# Fungal Biotechnology and Application

# Fungal Biotechnology and Application

Editor

**Baojun Xu**

MDPI

Basel • Beijing • Wuhan • Barcelona • Belgrade • Novi Sad • Cluj • Manchester

*Editor*
Baojun Xu
Beijing Normal
University-Hong Kong
Baptist University United
International College
Zhuhai, China

*Editorial Office*
MDPI
St. Alban-Anlage 66
4052 Basel, Switzerland

This is a reprint of articles from the Special Issue published online in the open access journal *Journal of Fungi* (ISSN 2309-608X) (available at: https://www.mdpi.com/journal/jof/special_issues/BFS2020).

For citation purposes, cite each article independently as indicated on the article page online and as indicated below:

| Lastname, A.A.; Lastname, B.B. Article Title. *Journal Name* **Year**, *Volume Number*, Page Range. |
|---|

**ISBN 978-3-0365-8810-0 (Hbk)**
**ISBN 978-3-0365-8811-7 (PDF)**
**doi.org/10.3390/books978-3-0365-8811-7**

# Contents

# About the Editor

**Baojun Xu**

Dr. Xu is a Chair Professor in Beijing Normal University–Hong Kong Baptist University United International College (UIC, a fully English teaching college in China), Fellow of the Royal Society of Chemistry, Zhuhai Scholar Distinguished Professor, Department Head of the Department of Life Sciences, Program Director of the Food Science and Technology Program, and author of over 300 peer-reviewed papers. Dr. Xu received his Ph.D. in Food Science from Chungnam National University, South Korea. He conducted postdoctoral research work in North Dakota State University (NDSU), Purdue University, and Gerald P. Murphy Cancer Foundation in the USA between 2005 and 2009. He undertook short-term visiting research at NDSU in 2012, the University Georgia in 2014, and undertook visiting research during his sabbatical leave (7 months) at Pennsylvania State University in the USA in 2016. Dr. Xu is currently serving as Associate Editor-in-Chief of *Food Science and Human Wellness*, Associate Editor of *Food Research International*, and Associate Editor of *Food Frontiers*, and is an editorial board member of approximately 10 international journals. He received the inaugural President's Award for Outstanding Research at UIC in 2016 and the President's Award for Outstanding Service to UIC in 2020. Dr. Xu was listed in the world's top 2% of scientists by Stanford University in 2020, 2021, 2022, and was listed as the Best Scientist in the world in the field of Biology and Biochemistry by Research.com in 2023.

# Preface

The demand for fossil fuels for utilization in industry, agriculture, transportation, and private sectors is sharply increasing globally. An insufficient fuel supply can be alleviated by employing alternative fuels. In the future bio-based society, an extensive range of challenges will have to be solved using novel biological solutions. For example, bioethanol is a liquid biofuel produced from biomass via the microbial fermentation of various biomass substrates, such as lignocelluloses. Millions of tons of agro-industrial waste byproducts are generated by agricultural practices using lignocellulose biomasses annually worldwide. The straws of crops are paradigms of agro-industrial waste byproducts containing lignocelluloses. Among the core technologies employed for these transformations are the fungal production of proteins, metabolites and chemicals, and biological processing. However, there remains a significant gap in our knowledge regarding the biological and chemical properties of fungi. Thus, this Special Issue discusses the latest knowledge and advances in fungal biotechnology and application. This Special Issue provides a platform for innovative academics, engineers, and industrial experts in the field of fungal biotechnology and food science to exchange new ideas and present their research results. This Special Issue includes nine original research papers and four reviews reflecting recent findings in the field.

With the increase in the world's population, the demand for industrialization has also enhanced humans' living standards. Fungi are considered a source of the essential constituents able to produce biocatalytic enzymes, including amylases, proteases, lipases, and cellulases, that offer broad-spectrum industrial and emerging applications. Emerging evidence has revealed that CRISPR/Cas9 technology has been successfully developed in various filamentous fungi and higher fungi for the purpose of editing specific genes. In addition to their medical importance, fungal proteases are extensively employed in various industries, such as in the preparation of butter, fruits, juices, and cheese, and in order to increase the shelf life of foods. It has been concluded that the hydrolysis of proteins in industries is one of the most significant applications of fungal enzymes, which has led to the substantial application of proteomics.

Overall, this Special Issue provides an up-to-date overview of research related to fungal biotechnologies and their application in the production of bioenergy and the utilization of agro-industrial waste, among others. We would like to take this opportunity to thank Prof. David S. Perlin, the Editor-in-Chief of the Journal of *Fungi*, for giving us this great opportunity to publish this Special Issue entitled "Fungal Biotechnology and Application". We also appreciate the contributions submitted by microbiological experts and congratulate them for their significant findings and hard work. We also owe our peer reviewers a huge debt of gratitude for their insightful comments, which helped to enhance the quality of the manuscripts published in this Special Issue.

**Baojun Xu**
*Editor*

Journal of
*Fungi*

MDPI

*Editorial*

# Fungal Biotechnology and Applications

**Baojun Xu**

Food Science and Technology Program, BNU-HKBU United International College, Zhuhai 519087, China; baojunxu@uic.edu.cn; Tel.: +86-756-3620636

The demand for fossil fuels for industry, agriculture, transportation, and private sectors is sharply increasing globally. An insufficient fuel supply can be alleviated using alternative fuels. In the future bio-based society, a wide range of challenges will have to be solved by new biological solutions. For example, bioethanol is a liquid biofuel produced from biomass via the microbial fermentation of various biomass substrates, such as lignocelluloses [1]. Millions of tons of agro-industrial waste byproducts are generated annually from agricultural practices on lignocellulose biomasses worldwide. Straws of crops are paradigms of agro-industrial waste byproducts containing lignocelluloses [2,3]. Among the core technologies for these transformations are the fungal production of proteins, metabolites and chemicals and biological processing. However, there is still a significant gap in our knowledge about the biological and chemical properties of fungal. Thus, this Special Issue discusses the latest knowledge and critical thinking on fungal biotechnology and its applications. This Special Issue provides an enabling platform for innovative academics, engineers and industrial experts in the field of fungal biotechnology and food science to exchange new ideas and present research results. This Special Issue includes nine original research papers and four reviews on recent findings in the field.

The thermotolerant yeast *Saccharomyces cerevisiae* has received considerable interest for industrial ethanol production. In a study conducted by Boonchuay et al. [1], an effective thermotolerant yeast strain S. cerevisiae TC-5 was selected for bioethanol production under elevated temperatures via simultaneous saccharification and fermentation SSF and the use of cellulose-rich corncob (CRC) residue. The maximum ethanol concentration and ethanol productivity values were 31.96 g/L and 0.222 g/L/h, respectively. In another study [4], a cDNA library was constructed from the mRNA of a methylotrophic yeast, *Ogataea polymorpha*, and then introduced into S. cerevisiae using horizontal gene transfer technology, a process through which an organism acquires genes from other organisms. The results confirmed that heat tolerance in yeast is a complex phenotype dependent on multiple quantitative loci. Artificial random gene transfer can emerge as a valuable tool in yeast bioengineering to investigate the background of complex phenotypes, such as heat tolerance.

As β-glucosidases represent the major bottleneck for the industrial degradation of plant biomass, great efforts are being devoted to developing efficient and inexpensive ways to produce them. In their study, Méndez-Líter et al. [5] utilized raw glycerol as an appropriate carbon source to produce efficient enzyme cocktails with high β-glucosidase levels using the fungus *Talaromyces amestolkiae*. The crude enzyme was successfully used to supplement a basal commercial cellulolytic cocktail for the saccharification of pretreated wheat straw, corroborating that even hardly exploitable industrial wastes, such as glycerol, can be used as secondary raw materials to produce valuable enzymatic preparations in a framework of the circular economy.

A research work [3] highlighted the valorization of the bulky recalcitrant lignocellulose byproduct of wheat straw (WS) for the enhanced production of value-added xylanase by the locally sourced novel *Penicillium chrysogenum* strain A3 DSM105774. The optimized production of xylanase via a submerged state of fermentation of WS was achieved using a three-step statistical and sequential approach.

**Citation:** Xu, B. Fungal Biotechnology and Applications. *J. Fungi* **2023**, *9*, 871. https://doi.org/10.3390/jof9090871

Received: 18 August 2023
Accepted: 22 August 2023
Published: 24 August 2023

Solid state fermentation (SsF) is recognized as a suitable process to produce enzymes using organic residues as substrates. However, only a few studies have integrated an evaluation of the feasibility of applying enzymes produced by SsF into subsequent hydrolyses followed by the production of target compounds, e.g., lactic acid (LA), through submerged-liquid fermentations (SmF). In another study [6], wheat bran (WB) was used as the substrate to produce enzymes via SsF by *Aspergillus awamori* DSM No. 63272. These enzymes were then used for the hydrolysis of the organic fraction of municipal solid waste (OFMSW). Subsequently, hydrolysates were fermented in SmF by *Bacillus coagulans* A166, increasing the LA concentration by 15.59 g/L. The data reported in this study provide an example of how SsF and SmF technologies can be combined for the valorization of WB and OFMSW.

The potential of rose flowers and lavender straw waste biomass was studied [7] as feed for lignocellulose substrates for the cultivation of the newly isolated *Ganoderma resinaceum* GA1M in Bulgaria, with the objective of obtaining mycelium-based bio-composites.

The bicarbonate ($HCO_3^-$) transporter family, including the anion exchanger (AE) group, is involved in multiple physiological processes through regulating acid–base homeostasis. $HCO_3^-$ transporters have been extensively studied in mammals, but fungal homologues of AEs are poorly understood. Dang et al. characterized the AE group member (MoAE4) in *Magnaporthe oryzae*. MoAE4 exhibits more sequence and structure homologies with the reported AE4 and BOR1 proteins [8]. Summarily, the data delineate a cytomembrane- and tonoplast-located $HCO_3^-$ transporter, which is required for development and pathogenicity in *M. oryzae*, revealing a potential drug target for blast disease control.

MicroRNA plays an important role in multifarious biological processes by regulating their corresponding target genes. However, the biological function and regulatory mechanism of fungal microRNA-like RNAs (milRNAs) remain poorly understood. In a study [9], combined with deep sequencing and bioinformatics analysis, milRNAs and their targets from *Trichoderma guizhouence* NJAU 4742 were isolated and identified under solid-state fermentation (SSF) using rice straw as the sole carbon source. The authors concluded that Tr-milRNA1 from NJAU 4742 improved lignocellulose utilization under heat stress by regulating the expression of the corresponding target gene Trvip36. These findings might open avenues for exploring the mechanism of lignocellulase secretion in filamentous fungi.

Light is perceived by photoreceptors in fungi and further integrated into the stress-activated MAPK HOG pathway, and thereby potentially activates the expression of genes for stress responses. The precise control of light conditions can likely improve the conidial yield and stress resistance to guarantee the low cost and long shelf life of Trichoderma-based biocontrol agents and biofertilizers. In a study [10], effects of wavelengths and intensities of light on conidial yield and stress tolerance to osmotic, oxidative and pH stresses in *Trichoderma guizhouense* were investigated. The authors found that blue light increased the conidial yield more than 1000 folds as compared to dark conditions and simultaneously enhanced conidial stress resistance.

Enzymatic catalysis is one of the main pillars of sustainability for industrial production. Enzyme application allows the use of toxic solvents to be minimized and the valorization of agro-industrial residues through reuse [11]. In addition, the enzymes are safe and energy-efficient. Nonetheless, their use in biotechnological processes is still hindered by their cost, stability and low rate of recycling and reuse. Among the many industrial enzymes, fungal laccases (LCs) are perfect candidates to serve as a biotechnological tool, as they are outstanding, versatile catalytic oxidants, only requiring molecular oxygen to function. Loi et al. [11] presented an overview of fungal LC (a promising green and sustainable enzyme), its mechanism of action, advantages, disadvantages, and solutions for its use as a tool to reduce the environmental and economic impact of industrial processes with a particular insight on the reuse of agro-wastes.

As a fundamental mineral element for cell growth and development, iron is available for uptake as ferric ions, which are usually oxidized into complex oxyhydroxide polymers, insoluble under aerobic conditions. In these conditions, the bioavailability of iron is dramatically reduced. As a result, microorganisms face problems of iron acquisition, especially under low concentrations of this element. However, some microbes have evolved mechanisms for obtaining ferric irons from the extracellular medium or environment by forming small molecules often regarded as siderophores. Pecoraro et al. systematically reviewed biosynthesis pathways, transport mechanisms and biotechnological applications of fungal siderophores [12].

With the increasing world population, demand for industrialization has also increased to meet humans' living standards. Fungi are considered a source of essential constituents to produce biocatalytic enzymes, including amylases, proteases, lipases, and cellulases, with broad-spectrum industrial and emerging applications. Naeem et al. discussed the origin, nature, mechanism of action, emerging aspects of genetic engineering for designing novel proteases, genome editing of fungal strains through CRISPR technology, present challenges and future recommendations of fungal proteases [13]. The emerging evidence revealed that CRISPR/Cas9 technology had been successfully developed in various filamentous fungi and higher fungi for the editing of specific genes. In addition to medical importance, fungal proteases are extensively used in different industries such as foods to prepare butter, fruits, juices and cheese, and to increase their shelf life. It is concluded that hydrolysis of proteins in industries is one of the most significant applications of fungal enzymes that led to massive usage of proteomics.

Alternaria is a ubiquitous fungal genus in many ecosystems, consisting of species and strains that can be saprophytic, endophytic or pathogenic to plants or animals, including humans. Alternaria species can produce a variety of secondary metabolites (SMs), especially low-molecular-weight toxins. Wang et al. [14] reviewed the recent advances in Alternaria phytotoxins, which covered the classification, chemical structure, occurrence, bioactivity and biosynthesis of the major Alternaria phytotoxins.

Overall, this Special Issue provides up-to-date findings related to fungal biotechnologies and their application in bioenergy production, agro-industrial waste utilization, etc. We would like to take this opportunity to thank Prof. David S. Perlin, the Editor-in-Chief of the *Journal of Fungi*, for giving us this great opportunity to publish a Special Issue "Fungal Biotechnology and Applications". We also appreciate the contributions from microbiological experts and congratulate them for their significant findings and hard work. We also owe our peer reviewers a huge debt of gratitude for their insightful comments that helped to improve the quality of manuscripts published in this issue.

**Funding:** This research received no external funding.

**Conflicts of Interest:** The author declares no conflict of interest.

## References

1. Boonchuay, P.; Techapun, C.; Leksawasdi, N.; Seesuriyachan, P.; Hanmoungjai, P.; Watanabe, M.; Srisupa, S.; Chaiyaso, T. Bioethanol production from cellulose-rich corncob residue by the thermotolerant *Saccharomyces cerevisiae* TC-5. *J. Fungi* **2021**, *7*, 547. [CrossRef] [PubMed]
2. Anwar, Z.; Gulfraz, M.; Irshad, M. Agro-industrial lignocellulosic biomass a key to unlock the future bioenergy: A brief review. *J. Radiat. Res. Appl. Sci.* **2014**, *7*, 163–173.
3. Matrawy, A.A.; Khalil, A.I.; Marey, H.S.; Embaby, A.M. Use of wheat straw for value-added product xylanase by *Penicillium chrysogenum* strain A3 DSM105774. *J. Fungi* **2021**, *7*, 696. [CrossRef] [PubMed]
4. Seike, T.; Narazaki, Y.; Kaneko, Y.; Shimizu, H.; Matsuda, F. Random transfer of *Ogataea polymorpha* genes into *Saccharomyces cerevisiae* reveals a complex background of heat tolerance. *J. Fungi* **2021**, *7*, 302. [CrossRef] [PubMed]
5. Méndez-Líter, J.A.; de Eugenio, L.I.; Hakalin, N.L.S.; Prieto, A.; Martínez, M.J. Production of a β-glucosidase-rich cocktail from *Talaromyces amestolkiae* using raw glycerol: Its role for lignocellulose waste valorization. *J. Fungi* **2021**, *7*, 363. [CrossRef] [PubMed]
6. Martău, G.-A.; Unger, P.; Schneider, R.; Venus, J.; Vodnar, D.C.; López-Gómez, J.P. Integration of solid state and submerged fermentations for the valorization of organic municipal solid waste. *J. Fungi* **2021**, *7*, 766. [CrossRef] [PubMed]
7. Angelova, G.; Brazkova, M.; Stefanova, P.; Blazheva, D.; Vladev, V.; Petkova, N.; Slavov, A.; Denev, P.; Karashanova, D.; Zaharieva, R.; et al. Waste rose flower and lavender straw biomass—An innovative lignocellulose feedstock for mycelium bio-materials development using newly isolated *Ganoderma resinaceum* GA1M. *J. Fungi* **2021**, *7*, 866. [CrossRef] [PubMed]

8. Dang, Y.; Wei, Y.; Zhang, P.; Liu, X.; Li, X.; Wang, S.; Liang, H.; Zhang, S.-H. The bicarbonate transporter (MoAE4) localized on both cytomembrane and tonoplast promotes pathogenesis in *Magnaporthe oryzae*. *J. Fungi* **2021**, *7*, 955. [CrossRef] [PubMed]
9. Li, T.; Liu, J.; Wang, Q.; Liu, Y.; Li, T.; Liu, D.; Shen, Q. *Tr*-milRNA1 contributes to lignocellulase secretion under heat stress by regulating the lectin-type cargo receptor gene *Trvip36* in *Trichoderma guizhouence* NJAU 4742. *J. Fungi* **2021**, *7*, 997. [CrossRef] [PubMed]
10. Li, Y.; Meng, X.; Guo, D.; Gao, J.; Huang, Q.; Zhang, J.; Fischer, R.; Shen, Q.; Yu, Z. A Simple and low-cost strategy to improve conidial yield and stress resistance of *Trichoderma guizhouense* through optimizing illumination conditions. *J. Fungi* **2022**, *8*, 50. [CrossRef] [PubMed]
11. Loi, M.; Glazunova, O.; Fedorova, T.; Logrieco, A.F.; Mulè, G. Fungal laccases: The forefront of enzymes for sustainability. *J. Fungi* **2021**, *7*, 1048. [CrossRef]
12. Pecoraro, L.; Wang, X.; Shah, D.; Song, X.; Kumar, V.; Shakoor, A.; Tripathi, K.; Ramteke, P.W.; Rani, R. Biosynthesis pathways, transport mechanisms and biotechnological applications of fungal siderophores. *J. Fungi* **2022**, *8*, 21. [CrossRef]
13. Naeem, M.; Manzoor, S.; Abid, M.-U.-H.; Tareen, M.B.K.; Asad, M.; Mushtaq, S.; Ehsan, N.; Amna, D.; Xu, B.; Hazafa, A. Fungal proteases as emerging biocatalysts to meet the current challenges and recent developments in biomedical therapies: An updated review. *J. Fungi* **2022**, *8*, 109. [CrossRef] [PubMed]
14. Wang, H.; Guo, Y.; Luo, Z.; Gao, L.; Li, R.; Zhang, Y.; Kalaji, H.M.; Qiang, S.; Chen, S. Recent advances in *Alternaria* phytotoxins: A review of their occurrence, structure, bioactivity, and biosynthesis. *J. Fungi* **2022**, *8*, 168. [CrossRef] [PubMed]

Journal of *Fungi*

MDPI

*Article*

# Random Transfer of *Ogataea polymorpha* Genes into *Saccharomyces cerevisiae* Reveals a Complex Background of Heat Tolerance

**Taisuke Seike [1], Yuki Narazaki [1], Yoshinobu Kaneko [2], Hiroshi Shimizu [1] and Fumio Matsuda [1,*]**

1 Department of Bioinformatics Engineering, Graduate School of Information Science and Technology, Osaka University, 1-5 Yamadaoka, Suita, Osaka 565-0871, Japan; taisuke_seike@ist.osaka-u.ac.jp (T.S.); navynarazaki0911@gmail.com (Y.N.); shimizu@ist.osaka-u.ac.jp (H.S.)
2 Yeast Genetic Resources Laboratory, Graduate School of Engineering, Osaka University, Osaka 565-0871, Japan; kaneko@bio.eng.osaka-u.ac.jp
* Correspondence: fmatsuda@ist.osaka-u.ac.jp; Tel.: +81-668-797-433

**Abstract:** Horizontal gene transfer, a process through which an organism acquires genes from other organisms, is a rare evolutionary event in yeasts. Artificial random gene transfer can emerge as a valuable tool in yeast bioengineering to investigate the background of complex phenotypes, such as heat tolerance. In this study, a cDNA library was constructed from the mRNA of a methylotrophic yeast, *Ogataea polymorpha*, and then introduced into *Saccharomyces cerevisiae*. *Ogataea polymorpha* was selected because it is one of the most heat-tolerant species among yeasts. Screening of *S. cerevisiae* populations expressing *O. polymorpha* genes at high temperatures identified 59 *O. polymorpha* genes that contribute to heat tolerance. Gene enrichment analysis indicated that certain *S. cerevisiae* functions, including protein synthesis, were highly temperature-sensitive. Additionally, the results confirmed that heat tolerance in yeast is a complex phenotype dependent on multiple quantitative loci. Random gene transfer would be a useful tool for future bioengineering studies on yeasts.

**Keywords:** *Saccharomyces cerevisiae*; cDNA library; heat tolerance; random gene transfer; *Ogataea polymorpha*

**Citation:** Seike, T.; Narazaki, Y.; Kaneko, Y.; Shimizu, H.; Matsuda, F. Random Transfer of *Ogataea polymorpha* Genes into *Saccharomyces cerevisiae* Reveals a Complex Background of Heat Tolerance. *J. Fungi* **2021**, *7*, 302. https://doi.org/10.3390/jof7040302

Academic Editor: Baojun Xu

Received: 19 March 2021
Accepted: 13 April 2021
Published: 15 April 2021



## 1. Introduction

Horizontal gene transfer is a rare evolutionary event in yeast, in which functional genes are acquired from other species [1]. Gene transfer among eukaryotes is a relatively rare event that is limited by unknown barriers [2–4]. Genome analysis revealed that the budding yeast *Saccharomyces cerevisiae* has acquired several genes from bacteria, such as those encoding metabolite enzymes and transporters [5–7].

Artificial random gene transfer is a technique used in conventional genetic engineering to identify valuable genes, such as those useful for metabolic engineering [8]. In addition, the technique can help investigate the genetic background of quantitative traits, as many loci, each with small effects, contribute to heat tolerance [9,10]. Transfer of a gene pool would provide a list of genes that contribute to heat tolerance in the host yeast, and this list of genes will help elucidate the genetic complexity behind the heat tolerance phenotype.

In this study, we used a cDNA library for a random gene transfer experiment [11]. The cDNA library was constructed from the mRNA of a methylotrophic yeast, *Ogataea polymorpha*, and then introduced into *S. cerevisiae*. *Saccharomyces cerevisiae* is an industrial host for bioethanol production. Since improved heat tolerance reduces the costs required for cooling during fermentation [12], many genetic analyses and adaptive evolutionary studies have been performed [13–15]. *Ogataea polymorpha* was selected because it can grow at temperatures close to 50 °C and is one of the most heat-tolerant species among yeasts [16,17]. Under high temperatures, screening the *S. cerevisiae* population identified 60 colonies showing improved heat tolerance and the corresponding *O. polymorpha* genes responsible for heat tolerance. Additionally, the list of genes uncovered the *S. cerevisiae* functions that are sensitive to high temperatures.

## 2. Materials and Methods

### *2.1. Strains, Plasmids, and Yeast Transformation*

The yeast strains and plasmids used in this study are listed in Table 1. Plasmids were derived from pGK413, pGK414, or pGK416, in which gene expression is controlled by the PGK1 promoter [18]. For the construction of cDNA libraries, *O. polymorpha* BY4329 cells were cultured till the exponential growth phase in 5 mL of yeast extract-peptone-adenine-dextrose (YPAD) medium and then harvested by centrifugation at 12,000× *g* for 5 min. mRNA was extracted from the cells using the Ribo-Pure Yeast Kit (Thermo Fisher Scientific, Waltham, MA, USA). Two overlapping regions for the In-Fusion method were added to the three plasmids using the inverse PCR method with pGK413, pGK414, and pGK416 as the templates and the primers pGK_inv_fw (TCTCATCGTACCCCGGAAATAAATT) and pGK_inv_rv (AACTATGGTGACGAAGTTTTATATTTGTTG) [11]. A cDNA library was constructed from the mRNA mixture and the amplicon of inverse PCR, using the In-Fusion SMARTer Directional cDNA Library Construction Kit (Takara Bio, Inc., Shiga, Japan). The In-Fusion mixture was introduced into *Escherichia coli* HST08 competent cells (Takara Bio, Inc.) by electroporation at 25 μF and 2 kV. The pulse controller was set to 200 Ω using a GenePulser (Bio-Rad Laboratories, Hercules, CA, USA). Ampicillin-resistant cells were recovered from agar plates, from which the plasmid pools were prepared. The growth conditions, DNA-related techniques, and the lithium-acetate method for transformation have been described previously [19].

**Table 1.** Strains and plasmids used in this study.

| Strain Name | Genotype | Source |
| --- | --- | --- |
| *Ogataea polymorpha* BY4329 | *Leu1-1* | Obtained from NBRP Yeast |
| *Saccharomyces cerevisiae* YPH499 | *MATa, ura3-52 lys2-801_amber ade2-101_ochre trp1-Δ63 his3-Δ200 leu2-Δ1* | Thermo Scientific |
| TT01 | YPH499 (pGK416_BY4329 cDNA library) | This study |
| TT02 | YPH499 (pGK413_BY4329 cDNA library) | This study |
| TT03 | YPH499 (pGK414_BY4329 cDNA library) | This study |
| TT01c | YPH499 (pGK416) | This study |
| TT02c | YPH499 (pGK413) | This study |
| TT03c | YPH499 (pGK414) | This study |
| *Escherichia coli* DH5α | *deoR endA1 gyrA96 hsdR17(rk-mk+) recA1 relA1 supE44 thi-1Δ(lacZYA-argFV169) φ80lacZΔM15* F- | |
| *Escherichia coli* HST08 | *F,endA1, supE44, thi-1, recA1, relA1, gyrA96, phoA, Φ80d lacZΔM15, Δ(lacZYA-argF) U169, Δ(mrr-hsdRMS-mcrBC), ΔmcrA,λ−* | |
| Plasmids | | |
| pGK413 | Yeast expression vector containing PGK1 promoter, origin, ARS4/CEN6 HIS3 marker, no expression (control plasmid) | [18] |
| pGK414 | Yeast expression vector containing PGK1 promoter, origin, ARS4/CEN6 TRP1 marker, no expression (control plasmid) | [18] |
| pGK416 | Yeast expression vector containing PGK1 promoter, origin, ARS4/CEN6 URA3 marker, no expression (control plasmid) | [18] |

### *2.2. Culture Conditions*

All strains were cultured in YPAD medium (1% Bacto yeast extract, 2% Bacto peptone, 2% glucose, and 0.004% adenine) and synthetic dextrose (SD) medium (0.67% yeast nitrogen base without amino acids and 2% or 0.5% glucose, as necessary, 0.006% leucine, 0.003% lysine hydrochloride, 0.002% histidine, 0.004% adenine, 0.004% tryptophan, and 0.002%

uracil). Yeast cells grown on the agar plate were cultured in 5 mL of SD medium containing the required amino acids overnight at 30 °C and 150 rpm. To screen for cell growth, the transformants were cultured overnight in SD agar medium containing amino acids at 39 °C or 39.5 °C in an incubator (TVA360DB, ADVANTEC, Tokyo, Japan).

### 2.3. Construction of Screening System for the Heat-Resistant Evolved Strain

*Saccharomyces cerevisiae* YPH499 was transformed with the cDNA library of *O. polymorpha* using the lithium-acetate method and then cultured for several days in SD agar medium to obtain colonies of transformants. From the original SD agar plates, replica plates were prepared on other SD agar plates using the replica plating method. The replicas were cultured at 39 °C or 39.5 °C. The plasmids in the selected transformants were extracted using the Easy Yeast Plasmid Isolation Kit (Takara). Each plasmid was introduced into *E. coli* HST08 competent cells and cultured in L medium containing 5 mL ampicillin. Plasmid purification was performed using LaboPass Mini (Hokkaido System Science), and sequence analysis was performed using the PGK 5′ primer (TAGTTTTTCAAGTTCTTAGA) and PGK 3′ primer (CTATTATTTTAGCGTAAAGG). For each plasmid, the corresponding *O. polymorpha* gene was identified using the BLAST search function in the UniProt database, including *O. polymorpha* genome information (http://www.uniprot.org/ accessed on 13 April 2021) [17]. *Saccharomyces cerevisiae* orthologs were identified using the BLAST search of the Saccharomyces Genome Database (SGD, https://www.yeastgenome.org/ accessed on 13 April 2021). Gene enrichment analysis was performed using the over-representation analysis function of the WebGestalt web tool (http://www.webgestalt.org/ accessed on 13 April 2021) [20]. The Gene Ontology (GO) dataset of all *O. polymorpha* proteins was retrieved from the UniProt database. The Benjamini-Hochberg (GH) method was used to evaluate the false discovery rate (FDR).

### 2.4. Confirmation of Reproducibility by Spot Method

Transformants were inoculated on SD agar medium containing 20 g/L glucose and cultured at 30 °C for two days. A single colony grown on the plate was inoculated into a test tube containing 5 mL of SD medium and precultured at 30 °C and 150 rpm. The preculture solution was then centrifuged at 3000 rpm and 4 °C. The collected transformants were suspended in sterile distilled water. Suspensions (6 μL) were then spotted onto SD agar medium supplemented with the appropriate amino acids and incubated at 39 °C or higher for five days.

## 3. Results

### 3.1. Comparison of Vectors for Artificial Random Gene Transfer

For the construction of an *O. polymorpha* cDNA library, three CEN/ARS plasmid vectors (single copy-type), namely pGK416 (possessing *URA3*), pGK413 (possessing *HIS3*), and pGK414 (possessing *TRP1*), were employed [18]. Because the relationship between amino acid auxotrophy and heat tolerance was expected, three control strains possessing pGK416, pGK413, and pGK414 (strains TT01c, TT02c, and TT03c, respectively) were constructed from the *S. cerevisiae* YPH499 strain and cultured on agar plates to compare their heat tolerance phenotypes (Figure 1A,B). The TT01c and TT02c strains were able to grow at 39 °C but failed to grow at 39.5 °C. However, many colonies that grew at 39 °C showed an abnormally wet phenotype. The upper growth limit of TT03c was 38 °C (Figure 1C). These results showed that the amino acid auxotrophy of *S. cerevisiae* affected the heat tolerance of yeast for as yet unknown reasons.

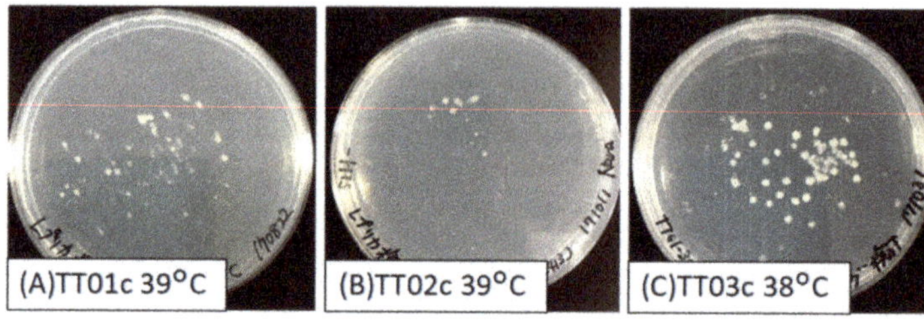


**Figure 1.** Colonies on replica plates of control strains cultivated under high temperature conditions for one week. (**A**) TT01c (YPH499 (pGK416)) cultured at 39 °C. (**B**) TT02c (YPH499 (pGK413)) cultured at 39 °C. (**C**) TT03c (YPH499 (pGK414)) cultured at 38 °C.

### 3.2. Screening of Heat-Tolerant S. cerevisiae Strains Expressing O. polymorpha cDNA

A cDNA library was constructed from the *O. polymorpha* BY4329 strain. The cDNA fragments were inserted into pGK416, pGK413, and pGK414, which were then introduced into the *S. cerevisiae* YPH499 strain to produce three populations possessing *O. polymorpha* cDNA (TT01, TT02, and TT03, respectively). Approximately 400 colonies grew on each selection plate, and a replica plate was prepared using the replica plating technique. A total of 50–80 replica plates consisting of approximately 20,000–32,000 colonies were prepared for each population (TT01, TT02, and TT03).

The replica plates were incubated at 39.5 °C for the TT01 and TT02 populations and at 39 °C for the TT03 population. After one week, no colonies were obtained from the TT02 population. In contrast, 11 and 49 colonies were obtained on the replica plates of the TT01 and TT03 populations, respectively. No colonies were identified after additional screening at higher temperatures.

Following the collection of plasmid vectors from the 60 colonies, sequences of open reading frames of the cDNAs were determined to identify the corresponding *O. polymorpha* genes, *S. cerevisiae* ortholog genes, and their putative functions using the BLAST search of UniProt and SGD databases (Tables 2 and S1) [17]. Among the 60 colonies, an identical gene (*OGAPODRAFT_52470*, an ortholog of *S. cerevisiae QCR8* ubiquinol-cytochrome c reductase subunit 8) was identified from two independent colonies (TT01-2 and TT01-8). The cDNAs obtained from four colonies (TT03-46, -47, -48, and -49) had poor homology to all *S. cerevisiae* ORFs (E-value < $1.0 \times 10^{-3}$), suggesting that these cDNAs were derived from *O. polymorpha*-specific genes.

**Table 2.** Annotation of *O. polymorpha* genes obtained from colonies of heat-tolerant *S. cerevisiae* expressing *O. polymorpha* cDNA [(1)].

| Colony ID | Gene ID of *O. polymorpha* [(2)] | *S. cerevisiae* Ortholog [(3)] | Functional Annotation of *S. cerevisiae* Ortholog |
|---|---|---|---|
| TT01-1 | *OGAPODRAFT_7331* | *CAF20* | cap-associated protein CAF20 |
| TT01-2 | *OGAPODRAFT_52470* | *QCR8* | ubiquinol-cytochrome c reductase subunit 8 |
| TT01-3 | *OGAPODRAFT_16764* | *ALD4* | aldehyde dehydrogenase |
| TT01-4 | *HPODL_02546* | *RPL16A* | 60S ribosomal protein L16-B |
| TT01-5 | *HPODL_00806* | *GUP1* | acyltransferase |
| TT01-6 | *OGAPODRAFT_17522* | *THO1* | SAP domain-containing ribonucleoprotein |
| TT01-7 | *OGAPODRAFT_12972* | *HSP10* | chaperonin GroES |
| TT01-8 | *OGAPODRAFT_52470* | *QCR8* | ubiquinol-cytochrome c reductase subunit 8 |
| TT01-9 | *HPODL_02610* | *CYT1* | cytochrome c1, heme protein, mitochondrial |
| TT01-10 | *HPODL_04437* | *FRK1* | serine/threonine protein kinase |
| TT01-11 | *OGAPODRAFT_15309* | *PAF1* | RNA polymerase II-associated factor 1 |
| TT03-1 | *HPODL_02637* | *GRS1* | glycine–tRNA ligase 1, mitochondrial |
| TT03-2 | *HPODL_00026* | *NAP1* | histone chaperone NAP1 |
| TT03-3 | *HPODL_05027* | *NAB2* | mRNA-binding protein NAB2 |

**Table 2.** *Cont.*

| Colony ID | Gene ID of *O. polymorpha* (2) | *S. cerevisiae* Ortholog (3) | Functional Annotation of *S. cerevisiae* Ortholog |
|---|---|---|---|
| TT03-4 | *HPODL_03235* | *ERV25* | p24 family protein delta-1 |
| TT03-5 | *HPODL_05028* | *RPS2* | ribosomal 40S subunit protein S2 |
| TT03-6 | *OGAPODRAFT_25583* | *RIB3* | 3,4-dihydroxy-2-butanone-4-phosphate synthase RIB3 |
| TT03-7 | *HPODL_03162* | *ACB1* | long-chain fatty acid transporter ACB1 |
| TT03-8 | *HPODL_01585* | *RAD4* | DNA repair protein RAD4 |
| TT03-9 | *HPODL_00194* | *MRP7* | mitochondrial 54S ribosomal protein YmL2 |
| TT03-10 | *HPODL_02367* | *RPS31* | ubiquitin-ribosomal 40S subunit protein S31 fusion protein |
| TT03-11 | *OGAPODRAFT_76806* | *CYT2* | cytochrome c1 heme lyase CYT2 |
| TT03-12 | *OGAPODRAFT_92206* | *PSA1* | mannose-1-phosphate guanylyltransferase |
| TT03-13 | *HPODL_01049* | *GRX6* | glutathione-disulfide reductase GRX6 |
| TT03-14 | *HPODL_00042* | *RPL7A* | ribosomal 60S subunit protein L7A |
| TT03-15 | *HPODL_04105* | *RPL42A* | ribosomal 60S subunit protein L42A |
| TT03-16 | *OGAPODRAFT_17069* | *PTI1* | cleavage polyadenylation factor subunit PTI1 |
| TT03-17 | *HPODL_01073* | *ANB1* | translation elongation factor eIF-5A |
| TT03-18 | *HPODL_02594* | *MMF1* | isoleucine biosynthesis protein MMF1 |
| TT03-19 | *OGAPODRAFT_102344* | *PGK1* | 3-phosphoglycerate kinase |
| TT03-20 (4) | *HPODL_02458* | *SOD2* | superoxide dismutase SOD2 |
| TT03-21 (4) | *HPODL_02693* | *PFK26* | 6-phosphofructo-2-kinase |
| TT03-22 | *HPODL_02169* | *TAF9* | transcription initiation factor TFIID subunit 9 |
| TT03-23 (4) | *HPODL_01966* | *RAD6* | E2 ubiquitin-conjugating protein RAD6 |
| TT03-24 | *HPODL_02705* | *RPL1A* | ribosomal 60S subunit protein L1A |
| TT03-25 (4) | *HPODL_01497* | *ASC1* | guanine nucleotide-binding protein subunit beta |
| TT03-26 | *HPODL_01957* | *MET5* | sulfite reductase (NADPH) subunit beta |
| TT03-27 | *OGAPODRAFT_75779* | *CEP3* | centromere DNA-binding protein complex CBF3 subunit B |
| TT03-28 | *HPODL_03364* | *RPL23B* | ribosomal 60S subunit protein L23B |
| TT03-29 (4) | *HPODL_00942* | *RPP2B* | ribosomal protein P2B |
| TT03-30 | *HPODL_01497* | *ASC1* | guanine nucleotide-binding protein subunit beta |
| TT03-31 | *HPODL_02465* | *SER2* | phosphoserine phosphatase |
| TT03-32 | *OGAPODRAFT_74529* | *STE5* | pheromone-responsive MAPK scaffold protein |
| TT03-33 (4) | *HPODL_03495* | *ACC1* | acetyl-CoA carboxylase |
| TT03-34 | *OGAPODRAFT_16247* | *DEG1* | pseudouridine synthase DEG1 |
| TT03-35 | *OGAPODRAFT_76195* | *STM1* | Uncharacterized protein |
| TT03-36 | *OGAPODRAFT_17428* | *SLM1* | phosphatidylinositol 4,5-bisphosphate-binding protein |
| TT03-37 | *OGAPODRAFT_15585* | *RPS26A* | ribosomal 40S subunit protein S26A |
| TT03-38 | *HPODL_03366* | *SNF3* | high-affinity glucose transporter SNF3 |
| TT03-39 | *HPODL_03527* | *IDP1* | isocitrate dehydrogenase (NADP(+)) |
| TT03-40 (4) | *OGAPODRAFT_7594* | *SOM1* | mitochondrial export protein Som1 |
| TT03-41 | *HPODL_02149* | *ETR1* | trans-2-enoyl-CoA reductase |
| TT03-42 | *HPODL_04585* | *MYO5* | myosin-5 |
| TT03-43 | *HPODL_01873* | *SBA1* | hsp90 cochaperone SBA1 |
| TT03-44 (4) | *HPODL_01380* | *PRY2* | sterol-binding protein |
| TT03-45 | *HPODL_01021* | *RPS27B* | ribosomal 40S subunit protein S27B |
| TT03-46 | *HPODL_02251* | n.d. | n.d. |
| TT03-47 | *HPODL_04413* | n.d. | n.d. |
| TT03-48 (4) | *OGAPODRAFT_16908* | n.d. | n.d. |
| TT03-49 (4) | *OGAPODRAFT_15905* | n.d. | n.d. |

(1) Full data are shown in Table S1. (2) *Ogataea polymorpha* genes were identified using the BLASTN function of UniProt. Partial nucleotide sequences were used as queries. (3) *Saccharomyces cerevisiae* orthologs were identified using the BLASTP function of SGD (E-value < $1.0 \times 10^{-4}$). The full amino acid sequences of the *O. polymorpha* gene products were used as queries. (4) Corresponding strain was reconstructed for confirmation, as shown in Figure 2.

Functional categorization of the annotation list revealed that the transferred cDNAs encoded genes involved in various functions such as metabolism (for example,

*HPODL_02693* encoding 6-phosphofructo-2-kinase, obtained from colony ID TT03-21), translation (*HPODL_00942* encoding ribosomal protein P2B, obtained from colony ID TT03-29), electron transport chain (*HPODL_02610* encoding cytochrome c1, obtained from colony ID TT01-9), and protein quality control (*OGAPODRAFT_12972* encoding the chaperonin GroES, obtained from colony ID TT01-7).

Gene enrichment analysis was performed using the GH method to control the FDR. The results showed that genes encoding ribosomal proteins and other proteins involved in translation were overrepresented in the list of 59 *O. polymorpha* genes, with 11 (19%) and 9 (15%) cDNAs encoding genes related to the GO terms "structural constituent of ribosome" and "translation", respectively (Table 3).

**Table 3.** Gene enrichment analysis of the list of 59 *O. polymorpha* genes.

| GO Term | | False Discovery Rate (FDR) | Number of Matches |
|---|---|---|---|
| structural constituent of ribosome | GO:0003735 | 0.000014579 | 11 |
| translation | GO:0006412 | 0.00028796 | 9 |
| ribosome | GO:0005840 | 0.047143 | 6 |

### 3.3. Reconstruction of Heat-Tolerant S. cerevisiae Strains

The *S. cerevisiae* strains listed in Table 2 were reconstructed to check the false-positive rate derived from the screening approach. For this purpose, 10 out of the 49 plasmid vectors were randomly selected from the TT03 populations and then introduced into the YPH499 strains. The heat tolerance of the reconstructed strains was investigated using a spot assay under high temperature conditions (Figure 2). Although 5-fold serial dilutions were employed to confirm differences in heat tolerance, we found that almost all reconstructed strains tended to be more tolerant to high temperatures (39 °C) than the control strain (TT03c). These results suggest that the false-positive rate was low enough in the screening approach.

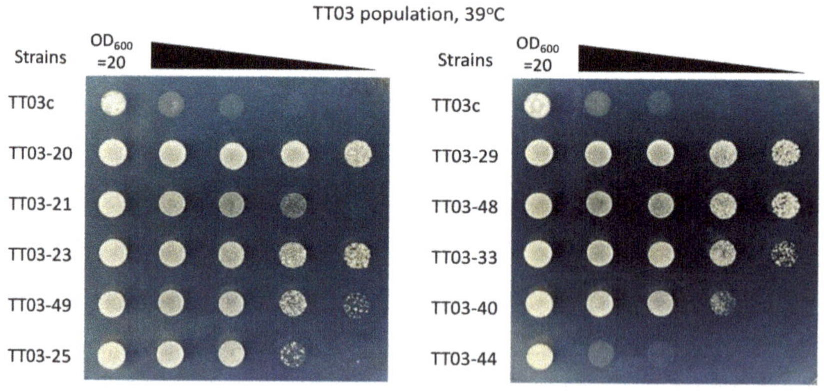


**Figure 2.** Heat tolerance of reconstructed strains. The dilution series was employed to confirm differences in heat tolerance. Samples were withdrawn from cultures, and their $OD_{600}$ values were adjusted to 20. Five-fold serial dilutions of these cultures were prepared in sterile distilled water, and 6 μL each of the cultures at $OD_{600}$ = 20 and their dilutions were spotted onto SD plates without tryptophan. The plates were then incubated at 39 °C for 5 days and photographed.

## 4. Discussion

In this study, we introduced a cDNA library derived from *O. polymorpha* into *S. cerevisiae*. Screening of the *S. cerevisiae* populations expressing the *O. polymorpha* cDNA library under high temperature conditions resulted in 60 colonies showing improved heat tolerance and identification of the *O. polymorpha* genes responsible for heat tolerance (Figure 2 and Table 2). These results reveal three aspects of the high temperature tolerance of *S. cerevisiae*.

First, we identified 59 candidate genes in *O. polymorpha* that contribute to heat tolerance. However, these results do not imply that the heat stability of proteins is derived from these genes. Since a strong promoter (PGK1 promoter) was used to express cDNA, a large amount of overexpressed proteins might have also contributed to heat tolerance. Further genetic and biochemical characterization is needed to examine the heat stability of the proteins expressed from the candidate genes. Moreover, this experiment failed to transfer all *O. polymorpha* genes to *S. cerevisiae* because the cDNA library used in this study was collected from *O. polymorpha* in exponential growth phase at 30 °C. More candidate genes are likely to be obtained using more comprehensive cDNA libraries prepared from *O. polymorpha*, for instance, under high temperature conditions.

Second, the genes identified in this study revealed the *S. cerevisiae* functions that are sensitive to high temperatures, because these functions were complemented by the expression of corresponding genes derived from *O. polymorpha*. Gene enrichment analysis showed that many of the *S. cerevisiae* proteins sensitive to high temperatures were ribosomal proteins and those involved in other steps of translation. Previous studies have reported that genes related to chaperonins [21], superoxide dismutase [22], ubiquitination [23], nitric oxide [24], $H^+$-ATPase [25], and trehalose biosynthesis [26,27] were responsible for the heat tolerance of *S. cerevisiae*. While genes encoding chaperonin (TT01-7) and superoxide dismutase (TT03-20) were found, genes responsible for other functions such as $H^+$-ATPase activity and trehalose biosynthesis were not found in the present study (Table 2). These results indicate that chaperonins, superoxide dismutase, ribosome, and translation may be additional targets for improving the heat tolerance of *S. cerevisiae*.

Thirdly, our results highlight that heat tolerance in yeasts is a complex phenotype that is controlled by multiple genes. This supports the idea that the improvement of heat tolerance in *S. cerevisiae* requires the expression of multiple heat-stable proteins. This study demonstrated that random gene transfer is a helpful laboratory evolution tool for investigating the genetic background of complex phenotypes, as well as for enabling future bioengineering studies.

**Supplementary Materials:** The following are available online at https://www.mdpi.com/article/10.3390/jof7040302/s1, Table S1: Annotation of *O. polymorpha* genes obtained from colonies of heat tolerant *S. cerevisiae* expressing *O. polymorpha* cDNA.

**Author Contributions:** Conceptualization, H.S. and F.M.; investigation, T.S. and Y.N.; resources, Y.K.; data curation, F.M.; writing—original draft preparation, Y.N.; writing—review and editing, T.S., Y.K., H.S. and F.M. All authors have read and agreed to the published version of the manuscript.

**Funding:** This research was supported in part by the Grants-in-Aid for Scientific Research (C) (Grant No.18K04851).

**Institutional Review Board Statement:** Not applicable.

**Informed Consent Statement:** Not applicable.

**Acknowledgments:** We thank Yoshihiro Toya and Minetaka Sugiyama for helpful comments to this manuscript. pGK series vectors were provided by the National Bio-Resource Project (NBRP) Yeast, Japan.

**Conflicts of Interest:** The authors declare no conflict of interest.

## References

1. Soucy, S.M.; Huang, J.; Gogarten, J.P. Horizontal gene transfer: Building the web of life. *Nat. Rev. Genet.* **2015**, *16*, 472–482. [CrossRef] [PubMed]
2. Yoshida, S.; Maruyama, S.; Nozaki, H.; Shirasu, K. Horizontal gene transfer by the parasitic plant *Striga hermonthica*. *Science* **2010**, *328*, 1128. [CrossRef] [PubMed]

3. Zhang, D.; Qi, J.; Yue, J.; Huang, J.; Sun, T.; Li, S.; Wen, J.F.; Hettenhausen, C.; Wu, J.; Wang, L.; et al. Root parasitic plant *Orobanche aegyptiaca* and shoot parasitic plant *Cuscuta australis* obtained Brassicaceae-specific strictosidine synthase-like genes by horizontal gene transfer. *BMC Plant Biol.* **2014**, *14*, 19. [CrossRef] [PubMed]
4. Novo, M.; Bigey, F.; Beyne, E.; Galeote, V.; Gavory, F.; Mallet, S.; Cambon, B.; Legras, J.L.; Wincker, P.; Casaregola, S.; et al. Eukaryote-to-eukaryote gene transfer events revealed by the genome sequence of the wine yeast Saccharomyces cerevisiae EC1118. *Proc. Natl. Acad. Sci. USA* **2009**, *106*, 16333–16338. [CrossRef] [PubMed]
5. Hall, C.; Brachat, S.; Dietrich, F.S. Contribution of horizontal gene transfer to the evolution of *Saccharomyces cerevisiae*. *Eukaryot. Cell* **2005**, *4*, 1102–1115. [CrossRef] [PubMed]
6. Moriguchi, K.; Yamamoto, S.; Tanaka, K.; Kurata, N.; Suzuki, K. Trans-kingdom horizontal DNA transfer from bacteria to yeast is highly plastic due to natural polymorphisms in auxiliary nonessential recipient genes. *PLoS ONE* **2013**, *8*, e74590. [CrossRef]
7. Milner, D.S.; Attah, V.; Cook, E.; Maguire, F.; Savory, F.R.; Morrison, M.; Muller, C.A.; Foster, P.G.; Talbot, N.J.; Leonard, G.; et al. Environment-dependent fitness gains can be driven by horizontal gene transfer of transporter-encoding genes. *Proc. Natl. Acad. Sci. USA* **2019**, *116*, 5613–5622. [CrossRef]
8. Shi, S.; Ji, H.; Siewers, V.; Nielsen, J. Improved production of fatty acids by Saccharomyces cerevisiae through screening a cDNA library from the oleaginous yeast Yarrowia lipolytica. *FEMS Yeast Res.* **2016**, *16*, fov108. [CrossRef] [PubMed]
9. Wang, Z.; Qi, Q.; Lin, Y.; Guo, Y.; Liu, Y.; Wang, Q. QTL analysis reveals genomic variants linked to high-temperature fermentation performance in the industrial yeast. *Biotechnol. Biofuels* **2019**, *12*, 59. [CrossRef] [PubMed]
10. Yang, Y.; Foulquie-Moreno, M.R.; Clement, L.; Erdei, E.; Tanghe, A.; Schaerlaekens, K.; Dumortier, F.; Thevelein, J.M. QTL analysis of high thermotolerance with superior and downgraded parental yeast strains reveals new minor QTLs and converges on novel causative alleles involved in RNA processing. *PLoS Genet.* **2013**, *9*, e1003693. [CrossRef]
11. Narazaki, Y.; Nomura, Y.; Morita, K.; Shimizu, H.; Matsuda, F. Expression of *Saccharomyces cerevisiae* cDNAs to enhance the growth of non-ethanol-producing *S. cerevisiae* strains lacking pyruvate decarboxylases. *J. Biosci. Bioeng.* **2018**, *126*, 317–321. [CrossRef]
12. Steensels, J.; Snoek, T.; Meersman, E.; Picca Nicolino, M.; Voordeckers, K.; Verstrepen, K.J. Improving industrial yeast strains: Exploiting natural and artificial diversity. *FEMS Microbiol. Rev.* **2014**, *38*, 947–995. [CrossRef]
13. Riles, L.; Fay, J.C. Genetic basis of variation in heat and ethanol tolerance in *Saccharomyces cerevisiae*. *G3* **2019**, *9*, 179–188. [CrossRef] [PubMed]
14. Caspeta, L.; Nielsen, J. Thermotolerant yeast strains adapted by laboratory evolution show trade-off at ancestral temperatures and preadaptation to other stresses. *MBio* **2015**, *6*, e00431. [CrossRef] [PubMed]
15. Kitichantaropas, Y.; Boonchird, C.; Sugiyama, M.; Kaneko, Y.; Harashima, S.; Auesukaree, C. Cellular mechanisms contributing to multiple stress tolerance in *Saccharomyces cerevisiae* strains with potential use in high-temperature ethanol fermentation. *AMB Express* **2016**, *6*, 107. [CrossRef]
16. Ishchuk, O.P.; Voronovsky, A.Y.; Abbas, C.A.; Sibirny, A.A. Construction of *Hansenula polymorpha* strains with improved thermotolerance. *Biotechnol. Bioeng.* **2009**, *104*, 911–919. [CrossRef] [PubMed]
17. Riley, R.; Haridas, S.; Wolfe, K.H.; Lopes, M.R.; Hittinger, C.T.; Goker, M.; Salamov, A.A.; Wisecaver, J.H.; Long, T.M.; Calvey, C.H.; et al. Comparative genomics of biotechnologically important yeasts. *Proc. Natl. Acad. Sci. USA* **2016**, *113*, 9882–9887. [CrossRef]
18. Ishii, J.; Izawa, K.; Matsumura, S.; Wakamura, K.; Tanino, T.; Tanaka, T.; Ogino, C.; Fukuda, H.; Kondo, A. A simple and immediate method for simultaneously evaluating expression level and plasmid maintenance in yeast. *J. Biochem.* **2009**, *145*, 701–708. [CrossRef] [PubMed]
19. Ishii, J.; Morita, K.; Ida, K.; Kato, H.; Kinoshita, S.; Hataya, S.; Shimizu, H.; Kondo, A.; Matsuda, F. A pyruvate carbon flux tugging strategy for increasing 2,3-butanediol production and reducing ethanol subgeneration in the yeast *Saccharomyces cerevisiae*. *Biotechnol. Biofuels* **2018**, *11*, 180. [CrossRef]
20. Wang, J.; Vasaikar, S.; Shi, Z.; Greer, M.; Zhang, B. WebGestalt 2017: A more comprehensive, powerful, flexible and interactive gene set enrichment analysis toolkit. *Nucleic Acids Res.* **2017**, *45*, W130–W137. [CrossRef]
21. Verghese, J.; Abrams, J.; Wang, Y.; Morano, K.A. Biology of the heat shock response and protein chaperones: Budding yeast (*Saccharomyces cerevisiae*) as a model system. *Microbiol. Mol. Biol. Rev.* **2012**, *76*, 115–158. [CrossRef] [PubMed]
22. Zyrina, A.N.; Smirnova, E.A.; Markova, O.V.; Severin, F.F.; Knorre, D.A. Mitochondrial superoxide dismutase and Yap1p act as a signaling module contributing to ethanol tolerance of the yeast *Saccharomyces cerevisiae*. *Appl. Environ. Microbiol.* **2017**, *83*, e02759-16. [CrossRef]
23. Uesugi, S.; Watanabe, D.; Kitajima, M.; Watanabe, R.; Kawamura, Y.; Ohnishi, M.; Takagi, H.; Kimura, K. Calcineurin inhibitors suppress the high-temperature stress sensitivity of the yeast ubiquitin ligase Rsp5 mutant: A new method of screening for calcineurin inhibitors. *FEMS Yeast Res.* **2014**, *14*, 567–574. [CrossRef] [PubMed]
24. Nishimura, A.; Kawahara, N.; Takagi, H. The flavoprotein Tah18-dependent NO synthesis confers high-temperature stress tolerance on yeast cells. *Biochem. Biophys. Res. Commun.* **2013**, *430*, 137–143. [CrossRef] [PubMed]
25. Coote, P.J.; Jones, M.V.; Seymour, I.J.; Rowe, D.L.; Ferdinando, D.P.; McArthur, A.J.; Cole, M.B. Activity of the plasma membrane $H^+$-ATPase is a key physiological determinant of thermotolerance in *Saccharomyces cerevisiae*. *Microbiology* **1994**, *140 (Pt 8)*, 1881–1890. [CrossRef]

26. Gibney, P.A.; Lu, C.; Caudy, A.A.; Hess, D.C.; Botstein, D. Yeast metabolic and signaling genes are required for heat-shock survival and have little overlap with the heat-induced genes. *Proc. Natl. Acad. Sci. USA* **2013**, *110*, E4393–E4402. [CrossRef] [PubMed]
27. Shima, J.; Hino, A.; Yamada-Iyo, C.; Suzuki, Y.; Nakajima, R.; Watanabe, H.; Mori, K.; Takano, H. Stress tolerance in doughs of *Saccharomyces cerevisiae* trehalase mutants derived from commercial Baker's yeast. *Appl. Environ. Microbiol.* **1999**, *65*, 2841–2846. [CrossRef] [PubMed]

Journal of *Fungi*

MDPI

*Article*

# Production of a β-Glucosidase-Rich Cocktail from *Talaromyces amestolkiae* Using Raw Glycerol: Its Role for Lignocellulose Waste Valorization

**Juan A. Méndez-Líter, Laura I. de Eugenio, Neumara L. S. Hakalin, Alicia Prieto * and María Jesús Martínez ***

Centro de Investigaciones Biológicas Margarita Salas (CIB), Consejo Superior de Investigaciones Científicas (CSIC), Ramiro de Maeztu 9, 28040 Madrid, Spain; jmendez@cib.csic.es (J.A.M.-L.); lidem@cib.csic.es (L.I.d.E.); nhakalin@gmail.com (N.L.S.H.)

* Correspondence: aliprieto@cib.csic.es (A.P.); mjmartinez@cib.csic.es (M.J.M.)

**Abstract:** As β-glucosidases represent the major bottleneck for the industrial degradation of plant biomass, great efforts are being devoted to discover both novel and robust versions of these enzymes, as well as to develop efficient and inexpensive ways to produce them. In this work, raw glycerol from chemical production of biodiesel was tested as carbon source for the fungus *Talaromyces amestolkiae* with the aim of producing enzyme β-glucosidase-enriched cocktails. Approximately 11 U/mL β-glucosidase was detected in these cultures, constituting the major cellulolytic activity. Proteomic analysis showed BGL-3 as the most abundant protein and the main β-glucosidase. This crude enzyme was successfully used to supplement a basal commercial cellulolytic cocktail (Celluclast 1.5 L) for saccharification of pretreated wheat straw, corroborating that even hardly exploitable industrial wastes, such as glycerol, can be used as secondary raw materials to produce valuable enzymatic preparations in a framework of the circular economy.

**Keywords:** biodiesel by-products; cellulases; fungi; hydrolases; wastes

**Citation:** Méndez-Líter, J.A.; de Eugenio, L.I.; Hakalin, N.L.S.; Prieto, A.; Martínez, M.J. Production of a β-Glucosidase-Rich Cocktail from *Talaromyces amestolkiae* Using Raw Glycerol: Its Role for Lignocellulose Waste Valorization. *J. Fungi* **2021**, *7*, 363. https://doi.org/10.3390/jof7050363

Academic Editor: Baojun Xu

Received: 14 April 2021
Accepted: 3 May 2021
Published: 6 May 2021



## 1. Introduction

The current generation of wastes at worldwide level is a direct consequence of the inefficient use of natural resources in human activities. Industry has traditionally focused on the production of goods and services, consuming raw materials and generating wastes that remain in the environment. Valorization of the wide variety of mankind-produced waste has become an essential task for a sustainable world [1]. On the other hand, the need for energy and commodities is continuously increasing, while fossil fuels reserves are declining and finite. Therefore, it is imperative to learn how to use waste from other processes as secondary raw materials to produce alternative energy sources and materials.

Many carbon-rich industrial or urban residual streams are susceptible to re-valorization into value added products. In this sense, biodiesel is a renewable biofuel that can be produced from vegetal oils, animal fat, synthetic algae, or cooking oil waste [2]. In fact, fuels from renewable biomass are the cleanest sources of energy, having the potential to reduce the amount of $CO_2$, nitrogen oxides (NOx), sulfur oxides (SOx), CO, and other hazardous particles released to the atmosphere compared with fossil diesel [3]. One of the main problems derived from industrial production of biodiesel is the generation of huge amounts of glycerol as byproduct (10:1 ratio). The global biodiesel production has increased during the last decade, reaching more than 30.8 $Mm^3$ in 2016, and its production will achieve 41 $Mm^3$ in 2022 [2], which will be translated into millions of $m^3$ of glycerol. Then, in the global market, glycerol has become an abundant raw material to be used, mainly in the chemical industry. However, the current industrial process for synthesis of biodiesel generates a by-product containing about 70–80% glycerol together with water, catalysts, fatty acids, and salts. This makes the commercialization and valorization of this by-product difficult, since the glycerol required as raw material for pharmaceutical, cosmetic, and food industries requires high purity standards.

As the structural component of many lipids, glycerol is widely found in nature and some microorganisms can naturally use it to grow in the absence of additional carbon sources. These organisms can be potent tools for bioconversion of glycerol from biodiesel manufacturing into valuable microbial products, such as enzymes with industrial interest, medicinal drugs, antibiotics, and other chemicals [4–6]. Glycerol has substituted typical carbohydrates, such as sucrose or glucose, in industrial fermentations for production of 1,3-propanediol, dihydroxyacetone, succinic acid, propionic acid, citric acid, pigments, polyhydroxyalkanoates, biosurfactants, docosahexaenoic acid, lipids, or syngas [7].

On the other hand, lignocellulosic biomass can be converted into fermentable sugars by the synergistic action of a battery of enzymes including, exo- and endoglucanases and β-glucosidases, as well as other auxiliary enzymes [8]. Among them, β-glucosidases are essential to release free glucose, but commercial cellulase preparations, especially those from *Trichoderma* spp., are usually deficient in this activity and must be supplemented to increase their efficiency in saccharification processes. In this context, *Talaromyces amestolkiae* has been studied for its ability to secrete a powerful cellulosic cocktail with several β-glucosidases depending on the carbon source in the culture media [9]. Analysis of the secretomes released in those media revealed two mayor β-glucosidases. One of them, named as BGL-2, was induced exclusively by cellulosic substrates [10], while the other, BGL-3, was produced independently of culture carbon source [11]. BGL-3 is a versatile 1,4-β-glucosidase that has proved to be very efficient for saccharification of wheat straw, but also highly active over 1,3-β-glucose bonds [11].

In this work we propose the use of raw glycerol as an appropriate carbon source to produce efficient enzyme cocktails with high β-glucosidase levels using the fungus *T. amestolkiae*. The efficiency of these cocktails for degradation of lignocellulosic residues was evaluated, and plant biomass valorization is discussed in the framework of a circular economy approach.

## 2. Materials and Methods

### 2.1. Chemicals, Microorganisms, Culture Media, and Crude Enzyme Extract Production

Raw glycerol was obtained by separation of biodiesel after chemical transesterification of waste cooking oil and sodium metoxide [12]. It was directly supplied as carbon source in a basal medium for fungal growth. All other chemicals used in this work were of reagent grade and purchased from Sigma-Aldrich.

*T. amestolkiae* was isolated from cereal waste and included in the collection of the Institute Jaime Ferrán of Microbiology (IJFM) with the number A795. For spore production, the fungus was grown in potato dextrose agar (PDA, Difco) for 5 days at 28 °C, and a suspension was obtained by placing 1 $cm^2$ agar in 5 mL of a 1% NaCl solution with 0.1% of Tween 80. Pre-inocula were prepared by inoculating 200 μL of this homogenous spore suspension in 250 mL flasks containing 50 mL of corn steep solid (CSS) medium (40 g/L glucose, 0.4 g/L $FeSO_4 \cdot 7H_2O$, 9 g/L $(NH_4)_2SO_4$, 4 g/L $K_2HPO_4$, 26.3 g/L corn steep solid, 7 g/L $CaCO_3$, and 2.8 mL/L soybean oil), previously autoclaved at 110 °C for 20 min. The cultures were incubated at 28 °C and 250 rpm for 5 days.

Enzyme production was carried out in 1 L flasks containing 200 mL of Mandels minimal medium (2.0 g/L $KH_2PO_4$, 1.3 g/L $(NH_4)_2SO_4$, 0.3 g/L urea, 0.3 g/L $MgSO_4 \cdot 7H_2O$, 0.3 g/L $CaCl_2$, 5 mg/L $FeSO_4 \cdot 7H_2O$, 1.6 mg/L $MnSO_4 \cdot H2O$, 1.4 mg/L $ZnSO_4 \cdot 7H_2O$, and 1 g/L Bacto peptone) [13], supplemented with 0.5%, 1%, and 2% (*w/v*) raw glycerol as the carbon source, both unbuffered (pH 4.5) and buffered with 100 mM sodium phosphate pH 6. After autoclaving at 121 °C for 20 min, which was also useful for glycerol sterilization, the flasks were inoculated with 2 mL of the pre-inoculum, and incubated at 28 °C and 250 rpm for 10 days, taking 1.5 mL aliquots daily for analytical determinations. After that time, cultures were filtered, concentrated, and dialyzed against a pH 4 sodium acetate buffer at 10 mM to produce the enzymatic cocktails. Concentration was achieved using an ultrafiltration cell (Amicon, Merck–Millipore, Kenilworth, NJ, USA) with a polysulfone membrane (3 kDa cut-off, Millipore). The process was carried out with gentle stirring, using nitrogen gas for pressure (2 bar).

### 2.2. Determination of Biomass, Proteins, Enzyme Activity, and Glycerol Consumption

Aliquots from the cultures were centrifuged and filtered. The content of *T. amestolkiae* biomass in the different samples was determined from the dry weight of the lyophilized pellets. The other parameters were measured in the supernatants. Protein concentration was evaluated by using the commercial Bio-Rad Bradford assay (dye reagent concentrate) (Bio-Rad, CA, USA) and bovine serum albumin as the standard. β-glucosidase activity was tested following the hydrolysis of 0.1% (*w/v*) *p*-nitrophenyl-β-D-glucopyranoside (*p*NPG, Sigma, St. Louis, MO, USA) at 50 °C, in a 100 mM sodium acetate buffer at pH 4. The reaction was stirred at 1200 rpm for 10 min in a Thermo shaker (T-100, SC-24 block), stopped by adding 2% (*w/v*) (1.42% final concentration) of $Na_2CO_3$, and the *p*-nitrophenol (*p*NP) released was spectrophotometrically measured at 410 nm ($\varepsilon_{410}$ = 15,200 $M^{-1}$ $cm^{-1}$). One unit of β-glucosidase activity was defined as the release of 1 µmol of *p*NP per minute. Avicelase (as indicative of exocellulase activity) and β-1,4-endoglucanase activities were measured by determining the release of reducing sugars by the Somogyi–Nelson method, as previously described [9], with 1% Avicel (Merck–Millipore) or 3% low viscosity carboxymethylcellulose (Sigma), respectively, as substrates. One unit of activity was defined as the corresponding to the release of 1 µmol of reducing sugar per minute. To determine glycerol consumption, the Glycerol Assay Kit (Sigma) was used according to the manufacturer's instructions.

All assays were performed in triplicate, and significant differences between samples were explored using Student's *t*-test, considering a $p$ value < 0.05 as the limit for considering differences.

### 2.3. Proteomic Analysis of T. amestolkiae Secretomes

For proteomic analysis, samples from culture supernatants of *T. amestolkiae* grown in Mandels medium plus 1% raw glycerol were analyzed. The process was carried out as previously described [9], with some minor variations. In brief, 2 mL of the liquid supernatants from 7-day-old cultures were centrifuged at 20,000× *g* for 10 min to remove mycelium. Then, aliquots containing 5 µg of proteins were first dissolved in SDS-PAGE loading buffer, denatured, loaded in a 12% SDS gel, and run for 10 min, in order to remove non-protein compounds before the proteomic analysis. The whole protein band was excised from the gel, cut in small pieces, and digested with trypsin overnight at 37 °C. Then the peptides mixture was extracted and analyzed by nano-HPLC–MS/MS.

A NanoEasy HPLC (Proxeon Biosystems, Odense, Denmark), coupled to a nano-electrospray ion source (Proxeon Biosystems), was used for all peptide separations. Samples were loaded onto a C18-A1 ASY-Column 2 cm precolumn (Thermo Scientific, MA, USA) and then eluted onto a Biosphere C18 column (C18, inner diameter 75 µm, 15 cm long, 3 µm particle size, Nano Separations, Nieuwkoop, Netherlands) as previously reported. Full-scan MS spectra (*m/z* 300–1800) were acquired in the LTQ-Orbitrap Velos in the positive ion mode (Figure S1).

Mass spectra files were searched in a specific database against the *T. amestolkiae* genome. Precursor and fragments mass tolerance were set to 10 ppm and 0.5 Da, respectively. Search parameters included a maximum of two missed cleavages allowed, carbamidomethylation of cysteines as a fixed modification, and oxidation of methionine as a variable modification. Peptides were validated through the algorithm Percolator [14], and only those with high and medium confidence were admitted (FDR 0.05). Protein identifications were only accepted if they contained at least two identified peptides. Results were inferred from data obtained from two technical replicates from two different biological samples. Relative quantification of the most abundant proteins in the samples analyzed was calculated from the number of peptide spectrum matches (PSMs) corresponding to each protein [15,16].

### 2.4. Saccharification of Pretreated Wheat Straw

The efficiency of *T. amestolkiae* enzyme cocktails as β-glucosidase supplements for saccharification was studied. For this, wheat straw subjected to three different pretreatments was used (Figure 1A): (i) acidic wheat straw (abbreviated as AcSE), obtained by steam explosion in the presence of dilute sulfuric acid (kindly provided by Abengoa Bioenergía, Babilafuente, Salamanca); (ii) neutral wheat straw (abbreviated as SE), obtained by steam explosion in the presence of water, also provided by Abengoa Bioenergía; (iii) and alkaline wheat straw (abbreviated as AP), prepared as follows: 5 g of wheat straw was incubated with 100 mL of 2% NaOH for 40 min and 110 °C using an autoclave, the solid material was then washed three times with water, the pH was adjusted to 5 with sulfuric acid, lyophilized, and stored. The composition of AP wheat straw was analyzed by sequential fractionation. In this process, the contents of cellulose, hemicellulose, and Klason lignin were determined through a quantitative acid hydrolysis [17]. In brief, the solid residue (300 mg) was treated for 1 h with 3 mL of 13.5 M sulfuric acid ($H_2SO_4$) at 30 °C and 150 rpm. Then, $H_2SO_4$ was diluted to 0.5 M and heated at 110 °C for 1 h. Sugars in the liquid fraction were identified and quantified by gas chromatography [18]. The slurry remaining after acid hydrolysis (Klason lignin) was washed until neutrality, oven dried at 55 °C, and weighted. All analyses were carried out in triplicate.

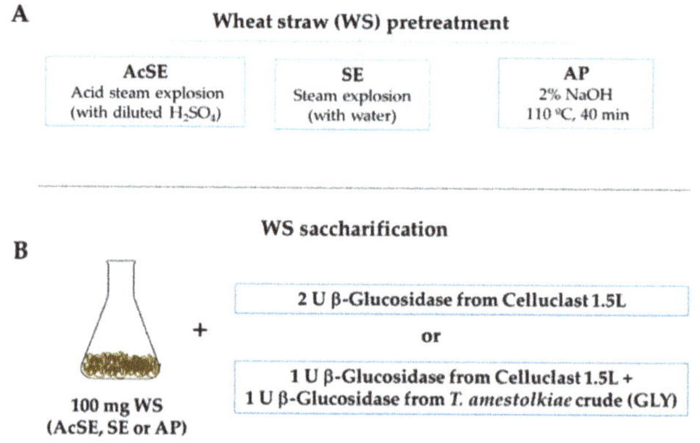


**Figure 1.** (**A**) Scheme of different wheat straw pretreatments used for enzymatic saccharification and (**B**) enzymatic doses used in the saccharification experiments.

For saccharification, 100 mg of AcSE, SE, or AP wheat straw was incubated in 2 mL 100 mM sodium acetate buffer at pH 4 for 96 h, at 40 °C and 1200 rpm in the presence of different enzymatic combinations (Figure 1B):

- 2 U of β-glucosidase activity from Celluclast 1.5 L (Novozymes, Copenhagen, Denmark). This is a commercial cocktail, rich in cellobiohydrolase and endoglucanase activities, and was used for comparison purposes;
- 1 U of β-glucosidase activity from Celluclast 1.5 L + 1 U of β-glucosidase activity from *T. amestolkiae* enzymatic crudes obtained in this work, using 1% glycerol as the carbon source.

## 3. Results

### 3.1. BGL Production and Growth of T. amestolkiae in Media with Raw Glycerol

The ascomycete *T. amestolkiae* has been proven as an excellent producer of robust and efficient cellulolytic enzymes in media with different carbon sources [9]. This finding suggested the convenience of testing cheaper carbon sources to obtain these cocktails, since they are rich in β-glucosidases, the key enzymes for cellulose saccharification.

In this work, we studied the use of different concentrations of raw glycerol, an abundant waste generated from biodiesel production, to produce added-value enzyme cocktails. To analyze the potential of this carbon source, we followed the evolution of fungal growth and the extracellular levels of β-glucosidase over 10-day-old cultures (240 h). As observed in Figure 2A, *T. amestolkiae* cultures in unbuffered media reached maximal biomass values around 24 h, regardless of the glycerol concentration assayed, although total biomass was higher as the glycerol concentration increased. The highest β-glucosidase levels were detected in cultures containing 1% or 2% glycerol (Figure 2B). BGL levels were low in the first 24–48 h, and then they started to rise uninterruptedly at a high rate between 72 h and 168 h. Surprisingly, both cultures produced around 8 U/mL at the final incubation time, despite that the fungal biomass was higher with 2% glycerol. Searching for an explanation to this result, it was observed that the pH was more acidic in the culture with 2% glycerol (pH 5) than in the other cultures (pH 6–6.5). The acidification was detected at 168 h and maintained until the end of the incubation period.

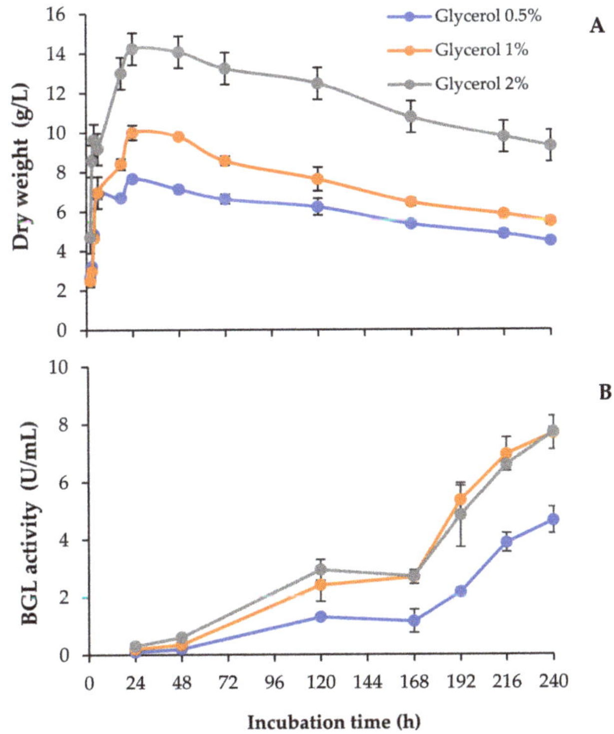


**Figure 2.** (**A**) Biomass and (**B**) β-glucosidase activity of *T. amestolkiae* growing in unbuffered cultures with raw glycerol (0.5%, 1%, and 2%), as the carbon source. All assays were performed in triplicate.

Thus, to investigate the effect of pH on BGL production, we repeated the experiment buffering all media with phosphate pH 6 to maintain the pH as constant across the incubation period. In these conditions, BGL activity and fungal growth went in parallel, and the highest values for both were detected in cultures with 2% glycerol (Figure 3A). It should also be noticed that the total amount of secreted proteins detected in 2% glycerol unbuffered cultures was considerably lower than in the buffered ones (0.10 and 0.17 mg/mL, respectively). These results suggest that pH should be strictly controlled in the cultures, since its value is related to the levels of secreted proteins. Buffered and unbuffered cultures with 0.5% and 1% glycerol followed a very similar behavior, where the pH was maintained in a pH range of 6–6.5.

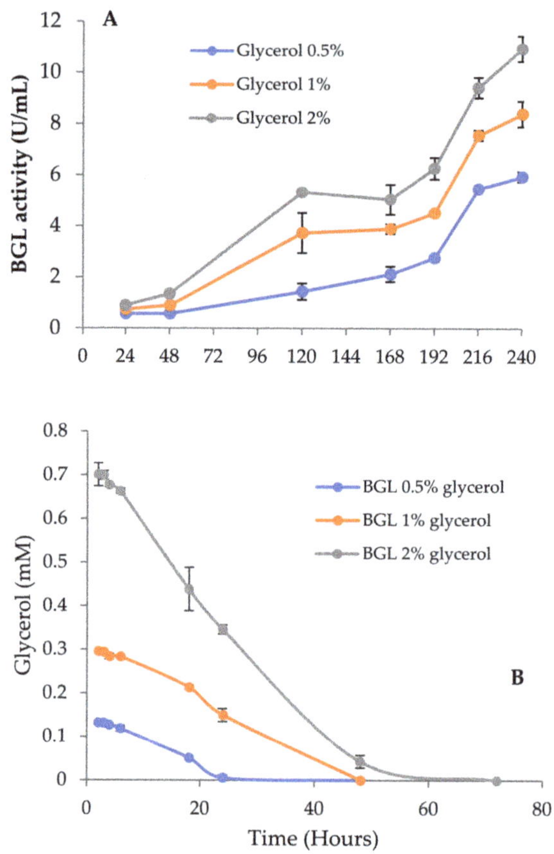


**Figure 3.** (**A**) β-glucosidase activity of *T. amestolkiae* growing in buffered cultures with raw glycerol (0.5%, 1%, and 2%), as the carbon source. (**B**) Glycerol consumption by *T. amestolkiae* in the buffered cultures. All assays were performed in triplicate.

On the other hand, these results confirm that *T. amestolkiae* does not require any specific inducer (cellulosic or non-cellulosic) to produce high amounts of β-glucosidases, as previously reported. This is probably due to the presence of BGL-3, an efficient β-glucosidase secreted under carbon starvation conditions [11]. BGL-3 is a β-1,4 glucosidase with a prominent activity on β-1,3 glucans. The high levels of this protein detected during secondary metabolism in the cultures may be due to its activity degrading the β-1,3-glucan released from the fungal cell wall during autolysis. As expected, the amount of other cellulolytic activities like avicelase or β-1,4-endoglucanase was negligible (data not shown), since these enzymes are usually released upon induction by cellulosic substrates.

Only a handful of studies have reported the production of fungal β-glucosidases using glycerol as the carbon source. For *Penicillium echinulatum*, glycerol was the preferred carbon source over cellulose, sugar cane bagasse (pretreated by steam explosion), or glucose [19]. *Aspergillus niger* NRRL 3112 could produce considerable amounts of this activity when grown on wheat bran and glycerol as co-substrates [20]. In other cases, the use of glycerol was not suitable for β-glucosidase production, as reported in *Penicillium funiculosum* [21]. *Trichoderma reesei* has been the most used fungus for producing cellulolytic cocktails, although it is well documented that the amount of β-glucosidase released by this fungus is insufficient for efficient hydrolysis of lignocellulosic biomass [22]. In this sense, one of the main advantages of the use of *Aspergillus* spp., *Penicillium* spp., or *Talaromyces* spp. lies in their higher secretion of β-glucosidase activity [22]. Many published reports allow us to compare its production among these fungal genera, mainly using crystalline cellulose as the carbon source. Different *Aspergillus* sp. were secreted between 0.5 U/mL and 2.5 U/mL of β-glucosidase [23,24]. *Penicillium* species, like *Penicillium brasilianum* or *Penicillium decumbens*, reached a total activity of 3.5 U/mL and 2.39 U/mL, respectively [25,26]. Higher β-glucosidase levels, from 10 to 150 U/mL, have been reported in strains of *P. funiculosum*, *Penicillium occitanis*, or *Penicillium verriculosum* [21,27,28] using cellulosic inducers. It should be noted that the amounts of BGL observed in *T. amestolkiae* cultures in media with raw glycerol as the sole carbon source are among the highest reported to date. The fact that this high production was achieved using a by-product of the biodiesel industry makes it even more interesting.

Glycerol consumption was monitored and related with the increase of BGL activity (Figure 3B). In cultures with 0.5% glycerol, the depletion of the carbon source occurred in 24 h, while in those with 1% and 2%, it took around 48 h. Low β-glucosidase levels were detected before glycerol exhaustion, but the activity started to rise slowly once consumed, and increased faster when carbon starvation conditions were fully established, expanding over time. The data in Figure 3B indicate that the maximal exploitation of glycerol for BGL production was obtained with 2% glycerol, so we chose this culture as the best model to produce BGL-rich cocktails.

The secretion of β-glucosidases in aged fungal cultures has been scarcely studied, although it would be interesting to know to what extent their production under carbon starvation is common among filamentous fungi. In the case of *T. amestolkiae*, it has been proved that induction of β-glucosidase BGL-3 was detected during carbon shortage. This enzyme is highly active on β-1,3 polysaccharides, suggesting its possible physiological role in cell wall metabolism during carbon starvation, releasing products that could be used as an alternative carbon source [11]. Some authors have reported the production of glycosyl hydrolases after carbon depletion in cultures of *A. niger* [29,30], but their finding was not discussed as a potential way to produce these enzymes. In any case, the versatility of *T. amestolkiae* to produce high levels of BGLs from different carbon sources, including some important industrial residues, should be exploited.

### 3.2. Fungal Secretome Analysis

An extracellular proteomic analysis of 7-day-old cultures was carried out to elucidate the protein profiles of *T. amestolkiae* cultures with 1% raw glycerol. The samples were subjected to tryptic digestion and LC–MS/MS of the whole peptide mixtures produced. The number of proteins identified was 148, similar to that reported in cultures with glucose or cellulosic substrates [9]. In the TIC scan of the proteomic analysis (Figure S1), an archive with al peptides was detected (Table S1), and a list with the 148 proteins detected (Table S2) can be found in the Supplementary Material Section of the article.

According to KOG, the functional role of most of the proteins identified in cultures with 1% raw glycerol was related to carbohydrate metabolism and transport (55.0%), followed by enzymes involved in amino acid metabolism and transport (Table 1). These data also agree with those previously published for *T. amestolkiae* secretomes with glucose as the carbon source [9]. As the main objective of this work was to produce BGLs from contaminated glycerol waste, we started our analysis classifying the CAZymes (carbohydrate active enzymes), finding 80 out of 141 detected proteins (54%). Among them, the most abundant were glycosyl hydrolases (GHs) from GH3, GH15, GH31, and GH55 families (Table 2). It is worth noting that GH15 enzymes were predominant in cultures with glucose as the carbon source [9], while in media with glycerol, GH3 proteins were the most abundant. The family GH3 is normally associated with enzymes degrading polysaccharides from lignocellulosic biomass, including many β-glucosidases, which supports the suitability of using this carbon source for β-glucosidase production. As expected, BGL-3 was found in the studied secretome (Table 3).

As already mentioned, this enzyme is secreted under carbon starvation [11], but other GHs potentially related with carbon depletion were also detected. This is the case of glucoamylases (GH15), α-glucosidases (GH31), and exo-β-1,3-glucanases (GH55), related to the fungal autophagy process, since 1,3-β-glucans and 1,4-α-glucan are components of the cell wall of many *Talaromyces* species [31]. Besides, it is noticeable that some of the most abundant proteins are related with the fungal autolysis processes. For example, cathepsin, which was detected in every condition, is a well-known protease that could be used by *T. amestolkiae* for degrading proteins and using its components as nutrients. The amino acids produced by cathepsin can be utilized by glutaminase, which has been related with glutamine metabolism under starvation conditions, when no other carbon source is present in the medium.

**Table 1.** Functional classification of the proteins identified in the secretome of 7-day-old *T. amestolkiae* cultures with raw glycerol as the carbon source, compared to those obtained in the same medium with glucose [9].

| | % PSM | |
|---|---|---|
| | Glycerol | Glucose |
| A—RNA processing and modification | 0.63 | 0.22 |
| C—Energy production and conversion | 2.72 | 4.90 |
| E—Amino acid metabolism and transport | 13.11 | 10.38 |
| F—Nucleotide metabolism and transport | 0.74 | 1.42 |
| G—Carbohydrate metabolism and transport | 55.01 | 65.16 |
| I—Lipid metabolism | 0.09 | 0.17 |
| M—Cell wall/membrane/envelop biogenesis | 0.77 | 3.61 |
| O—Post-translational modification. protein turnover. chaperone functions | 1.96 | 1.71 |
| Q—Secondary structure | 2.55 | 0.91 |
| R—General functional prediction only | 4.81 | 1.89 |
| S—Function unknown | 5.79 | 3.71 |
| T—Signal transduction | 4.13 | 4.86 |

**Table 2.** Glycosyl hydrolase families identified in secretomes from 7-day-old *T. amestolkiae* cultures growing in raw glycerol, compared to those obtained in the same medium with glucose as the carbon source [9].

| | % PSM | |
|---|---|---|
| GH Family | Glycerol | Glucose |
| GH2 | 3.2 | 1.7 |
| GH3 | 16.6 | 16.3 |
| GH13 | 3.0 | 4.3 |
| GH15 | 10.1 | 28.4 |
| GH18 | 2.8 | 0.6 |
| GH20 | 3.7 | 4.8 |
| GH27 | 2.1 | 1.7 |
| GH31 | 8.6 | 11.8 |
| GH35 | 2.7 | 1.8 |
| GH55 | 8.1 | 3.9 |
| GH71 | 4.0 | 0.1 |
| GH72 | 3.2 | 1.3 |
| GH92 | 5.3 | 1.1 |
| GH127 | 3.6 | 3.2 |

**Table 3.** Most abundant extracellular proteins identified in the *T. amestolkiae* secretome obtained from 7-day-old cultures growing in raw glycerol. BGL-3 is indicated in bold.

| Accession ID | % PSM (Average) | Predicted Protein Function | Cazyme Family | $M_w$ (kDa) |
|---|---|---|---|---|
| **g377 (BGL-3)** | **7.09** | **beta-glucosidase** | **GH3** | **88.7** |
| g3995 | 6.47 | Glutaminase | - | 76.4 |
| g8295 | 3.59 | alpha-glucosidase | GH31 | 98.6 |
| g2158 | 3.28 | Glucoamylase | GH15 | 65.2 |
| g9324 | 3.18 | Exo-beta-1,3-glucanase | GH55 | 84.3 |
| g2140 | 2.50 | Glucoamylase | GH15 | 67.7 |
| g5915 | 2.23 | non-reducing end β-L-arabinofuranosidase | GH127 | 68.8 |
| g4076 | 2.00 | hexosaminidase | GH20 | 67.9 |
| g216 | 1.77 | neutral/alkaline nonlysosomal ceramidase | - | 160.0 |
| g9148 | 1.58 | catalase | - | 79.1 |

Regarding β-glucosidases, the most important result was the confirmation of the presence of BGL-3 as the most abundant extracellular protein. This is in good agreement with our previous results describing its production upon carbon exhaustion [9]. Besides, one GH1, recently characterized [32], and four GH3 β-glucosidases were also identified (Table 4). This profusion of β-glucosidases could contribute to explain why this activity was so high in this medium, as compared with other conditions

**Table 4.** Main hypothetical BGLs detected identified in the *T. amestolkiae* secretome obtained from 7-day-old cultures with raw glycerol as carbon source.

| Accession ID | % PSM (Average) | Cazyme Family | $M_w$ (kDa) |
|---|---|---|---|
| g377 (BGL-3) | 7.09 | GH3 | 88.7 |
| g9150 | 1.54 | GH3 | 86.5 |
| g8384 | 0.85 | GH1 | 68.1 |
| g6857 | 0.79 | GH3 | 109 |
| g3139 | 0.30 | GH3 | 93.6 |
| g6753 | 0.09 | GH3 | 81.8 |

In summary, the proteomic analysis of *T. amestolkiae* growing in raw glycerol confirmed the presence of BGL-3 and revealed that other interesting enzymes, including different β-glucosidases, can also be obtained for different biotechnological applications by exploiting the carbon starvation metabolism of *T. amestolkiae*.

### 3.3. Wheat Straw Saccharification

To check the efficiency of the novel enzyme cocktail obtained in this work for saccharification of lignocellulosic biomass, we used it as supplement of β-glucosidase activity for Celluclast 1.5 L (a basal cellulolytic preparation, deficient in this activity). Since lignocellulosic substrates have different characteristics, the most effective pretreatment and enzyme cocktail should be used, considering the properties of each raw material. Wheat straw is one of the most important agricultural residues worldwide. According to Iskalieva et al. [33], it is mainly composed of cellulose (38.8%), hemicelluloses (39.5%), and lignin (17.1%). In the present work, three different pretreatments were used to make wheat straw cellulose more accessible: steam explosion (SE), steam explosion in the presence of dilute sulfuric acid (AcSE), and alkaline pretreatment (AP) (Figure 1). Steam explosion is one of the most commonly used methods for lignocellulose pretreatment, deconstructing and modifying part of the lignin, and solubilizing hemicellulose [34,35]. However, although cellulose is more accessible in the steam-exploded biomass, compounds derived from the partial hydrolysis of sugars and lignin remain embedded, producing adverse effects on downstream processes, including enzyme inhibition [36]. The use of dilute sulfuric acid could theoretically avoid this problem, since it helps to remove a bigger part of the lignin, but some of the polymer is always present in the material. On the other hand, the alkaline pretreatment with NaOH has been reported to increase cellulose digestibility by inducing a big reduction of the lignin content [37]. These pretreatments affect the composition of the material used for saccharification differently (Table 5).

**Table 5.** Main components of wheat straw pretreated by steam explosion in water (SE), steam explosion with dilute sulfuric acid (AcSE), or alkaline pretreatment (AP).

| | SE | AcSE | AP |
|---|---|---|---|
| Cellulose | 49.0% | 43.6% | 71.8% |
| Hemicellulose | 15.4% | 17.1% | 24.1% |
| Lignin | 35.6% | 39.3% | 4.1% |

In the case of alkaline pretreatment, most lignin was solubilized and removed, leaving a solid fraction enriched in polysaccharides. The two steam explosion treatments gave wheat straw slurries with different cellulose and hemicellulose contents. The AcSE treatment (dilute sulfuric acid) more strongly affected hemicellulose and lignin, forming compounds that produce adverse effects on downstream processes [38]. The results of the saccharification of pretreated wheat straw can be seen in Figure 4.

The combination of Celluclast 1.5 L with the *T. amestolkiae* cocktail improved the saccharification yield of steam-exploded wheat straw (with or without sulfuric acid), compared with Celluclast 1.5 L alone, reaching a cellulose conversion of 65% for SE and of 86% for AcSE. The commercial Celluclast 1.5 L contains an array of cellulolytic and hemicellulolytic activities in addition to BGL, which could theoretically improve saccharification. However, cellulose degradation with this cocktail was 41% for SE and 53% for AcSE.

On the other hand, the saccharification of alkali-pretreated wheat straw was similar with both cocktails (86% for only Celluclast and 89% when supplemented with β-glucosidase from *T. amestolkiae*). One possible reason for this is the virtual lack of lignin in the alkali-pretreated wheat straw. It has been reported that lignin could trigger an irreversible cellulase inactivation, even when using pretreated lignocellulosic biomass, by two possible mechanisms: (i) forming a physical barrier that prevents enzyme access, and (ii) by non-productively binding of cellulolytic enzymes [34,39]. The non-productive binding of enzymes to lignin is recognized as a problem to overcome for improving saccharification efficiency of pretreated lignocellulosic biomass to fermentable sugars. A feasible explanation for our results is that the enzymes from the BGL-rich cocktails from *T. amestolkiae* were better than those contained in Celluclast 1.5 L, avoiding the non-productive binding to lignin. This observation perfectly correlates with some studies that suggest that β-glucosidases and xylanases could avoid non-productive binding to lignin better than other cellulolytic enzymes [40]. Besides, it is remarkable that BGL-3, the most abundant cellulase in *T. amestolkiae* cocktails, possesses a fibronectin III domain (FnIII) [11]. Lima et al. [41], postulated that the FnIII domain strongly interacts with lignin fragments, preventing unproductive binding of cellulases to the lignin. Taking this into consideration, BGL-3 rich cocktails may be advantageous for the saccharification of pretreated residues where lignin is present, representing an effective and cheap alternative to current commercial enzymatic preparations.

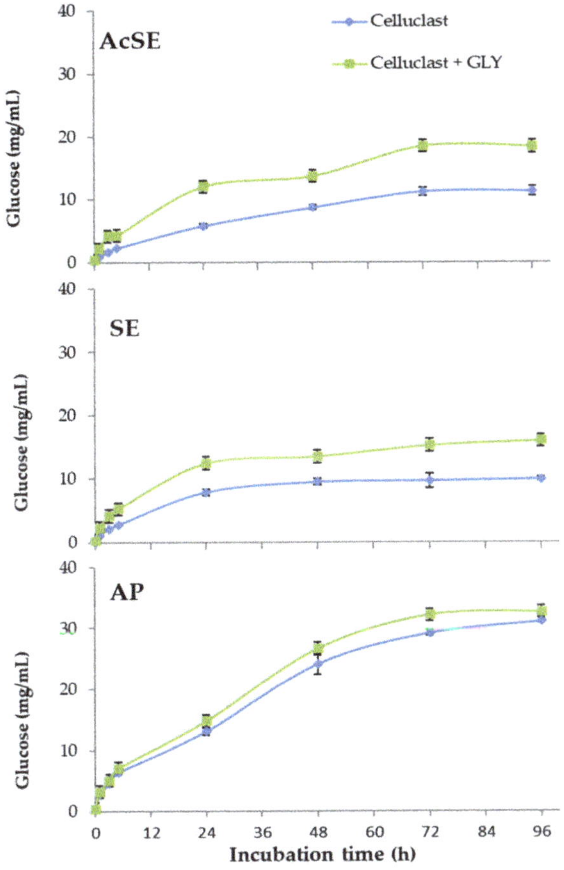


**Figure 4.** Saccharification of pretreated wheat straw (AcSE: acidic steam explosion; SE: steam explosion; AP: alkali pretreatment) with Celluclast 1.5 L (basal cocktail) and Celluclast 1.5 L supplemented with the *T. amestolkiae* cocktail obtained with raw glycerol as the carbon source (GLY). All assays were performed in triplicate.

## 4. Conclusions

In this work, we demonstrated the potential of the fungus *T. amestolkiae* for producing high levels of β-glucosidase activity using raw glycerol as the carbon source. Proteomic analysis of the crudes revealed that, although other minor glycosyl hydrolases were also found in the secretomes, BGL-3 was the main enzyme responsible for this behavior. This novel enzymatic cocktail was efficiently used to supplement the commercial cellulolytic preparation Celluclast 1.5 L, increasing the glucose yield from wheat straw pretreated by steam explosion. This work set the foundations for expanding the industrial applications of *T. amestolkiae* and exploiting its efficient metabolism, able to use inexpensive carbon sources and to produce β-glucosidases and other value-added enzymes under carbon starvation.

**Supplementary Materials:** Supplementary materials are available online at https://www.mdpi.com/article/10.3390/jof7050363/s1.

**Author Contributions:** J.A.M.-L. and L.I.d.E. performed the experiments, the analysis of the results, and drafted the manuscript. N.L.S.H. performed the pretreatments of the wheat straw. M.J.M. and A.P. planned, reviewed, and edited the manuscript. All authors have read and agreed to the published version of the manuscript.

**Funding:** This work has been funded by Projects GLYSUS RTI2018-093683-B-I00 (MCIU/AEI/FEDER), RETOPROSOST2 S2018/EMT-4459 (Comunidad de Madrid) and BioSFerA EU-H2020-NMBP-TR-IND.

**Institutional Review Board Statement:** Not applicable.

**Informed Consent Statement:** Not applicable.

**Data Availability Statement:** *T. amestolkiae* whole genome shotgun project is deposited at DDBJ/ENA/GenBank under the accession MIKG00000000.

**Acknowledgments:** The authors thank the Proteomics and Genomics facility at CIB, Novozymes for providing the commercial cocktails, and Abengoa for supplying pretreated wheat straw. The authors acknowledge support of the publication fee by the CSIC Open Access Publication Support Initiative through its Unit of Information Resources for Research (URICI).

**Conflicts of Interest:** The authors declare that they have no competing interests.

## References


1. Giusti, L. A review of waste management practices and their impact on human health. *WASTE Manag.* **2009**, *29*, 2227–2239. [CrossRef] [PubMed]
2. Monteiro, M.R.; Kugelmeier, C.L.; Pinheiro, R.S.; Batalha, M.O.; da Silva César, A. Glycerol from biodiesel production: Technological paths for sustainability. *Renew. Sustain. Energy Rev.* **2018**, *88*, 109–122. [CrossRef]
3. Lapinskiene, A.; Martinkus, P.; Rebzdaite, V. Eco-toxicological studies of diesel and biodiesel fuels in aerated soil. *Environ. Pollut.* **2006**, *142*, 432–437. [CrossRef] [PubMed]
4. Tang, S.; Boehme, L.; Lam, H.; Zhang, Z. Pichia pastoris fermentation for phytase production using crude glycerol from biodiesel production as the sole carbon source. *Biochem. Eng. J.* **2009**, *43*, 157–162. [CrossRef]
5. Luo, X.; Ge, X.; Cui, S.; Li, Y. Value-added processing of crude glycerol into chemicals and polymers. *Bioresour. Technol.* **2016**, *215*, 144–154. [CrossRef]
6. Magdouli, S.; Guedri, T.; Tarek, R.; Brar, S.K.; Blais, J.F. Valorization of raw glycerol and crustacean waste into value added products by Yarrowia lipolytica. *Bioresour. Technol.* **2017**, *243*, 57–68. [CrossRef]
7. Yang, F.; Hanna, M.A.; Sun, R. Value added use crude glycerol-a byproduct of biodiesel production. *Biotechnol. Biofuels* **2012**, *5*, 1–10.
8. Guo, H.; Chang, Y.; Lee, D.J. Enzymatic saccharification of lignocellulosic biorefinery: Research focuses. *Bioresour. Technol.* **2018**, *252*, 198–215. [CrossRef]
9. de Eugenio, L.; Méndez-Líter, J.; Nieto-Domínguez, M.; Alonso, L.; Gil-Muñoz, J.; Barriuso, J.; Prieto, A.; Martínez, M. Differential β-glucosidase expression as a function of carbon source availability in Talaromyces amestolkiae: A genomic and proteomic approach. *Biotechnol. Biofuels* **2017**, *10*, 1–14. [CrossRef]
10. Méndez-Líter, J.A.; Gil-Muñoz, J.; Nieto-Domínguez, M.; Barriuso, J.; De Eugenio, L.I.; Martínez, M.J. A novel, highly efficient β-glucosidase with a cellulose-binding domain: Characterization and properties of native and recombinant proteins. *Biotechnol. Biofuels* **2017**, *10*, 1–15. [CrossRef]
11. Méndez-Líter, J.A.; de Eugenio, L.I.; Prieto, A.; Martínez, M.J. The β-glucosidase secreted by Talaromyces amestolkiae under carbon starvation: A versatile catalyst for biofuel production from plant and algal biomass. *Biotechnol. Biofuels* **2018**, *11*, 1–14. [CrossRef]
12. de Sousa, M.H.; da Silva, A.S.F.; Correia, R.C.; Leite, N.P.; Bueno, C.E.G.; dos Santos Pinheiro, R.L.; de Santana, J.S.; da Silva, J.L.; Sales, A.T.; de Souza, C.C.; et al. Valorizing municipal organic waste to produce biodiesel, biogas, organic fertilizer, and value-added chemicals: An integrated biorefinery approach. *Biomass Convers. Biorefinery* **2021**. [CrossRef]
13. Mandels, M.; Weber, J. The production of cellulases. In *Cellulases and Their Applications*; American Chemistry Society: Washington, DC, USA, 1969; pp. 391–414. ISBN 0-8412-0095-5.
14. Kall, L.; Krogh, A.; Sonnhammer, E.L.L. A combined transmembrane topology and signal peptide prediction method. *J. Mol. Biol.* **2004**, *338*, 1027–1036. [CrossRef]
15. Liu, H.B.; Sadygov, R.G.; Yates, J.R. A model for random sampling and estimation of relative protein abundance in shotgun proteomics. *Anal. Chem.* **2004**, *76*, 4193–4201. [CrossRef]
16. Schulze, W.X.; Usadel, B. Quantitation in Mass-Spectrometry-Based Proteomics. *Annu. Rev. Plant Biol.* **2010**, *61*, 491–516. [CrossRef]
17. Mardones, W.; Di Genova, A.; Cortés, M.P.; Travisany, D.; Maass, A.; Eyzaguirre, J. The genome sequence of the soft-rot fungus Penicillium purpurogenum reveals a high gene dosage for lignocellulolytic enzymes. *Mycology* **2018**, *9*, 59–69. [CrossRef]
18. Notararigo, S.; Nacher-Vazquez, M.; Ibarburu, I.; Laura Werning, M.; de Palencia, P.; Teresa Duenas, M.; Aznar, R.; Lopez, P.; Prieto, A. Comparative analysis of production and purification of homo- and hetero-polysaccharides produced by lactic acid bacteria. *Carbohydr. Polym.* **2013**, *93*, 57–64. [CrossRef]

19. Schneider, W.D.H.; Goncalves, T.A.; Uchima, C.A.; Couger, M.B.; Prade, R.; Squina, F.M.; Pinheiro Dillon, A.J.; Camassola, M. Penicillium echinulatum secretome analysis reveals the fungi potential for degradation of lignocellulosic biomass. *Biotechnol. Biofuels* **2016**, *9*. [CrossRef]
20. Abdella, A.; Mazeed, T.E.-S.; El-Baz, A.F.; Yang, S.-T. Production of beta-glucosidase from wheat bran and glycerol by Aspergillus niger in stirred tank and rotating fibrous bed bioreactors. *Process Biochem.* **2016**, *51*, 1331–1337. [CrossRef]
21. Ramani, G.; Meera, B.; Vanitha, C.; Rao, M.; Gunasekaran, P. Production, Purification, and Characterization of a beta-Glucosidase of Penicillium funiculosum NCL1. *Appl. Biochem. Biotechnol.* **2012**, *167*, 959–972. [CrossRef]
22. Vaishnav, N.; Singh, A.; Adsul, M.; Dixit, P.; Sandhu, S.K.; Mathur, A.; Puri, S.K.; Singhania, R.R. Penicillium: The next emerging champion for cellulase production. *Bioresour. Technol. Rep.* **2018**, *2*, 131–140. [CrossRef]
23. Decker, C.H.; Visser, J.; Schreier, P. beta-glucosidases from five black Aspergillus species: Study of their physico-chemical and biocatalytic properties. *J. Agicult. Food Chem.* **2000**, *48*, 4929–4936. [CrossRef] [PubMed]
24. Riou, C.; Salmon, J.M.; Vallier, M.J.; Gunata, Z.; Barre, P. Purification, characterization, and substrate specificity of a novel highly glucose-tolerant beta-glucosidase from Aspergillus oryzae. *Appl. Environ. Microbiol.* **1998**, *64*, 3607–3614. [CrossRef] [PubMed]
25. Jorgensen, H.; Olsson, L. Production of cellulases by Penicillium brasilianum IBT 20888 - Effect of substrate on hydrolytic performance. *Enzyme Microb. Technol.* **2006**, *38*, 381–390. [CrossRef]
26. Sun, X.; Liu, Z.; Zheng, K.; Song, X.; Qu, Y. The composition of basal and induced cellulase system in Penicillium decumbens under induction or repression conditions. *Enzym. Microb. Technol.* **2008**, *42*, 560–567. [CrossRef]
27. Chaabouni, S.E.; Hadjtaieb, N.; Mosrati, R.; Ellouz, R. Preliminary assessment of Penicillium occitanis cellulase: A futher useful system. *Enzyme Microb. Technol.* **1994**, *16*, 538–542. [CrossRef]
28. Solov'eva, I.V.; Okunev, O.N.; Vel'kov, V.V.; Koshelev, A.V.; Bubnova, T.V.; Kondrat'eva, E.G.; Skomarovskii, A.A.; Sinitsyn, A.P. The selection and properties of Penicillium verruculosum mutants with enhanced production of cellulases and xylanases. *Microbioloy* **2005**, *74*, 141–146. [CrossRef]
29. Nitsche, B.M.; Jorgensen, T.R.; Akeroyd, M.; Meyer, V.; Ram, A.F.J. The carbon starvation response of Aspergillus niger during submerged cultivation: Insights from the transcriptome and secretome. *BMC Genom.* **2012**, *13*. [CrossRef]
30. van Munster, J.M.; Daly, P.; Delmas, S.; Pullan, S.T.; Blythe, M.J.; Malla, S.; Kokolski, M.; Noltorp, E.C.M.; Wennberg, K.; Fetherston, R.; et al. The role of carbon starvation in the induction of enzymes that degrade plant-derived carbohydrates in Aspergillus niger. *Fungal Genet. Biol.* **2014**, *72*, 34–47. [CrossRef]
31. Prieto, A.; Bernabe, M.; Leal, J.A. Isolation, purification and chemical characterization of alkali-extractable polysacharides from the cell-walls of Talaromyces species. *Mycol. Res.* **1995**, *99*, 69–75. [CrossRef]
32. Méndez-Líter, J.A.; Nieto-Domínguez, M.; Fernández De Toro, B.; González Santana, A.; Prieto, A.; Asensio, J.L.; Cañada, F.J.; De Eugenio, L.I.; Martínez, M.J. A glucotolerant β-glucosidase from the fungus Talaromyces amestolkiae and its conversion into a glycosynthase for glycosylation of phenolic compounds. *Microb. Cell Fact.* **2020**, *19*, 1–13. [CrossRef]
33. Iskalieva, A.; Yimmou, B.M.; Gogate, P.R.; Horvath, M.; Horvath, P.G.; Csoka, L. Cavitation assisted delignification of wheat straw: A review. *Ultrason. Sonochem.* **2012**, *19*, 984–993. [CrossRef]
34. Pan, X.J.; Xie, D.; Gilkes, N.; Gregg, D.J.; Saddler, J.N. Strategies to enhance the enzymatic hydrolysis of pretreated softwood with high residual lignin content. *Appl. Biochem. Biotechnol.* **2005**, *121*, 1069–1079. [CrossRef]
35. Galbe, M.; Wallberg, O. Pretreatment for biorefineries: A review of common methods for efficient utilisation of lignocellulosic materials. *Biotechnol. Biofuels* **2019**, *12*, 1–26. [CrossRef] [PubMed]
36. Garcia-Aparicio, M.; Ballesteros, I.; Gonzalez, A.; Oliva, J.; Ballesteros, M.; Negro, M. Effect of inhibitors released during steam-explosion pretreatment of barley straw on enzymatic hydrolysis. *Appl. Biochem. Biotechnol.* **2006**, *129*, 278–288. [CrossRef]
37. Kumar, R.; Wyman, C.E. Effects of cellulase and xylanase enzymes on the deconstruction of solids from pretreatment of poplar by leading technologies. *Biotechnol. Prog.* **2009**, *25*, 302–314. [CrossRef]
38. Moreno, A.D.; Ibarra, D.; Alvira, P.; Tomás-Pejó, E.; Ballesteros, M. A review of biological delignification and detoxification methods for lignocellulosic bioethanol production. *Crit. Rev. Biotechnol.* **2015**, *35*, 342–354. [CrossRef]
39. Gao, D.; Haarmeyer, C.; Balan, V.; Whitehead, T.A.; Dale, B.E.; Chundawat, S.P.S. Lignin triggers irreversible cellulase loss during pretreated lignocellulosic biomass saccharification. *Biotechnol. Biofuels* **2014**, *7*. [CrossRef]
40. Yarbrough, J.M.; Mittal, A.; Mansfield, E.; Taylor II, L.E.; Hobdey, S.E.; Sammond, D.W.; Bomble, Y.J.; Crowley, M.F.; Decker, S.R.; Himmel, M.E.; et al. New perspective on glycoside hydrolase binding to lignin from pretreated corn stover. *Biotechnol. Biofuels* **2015**, *8*. [CrossRef]
41. Lima, M.A.; Oliveira-Neto, M.; Kadowaki, M.A.S.; Rosseto, F.R.; Prates, E.T.; Squina, F.M.; Leme, A.F.P.; Skaf, M.S.; Polikarpov, I. Aspergillus niger beta-Glucosidase Has a cellulase-like tadpole molecular shape: Insights into gycoside hydrolase famyly 3 (GH3) beta-glucosidase structure and function. *J. Biol. Chem.* **2013**, *288*, 32991–33005. [CrossRef]

Journal of *Fungi*

MDPI

*Article*

# Bioethanol Production from Cellulose-Rich Corncob Residue by the Thermotolerant *Saccharomyces cerevisiae* TC-5

**Pinpanit Boonchuay [1], Charin Techapun [1], Noppol Leksawasdi [2], Phisit Seesuriyachan [1], Prasert Hanmoungjai [1], Masanori Watanabe [3], Siraprapa Srisupa [1] and Thanongsak Chaiyaso [1,*]**

1 Division of Biotechnology, Faculty of Agro-Industry, Chiang Mai University, Chiang Mai 50100, Thailand; pinpanit_boonchuay@cmu.ac.th (P.B.); charin.t@cmu.ac.th (C.T.); phisit.s@cmu.ac.th (P.S.); prasert.h@cmu.ac.th (P.H.); Siraprapa_sri@cmu.ac.th (S.S.)

2 Division of Food Engineering, Faculty of Agro-Industry, Chiang Mai University, Chiang Mai 50100, Thailand; noppol.l@cmu.ac.th

3 Department of Food, Life and Environmental Sciences, Faculty of Agriculture, Yamagata University, Tsuruoka, Yamagata 9978555, Japan; mwata@tds1.tr.yamagata-u.ac.jp

* Correspondence: thanongsak.c@cmu.ac.th or thachaiyaso@hotmail.com; Tel.: +66-5394-8262; Fax: +66-5394-8219

**Abstract:** This study aimed to select thermotolerant yeast for bioethanol production from cellulose-rich corncob (CRC) residue. An effective yeast strain was identified as *Saccharomyces cerevisiae* TC-5. Bioethanol production from CRC residue via separate hydrolysis and fermentation (SHF), simultaneous saccharification and fermentation (SSF), and prehydrolysis-SSF (pre-SSF) using this strain were examined at 35–42 °C compared with the use of commercial *S. cerevisiae*. Temperatures up to 40 °C did not affect ethanol production by TC-5. The ethanol concentration obtained via the commercial *S. cerevisiae* decreased with increasing temperatures. The highest bioethanol concentrations obtained via SHF, SSF, and pre-SSF at 35–40 °C of strain TC-5 were not significantly different (20.13–21.64 g/L). The SSF process, with the highest ethanol productivity (0.291 g/L/h), was chosen to study the effect of solid loading at 40 °C. A CRC level of 12.5% (*w*/*v*) via fed-batch SSF resulted in the highest ethanol concentrations of 38.23 g/L. Thereafter, bioethanol production via fed-batch SSF with 12.5% (*w*/*v*) CRC was performed in 5-L bioreactor. The maximum ethanol concentration and ethanol productivity values were 31.96 g/L and 0.222 g/L/h, respectively. The thermotolerant *S. cerevisiae* TC-5 is promising yeast for bioethanol production under elevated temperatures via SSF and the use of second-generation substrates.

**Keywords:** bioethanol; cellulose-rich corncob; thermotolerant *Saccharomyces cerevisiae* TC-5; simultaneous saccharification and fermentation; thermotolerant yeast

**Citation:** Boonchuay, P.; Techapun, C.; Leksawasdi, N.; Seesuriyachan, P.; Hanmoungjai, P.; Watanabe, M.; Srisupa, S.; Chaiyaso, T. Bioethanol Production from Cellulose-Rich Corncob Residue by the Thermotolerant *Saccharomyces cerevisiae* TC-5. *J. Fungi* **2021**, *7*, 547. https://doi.org/10.3390/jof7070547

Academic Editor: Baojun Xu

Received: 16 June 2021
Accepted: 8 July 2021
Published: 9 July 2021



## 1. Introduction

Globally, the demand for petroleum-based fuels for industry, agriculture, transportation, and private sectors is sharply increasing. An insufficient fuel supply can be alleviated by using alternative fuels. For example, bioethanol is a liquid biofuel produced from biomass by microbial fermentation of various substrates, such as sugar-based substrate, starchy biomass, and lignocellulosic biomass [1–3]. Among them, lignocellulosic biomass, a renewable substrate, is the cheapest and most sustainable substrate for bioethanol production [4,5]. Cellulose-rich corncob (CRC) residue is a solid waste obtained from corncob-xylooligosaccharides (corncob-XO) production. During XO production, most of the xylan in KOH-treated corncob is hydrolyzed by cellulase-free endo-xylanase, whereas the cellulose remains. The CRC residue contains a relatively high cellulose content in the range of 74–80% (*w*/*w*), with 10–13% (*w*/*w*) hemicellulose and 2–6% (*w*/*w*) lignin, making it an attractive substrate for fermentable sugar and bioethanol production [6].

Biethanol production from lignocellulosic biomass can be categorized into three main processes, namely separate hydrolysis fermentation (SHF), simultaneous saccharification and fermentation (SSF), and prehydrolysis-simultaneous saccharification and fermentation (pre-SSF) [7,8]. During SHF, enzymatic hydrolysis and ethanol fermentation are carried out separately, using two bioreactors. The hydrolysis and fermentation step can be carried out independently at their optimum conditions regarding temperature, pH, and time. However, this process requires long operation periods and complex equipment. The bioethanol production rate of SHF might be limited by the high concentration of initial glucose and cellobiose in the hydrolysate [9,10]. In contrast, in the SSF process, enzymatic hydrolysis and ethanol fermentation are carried out simultaneously in a single bioreactor [11]. Consequently, the overall process time, investment cost, substrate feedback inhibition, and contamination risk are decreased; SSF is therefore preferably employed in bioethanol production from lignocellulosic biomass [12,13]. One of the major limitations of SSF is the operation temperature. The optimal temperature for enzymatic hydrolysis is around 40–55 °C [14], whereas that for fermentation is around 20–35 °C [15]. Thus, screening of thermotolerant ethanolic microorganisms, which can grow at 37–50 °C, is required to overcome these limiting factors [5]. As ethanol-producing yeast, *S. cerevisiae* is outstanding candidate suitable for industrial application. Several desirable characteristics of *S. cerevisiae* have been reported, including generally regarded as safe (GRAS) status, high fermentation rate, genetic tractability, and stress tolerant (robustness) [16,17]. Some strains of *S. cerevisiae* are able to grow and produce ethanol at high temperature. Hence, thermotolerant *S. cerevisiae* receives considerable interest for industrial ethanol production [18].

In this sense, an effective thermotolerant yeast strain that is suitable for second-generation bioethanol production from CRC at high temperatures was searched. The effects of temperature and solid loading on bioethanol production by SHF, SSF, and pre-SSF processes, using the selected thermotolerant yeasts, were investigated. To determine the suitability of the selected thermotolerant yeast in ethanol production at the pilot and the industrial scale, the scale-up of bioethanol production in a 5-L stirred tank bioreactor was investigated. The thermotolerant yeast strain *Saccharomyces cerevisiae* TC-5 could grow well and produce bioethanol via CRC, even under low nutrient and high temperature conditions, indicating minimized production costs. High ethanol yields were achieved by strain TC-5 via the fed-batch SSF at elevated temperatures. Using this thermotolerant yeast under optimal conditions, the overall process time for bioethanol production could be reduced and the production of ethanol increased.

## 2. Materials and Methods

### 2.1. Chemicals and Materials

Analytical-grade *p*-nitrophenyl-β-D-glucopyranoside (*p*NPG), arabinose, dinitrosalicylic acid (DNS), and cellobiose were purchased from Sigma (St. Louis, MO, USA). Glucose was purchased from Ajax FineChem (Australia). Malt extract, peptone (Bacto™ Peptone), and yeast nitrogen base without amino acid were purchased from BD Difco™ (Franklin Lakes, NJ, USA). Xylan (Beechwood) was purchased from Megazyme (Wicklow, Ireland). Trehalose and xylose were purchased from Loba Chemie (Mumbai, India). Yeast extract was purchased from HiMedia (Mumbai, India). Lactose, maltose, sulfuric acid, sucrose, and water (HPLC grade) were purchased from RCI Labscan (Bangkok, Thailand). Ethanol (99.8%) was purchased from a Liquor Distillery Organization (Bangkok, Thailand). All other chemicals used in this study were of analytical grade.

A commercial cellulase cocktail (iKnowZyMe AC cellulase) was purchased from Reach Biotechnology Co. Ltd. (Bangkok, Thailand). Endo-glucanase, total cellulase activity (FPase), β-glucosidase, and xylanase were used at concentrations of commercial enzyme were 1200, 50, 140, and 2100 U/mL, respectively.

The CRC residue was obtained from the corncob-XO production process according to the method described by Boonchuay et al. [6]. The solid CRC residue was recovered by filtration using filter cloth and rinsed with tap water to remove other products. Subsequently, CRC residue was dried at 80 °C in a hot air oven (350i, Schwabach, Memmert, Germany) for 48 h. The cellulose, hemicellulose, and lignin contents of the CRC residue were determined by the TAPPI method and analyzed by the Animal Nutrition Laboratory, Department of Animal and Aquatic Sciences, Faculty of Agriculture, Chiang Mai University, Thailand [19,20]. The CRC residue was either used as substrate for hydrolysate (fermentable sugars) production or for bioethanol production via SSF and pre-SSF processes.

### 2.2. Cellulose-Rich Corncob (CRC) Hydrolysate Production

The reaction mixture for CRC hydrolysate production was composed of 7.5% (*w*/*v*) CRC and 22.04 FPU/$g_{CRC}$ of commercial cellulase cocktail in 0.1 M sodium-citrate buffer pH 5.0. The reaction was conducted at 46 °C with a shaking speed of 150 rpm for 96 h, in accordance with the optimal condition for CRC hydrolysate production that reported by Boonchuay et al. [6] and the manual guide for iKnowZyMe AC cellulase. After 96 h, the CRC hydrolysate was recovered by centrifugation (Z236K, Wehingen, Hermle, Germany) at 6000 rpm (4430× *g*) and 4 °C for 10 min and kept at −20 °C until use. The fermentable sugar (glucose, xylose, and arabinose) concentration of the CRC hydrolysate was determined by high-performance liquid chromatography (HPLC).

### 2.3. Medium and Medium Preparation

#### 2.3.1. Yeast Extract–Malt Extract (YM) Medium

The yeast extract–malt extract (YM) medium was composed of yeast extract 4 g/L, malt extract 10 g/L, and glucose 4 g/L. All components were dissolved in distilled water. The initial pH of YM medium was adjusted to 6.5 with 0.1 N NaOH or 0.1 N HCl and autoclaved at 121 °C for 15 min [21].

#### 2.3.2. Separate Hydrolysis and Fermentation (SHF)-Bioethanol Production Medium

Briefly, $(NH_4)_2SO_4$ 4 g/L, yeast extract 1 g/L, $NH_4H_2PO_4$ 1 g/L, and $MgSO_4 \cdot 7H_2O$ 0.1 g/L were dissolved in CRC hydrolysate pH 5.0. This CRC hydrolysate produced under optimal condition contained glucose 54.0 g/L, xylose 14.18 g/L, and arabinose 0.65 g/L. The SHF-bioethanol production medium was autoclaved at 121 °C for 15 min before inoculation [6].

#### 2.3.3. Simultaneous Saccharification and Fermentation (SSF)-Bioethanol Production Medium

Sodium citrate buffer pH 5.0 (0.1 M) supplemented with $(NH_4)_2SO_4$ 4 g/L, yeast extract 1 g/L, $NH_4H_2PO_4$ 1 g/L, and $MgSO_4 \cdot 7H_2O$ 0.1 g/L was mixed with 7.5, 10, 12.5, and 15% (*w*/*v*) CRC residue. The SSF bioethanol production medium (pH 5.0) was autoclaved at 121 °C for 15 min. After the medium was cooled down, 22.04 FPU/$g_{CRC}$ of commercial cellulase cocktail and 10% (*v*/*v*) prepared seed culture of yeast were transferred into the medium [6].

### 2.4. Microorganisms and Inoculum Preparation

Commercial active dry yeast (*S. cerevisiae*) was purchased from Danstil, Lallemand Inc. (Fredericia, Denmark), and used as a benchmark strain. This commercial strain demonstrates good tolerance to high fermentation temperatures (26–36 °C) and suitable for application both in batch and semi-continuous fermentations. Thermotolerant yeasts, including isolates SB1, SC10, G3, and TC-5, were obtained from the culture collection of the Division of Biotechnology, Faculty of Agro-Industry, Chiang Mai University, Thailand. Based on their 26S rDNA sequences and sugar assimilation test, isolates SB1, SC10, and G3 were identified as *Candida glabrata*. The accession number of these strains are MN784460, MN784462, and KY618710, respectively. Whereas isolate TC-5 was identified as *S. cerevisiae* with the accession number KY681804 (Figure S1). The isolates were first selected based on their ability to grow and produce bioethanol at 42 °C [22]. Stock cultures were maintained in glycerol (30%, *w*/*w*) at −30 °C until use.

For all yeast inoculum preparations, 1.0 mL of glycerol stock was transferred to a 250-mL Erlenmeyer flask containing 50 mL of YM broth and cultivated in an incubator shaker (LSI-3016R, Gyeonggi, Republic of Korea) at 37 °C with a shaking speed of 200 rpm for 24–48 h until the optical density at 600 nm ($OD_{600}$) = 6.0 was reached. Thereafter, the cultivation broth was centrifuged at 6000 rpm (4430× *g*) and 4 °C for 10 min. The obtained cell pellet was washed twice and resuspended in 50 mL of sterile 0.85% (*w*/*v*) NaCl, and the $OD_{600}$ was adjusted to 6.0 before inoculation. Then, the 10% (*v*/*v*) resuspended yeast cell was inoculated to the bioethanol production medium.

### 2.5. *Bioethanol Production in the Laboratory Bottle*

All experiments were conducted in 100-mL laboratory bottles (Duran, Mainz, Germany) equipped with an airlock, containing 90 mL of each bioethanol production medium. The experiments conducted under static condition were operated in an incubator (SMART i250, Accuplus, Bangkok, Thailand) and those conducted under shaking were operated at 150 rpm in the incubator shaker (LSI-3016R, Labtech, Gyeonggi, Republic of Korea).

#### 2.5.1. Effects of Temperature and Process on Bioethanol Production

Separate hydrolysis and fermentation (SHF) process. The effects of temperature on bioethanol production via SHF, using the selected thermotolerant yeast *S. cerevisiae* TC-5 and commercial *S. cerevisiae* (control), were determined using SHF-bioethanol production medium. The effect of temperature on bioethanol production of all tested strains was investigated at 35, 37, 40, or 42 °C for 120 h under static and limited $O_2$ conditions. Samples were periodically taken at 12-h intervals, and the levels of remaining sugars and ethanol were determined by HPLC.

Simultaneous saccharification and fermentation (SSF) process. After the autoclaved SSF-bioethanol production medium was cooled down, 22.04 FPU/$g_{CRC}$ of commercial cellulase cocktail and 10% (*v*/*v*) of each seed culture of the thermotolerant strain *S. cerevisiae* TC-5 or commercial *S. cerevisiae* (control) were inoculated into the medium. The effect of temperature on bioethanol production of all tested strains was investigated at 35, 37, 40, or 42 °C at 150 rpm for 120 h, under limited $O_2$ conditions in the incubator shaker. Samples were periodically taken at 12-h intervals, and the concentrations of residual sugars and ethanol were determined by HPLC.

Prehydrolysis-simultaneous saccharification and fermentation (pre-SSF) process; Bioethanol production via pre-SSF was performed using SSF-bioethanol production medium. Commercial cellulase cocktail (22.04 FPU/$g_{CRC}$) was added to the medium, and the prehydrolysis step was conducted at 46 °C, 150 rpm, for 24 h. After that, 10% (*v*/*v*) of prepared seed culture of the thermotolerant yeast strain TC-5 or commercial *S. cerevisiae* (control) was subsequently inoculated. The effect of temperature on bioethanol production of all tested strains was investigated at 35, 37, 40, or 42 °C, keeping the other parameters identical to those in the SSF process. Samples were periodically taken at 12-h intervals. The concentrations of residual sugars and ethanol in the samples were determined by HPLC. The optimal temperature and fermentation conditions, resulting in the highest ethanol concentration, yield, and productivity, were selected for the following experiment.

#### 2.5.2. Effect of Solid Loading on Bioethanol Production by *S. cerevisiae* TC-5

Batch simultaneous saccharification and fermentation (batch SSF) process. The effect of solid loading on bioethanol production by *S. cerevisiae* TC-5 via the batch SSF process was studied at 7.5, 10, 12.5, and 15% (*w*/*v*) CRC residue, using SSF-ethanol production medium. The experimental conditions were 40 °C, 150 rpm, for 168 h. The inoculum size and enzyme dosage were fixed at 10% (*v*/*v*) and 22.04 FPU/$g_{CRC}$, respectively. Samples were periodically taken at 12-h intervals. The concentration of residual sugar and ethanol was determined by HPLC.

Fed-batch simultaneous saccharification and fermentation (fed-batch SSF) process; Fed-batch SSF experiments with 10, 12.5, and 15% (*w*/*v*) CRC residue were performed by keeping the other parameters constant. According to our previous study [6], the effect of the CRC concentration on hydrolysate production was studied using a statistical design at different levels from 2.5–17.5% (*w*/*v*). The highest fermentable sugar concentration was attained when using the 7.8% (*w*/*v*). However, when the CRC concentration was increased to 17.5% (*w*/*v*), the fermentable sugar yields ($g/g_{CRC}$) was lower than that obtained from other experiments. Whereas, at low substrate loading experiment (2.5–5.0%, *w*/*v*), the glucose and total sugar concentration (g/L) was lower than that of high substrate loading experiment. The initial CRC residue concentration of all experiments was 7.5% (*w*/*v*). The fed-batch SSF reactors were fed with 2.25 g of CRC residue at 48, 72, and 96 h to increase the final substrate loading to 10, 12.5, and 15% (*w*/*v*) CRC residue, respectively. Samples were periodically taken at 12-h intervals. The concentrations of residual sugars and ethanol were determined by HPLC. The optimal solid loading and fermentation conditions, resulting in the maximum bioethanol concentration, yield, and productivity, were then selected for further experiments.

### 2.6. *Bioethanol Production via Fed-Batch SSF by S. cerevisiae TC-5 in a Bioreactor*

For the scale-up of bioethanol production employing *S. cerevisiae* TC-5, the fed-batch SSF process was performed in a 5-L stirred tank bioreactor (MDFT-N-5L, BE Marubishi, Bangkok, Thailand) with a 3-L working volume. The experiment was performed with 7.5% (*w*/*v*) CRC residue as the initial solid loading. The 75 g of CRC residue was fed after 48 and 72 h to increase the final solid loading to 12.5% (*w*/*v*). Samples were periodically taken at 12-h intervals. The concentration of residual sugar and ethanol was determined by HPLC.

### 2.7. *Analytical Methods*

#### 2.7.1. Sugar and Ethanol Determination

Samples were centrifuged at 13,000 rpm (16,060× *g*) and 4 °C for 15 min. The obtained supernatant was filtered through a nylon membrane filter (0.2 µm, FiltrEX, USA) and subjected to HPLC analysis (SCL-10Avp, Kyoto, Shimadzu, Japan) with an Aminex HPX 87H column (300 × 7.8 mm; Bio-Rad, Hercules, CA, USA). The mobile phase consisted of 5.0 mM $H_2SO_4$ as an eluent at a flow rate of 0.60 mL/min. The column thermostat was set at 40 °C. Sugar and ethanol were detected using a refractive index (RI) detector (RID-10A, Shimadzu, Kyoto, Japan) over 25 min [6].

#### 2.7.2. Enzyme Assay

Endo-glucanase activity was measured using 0.5% (*w*/*v*) sodium carboxymethyl cellulose solution in 0.1 M sodium-citrate buffer (pH 5.0) as a substrate, according to the modified method of Zhang et al. [23]. Total cellulase activity (FPase; filter paper unit or FPU) was measured using filter paper as substrate, according to the modified method of Ghose [24]. The β-glucosidase activity was measured using *p*-nitrophenyl-β-D-glucopyranoside (*p*NPG) in 0.1 M sodium-citrate buffer (pH 5.0) as substrate and calculated using the molar extinction coefficient, ε400 = 18,300 $M^{-1}cm^{-1}$, according to the methods of Salma [25] and Boonchuay et al. [6]. Endo-xylanase activity was measured using 1.0% (*w*/*v*) beechwood xylan solution in 0.1 M sodium-citrate (pH 5.0) as substrate [26].

#### 2.7.3. Calculation and Statistical Analysis

All experiments were carried out in triplicate. The data were analyzed for statistical significance using one-way analysis of variance (ANOVA), followed by Duncan's multiple range test ($p < 0.05$). The statistical software package SPSS v. 17 was used in the analysis of the experimental data. The fermentable sugar yield, the conversion of cellulose to glucose (hydrolysis efficiency, %), the initial glucose in the SSF-bioethanol production medium, the theoretical ethanol yield (Y, %), the conversion of cellulose to ethanol (%), ethanol yield ($Y_{EtOH}$, g/g), and ethanol productivity ($Q_p$, g/L/h) were calculated according to Boonchuay et al. [6].

## 3. Results

### 3.1. Composition of Cellulose-Rich Corncob (CRC) Residue and CRC Hydrolysate Production

The cellulose, hemicellulose, and lignin contents of CRC residue were 74.36 ± 1.11%, 18.19 ± 0.90%, and 3.96 ± 0.07%, respectively. The CRC residue was a solid waste obtained from corncob-XO production. After hydrolysis by commercial cellulase cocktails, CRC hydrolysate consisted of glucose 54.04 ± 0.91 g/L, xylose 14.18 ± 0.90 g/L, and arabinose 0.65 ± 0.04 g/L. The hydrolysis efficiency (85.69 ± 2.03%) was calculated, and the produced CRC hydrolysate was further applied as a carbon source for bioethanol production by the thermotolerant yeast via SHF.

### 3.2. Selection of the Thermotolerant Yeast for Bioethanol Production from CRC Hydrolysate

The ability of strain SB1, SC10, G3, and TC-5 to produce bioethanol from CRC hydrolysate was studied at a high temperature (42 °C). Figure 1 shows that *C. glabrata* G3 produced the highest ethanol concentration ($C_{EtOH}$), ethanol productivity ($Q_P$), ethanol yield ($Y_{p/s}$), and theoretical ethanol yield (Y) of 19.05 g/L, 0.256 g/L/h, 0.349 $g_{EtOH}/g_{glucose}$, and 68.45%, respectively. However, *S. cerevisiae* TC-5 also showed the highest and comparable ethanol concentration, ethanol productivity, ethanol yield, and theoretical ethanol yield of 18.38 g/L, 0.252 g/L/h, 0.334 $g_{EtOH}/g_{glucose}$, and 66.05%, respectively. In addition, ethanol production using both of thermotolerant yeast strains G3 and TC-5 were not significantly difference.

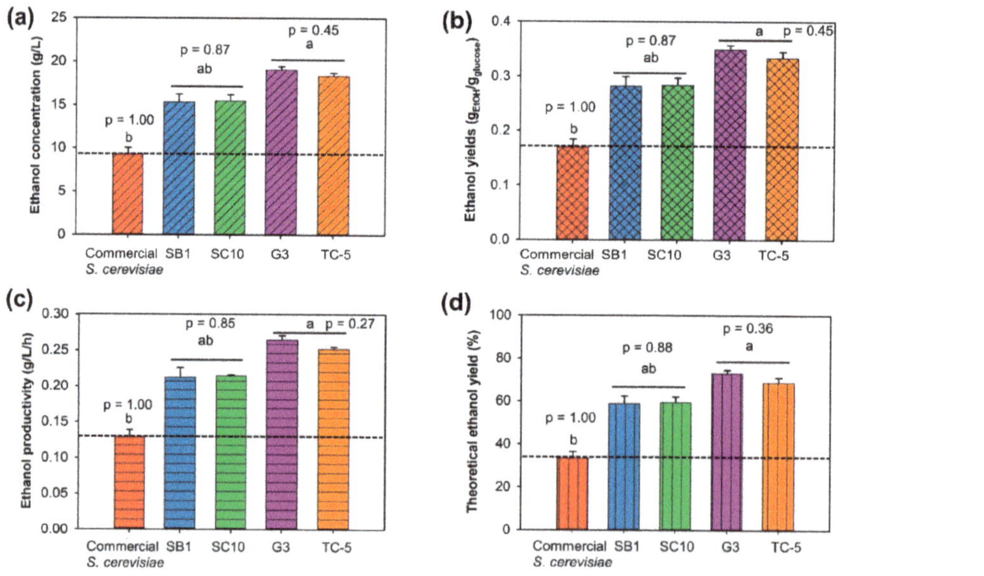


**Figure 1.** Comparison of ethanol concentration (**a**), ethanol yield (**b**), ethanol productivity (**c**), and theoretical ethanol yield (**d**) from cellulose-rich corncob hydrolysate by four thermotolerant yeasts and commercial *S. cerevisiae* at 42 °C for 72 h. Note: values are presented as mean ± standard deviation. Different letters above the bars denote statistically significant different. Bar chart with the same pattern and letter are not significantly different. The level of significance was tested by Duncan's multiple range test. (Each kinetic value (ethanol concentration, ethanol yield, ethanol productivity, or theoretical ethanol yield) was compared between yeast strains).

### 3.3. Bioethanol Production in the Laboratory Bottle

#### 3.3.1. Effects of Temperature and Processes on Bioethanol Production by *S. cerevisiae* TC-5

Separate hydrolysis and fermentation (SHF) process. The effect of temperature on bioethanol production from CRC hydrolysate by *S. cerevisiae* TC-5 and commercial *S. cerevisiae* (control) via SHF was investigated at 35–42 °C. When the temperature was set at 35 °C, glucose was completely consumed by the thermotolerant yeast strain TC-5 and the commercial *S. cerevisiae* (Figures S2 and S3). The ethanol concentration and ethanol yield of both tested strains ranged from 20.33–20.59 g/L and 0.373–0.378 $g_{EtOH}/g_{glucose}$, respectively (Figure 2a and Table S1). In addition, the range of theoretical ethanol yield at 35 °C was 78.01–79.01%. The ethanol production at 35 °C of both tested strains did not significantly differ. However, elevated temperature (37–42 °C) had a negative effect on ethanol fermentation by commercial *S. cerevisiae*. At 40 and 42 °C, commercial *S. cerevisiae* could not metabolize glucose completely, although the cultivation period was extended to 120 h (Figure S2). When the temperature ranged from 35 to 42 °C, the ethanol concentration of this strain decreased continuously from 20.59 to 9.81 g/L, theoretical ethanol yield decreased from 79.01 to 37.64%, and ethanol yield dramatically decreased from of 0.378 to 0.180 $g_{EtOH}/g_{glucose}$. In contrast, temperatures of 37 and 40 °C did not affect the ethanol production by the thermotolerant yeast strain TC-5. At these temperatures, glucose was rapidly and completely consumed within 48 h. When the SHF process was investigated at 42 °C, the glucose consumption of strain TC-5 was slightly impeded until 72 h of cultivation time (Figure S3). After that, the consumption was steady, and a small amount of glucose remained in the fermentation medium. The significantly highest ethanol concentration of 20.50 g/L was obtained from strain TC-5 when the cultivation temperature was 40 °C (Figure 2a). Moreover, the ethanol productivity of strain TC-5 at 35–40 °C was not significantly different (Figure 2d). However, ethanol concentration (Figure 2a), ethanol productivity (Figure 2d), ethanol yield (Table S1), and theoretical ethanol yield (Table S1) of strain TC-5 slightly decreased when the temperature increased to 42 °C.

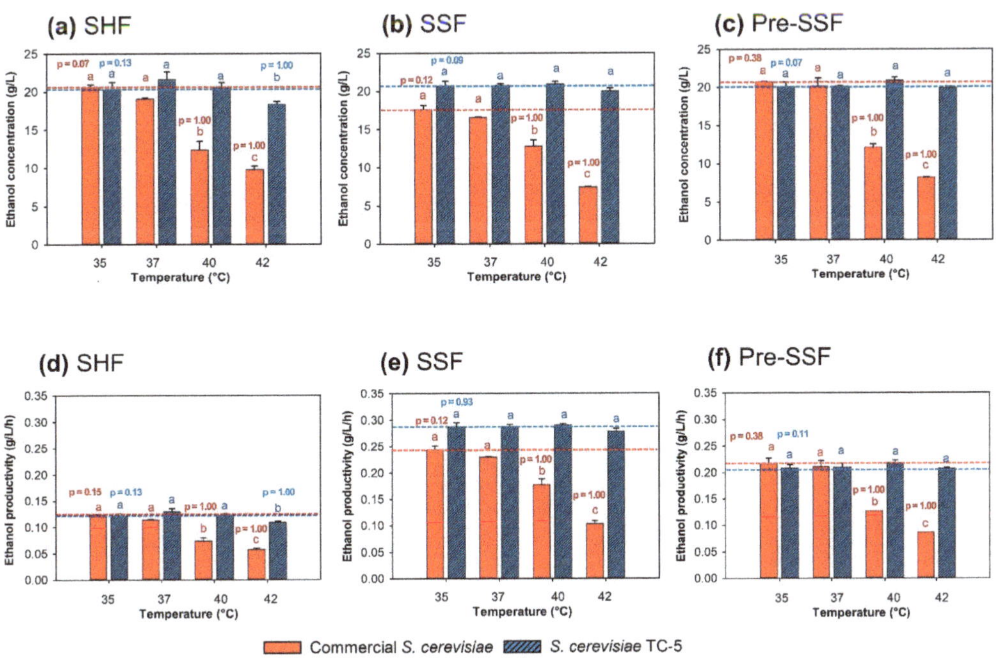


**Figure 2.** Ethanol production via separate hydrolysis and fermentation (SHF) (**a**), simultaneous saccharification and fermentation (SSF) (**b**), and prehydrolysis-simultaneous saccharification (pre-SSF) (**c**); ethanol productivity via SHF (**d**), SSF (**e**), and pre-SSF (**f**) by commercial *S. cerevisiae* and thermotolerant *S. cerevisiae* TC-5 at 35, 37, 40, and 42 °C for 72 h. Note: values are presented as mean ± standard deviation. Different letters above the bars denote statistically significant different. The level of significance was tested by Duncan's multiple range test. Bar chart with the same pattern and letter are not significantly different (ethanol concentration or productivity of each yeast strain were compared between temperature).

Simultaneous saccharification and fermentation (SSF) process; To determine the feasibility to produce bioethanol from CRC by using the thermotolerant TC-5 via SSF at high temperatures, fermentation was carried out at 35, 37, 40, or 42 °C, 150 rpm, under limited $O_2$ condition. The effect of temperature on ethanol production was examined and is shown in Figure 2b,e, Figures S4 and S5, and Table S1. At 35 and 37 °C, the released glucose was rapidly consumed within 72 h by commercial *S. cerevisiae* since the sugar consumption by yeast might be equal to the rate of enzymatic hydrolysis; glucose in the fermentation medium was not detected (Figures S4 and S5). The final ethanol concentration of the commercial strain at 35 and 37 °C was 17.58 and 16.58 g/L, respectively. Theoretical ethanol yields of 67.46 and 63.62% were obtained, respectively (Table S1). On the contrary, at 40 and 42 °C, the glucose concentration continuously increased because high temperature might inhibit the ethanol fermentation and growth of the commercial strain. Under high temperatures (40 and 42 °C), the ethanol production of the commercial yeast strain was relatively low at 7.45–12.81 g/L (Figure 2b,e, Table S1). The ethanol yield of commercial *S. cerevisiae* also considerably decreased with increasing temperatures. In the SSF process, the high temperatures at 40–42 °C did not affect the ethanol fermentation of strain TC-5 (Figure 2b,e). The ethanol concentration at 40 °C was 20.92 g/L, with an ethanol yield of 0.384 $g_{EtOH}/g_{glucose}$ and an ethanol productivity of 0.291 g/L/h. When the temperature was elevated to 42 °C, strain TC-5 obtained an ethanol concentration, ethanol yield, and ethanol productivity of 20.02 g/L, 0.368 $g_{EtOH}/g_{glucose}$, and 0.278 g/L/h, respectively.

Prehydrolysis-simultaneous saccharification and fermentation (pre-SSF) process; The effect of temperature on ethanol production during the pre-SSF process by thermotolerant *S. cerevisiae* TC-5 and commercial *S. cerevisiae* was examined (Figure 2c,f; Table S1). After prehydrolysis at 46 °C for 24 h, the glucose concentration was rapidly increased (~50 g/L). Subsequently, each yeast inoculum was added to the SSF-bioethanol production medium, and the temperature was adjusted to 35, 37, 40, and 42 °C. The glucose consumption of commercial *S. cerevisiae* at 35 and 37 °C increased within 24–72 h (Figure S6). At the same time and under high temperatures (40 and 42 °C), the glucose consumption decreased dramatically. Thereafter, the consumption rate was steady and glucose was not consumed completely, even when the cultivation period was extended to 144 h (Figure S6). When the temperature was increased from 35 to 42 °C, the ethanol concentration and theoretical ethanol yield of commercial *S. cerevisiae* decreased from 20.71 to 8.24 g/L, and from 79.47 to 31.62%, respectively. Glucose was rapidly metabolized by strain TC-5 at 35 and 37 °C. This strain also showed effective glucose use in a temperature range of 35–42 °C (Figure S7). An ethanol concentration in the range of 19.91–20.89 g/L was obtained, with a theoretical ethanol yield of 76.40–80.16% (Table S1).

### 3.3.2. Effect of Solid Loading on Bioethanol Production by *S. cerevisiae* TC-5

Batch simultaneous saccharification and fermentation (batch SSF) process. The bioethanol production by strain TC-5 via batch SSF was investigated using CRC as substrate at 40 °C, 150 rpm, for 168 h. Four concentrations of solid loading, namely 7.5, 10, 12.5, and 15% (*w*/*v*) CRC, were studied. At 7.5% (*w*/*v*) CRC residue, glucose was rapidly consumed within 12 h as the sugar consumption by yeast might equal the rate of enzymatic hydrolysis (Figure 3a). Therefore, glucose was not detected in the fermentation medium. The ethanol concentration and theoretical ethanol yield were 20.92 g/L and 80.26%, respectively (Table 1). At 10% (*w*/*v*) solid loading, glucose was completely consumed within 120 h (Figure 3b), with ethanol concentration and theoretical ethanol yield values of 28.90 g/L and 88.37%, respectively. However, at 12.5 and 15% (*w*/*v*) CRC loading, glucose was not completely used (Figure 3c,d). The ethanol concentrations were 35.91 and 34.90 g/L, with theoretical ethanol yields of 87.86% and 71.13%, respectively (Table 1).

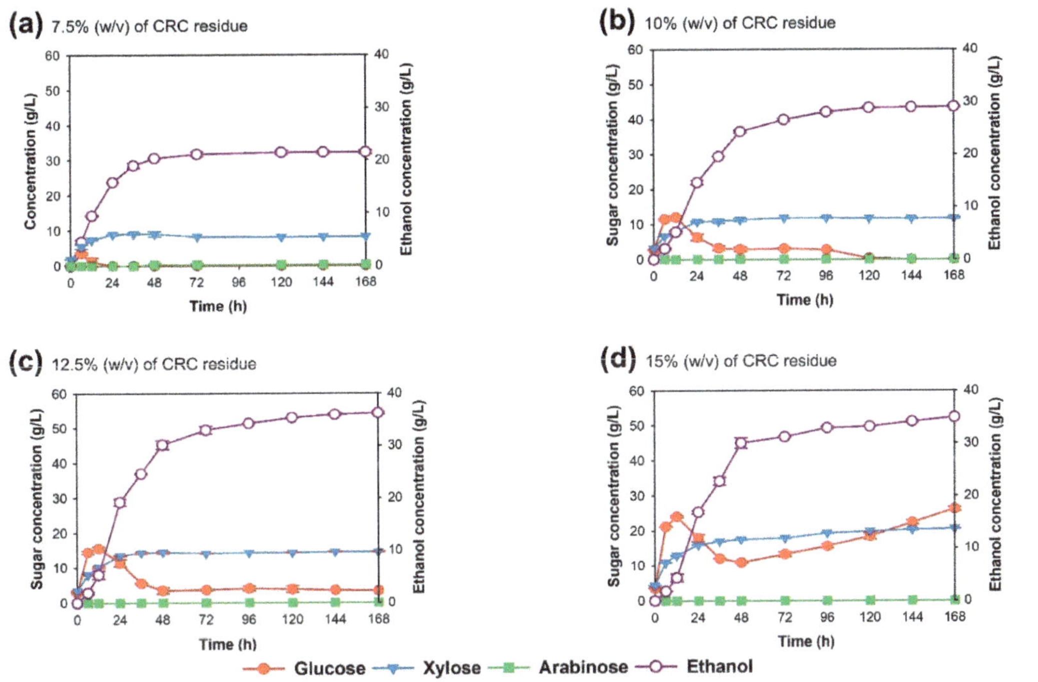


**Figure 3.** Effect of solid loading at 7.5 (**a**), 10 (**b**), 12.5 (**c**), and 15% (**d**) (*w*/*v*) cellulose-rich corncob residue on ethanol fermentation by the thermotolerant *S. cerevisiae* TC-5 via simultaneous saccharification and fermentation conducted in a 100-mL laboratory bottle at 40 °C for 168 h.

**Table 1.** Ethanol production from CRC via batch and fed-batch simultaneous saccharification and fermentation (SSF) at different cellulose-rich corncob (CRC) solid loadings by the thermotolerant *S. cerevisiae* TC-5 at 40 °C.

| Processes | CRC Loading (%, *w*/*v*) | Time (h) | $C_{EtOH}$ * (g/L) | $Y_{p/s}$ ** | | $Q_P$ *** (g/L/h) | Y **** (%) |
|---|---|---|---|---|---|---|---|
| | | | | $g_{EtOH}/g_{glucose}$ | $g_{EtOH}/g_{CRC}$ | | |
| **100-mL laboratory Bottle** | | | | | | | |
| Batch | 7.5 | 72 | 20.92 ± 0.16 $^{Gd}$ | 0.436 ± 0.003 $^{Cb}$ | 0.269 ± 0.002 $^{Cb}$ | 0.291 ± 0.002 $^{Ab}$ | 80.26 ± 0.63 $^{Cb}$ |
| | 10 | 120 | 28.90 ± 0.17 $^{Fc}$ | 0.452 ± 0.003 $^{Ba}$ | 0.289 ± 0.001 $^{Ba}$ | 0.241 ± 0.001 $^{Da}$ | 88.37 ± 0.52 $^{Ba}$ |
| | 12.5 | 144 | 35.91 ± 0.30 $^{Ca}$ | 0.449 ± 0.004 $^{Ba}$ | 0.288 ± 0.002 $^{Ba}$ | 0.249 ± 0.002 $^{Ca}$ | 87.86 ± 0.73 $^{Ba}$ |
| | 15 | 168 | 34.90 ± 0.01 $^{Db}$ | 0.364 ± 0.000 $^{Fc}$ | 0.233 ± 0.000 $^{Fc}$ | 0.208 ± 0.000 $^{Fc}$ | 71.13 ± 0.01 $^{Fc}$ |
| Fed-batch | 10 | 120 | 28.70 ± 0.11 $^{Fb}$ | 0.448 ± 0.002 $^{Bb}$ | 0.287 ± 0.001 $^{Bb}$ | 0.239 ± 0.001 $^{Db}$ | 87.74 ± 0.32 $^{Bb}$ |
| | 12.5 | 144 | 38.23 ± 0.19 $^{Aa}$ | 0.478 ± 0.002 $^{Aa}$ | 0.306 ± 0.001 $^{Aa}$ | 0.265 ± 0.001 $^{Ba}$ | 93.51 ± 0.47 $^{Aa}$ |
| | 15 | 168 | 37.55 ± 0.35 $^{Bb}$ | 0.391 ± 0.004 $^{Ec}$ | 0.251 ± 0.002 $^{Ec}$ | 0.224 ± 0.002 $^{Ec}$ | 76.56 ± 0.70 $^{Ec}$ |
| **5-L bioreactor** | | | | | | | |
| Fed-batch | 12.5 | 144 | 31.96 ± 0.78 $^{E}$ | 0.400 ± 0.000 $^{D}$ | 0.256 ± 0.001 $^{D}$ | 0.222 ± 0.001 $^{E}$ | 78.20 ± 0.19 $^{D}$ |

Note: values are presented as mean ± standard deviation. Data with the same superscript (capital letter) in the same column are not significantly different at $p \leq 0.05$ (different kinetic values were compared among all fermentation processes and CRC loadings). Data with the same superscript (small letter) are not significantly different at $p \leq 0.05$ (different kinetic values were compared among all fermentation processes and CRC loadings). The level of significance was tested by Duncan's multiple range test at $p \leq 0.05$. * $C_{EtOH}$: ethanol concentration; ** $Y_{p/s}$: ethanol yield; *** $Q_P$: ethanol productivity; **** Y: theoretical ethanol yield.

Fed-batch simultaneous saccharification and fermentation (fed-batch SSF) process. To investigate the possibility of bioethanol production by *S. cerevisiae* TC-5 via fed-batch SSF process at 40 °C, 150 rpm, for 168 h, fed-batch SSF was performed by using 10, 12.5, and 15% (*w*/*v*) final solid loading. The obtained ethanol concentration, ethanol productivity, ethanol yield, and theoretical ethanol yield are shown in Table 1. As depicted in Figure 4a–c, with a higher CRC solid loading, longer fermentation periods were required. At the CRC solid loading of 10% (*w*/*v*), ethanol concentration, ethanol productivity, and theoretical ethanol yield values of 28.70 g/L, 0.239 g/L/h, and 87.74%, respectively, were achieved at 120 h (Figure 4a and Table 1). At 12.5% (*w*/*v*) CRC solid loading, the highest ethanol concentration and theoretical ethanol yield values of 38.23 g/L and 93.51%, respectively, were obtained within 144 h (Figure 4b and Table 1). Ethanol yield and ethanol productivity were 0.478 $g_{EtOH}/g_{glucose}$ and 0.265 g/L/h, respectively. Surprisingly, ethanol production from 12.5% loading via fed-batch SSF process was significantly higher than that from the batch SSF process at 12.5% (*w*/*v*) CRC solid loading (Table 1). At the same solid loading, glucose was accumulated and was not metabolized completely in the batch SSF (Figure 3c), whereas the released glucose was completely used in the fed-batch SSF (Figure 4b). In contrast, bioethanol production via fed-batch SSF with 15% (*w*/*v*) CRC solid loading showed that the sugar concentration continuously increased, and sugar was accumulated in the fermentation medium (Figure 4c). The ethanol concentration was 37.55 g/L, with ethanol yield and theoretical ethanol yield values of 0.391 $g_{EtOH}/g_{glucose}$ and 76.56%, respectively (Table 1). However, the values obtained for the fed-batch SSF at 15% (*w*/*v*) CRC solid loading were higher than those obtained for the batch SSF at 15% (*w*/*v*) CRC solid loading (Figure 3d,c, and Table 1).

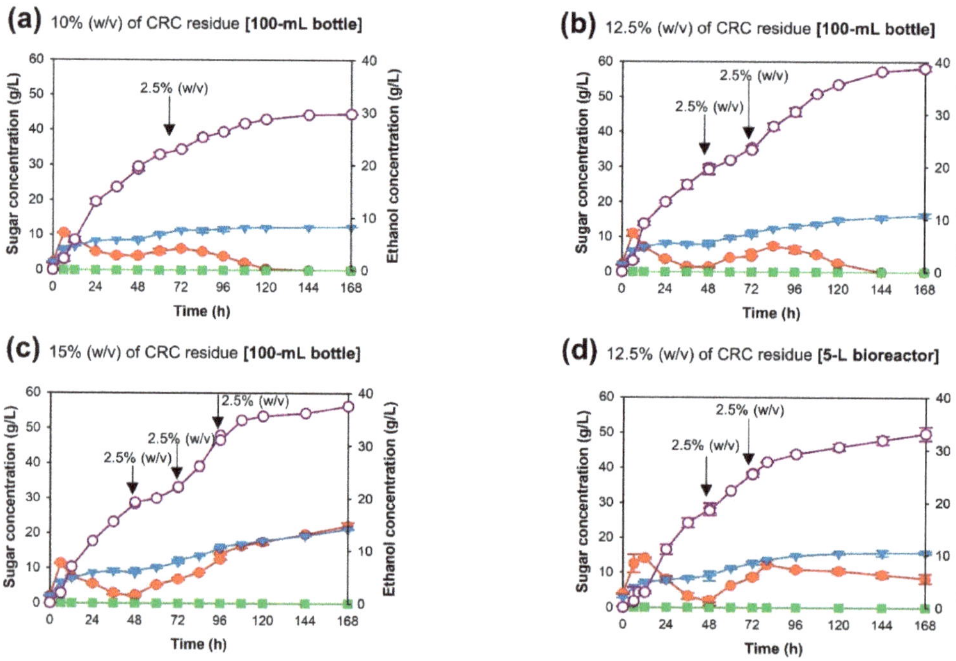


**Figure 4.** Time course of ethanol production from cellulose-rich corncob residue by the thermotolerant *S. cerevisiae* TC-5 via fed-batch simultaneous saccharification and fermentation, using 10 (**a**), 12.5 (**b**), and 15% (**c**) (*w*/*v*) cellulose-rich corncob residue, conducted in a 100-mL laboratory bottle at 40 °C for 168 h; time course of ethanol production via fed-batch simultaneous saccharification and fermentation using 12.5% (*w*/*v*) cellulose-rich corncob residue by the thermotolerant *S. cerevisiae* TC-5, conducted in a 5-L bioreactor at 40 °C for 168 h (**d**).

### *3.4. Bioethanol Production via Fed-Batch Simultaneous Saccharification and Fermentation (Fed-Batch SSF) by S. cerevisiae TC-5 in a Bioreactor*

For the scale-up of bioethanol production by *S. cerevisiae* TC-5, fed-batch SSF was performed in a 5-L stirred tank bioreactor with a 3-L working volume. The experiment was performed with the final solid loading of 12.5% (*w*/*v*); the glucose concentration decreased slowly, and glucose remained in the fermentation medium (Figure 4d). An ethanol concentration of 31.96 g/L, with an ethanol yield and a theoretical ethanol yield of 0.400 $g_{EtOH}/g_{glucose}$ and 78.20%, respectively, could be obtained.

### 3.5. Mass Balance of Bioethanol Production by the Thermotolerant S. cerevisiae TC-5

Figure 5 shows the mass balance for bioethanol production (based on 1 kg of raw corncob). One kg of raw corncob contained 426 g of cellulose (42.6%), 388 g of hemicellulose (38.82%), 87 g of lignin (8.69%), and 99 g of other components (9.88%). After KOH pretreatment, 629 g of treated corncob could be obtained.

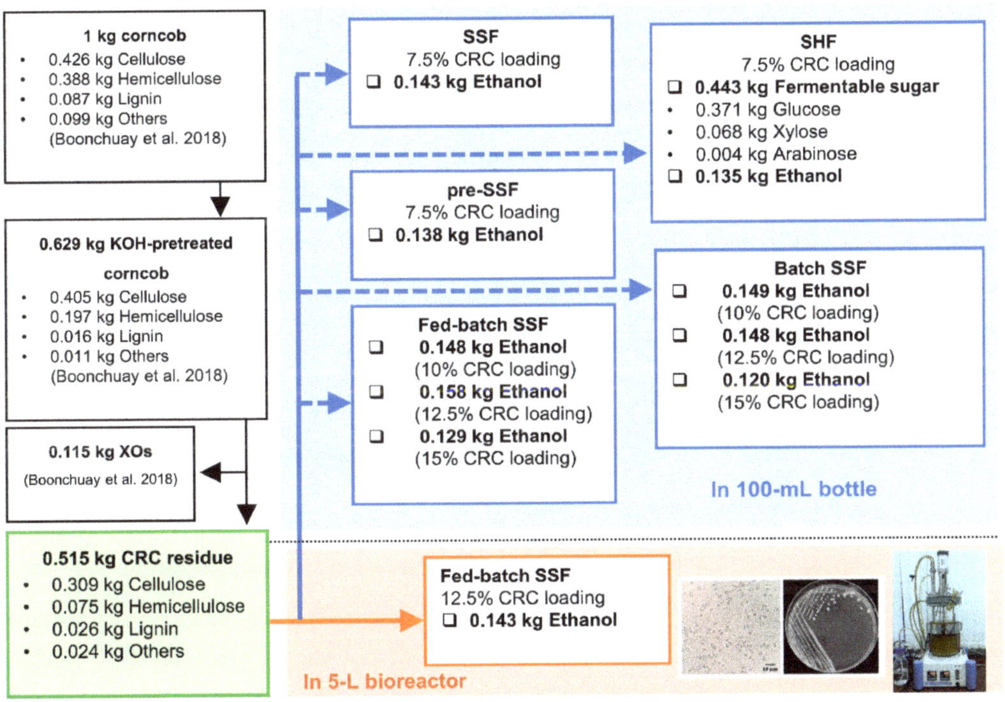


**Figure 5.** Mass balance of bioethanol production from cellulose-rich corncob residue by the thermotolerant *S. cerevisiae* TC-5.

The bioethanol production in the laboratory bottle by strain TC-5 via SHF, SSF, and pre-SSF with 7.5% (*w*/*v*) CRC residue was performed at 40 °C. When applying SHF, 443 g of fermentable sugar, 371 g of glucose (83.73%),68 g of xylose (15.26%), and 4 g of arabinose (1.01%) were obtained, with an ethanol yield of 135 g (0.263 $g_{EtOH}/g_{CRC}$ or 0.376 $g_{EtOH}/g_{glucose}$). The SSF process resulted in the highest ethanol yield of 143 g (0.277 $g_{EtOH}/g_{CRC}$ or 0.384 $g_{EtOH}/g_{glucose}$). With regard to the pre-SSF process, an ethanol yield of 138 g (0.268 $g_{EtOH}/g_{CRC}$ or 0.383 $g_{EtOH}/g_{glucose}$) was obtained. Hence, we selected SSF for the bioethanol production in laboratory bottles by strain TC-5 at 40 °C via batch and fed-batch SSF. For batch SSF, the ethanol yields from 10, 12.5 and 15% (*w*/*v*) CRC residue were 149, 148, and 120 g or 0.289, 0.287, and 0.233 $g_{EtOH}/g_{CRC}$, respectively. Moreover, the highest ethanol yields of 148, 158, and 129 g or 0.287, 0.306, and 0.250 $g_{EtOH}/g_{CRC}$ were obtained from fed-batch SSF with 10, 12.5, and 15% (*w*/*v*) CRC residue. The scaled-up fed batch SSF with 12.5% (*w*/*v*) CRC residue provided a slightly lower ethanol yield of 143 g (0.256 $g_{EtOH}/g_{CRC}$ or 0.400 $g_{EtOH}/g_{glucose}$) than the bioethanol production in laboratory bottles.

## 4. Discussion

During the XO production process, almost all of the xylan, which is the main hemicellulose found in corncob, was hydrolyzed by endo-xylanase, and released into the liquid phase. In the solid phase of the CRC residue, the cellulose content remained constant [6]. Hence, the cellulose ratio in CRC was relatively high. Cellulose is a polysaccharide composed of linear homopolymers of glucose units linked together by β-1,4 glycosidic bonds [27], making CRC a promising substrate for lignocellulosic hydrolysate (fermentable sugar) production.

The species *C. glabrata* is a thermotolerant ethanol-producing yeast which can tolerate both high temperatures and high acid concentrations [6,28]. However, it is also an opportunistic human pathogen [29]. The species *S. cerevisiae* is the most widely employed yeast in industrial ethanol production because it can tolerate at wide range of pH and temperature levels (30–42 °C). Moreover, *S. cerevisiae* can produce high ethanol yields even under the limited $O_2$ condition [15,30–32]. Regarding the safety aspect, *S. cerevisiae* is generally regarded as safe (GRAS) and frequently employed, with high potential in the industrial ethanol production [33]. Therefore, the isolated thermotolerant *S. cerevisiae* TC-5 was selected for bioethanol production from CRC residue.

In the SHF process, enzymatic hydrolysis and ethanol fermentation are carried out separately using two separate vessels, making this process time- and labor-consuming [10,11]. These limitations can be overcome by employing the SSF process, which is a combination of both enzymatic hydrolysis and ethanol fermentation in a single bioreactor [34]. Hence, the effect of temperature on bioethanol production from CRC via SSF by the thermotolerant yeast strain TC-5 was also considered. The thermotolerant yeasts strain TC-5 showed satisfactory performance regarding bioethanol production. The glucose was completely consumed at all studied temperatures (35–42 °C), and xylose and arabinose were accumulated. Generally, *S. cerevisiae* can ferment glucose (hexose) into ethanol, but cannot ferment the pentoses, especially xylose and arabinose [15]. This limitation might be overcome by co-fermentation with pentose-fermenting yeasts. However, CRC hydrolysate contains low xylose and arabinose concentrations of only 14.18 and 0.65 g/L, respectively. A higher concentration of xylose has been detected in hydrolysate produced from various types of lignocellulosic biomass e.g., sugarcane bagasse (56 g/L), rice straw (44 g/L), and wheat straw (25 g/L) [35–37]. Generally, almost of pentose-fermenting yeasts cannot grow under high ethanol and inhibitor concentrations as well as under strictly anaerobic conditions. Besides, they can generate unwanted products such as xylitol [8]. From an economical perspective, along with the purity of the produced bioethanol, the use of liquid waste from distillation, which contains high concentrations of xylose and arabinose, applying pentose-fermenting yeasts or other potential microorganisms should be investigated in further study.

The uncompleted use of glucose in SSF might be related to stress conditions such as temperature, osmotic pressure, and inhibitor formation, which can inhibit the growth of yeasts and their ethanol production capacity [34]. Thermotolerant *S. cerevisiae* strains have been widely employed for bioethanol production from lignocellulosic biomasses via SSF. For example, Mendes et al. [38] reported that *S. cerevisiae* strain ATCC 26602 produced 41.7 g/L of ethanol, with a theoretical ethanol yield of 49.8% at 38 °C, when primary sludge was used as substrate. Da Silva et al. [39] studied ethanol production at 40 °C, using the *S. cerevisiae* strain CAT-1 and carnauba straw as substrate; ethanol concentration and theoretical yield were 5.47 g/L and 54.81%, respectively. Regarding this study, the bioethanol production by strain TC-5 at 42 °C was not significantly difference compared with that at 35–40 °C and of previous reports. In this sense, the thermotolerant strain TC-5 is a candidate yeast strain for bioethanol production under elevated temperature via SSF.

Ethanol production from various types of lignocellulosic biomass via the pre-SSF process has previously been investigated. For example, Gladis et al. [40] reported that bioethanol production from corn stover using baker's yeast. Prehydrolysis was performed at 50 °C for 24 h, and after the prehydrolysis step, the glucose concentration was 111.5 g/L.

Subsequently, the temperature was adjusted to 35 °C for yeast fermentation. An ethanol concentration and a theoretical ethanol yield of 53.5 g/L and 75% were attained, respectively. Öhgren et al. [41] reported that an ethanol concentration and theoretical ethanol yield of 26.4 g/L and 75%, obtained from corn stover by using baker's yeast. Prehydrolysis was performed at 55 °C for 16 h, followed by SSF at 30 °C. Pre-SSF is one type of SSF process that provides the short period needed for lignocellulose material to be partly hydrolyzed before fermentation. This way, the prepared fermentable sugars in the fermentation medium can readily be used by the yeast. This process enables the use of higher temperatures during the initial enzymatic hydrolysis, potentially increasing enzymatic activity. Another advantage of this process is the reduction of the ethanol production time, potentially increasing the overall ethanol productivity [41].

Regarding the overall process time, to obtain the highest ethanol concentrations of SHF, pre-SSF, and SSF processes at 40 °C, periods of 168, 96, and 72 h, respectively, were needed. The bioethanol production via SSF and pre-SSF was not significantly different, whereas the maximum bioethanol productivity was obtained via SSF. In this sense, single-step SSF is suitable for bioethanol production from CRC at 40 °C because of the short fermentation and processing periods and the high ethanol productivity.

The ethanol production from various types of feedstocks via SSF at different solid loadings has been investigated. For example, Qin et al. [42] studied the effect of solid loading (6, 7, and 9%, *w*/*v*) on the bioethanol production from ethylenediamine-pretreated corn stover at 34 °C. The results showed that ethanol yield decreased with an increase in solid loading, and the authors suggested that this is caused by the lack of enzyme activity, cell viability, or other unknown factors. In addition, the challenges in bioethanol production via SSF at high solid concentrations are high viscosity and high energy consumption [43]. A high substrate concentration also results in a high content of inhibitors, impeding ethanol concentration and yield. These issues can be solved by using fed-batch fermentation [37,44].

Fed-batch SSF is a combination of batch and continuous modes by the periodical addition of the substrate into the fermentation medium and has been used to overcome substrate inhibition in batch SSF [37,40]. Ethanol production from various types of feedstocks via fed-batch SSF has been reported. For example, Gao et al. [37] reported bioethanol production from sugarcane bagasse with 33% (*w*/*v*) solid loading via fed-batch SSF. The highest ethanol concentration of 76 g/L, with a theoretical ethanol yield of 66%, was obtained. Fed-batch SSF using 20% (*w*/*v*) corn stover has been studied by Gladis et al. [40], and an ethanol concentration and a theoretical ethanol yield of 58 g/L and 81% were obtained. In this study, bioethanol production via fed-batch SSF with 12.5% (*w*/*v*) CRC solid loading at 40 °C provided the highest ethanol yield compared with other processes (Table 1). Therefore, fed-batch SSF at 12.5% (*w*/*v*) solid loading was also simulated for bioethanol production in a bioreactor.

The ethanol concentration in the bioreactor was lower than that obtained from via the 100-mL laboratory bottle (Table 1). This result is in agreement with previous findings. For example, Orrego et al. [45] found that the ethanol concentration attained from a 5-L bioreactor was lower than that obtained via a 250-mL Erlenmeyer flask. Dwiarti et al. [46] also found that the ethanol production in 2-L Erlenmeyer flasks was decreased compared with that of 200-mL Erlenmeyer flasks (11.80 to 8 g/L). Most likely, this can be explained by insufficient oxygen availability, mass, and heat transfer coefficients, absence of free water, accumulation of unused sugar (xylose and arabinose), viscosity, and mixing capacity [45–47].

During alkali pretreatment, the intermolecular ester bonds between lignin and hemicellulose are saponified, and the hemicellulose and lignin are dissolved in the alkali solution [48]. Therefore, the cellulose contents in KOH-treated corncob increased to 64.46%, whereas the hemicellulose and lignin contents decreased to 31.27 and 2.52%, respectively. After hydrolysis by endo-xylanase, 115 g of XOs and 515 g of CRC residue were obtained. Endo-xylanase is the key enzyme for hemicellulose hydrolysis and can randomly break down xylan, the main hemicellulose found in corncob, to produce XOs [49]; in treated corncob, the cellulose remained. Consequently, the cellulose content in CRC residue was relatively high (75.69%), and the hemicellulose content decreased to 14.53% (Figure 4). Hence, CRC residue with a high cellulose content could be a potential substrate for bioethanol production.

Table 2 shows the comparison of bioethanol production by different strains of *S. cerevisiae* from various types of lignocellulosic biomass via SSF at higher temperatures. Saravanikumar and Kathiresan [50], in their studies on sawdust, reported that sawdust hydrolysate was produced by consecutive acid treatment and enzymatic hydrolysis, using cellulase from *Trichoderma estonicum* SKS1. Bioethanol was produced via SHF by *S. cerevisiae* JN387604 at 36.5 °C, obtaining an ethanol concentration of 55.2 g/L and a theoretical yield of 85.6% at a fermentation period of 102 h. Recently, waste jasmine flower has been used as substrate for bioethanol production. The biomass was first treated using alkaline solutions and temperature, followed by enzymatic hydrolysis using cellulase for 24 h to obtain fermentable sugars. Then, bioethanol was produced using free cells of *S. cerevisiae* TISTR5020 in a 2-L bioreactor at 30–35 °C via SHF. Ethanol concentration, productivity, and yield were 14.39 g/L, 0.029 g/L/h, and 0.029 $g_{EtOH}/g_{biomass}$, obtained within 120 h [51].

**Table 2.** Comparison of bioethanol production from *S. cerevisiae* using lignocellulosic biomass via simultaneous saccharification and fermentation (SSF).

| Strain | Solid Loading; Scale | Fermentation Temperature (°C) | Substrate | $C_{EtOH}$ * (g/L) | $Y_{p/s}$ ** $g_{EtOH}/g_{substrate}$ | $Q_P$ *** (g/L/h) | Y **** (%) | Reference |
|---|---|---|---|---|---|---|---|---|
| *S. cerevisiae* mbc2 | 9% (*w*/*w*) glucan; Flask-scale | 42 | Silvergrass (*Miscanthus* sp.) | 15.30 | N/A | 0.32 | 90.10 | Cha et al. [12] |
| *S. cerevisiae* TJ14 | 5% (*w*/*v*) cellulose; Flask-scale | 42 | Paper sludge | 11.80 | N/A | 0.12 | 80.00 | Dwiarti et al. [46] |
| *S. cerevisiae* SC90 | 10% (*w*/*v*); Flask-scale | 40 | Oil palm trunk | 44.25 | 0.443 | 0.46 | 90.34 | Tareen et al. [52] |
| *S. cerevisiae* TC-5 | 12.5% (*w*/*v*); Flask-scale | 40 | Cellulose-rich corncob residue | 35.91 | 0.251 | 0.25 | 87.86 | This study |

Note: N/A: not available; * $C_{EtOH}$: ethanol concentration; ** $Y_{p/s}$: ethanol yield; *** $Q_P$: ethanol productivity; **** Y: theoretical ethanol yield.

With regard to fed-batch SSF (Table 3), another bioethanol production process has been reported by Gao et al. [37], who investigated bioethanol production from sugarcane bagasse via fed-batch SSF by using *S. cerevisiae* Y-2034 and commercial cellulase (Novozymes A/S). Fermentation was carried out in 250-mL Erlenmeyer flasks (100 mL of reaction volume) at 37 °C, an ethanol concentration and theoretical ethanol yield of 46.13 g/L and 70.06%, respectively, were obtained. In another study, bioethanol production from empty palm fruit bunch fiber in a 5-L stirring bioreactor (1 L of reaction volume) via fed-batch SSF has been investigated by Park et al. [53]. The strain *S. cerevisiae* L3262a was employed to produce ethanol at 30 °C, using fermentation medium with 10 g/L of yeast extract and 20 g/L of peptone. Under optimal conditions, an ethanol concentration of 62.5 g/L was obtained, with a theoretical yield of 70.6%. In a study by Tareen et al. [52], an oil palm trunk was treated by steam explosion and alkali extraction, followed by SSF. Fermentation was carried out in 500-mL flasks containing 300 mL of SSF production medium (10% treated oil palm trunk, 20 g/L of peptone, and 10 g/L of yeast extract). Two types of commercial cellulase, namely Celluclast 1.5 L and Novozymes 188 were employed in the SSF process, which was carried out at 40 °C using *S. cerevisiae* SC90. The highest ethanol concentration was 44.25 g/L, with a productivity of 0.45 g/L/h.

**Table 3.** Comparison of bioethanol production from *S. cerevisiae* using lignocellulosic biomass via fed-batch simultaneous saccharification and fermentation (fed-batch SSF).

| Strain | Solid Loading; Scale | Fermentation Temperature (°C) | Substrate | $C_{EtOH}$ * (g/L) | $Y_{p/s}$ ** $g_{EtOH}/g_{substrate}$ | $Q_P$ *** (g/L/h) | Y **** (%) | Reference |
|---|---|---|---|---|---|---|---|---|
| *S. cerevisiae* Y-2034 | 19% (*w*/*v*); Flask-scale | 37 | Sugarcane bagasse | 46.13 | 0.243 | 0.64 | 70.06 | Gao et al. [37] |
| *S. cerevisiae* L2524a | 30% (*w*/*v*); Bioreactor-scale | 30 | Empty palm fruit bunch fibers | 62.5 | 0.208 | 0.66 | 70.60 | Park et al. [53] |
| *S. cerevisiae* TC-5 | 12.5% (*w*/*v*); Bioreactor-scale | 40 | Cellulose-rich corncob residue | 31.96 | 0.256 | 0.22 | 78.20 | This study |

Note: N/A: not available; * $C_{EtOH}$: ethanol concentration; ** Yp/s: ethanol yield; *** QP: ethanol productivity; **** Y: theoretical ethanol yield.

In the present study, ethanol concentration, ethanol yield, and theoretical ethanol yield values of 31.96 g/L (143 $g/kg_{raw\ corncob}$), 0.256 $g_{EtOH}/g_{CRC}$, and 78.20%, respectively, were achieved by fed-batch SSF in a 5-L bioreactor. Ethanol concentration and productivity were not only comparable to those obtained in previous studies, but the cost of the SSF production medium used in this study was also remarkably low. Only 1 g/L of yeast extract was used, and inexpensive inorganic nutrients ($(NH_4)_2SO_4$, $NH_4H_2PO_4$, and $MgSO_4{\cdot}7H_2O$) were supplemented. Alternatively, *S. cerevisiae* TC-5 could grow and effectively produce bioethanol even under low nutrient conditions. The cost of these supplemental ingredients is a critical issue for industrial processes or commercial ethanol plants [54]. Yeast extract alone is estimated to account for approximately 20% of the raw material costs [55]. When medium with urea 3 g/L was used to produce bioethanol from carob waste, compared with the use of yeast extract-peptone medium (10 g/L of yeast extract and 20 g/L of peptone), the production costs could be reduced by up to 50% [56]. Another aspect for ethanol production is the minimization of the overall process time. Conventional SHF is considered a time-consuming process [57]. In this study, the overall process time of the up-scaled ethanol production, using thermotolerant *S. cerevisiae* TC-5, via SSF was only 144 h. In this sense, the newly isolated *S. cerevisiae* TC-5 might be an effective thermotolerant yeast strain suitable for bioethanol production from lignocellulosic material, minimizing processing time, and production costs. Regarding these concerns, this ethanol production procedure using *S. cerevisiae* TC-5 is effective, practical, and competitive on a cost basis.

## 5. Conclusions

This study shows that CRC residue can be used as a potential second-generation substrate for bioethanol production because of its high cellulose content. The isolated thermotolerant strain *S. cerevisiae* TC-5 could grow and produce bioethanol at 40 °C and required less supplemented nutrients and minerals. In addition, strain TC-5 could be employed in either batch SSF or fed-batch SSF at bioreactor scale. These characteristics make it an attractive thermotolerant yeast strain suitable for use in industrial bioethanol production from various lignocellulosic biomasses.

**Supplementary Materials:** The following are available online at https://www.mdpi.com/article/10.3390/jof7070547/s1, Table S1: Comparison of ethanol concentration, ethanol yield, ethanol productivity, and theoretical ethanol yield produced by commercial *S. cerevisiae* (control) and the thermotolerant *S. cerevisiae* TC-5 and via separate hydrolysis and fermentation (SHF), simultaneous saccharification and fermentation (SSF), and prehydrolysis-simultaneous saccharification and fermentation (pre-SSF) at 35, 37, 40, and 42 °C at 72 h of fermentation time. Figure S1: Phylogenetic tree of the thermotolerant *S. cerevisiae* TC-5, commercial *S. cerevisiae*, and the related species in GenBank database. Figure S2: Time course of ethanol production from cellulose-rich corncob hydrolysate by commercial *S. cerevisiae* via separate hydrolysis and fermentation (SHF) process at 35 (a), 37 (b), 40 (c), and 42 °C (d). Figure S3: Time course of ethanol production from cellulose-rich corncob hydrolysate by *S. cerevisiae* TC-5 via separate hydrolysis and fermentation (SHF) process at 35 (a), 37 (b), 40 (c), and 42 °C (d).

Figure S4: Time course of ethanol production from cellulose-rich corncob residue by commercial *S. cerevisiae* via simultaneous saccharification and fermentation (SSF) process at 35 (a), 37 (b), 40 (c), and 42 °C (d). Figure S5: Time course of ethanol production from cellulose-rich corncob residue by *S. cerevisiae* TC-5 via simultaneous saccharification and fermentation (SSF) process at 35 (a), 37 (b), 40 (c), and 42 °C (d). Figure S6: Time course of ethanol production from cellulose-rich corncob residue by commercial *S. cerevisiae* via prehydrolysis-simultaneous saccharification and fermentation (pre-SSF) process at 35 (a), 37 (b), 40 (c), and 42 °C (d). Figure S7: Time course of ethanol production from cellulose-rich corncob residue by *S. cerevisiae* TC-5 via prehydrolysis-simultaneous saccharification and fermentation (pre-SSF) process at 35 (a), 37 (b), 40 (c), and 42 °C (d).

**Author Contributions:** Conceptualization, T.C.; methodology, S.S.; software, P.B. and S.S.; validation, P.B. and T.C.; formal analysis, P.B.; investigation, S.S.; resources, S.S.; data curation, P.B.; writing—original draft preparation, S.S.; writing—review and editing, P.B., N.L., P.S., P.H., M.W. and T.C.; visualization, P.B. and S.S.; supervision, C.T.; project administration, T.C. All authors have read and agreed to the published version of the manuscript.

**Funding:** This research was funded by Chiang Mai University (CMU); the Faculty of Agro-Industry, CMU; the Functional Food Research Center for Wellbeing, CMU; and National Research Council of Thailand (NRCT).

**Institutional Review Board Statement:** Not applicable.

**Informed Consent Statement:** Not applicable.

**Data Availability Statement:** The 26S rDNA sequences data obtained in this work have been deposited in NCBI and are available on request from the corresponding author (thanongsak.c@cmu.ac.th or thachaiyaso@hotmail.com (accessed on 9 July 2021)).

**Acknowledgments:** This research was supported by Chiang Mai University (CMU); the Faculty of Agro-Industry, CMU; the Functional Food Research Center for Well-being, CMU; and National Research Council of Thailand (NRCT).

**Conflicts of Interest:** The authors declare no conflict of interest.

## References

1. Viikari, L.; Vehmaanperä, J.; Koivula, A. Lignocellulosic ethanol: From science to industry. *Biomass Bioenergy* **2012**, *46*, 13–24. [CrossRef]
2. Kumar, S.; Dheeran, P.; Singh, S.P.; Mishra, I.M.; Adhikari, D.K. Bioprocessing of bagasse hydrolysate for ethanol and xylitol production using thermotolerant yeast. *Bioprocess Biosyst. Eng.* **2015**, *38*, 39–47. [CrossRef]
3. Behera, B.C.; Sethi, B.K.; Mishra, R.R.; Dutta, S.K.; Thatoi, H.N. Microbial cellulases—Diversity & biotechnology with reference to mangrove environment: A review. *J. Genet. Eng. Biotechnol.* **2017**, *15*, 197–210. [CrossRef]
4. Wi, S.G.; Choi, I.S.; Kim, K.H.; Kim, H.M.; Bae, H.-J. Bioethanol production from rice straw by popping pretreatment. *Biotechnol. Biofuels* **2013**, *6*, 166. [CrossRef]
5. Talukder, A.A.; Easmin, F.; Mahmud, S.A.; Yamada, M. Thermotolerant yeasts capable of producing bioethanol: Isolation from natural fermented sources, identification and characterization. *Biotechnol. Biotechnol. Equip.* **2016**, *30*, 1106–1114. [CrossRef]
6. Boonchuay, P.; Techapun, C.; Leksawasdi, N.; Seesuriyachan, P.; Hanmoungjai, P.; Watanabe, M.; Takenaka, S.; Chaiyaso, T. An integrated process for xylooligosaccharide and bioethanol production from corncob. *Bioresour. Technol.* **2018**, *256*, 399–407. [CrossRef] [PubMed]
7. Liu, Y.; Xu, J.; Zhang, Y.; Yuan, Z.; He, M.; Liang, C.; Zhuang, X.; Xie, J. Sequential bioethanol and biogas production from sugarcane bagasse based on high solids fed-batch SSF. *Energy* **2015**, *90*, 1199–1205. [CrossRef]
8. Aditiya, H.B.; Mahlia, T.M.I.; Chong, W.T.; Nur, H.; Sebayang, A.H. Second generation bioethanol production: A critical review. *Renew. Sustain. Energy Rev.* **2016**, *66*, 631–653. [CrossRef]
9. Balat, M.; Balat, H. Recent trends in global production and utilization of bio-ethanol fuel. *Appl. Energy* **2009**, *86*, 2273–2282. [CrossRef]
10. Xiros, C.; Topakas, E.; Christakopoulos, P. Hydrolysis and fermentation for cellulosic ethanol production. *WIREs Energy Environ.* **2013**, *2*, 633–654. [CrossRef]
11. Olofsson, K.; Bertilsson, M.; Lidén, G. A short review on SSF—An interesting process option for ethanol production from lignocellulosic feedstocks. *Biotechnol. Biofuels* **2008**, *1*, 7. [CrossRef]
12. Cha, Y.-L.; An, G.H.; Yang, J.; Moon, Y.-H.; Yu, G.-D.; Ahn, J.-W. Bioethanol production from *Miscanthus* using thermotolerant *Saccharomyces cerevisiae* mbc 2 isolated from the respiration-deficient mutants. *Renew. Energy* **2015**, *80*, 259–265. [CrossRef]
13. Jahnavi, G.; Prashanthi, G.S.; Sravanthi, K.; Rao, L.V. Status of availability of lignocellulosic feed stocks in India: Biotechnological strategies involved in the production of Bioethanol. *Renew. Sustain. Energy Rev.* **2017**, *73*, 798–820. [CrossRef]

14. Baig, K.S. Interaction of enzymes with lignocellulosic materials: Causes, mechanism and influencing factors. *Bioresour. Bioprocess.* **2020**, *7*, 21. [CrossRef]
15. Mohd Azhar, S.H.; Abdulla, R.; Jambo, S.A.; Marbawi, H.; Gansau, J.A.; Mohd Faik, A.A.; Rodrigues, K.F. Yeasts in sustainable bioethanol production: A review. *Biochem. Biophys. Rep.* **2017**, *10*, 52–61. [CrossRef]
16. Demeke, M.M.; Dietz, H.; Li, Y.; Foulquié-Moreno, M.R.; Mutturi, S.; Deprez, S.; Den Abt, T.; Bonini, B.M.; Liden, G.; Dumortier, F.; et al. Development of a D-xylose fermenting and inhibitor tolerant industrial *Saccharomyces cerevisiae* strain with high performance in lignocellulose hydrolysates using metabolic and evolutionary engineering. *Biotechnol. Biofuels* **2013**, *6*, 89. [CrossRef] [PubMed]
17. Walker, G.M.; Walker, R.S.K. Chapter Three—Enhancing Yeast Alcoholic Fermentations. In *Advances in Applied Microbiology*; Gadd, G.M., Sariaslani, S., Eds.; Academic Press: Cambridge, MA, USA, 2018; Volume 105, pp. 87–129. ISBN 0065-2164.
18. Pattanakittivorakul, S.; Lertwattanasakul, N.; Yamada, M.; Limtong, S. Selection of thermotolerant *Saccharomyces cerevisiae* for high temperature ethanol production from molasses and increasing ethanol production by strain improvement. *Antonie Leeuwenhoek* **2019**, *112*, 975–990. [CrossRef] [PubMed]
19. Boonchuay, P.; Wongpoomchai, R.; Jaturasitha, S.; Mahatheeranont, S.; Watanabe, M.; Chaiyaso, T. Prebiotic properties, antioxidant activity, and acute oral toxicity of xylooligosaccharides derived enzymatically from corncob. *Food Biosci.* **2021**, *40*, 100895. [CrossRef]
20. Chaiyaso, T.; Boonchuay, P.; Takenaka, S.; Techapun, C.; Rachtanapun, P.; Jantanasakulwong, K.; Watanabe, M. Efficient enzymatic process for mulberry paper production: An approach for xylooligosaccharide production coupled with minimizing bleaching agent doses. *Waste Biomass Valorization* **2021**. [CrossRef]
21. Manowattana, A.; Techapun, C.; Laokuldilok, T.; Phimolsiripol, Y.; Chaiyaso, T. Enhancement of β-carotene-rich carotenoid production by a mutant *Sporidiobolus pararoseus* and stabilization of its antioxidant activity by microencapsulation. *J. Food Process. Preserv.* **2020**, *44*, e14596. [CrossRef]
22. Srisupa, S.; Boonchuay, P.; Hanmoungjai, P.; Chaiyaso, T. Bioethanol production using cellulose-rich corncob residue by thermotolerant yeasts. In Proceedings of the International Conference on Food and Applied Bioscience (FAB 2020), Chiangmai, Thailand, 6–7 February 2020; pp. 391–402.
23. Zhang, Y.H.P.; Hong, J.; Ye, X. Cellulase Assays. In *Biofuels: Methods in Molecular Biology (Methods and Protocols)*; Mielenz, J.R., Ed.; Humana Press: Totowa, NJ, USA, 2009; pp. 213–231.
24. Ghose, T.K. Measurement of cellulase activities. *Pure Appl. Chem.* **1987**, *59*, 257–268. [CrossRef]
25. Salma, F. Investigation of β-Xylosidase, α-L-Arabinofuranosidase and Acetylesterase from *Thermotoga hypogea*. Master's Thesis, University of Waterloo, Waterloo, ON, Canada, 2008.
26. Sinjaroonsak, S.; Chaiyaso, T.; H-Kittikun, A. Optimization of cellulase and xylanase productions by *Streptomyces thermocoprophilus* TC13W using low cost pretreated oil palm empty fruit bunch. *Waste Biomass Valorization* **2019**, *11*, 3925–3936. [CrossRef]
27. Bai, F.-W.; Yang, S.; Ho, N.W.Y. 3.05—Fuel ethanol production from lignocellulosic biomass. In *Comprehensive Biotechnology*, 3rd ed.; Moo-Young, M., Ed.; Pergamon: Oxford, UK, 2019; pp. 49–65. ISBN 978-0-444-64047-5.
28. Watanabe, I.; Nakamura, T.; Shima, J. Characterization of a spontaneous flocculation mutant derived from *Candida glabrata*: A useful strain for bioethanol production. *J. Biosci. Bioeng.* **2009**, *107*, 379–382. [CrossRef]
29. Tam, P.; Gee, K.; Piechocinski, M.; Macreadie, I. Candida glabrata, Friend and Foe. *J. Fungi* **2015**, *1*, 277–292. [CrossRef]
30. Chandel, A.K.; Lakshmi Narasu, M.; Chandrasekhar, G.; Manikyam, A.; Venkateswar Rao, L. Use of *Saccharum spontaneum* (wild sugarcane) as biomaterial for cell immobilization and modulated ethanol production by thermotolerant *Saccharomyces cerevisiae* VS3. *Bioresour. Technol.* **2009**, *100*, 2404–2410. [CrossRef] [PubMed]
31. Kasavi, C.; Finore, I.; Lama, L.; Nicolaus, B.; Oliver, S.G.; Toksoy Oner, E.; Kirdar, B. Evaluation of industrial *Saccharomyces cerevisiae* strains for ethanol production from biomass. *Biomass Bioenergy* **2012**, *45*, 230–238. [CrossRef]
32. Nuanpeng, S.; Thanonkeo, S.; Yamada, M.; Thanonkeo, P. Ethanol production from sweet sorghum juice at high temperatures using a newly isolated thermotolerant yeast *Saccharomyces cerevisiae* DBKKU Y-53. *Energies* **2016**, *9*, 253. [CrossRef]
33. Tesfaw, A.; Assefa, F. Current Trends in Bioethanol Production by *Saccharomyces cerevisiae*: Substrate, Inhibitor Reduction, Growth Variables, Coculture, and Immobilization. *Int. Sch. Res. Not.* **2014**, *2014*, 532852. [CrossRef]
34. Cheng, N.; Koda, K.; Tamai, Y.; Yamamoto, Y.; Takasuka, T.E.; Uraki, Y. Optimization of simultaneous saccharification and fermentation conditions with amphipathic lignin derivatives for concentrated bioethanol production. *Bioresour. Technol.* **2017**, *232*, 126–132. [CrossRef]
35. Tomás-Pejó, E.; Oliva, J.M.; Ballesteros, M.; Olsson, L. Comparison of SHF and SSF processes from steam-exploded wheat straw for ethanol production by xylose-fermenting and robust glucose-fermenting *Saccharomyces cerevisiae* strains. *Biotechnol. Bioeng.* **2008**, *100*, 1122–1131. [CrossRef]
36. Ma, K.; He, M.; You, H.; Pan, L.; Hu, G.; Cui, Y.; Maeda, T. Enhanced fuel ethanol production from rice straw hydrolysate by an inhibitor-tolerant mutant strain of *Scheffersomyces stipitis*. *RSC Adv.* **2017**, *7*, 31180–31188. [CrossRef]
37. Gao, Y.; Xu, J.; Yuan, Z.; Jiang, J.; Zhang, Z.; Li, C. Ethanol production from sugarcane bagasse by fed-batch simultaneous saccharification and fermentation at high solids loading. *Energy Sci. Eng.* **2018**, *6*, 810–818. [CrossRef]
38. Mendes, C.V.T.; Cruz, C.H.G.; Reis, D.F.N.; Carvalho, M.G.V.S.; Rocha, J.M.S. Integrated bioconversion of pulp and paper primary sludge to second generation bioethanol using *Saccharomyces cerevisiae* ATCC 26602. *Bioresour. Technol.* **2016**, *220*, 161–167. [CrossRef] [PubMed]

39. da Silva, F.L.; de Oliveira Campos, A.; dos Santos, D.A.; Batista Magalhães, E.R.; de Macedo, G.R.; dos Santos, E.S. Valorization of an agroextractive residue—Carnauba straw—For the production of bioethanol by simultaneous saccharification and fermentation (SSF). *Renew. Energy* **2018**, *127*, 661–669. [CrossRef]
40. Gladis, A.; Bondesson, P.-M.; Galbe, M.; Zacchi, G. Influence of different SSF conditions on ethanol production from corn stover at high solids loadings. *Energy Sci. Eng.* **2015**, *3*, 481–489. [CrossRef]
41. Öhgren, K.; Vehmaanperä, J.; Siika-Aho, M.; Galbe, M.; Viikari, L.; Zacchi, G. High temperature enzymatic prehydrolysis prior to simultaneous saccharification and fermentation of steam pretreated corn stover for ethanol production. *Enzym. Microb. Technol.* **2007**, *40*, 607–613. [CrossRef]
42. Qin, L.; Zhao, X.; Li, W.-C.; Zhu, J.-Q.; Liu, L.; Li, B.-Z.; Yuan, Y.-J. Process analysis and optimization of simultaneous saccharification and co-fermentation of ethylenediamine-pretreated corn stover for ethanol production. *Biotechnol. Biofuels* **2018**, *11*, 118. [CrossRef]
43. Roberto, I.C.; Castro, R.C.A.; Silva, J.P.A.; Mussatto, S.I. Ethanol production from high solid loading of rice straw by simultaneous saccharification and fermentation in a non-conventional reactor. *Energies* **2020**, *13*, 2090. [CrossRef]
44. Persson, M.; Galbe, M.; Wallberg, O. A strategy for synergistic ethanol yield and improved production predictability through blending feedstocks. *Biotechnol. Biofuels* **2020**, *13*, 156. [CrossRef]
45. Orrego, D.; Zapata-Zapata, A.D.; Kim, D. Optimization and scale-up of coffee mucilage fermentation for ethanol production. *Energies* **2018**, *11*, 786. [CrossRef]
46. Dwiarti, L.; Boonchird, C.; Harashima, S.; Park, E.Y. Simultaneous saccharification and fermentation of paper sludge without pretreatment using cellulase from *Acremonium cellulolyticus* and thermotolerant *Saccharomyces cerevisiae*. *Biomass Bioenergy* **2012**, *42*, 114–122. [CrossRef]
47. Kadhum, H.J.; Rajendran, K.; Murthy, G.S. Effect of solids loading on ethanol production: Experimental, economic and environmental analysis. *Bioresour. Technol.* **2017**, *244*, 108–116. [CrossRef] [PubMed]
48. Kim, J.S.; Lee, Y.Y.; Kim, T.H. A review on alkaline pretreatment technology for bioconversion of lignocellulosic biomass. *Bioresour. Technol.* **2016**, *199*, 42–48. [CrossRef]
49. Santibáñez, L.; Henríquez, C.; Corro-Tejeda, R.; Bernal, S.; Armijo, B.; Salazar, O. Xylooligosaccharides from lignocellulosic biomass: A comprehensive review. *Carbohydr. Polym.* **2021**, *251*, 117118. [CrossRef]
50. Saravanakumar, K.; Kathiresan, K. Bioconversion of lignocellulosic waste to bioethanol by *Trichoderma* and yeast fermentation. *3 Biotech* **2014**, *4*, 493–499. [CrossRef]
51. Khammee, P.; Unpaprom, Y.; Chaichompoo, C.; Khonkaen, P.; Ramaraj, R. Appropriateness of waste jasmine flower for bioethanol conversion with enzymatic hydrolysis: Sustainable development on green fuel production. *3 Biotech* **2021**, *11*, 216. [CrossRef]
52. Tareen, A.K.; Sultan, I.N.; Songprom, K.; Laemsak, N.; Sirisansaneeyakul, S.; Vanichsriratana, W.; Parakulsuksatid, P. Two-step pretreatment of oil palm trunk for ethanol production by thermotolerent *Saccharomyces cerevisiae* SC90. *Bioresour. Technol.* **2021**, *320*, 124298. [CrossRef] [PubMed]
53. Park, J.M.; Oh, B.-R.; Seo, J.-W.; Hong, W.-K.; Yu, A.; Sohn, J.-H.; Kim, C.H. Efficient production of ethanol from empty palm fruit bunch fibers by fed-batch simultaneous saccharification and fermentation using *Saccharomyces cerevisiae*. *Appl. Biochem. Biotechnol.* **2013**, *170*, 1807–1814. [CrossRef]
54. Kadam, K.L.; Newman, M.M. Development of a low-cost fermentation medium for ethanol production from biomass. *Appl. Microbiol. Biotechnol.* **1997**, *47*, 625–629. [CrossRef]
55. Unrean, P.; Khajeeram, S. Optimization and techno-economic assessment of high-solid fed-batch saccharification and ethanol fermentation by *Scheffersomyces stipitis* and *Saccharomyces cerevisiae* consortium. *Renew. Energy* **2016**, *99*, 1062–1072. [CrossRef]
56. Raposo, S.; Constantino, A.; Rodrigues, F.; Rodrigues, B.; Lima-Costa, M.E. Nitrogen sources screening for ethanol production using carob industrial wastes. *Appl. Biochem. Biotechnol.* **2017**, *181*, 827–843. [CrossRef]
57. Dey, P.; Pal, P.; Kevin, J.D.; Das, D.B. Lignocellulosic bioethanol production: Prospects of emerging membrane technologies to improve the process—A critical review. *Rev. Chem. Eng.* **2020**, *36*, 333–367. [CrossRef]

Journal of *Fungi*

MDPI

*Article*

# Use of Wheat Straw for Value-Added Product Xylanase by *Penicillium chrysogenum* Strain A3 DSM105774

**Amira A. Matrawy [1], Ahmed I. Khalil [1], Heba S. Marey [1] and Amira M. Embaby [2,*]**

1 Environmental Studies Department, Institute of Graduate Studies and Research, Alexandria University, Alexandria 21526, Egypt; amira.matrawy30@alexu.edu.eg (A.A.M.); airkhalil@gmail.com (A.I.K.); hebasaid06@gmail.com (H.S.M.)
2 Biotechnology Department, Institute of Graduate Studies and Research, Alexandria University, Alexandria 21526, Egypt
* Correspondence: Amira.Embaby@alexu.edu.eg

**Abstract:** The present work highlights the valorization of the bulky recalcitrant lignocellulose byproduct wheat straw (WS) for the enhanced production of value-added xylanase by the locally sourced novel *Penicillium chrysogenum* strain A3 DSM105774 for the first time. The optimized production of xylanase by submerged state of fermentation of WS was achieved using a three-step statistical and sequential approach: one factor at a time (OFAT), Plackett–Burman design (PBD), and Box Behnken design (BBD). Incubation temperature (30 °C), WS, and ammonium sulphate were the key determinants prompting xylanase production; inferred from OFAT. The WS concentration (%($w/v$)), yeast extract concentration (%($w/v$)), and initial pH of the production medium imposed significant effects ($p \leq 0.05$) on the produced xylanase, realized from PBD. The predicted levels of WS concentration, initial pH of the production medium, and yeast extract concentration provoking the ultimate xylanase levels (53.7 U/mL) with an 8.95-fold enhancement, localized by the estimated ridge of the steepest ascent of the ridge analysis path, were 3.8% ($w/v$), 5.1, and 0.098% ($w/v$), respectively; 94.7% lab validation. The current data underpin the up-scaling of xylanase production using this eco-friendly, cheap, and robust methodology for the valorization of WS into the value-added product xylanase.

**Keywords:** wheat straw; *Penicillium chrysogenum* strain A3 DSM105774; xylanase; submerged state of fermentation; low cost eco-friendly methodology

**Citation:** Matrawy, A.A.; Khalil, A.I.; Marey, H.S.; Embaby, A.M. Use of Wheat Straw for Value-Added Product Xylanase by *Penicillium chrysogenum* Strain A3 DSM105774. *J. Fungi* **2021**, *7*, 696. https://doi.org/10.3390/jof7090696

Academic Editor: Baojun Xu

Received: 14 July 2021
Accepted: 20 August 2021
Published: 27 August 2021



## 1. Introduction

Millions of tons of agro-industrial waste byproducts do generate from agricultural practices on lignocellulose biomasses annually worldwide. Wheat straw, wheat bran, wheat husk, corncob, rice straw, rice bran, rice husk, oat husk, oat straw, barley straw, sorghum straw, paddy straw, sugarcane straw, and maize straw are paradigms of agro-industrial waste byproducts containing lignocelluloses [1].

Wheat straw (designated hereafter as WS) is a recalcitrant agro-industrial waste byproduct containing lignocelluloses. Wheat straw, the dry stalk of wheat after the removal of the grain and the chaff [2], is composed of carbohydrates as a major constituent (27% cellulose, 21% hemicellulose, and 23% lignin on dry weight basis), minerals, proteins, silica, ash, and acid detergent fibers [3–5]. The worldwide consumption of wheat was estimated to reach 652.18 million ton in 2010 [6]. Normally, WS is collected and conserved in straw bales. Despite the involvement of WS in an extensive range of applications in medicine [7], fermentation industry [8], soil fertility [9], bio charcoal [10], basket making, pulp industry [11], and bioremediation [12,13], novel usages are urgently mandatory for effective management and economic valorization of the bulky amounts of this byproduct waste.

Wheat straw is a potential zero-cost alternate for the highly expensive synthetic substrates included in the manufacturing of value-added bioproducts like enzymes by the aid of a panel of filamentous fungi. For instance, *Penicillium janthinellum* [14], 2015), *Trichoderma harzianum* ZF-2 [15], *P. chrysogenum* [16], *Paecilomyces thermophila* J18 [17], *Aspergillus ochraceus* [18], and *A. niger* strain [19] were reported as potential fungi not only for the management of the lignocellulose byproduct WS but also for the WS valorization into value added products in the form of cellulases, laccases, α-amylases, xylanases, and β-xylosoxidases, respectively.

β-1,4-Xylanases (EC 3.2.1.8) of microbial origin are industrial enzymes with immense biotechnological applications for mankind such as waste treatment, paper-pulp, feed, and food industries [9,20] as well as the manufacturing of biofuel from cellulosic biomass [21]. Over the few last decades, numerous fungal strains were tried for the production of live xylanases using lignocelluloses agro-industrial waste byproducts such as sorghum straw by *Thermomyces lanuginosus* ($D_2W_3$) [22], rice husk by *T. lanuginosus* strain A3-1 DSM 105773 [23], rice bran by *Humicola lanuginosa* [24], rice straw by *Aspergillus fumigatus* NIT-DGPKA3 [25], wheat bran by *P. chrysogenum* PCL501 [26], wheat straw by *Paecilomyces thermophia* [27] and *Melanocarpus albomyces* IITD3A [28], and maize straw by *Trichoderma viride* [29]. Several *Penicillum* spp (e.g., *Penicillium oxalicum* ZH-30 [30], *P. oxalicum* T3.3 [31], *P. pinophilum* NTG1 II/6 [32], *P. decumbens* [33], *P. janthinelum* CRC 87M-115 [34], *P. purpurogenum* [35], *P. echiulatum* [36], *P. expansum* [37], *P. chrysogenum* PCL501 [26], and *P. canescens* (10-10c) [38] have been intensely studied for their capability to manufacture live xylanases upon their cultivation on wheat bran, wheat straw, rice straw, corncob, oat husk, and barely straw. However, the valorization of WS to a value-added product xylanase by *P. chrysogenum* (recently designated as *P. rubens*) has not been attempted yet. Currently, the xylanase producers at the industrial scale are mostly species of *Aspergillus* and *Trichoderma*. Nonetheless, the search of novel xylanase hyper-producer fungal strains alongside with the selection of zero-cost substrates has captured increasing interest to cover the needs of enzyme markets globally.

In the light of the mentioned above, this work was conducted: (a) to beneficiate the WS through an eco-friendly technology to a valuable end product xylanase, (b) to statistically optimize the xylanase production by the locally novel *Penicillium chrysogenum* strain A3 DSM105774 using WS-based production medium on the flask scale for the first time ever, (b) to customize an inexpensive xylanase production medium containing the zero-cost substrate WS as a co-sole carbon source and xylanase inducer.

## 2. Materials and Methods

### 2.1. Wheat Straw

The wheat straw waste was obtained from a regional farm in Alexandria City. It was subjected to drying in an oven at 60 °C, and was milled in a Devika commercial mixer grinder, Classic Model, Devika Industries Inc., Gujarat, India) into fluffy flakes (1–2 mm). Then it was preserved in airtight plastic bags at ambient temperature.

### 2.2. Reagents and Chemicals

Beechwood xylan was obtained from LOBAChemie, PVT., Ltd., Mumbai, India. The 3,5-dinitrosalicylic acid (DNS), xylose was purchased from Sigma-Aldrich Co., Missouri, Saint Louis, USA. The PDA (potato dextrose agar) was purchased from HiMedia Labs, Ltd., Mumbai, India. The18S rRNA fungal-specific primers were manufactured in Macrogen Co., Ltd., Seoul, Korea.

### 2.3. Media

The PDA was used for the purposes of activation and short-term maintenance of fungal strains. The minimal medium [39], in g/L: $(NH_4)_2SO_4$, 1.3; $KH_2PO_4$, 0.37; $MgSO_4{\cdot}7H_2O$, 0.25; $CaCl_2{\cdot}2H_2O$, 0.07; $FeCl_3$, 0.02; yeast extract 1.0; and 0.5% (*w*/*v*) beechwood xylan, was used as a core xylanase production medium by fungal candidates.

### 2.4. Isolation of Xylanase-Producing Fungal Strains

The fungal strain enrolled in this work was isolated from a sample gathered from Abbis-2 compost plant, Alexandria City, Egypt. Succinctly, 10 g compost sample was added to 90 mL sterilized saline solution and agitated for 30 min. Ten-fold serial dilution was done followed by plating 0.1 mL of an appropriately selected dilution on PDA plates complemented with streptomycin at a final concentration of 100 mg/L to suppress the growth of bacteria [40]. The incubation temperature for the inoculated PDA plates was 30 °C. After incubation for 6 days, single fungal colonies of distinct colorizations were picked for sub-culturing on PDA plates for further purification. All purified fungal isolates were preserved on PDA slants at 4 °C and sub-cultured monthly. The capability of the tested fungal isolates regarding xylanase production was assessed on a minimal medium complemented with beechwood xylan (0.5% (*w*/*v*)) and 2% (*w*/*v*) agar. The plates were incubated for 6 days at 30 °C followed by overflowing with Congo red (0.1% (*w*/*v*)) for 15 min. The zone of hydrolysis was developed by flooding the plates with 1.5 M NaCl for 30 min. The highest xylanase-producing fungal isolate, selected based on its largest hydrolysis zone surrounding the fungal colony on a xylan agar plate, was selected as the promising fungal strain to conduct this study (Figure S1).

### 2.5. Scnnaing Electron Microscopy

The microscopic features of the most promising xylanase-producing fungal strain were investigated under scanning electron microscope (SEM). Dry smears of the fungal strain were coated with approximately 15 nm gold (JEC-1100 E Sputter Coater, JEOL Inc., Pleasanton, CA, USA). After that, the golden coated samples were scanned using SEM (JEOL JSM-5300, JEOL Inc., Pleasanton, CA, USA). The coated slide was accelerated at 30 KV (at room temperature). The digital images of the samples were adjusted and saved for further investigation.

### 2.6. 18S rDNA Sequencing Approach

Genomic DNA of the quest fungal strain was extracted using the ZR Fungal/Bacterial DNA Miniprep™ (Zymo Co., Irvine, CA, USA) in accordance with the manufacturer's instructions. The *18S ribosomal DNA* (*18S rDNA*) gene full length of quest fungal strain was amplified by PCR using the fungal specific primer set EF4 (5′-GGAAGGGRTGTATTTATT AG-3′) and EF3 (5′-TCCTCTAAATGACAAGTTTG-3′) [41]. The reaction was conducted at a final volume of 50 µL: 5 µL (30 ng) of genomic DNA, (2.5 µL) 25 pmol of each primer (EF4 and EF3), 25 µL of 1X PCR master mix (iNTRON, Gyeonggi, Korea), and 15 µL nuclease free water. The PCR reaction was executed in PCR thermocycler (Biometra, Göttingen, Germany). PCR conditions were set as follow: initial denaturation at 95 °C for 5 min, 30 cycles (each cycle: denaturation at 94 °C for 45 s, annealing at 48 °C for 45 s and extension at 72 °C for 1.5 min) and final extension at 72 °C for 10 min. The presence of PCR products was checked on 1.0% (*w*/*v*) agarose gel. The 18S rDNA PCR product was purified using the GeneJET PCR Purification kit (Thermo Fisher Scientific Co., Waltham, Massachusetts, MA, USA). Then, the 18S rDNA nucleotide sequence of the fungal strain was searched against the international nucleotide databases (EMBL, GenBank, and DDBJ) using the BLASTN algorithm (http://blast.ncbi.nlm.nih.gov/Blast.cgi, accessed on 15 May 2017). The phylogenetic tree was constructed out via CLC Sequence Viewer 8.0 software to define the evolutionary relationship between the quest fungal isolate and other fungal members. The 18S rDNA nucleotide sequence of the quest fungal strain was deposited in the EMBL database.

### 2.7. Xylanase Assay

The xylanase activity of the fungal-free supernatant was determined by estimating the reducing sugars derived from the substrate beechwood xylan by the 3, 5 dinitrosalicylic acid (DNS) [42]. Concisely, 0.5 mL beechwood xylan (1.0% (*w*/*v*)) in 100 mL (0.1M sodium acetate buffer, pH 5.0) was added to crude enzyme preparation (0.5 mL) and incubated at 50 °C for 30 min. The reaction was stopped by the addition of DNS reagent (2.0 mL) followed by boiling at 100 °C for 5 min. Unlike the test reaction, the crude enzyme preparation was boiled for 5 min prior to its addition to the reaction mixture in the control reaction. All reaction tubes were cooled down followed by the addition of 7.0 mL distilled water to each tube with thorough mixing. The absorbance of the developed color was estimated spectrophotometrically at 540 nm. The standard curve was established using xylose. All assays were executed in 4 replicates. One arbitrary unit (U) of enzyme activity was described as the quantity of enzyme liberating 1 μmol xylose from the substrate beechwood xylan after one min.

### 2.8. Optimization of Xylanase Production

A three-step optimizing plan, OFAT, PBD, and BBD, was employed to optimize the xylanase production directed by the most potent xylanase-producing fungal strain under submerged state fermentation.

### 2.9. One Factor at a Time (OFAT)

The initial selection of the physicochemical determinants affecting the xylanase activity by the most potent xylanase-producing fungal strain was done through OFAT methodology. The theory behind this approach is the alteration of one factor at a time regardless of the likely interactions among the tried factors. In this study, the checked physiochemical factors were wheat straw as a carbon source instead of beechwood xylan, inorganic nitrogen source (i.e., sodium nitrate, ammonium sulphate, ammonium citrate, ammonium chloride, ammonium phosphate, and ammonium nitrate), and incubation temperature. Each inorganic nitrogen source was tested at a final concentration of 1.3 g/L at 35 °C. The following different incubation temperatures 25, 30, 35, and 40 °C were tried out. A 50 mL production medium was dispended in Erlenmeyer flasks (100 mL). Two discs of 4 days activated fungal strain were used to inoculate the Erlenmeyer flask-containing production medium. All cultures were shaken at 120 rpm in an incubator shaker for 6 days. The experimental runs were operated in triplicates.

The proper incubation temperature, deduced from OFAT trials, prompted the maximal levels of xylanase was chosen in order to conduct the subsequent optimization trials. The two independent factors imposing significant consequences on xylanase production deduced from OFAT: wheat straw concentration (% ($w/v$)) and ammonium sulphate concentration (% ($w/v$)) as well as three independent factors; incubation time, yeast extract concentration (%($w/v$)), and initial pH of the production medium were further studied by PDB; the next step in the optimization strategy.

### 2.10. Plackett–Burman Design (PBD)

In this study, Plackett–Burman (a full factorial design) [43] was employed in order to evaluate the likely considerable linear impact of the five independent factors: yeast extract concentration ($w/v$)), initial pH of the production medium, wheat straw concentration ($w/v$)), incubation time (days), and ammonium sulphate concentration ($w/v$)) on xylanase production. A matrix of twelve trials, full factorial design, was designed by Minitab 17.3 software as presented in Table 1a. Each factor was coded (−1) and (+1) in two levels low level and (high level, respectively. The possible linear impact enforced by the tried independent factors on the production of xylanase was portrayed through a first-order polynomial equation (Equation (1)).

$$Y = \beta o + \sum \beta i x i \quad (1)$$

A fifty mL of the production medium was dispensed in 100 mL Erlenmeyer flasks. All cultures were incubated at 120 rpm in an incubator shaker. The independent factors imposing significant influences on the xylanase production [i.e., initial pH of the production medium, yeast extract concentration (% ($w/v$)), and wheat straw concentration (% ($w/v$))] were tested further in the final stage in the optimization strategy, BBD. However, the independent factors displaying non-significant concerns on the production of xylanase were used at their initial values in the subsequent experiments.

**Table 1.** (**a**) Coded-real values of five factors in PBD for observing top significant factors regulating the production of xylanase by *P. chrysogenum* strain A3 DSM105774. (**b**) An outline for multiple linear regression of PBD data for assessment of the production of xylanase by *P. chrysogenum* strain A3 DSM105774.

(a)

| | Independent Factors | | | | | Y Xylanase Activity (U/mL) | |
|---|---|---|---|---|---|---|---|
| **Trial #** | **X1** | **X2** | **X3** | **X4** | **X5** | **Exp. [a]** | **Pred. [b]** |
| 1 | 1(1.0) | −1(0.065) | 1(7.0) | −1(4.0) | −1(0.025) | 14.43 | 14.39 |
| 2 | 1(1.0) | 1(0.25) | −1(5.0) | 1(8.0) | −1(0.025) | 15.00 | 15.54 |
| 3 | −1(0.25) | 1(0.25) | 1(7.0) | −1(4.0) | 1(0.15) | 6.54 | 5.92 |
| 4 | 1(1.0) | −1(0.065) | 1(7.0) | 1(8.0) | −1(0.025) | 13.47 | 13.98 |
| 5 | 1(1.0) | 1(0.25) | −1(5.0) | 1(8.0) | 1(0.15) | 14.32 | 13.04 |
| 6 | 1(1.0) | 1(0.25) | 1(7.0) | −1(4.0) | 1(0.15) | 9.63 | 11.98 |
| 7 | −1(0.25) | 1(0.25) | 1(7.0) | 1(8.0) | −1(0.025) | 9.39 | 7.99 |
| 8 | −1(0.25) | −1(0.065) | 1(7.0) | 1(8.0) | 1(0.15) | 6.21 | 5.42 |
| 9 | −1(0.25) | −1(0.065) | −1(5.0) | 1(8.0) | 1(0.15) | 4.47 | 6.89 |
| 10 | 1(1.0) | −1(0.065) | −1(5.0) | −1(4.0) | 1(0.15) | 15.45 | 13.38 |
| 11 | −1(0.25) | 1(0.25) | −1(5.0) | −1(4.0) | −1(0.025) | 9.48 | 9.89 |
| 12 | −1(0.25) | −1(0.065) | −1(5.0) | −1(4.0) | −1(0.025) | 9.84 | 9.81 |

(b)

| Independent Factor | Estimate Symbol | Estimate | t-Value | *p*-Value | Confidence Level (%) |
|---|---|---|---|---|---|
| Intercept | B° | 10.68575 | 34.2592 | $7.64 \times 10^{-18}$ | 100.00 * |
| X1 | B1 | 3.030083 | 9.714641 | $1.39 \times 10^{-8}$ | 99.99 * |
| X2 | B2 | 0.040083 | 0.12851 | 0.899171 | |
| X3 | B3 | −0.74008 | −2.37275 | 0.029001 | 97.10 * |
| X4 | B4 | −0.20925 | −0.67087 | 0.510822 | |
| X5 | B5 | −1.24925 | −4.00518 | 0.000830 | 99.91 * |

Values between brackets: real values for the independent factors. X1: WS concentration (%($w/v$)), X2: ammonium sulphate concentration (%($w/v$)), X3: initial pH of the production medium, X4: incubation time (days) and X5: yeast extract concentration (%($w/v$)). [a]: experimental values and [b]: predicted values. Values: mean of 3 readings. * Significant *p*-value: $\leq 0.05$.

*2.11. Box–Behnken Design (BBD)*

The key factors, showing significant influences on the production of xylanase were studied through the BBD [44] to determine the optimum level of each key factor in combination with the maximum level of xylanase. Each factor was coded (−1), (0), and (+1) indicating three levels low level, center level, and high level, respectively in a matrix of fifteen trials (Table 2a). The three studied independent factors were wheat straw concentration (% ($w/v$)), initial pH of the production medium, and yeast extract concentration (% ($w/v$)). All likely forms of interactions among the studied independent factors triggering significant consequences on the production of xylanase were depicted in Equation (2); a second-order polynomial equation.

$$Y = \beta o + \sum_{i=1}^{k} \beta ixi + \sum_{i=1}^{k} \beta iixixi + \sum_{i=1}^{k-1} \sum_{j=2}^{k} \beta ijxixj + \epsilon \quad (2)$$

All experimental trials had performed in Erlenmeyer flasks (100 mL) containing a 50 mL production medium. All experiments of BBD were incubated in an incubator shaker, with agitation at 120 rpm.

**Table 2.** (**a**) The levels of coded and real values of three independent factors in BBD for maximizing the xylanase production by *P. chrysogenum* strain A3 DSM105774. (**b**) Regression synopsis of BBD for assessing xylanase levels produced by *P. chrysogenum* strain A3 DSM105774.

**(a)**

| Trial # | X1 | X3 | X5 | Y6 Xylanase Activity (U/mL) Exp. [a] | Y6 Xylanase Activity (U/mL) Pred. [b] |
|---|---|---|---|---|---|
| 1 | −1(0.1) | −1(4.5) | 0(0.056) | 10.77 | 8.45 |
| 2 | 1(1.9) | −1(4.5) | 0(0.056) | 51.42 | 56.52 |
| 3 | −1(0.1) | 1(9.5) | 0(0.056) | 3.84 | −1.26 |
| 4 | 1(1.9) | 1(9.5) | 0(0.056) | 12.03 | 14.35 |
| 5 | −1(0.1) | 0(7.0) | −1(0.012) | 7.26 | 13.32 |
| 6 | 1(1.9) | 0(7.0) | −1(0.012) | 45.48 | 44.13 |
| 7 | −1(0.1) | 0(7.0) | 1(0.1) | 9.45 | 10.80 |
| 8 | 1(1.9) | 0(7.0) | 1(0.1) | 49.74 | 43.68 |
| 9 | 0(1.0) | −1(4.5) | −1(0.012) | 43.50 | 39.75 |
| 10 | 0(1.0) | 1(9.5) | −1(0.012) | 11.13 | 10.17 |
| 11 | 0(1.0) | −1(4.5) | 1(0.1) | 33.66 | 34.62 |
| 12 | 0(1.0) | 1(9.5) | 1(0.1) | 8.58 | 12.33 |
| 13 | 0(1.0) | 0(7.0) | 0(0.56) | 37.47 | 38.45 |
| 14 | 0(1.0) | 0(7.0) | 0(0.56) | 38.82 | 38.45 |
| 15 | 0(1.0) | 0(7.0) | 0(0.56) | 39.06 | 38.45 |

**(b)**

| Independent Factor | Estimate Symbol | Estimate | t-Value | *p*-Value | Confidence Level (%) |
|---|---|---|---|---|---|
| Intercept | B° | 38.45 | 22.75182 | $9.1 \times 10^{-6}$ | 100.00 * |
| X1 | B1 | 15.92 | 15.38201 | $1.51 \times 10^{-12}$ | 99.99 * |
| X3 | B2 | −12.97 | −12.5339 | $6.28 \times 10^{-11}$ | 99.999 * |
| X5 | B3 | −0.74 | −0.71746 | 0.481384 | - |
| X1.X1 | B11 | −7.59 | −4.97925 | $7.21 \times 10^{-5}$ | 99.99 * |
| X3.X3 | B22 | −11.35 | −7.45082 | $3.44 \times 10^{-7}$ | 99.999 * |
| X5.X5 | B33 | −2.88 | −1.89225 | 0.073025 | - |
| X1.X3 | B12 | −8.11 | −5.5447 | $1.99 \times 10^{-5}$ | 99.99 * |
| X1.X5 | B13 | 0.52 | 0.35359 | 0.727346 | - |
| X3.X5 | B23 | 1.82 | 1.245251 | 0.227433 | - |

alues between brackets: real values for the independent factors. X1: WS concentration (%($w/v$)), X3: initial pH of the production medium and X5: yeast extract concentration (%($w/v$)). [a]: experimental values and [b]: predicted values. Values: mean of 3 readings.* Significant *p*-values: ≤ 0.05.

### 2.12. Statistical Analyses and Softwares

The PBD and BBD used in this study had been designed using the Minitab software 17.3. All regression analyses were performed using Minitab software 17.3. The three-dimensional plots were drawn using the Statistica software 13.1. The RSM package (R Development Core team 2016), Comprehensive R Archive Network (http://CRAN.R-project.org/package=rsm, accessed on 17 March 2016), was used in this study for performing statistical, canonical, and ridge analyses.

## 3. Results

### 3.1. Penicillium Chrysogenum Strain A3 DSM105774

The full length of the *18S rDNA* gene (1500 bp) of the quest fungal strain was successfully amplified by PCR as depicted in Figure 1A. The analysis of the obtained 18S rDNA nucleotide sequence by BLASTn followed by a phylogenetic tree construction did reveal that the promising xylanase producer isolate affiliated as *Penicillium chrysogenum* (Figure 1B). To help deposit its 18S rRNA (18S rDNA) nucleotide sequence in GenBank, the fungus was assigned a strain nomenclature strain A3. Thus, the fungal isolate was named *Penicillium chrysogenum* strain A3. The deposited 18S rDNA nucleotide sequence was given the accession number KY010602. To guarantee the accessibility of the fungal strain for the public, DSMZ was chosen to deposit the fungal strain with the accession number DSM105774. Moreover, the microscopic features of the fungal strain, mainly the fungal conidia, were depicted in Figure 2 using a scanning electron microscope.

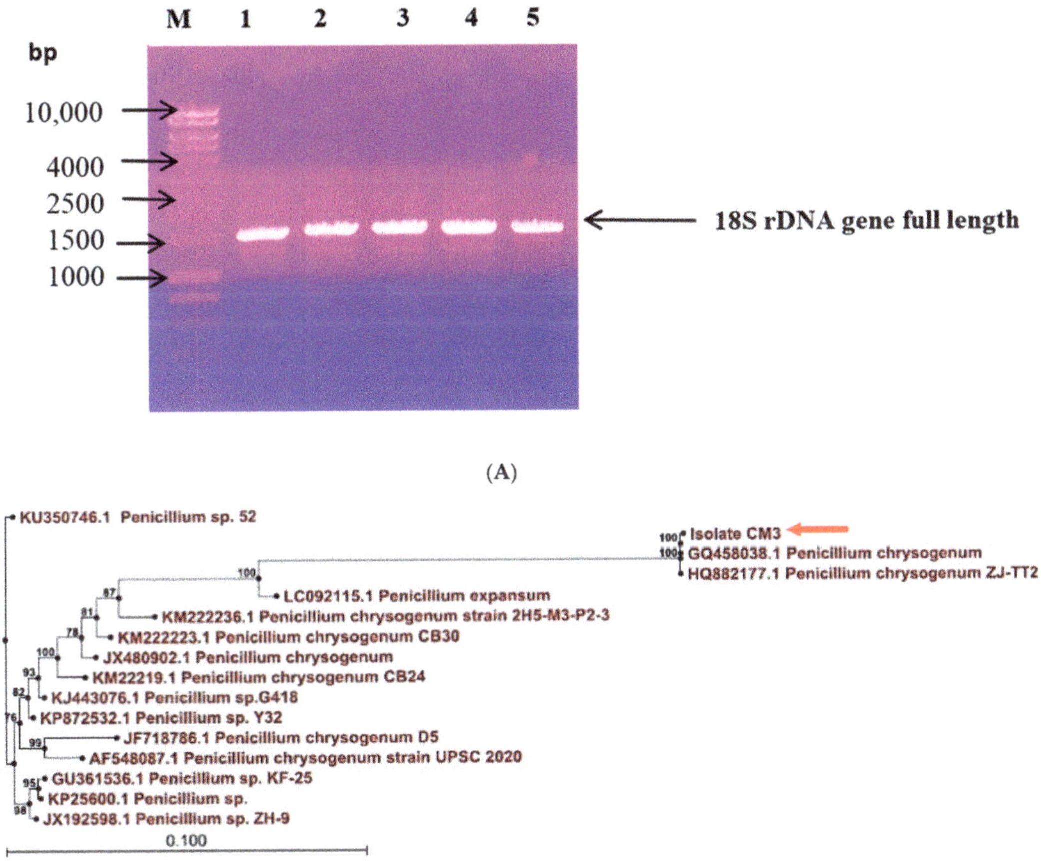


(A)

(B)

**Figure 1.** (**A**) Agarose gel (1%$w/v$) electrophoresis showing the PCR product for the full length *18S rDNA* gene amplified by PCR from the quest fungal strain. M: 1kbp DNA ladder. Lanes 1–5: amplified PCR product of 18S rDNA full length from the quest fungal strain from five PCR reactions Eppendorf tubes. (**B**) Neighbor-joining tree (constructed by CLC Sequence Viewer 8.0) depicting the phylogenetic relationship between 18S ribosomal DNA sequence of the Figure 3, (the fungal isolate in this study) and other 18S ribosomal DNA sequences assigning to closely related fungi. Bootstrap values were presented on branch nodes (100 re-samplings). The fungal isolate CM3 is the quest fungal strain *P. chrysogenum* strain A3 DSM105774 and is indicated by a red arrow.

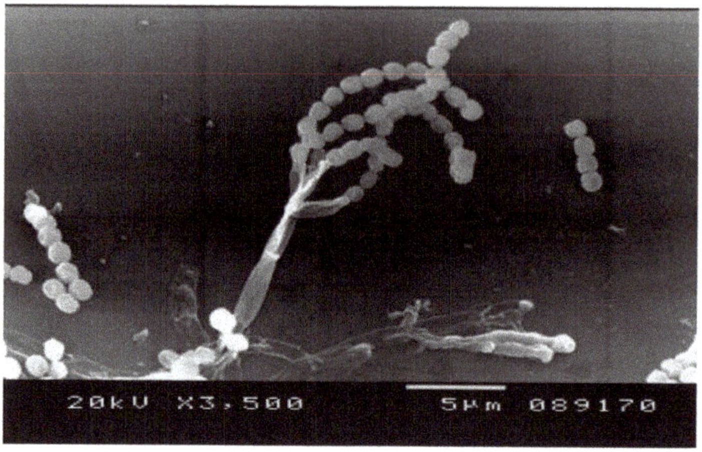


**Figure 2.** A photo of scanning electron microscope showing the conidia of *P. chrysogenum* A3 DSM105774. The magnification is ×3500.

### 3.2. Key Factors Directing Xylanase Production

With regard to the influence of nitrogen source on xylanase production via *P. chrysogenum* strain A3 DSM1057, the highest significant levels of xylanase activity ($p \leq 0.05$) were observed upon the incorporation of ammonium sulphate and sodium nitrate separately in the production medium (Figure 3). No significant difference ($p \geq 0.05$) in the level of xylanase activity was noticed as a result of the incorporation of ammonium sulphate and sodium nitrate separately in the production medium. Based on the present data, ammonium sulphate was chosen as the appropriate nitrogen source triggering xylanase production in the subsequent experimental trials.

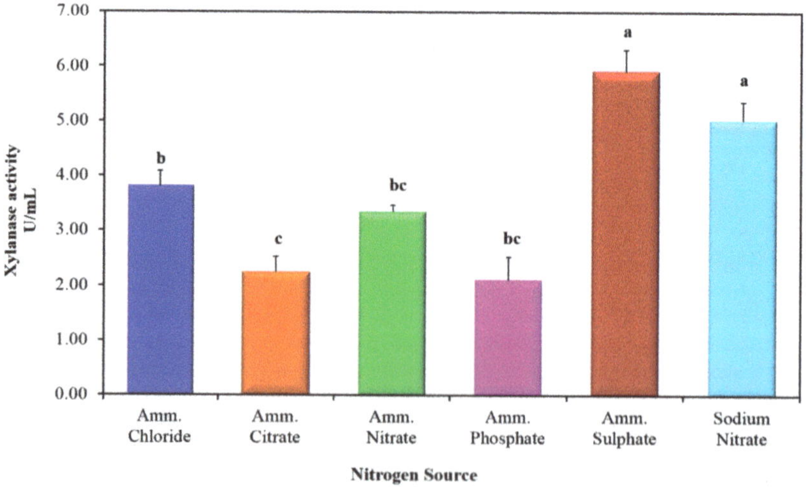


**Figure 3.** Influence of the nitrogen source on the production of xylanase from *P. chrysogenum* A3 DSM105774 at 35 °C. Values are mean of 4 readings with standard error (SE). Symbols (a, b, and c) displayed on histogram bars were used to distinguish the significance in the xylanase levels imposed by different inorganic nitrogen sources. The histogram bars bearing the same symbol letter indicated non-significant difference. The histogram bars bearing different symbol letters indicated a significant difference at $p$-value $\leq 0.05$.

Concerning the impact of the incubation temperature on xylanase production via *P. chrysogenum* strain A3 DSM105774, four incubation temperatures (25, 30, 35 and 40 °C) were studied. Data conferred a dramatic significant reduction ($p \leq 0.05$) in xylanase activity (1.09637 U/mL) when the fungal strain was cultivated at incubation temperatures higher than 30 °C (i.e., 35 and 40 °C) (Figure 4). Moreover, there was an appreciable variation in xylanase activity (34.4625 and 23.0393 U/mL) (Figure 4) at $p \leq 0.05$ when the fungus was cultivated at 30 and 25 °C, respectively. Consequently, 30 °C had been chosen as the applicable incubation temperature to carry out the next optimization experiments.

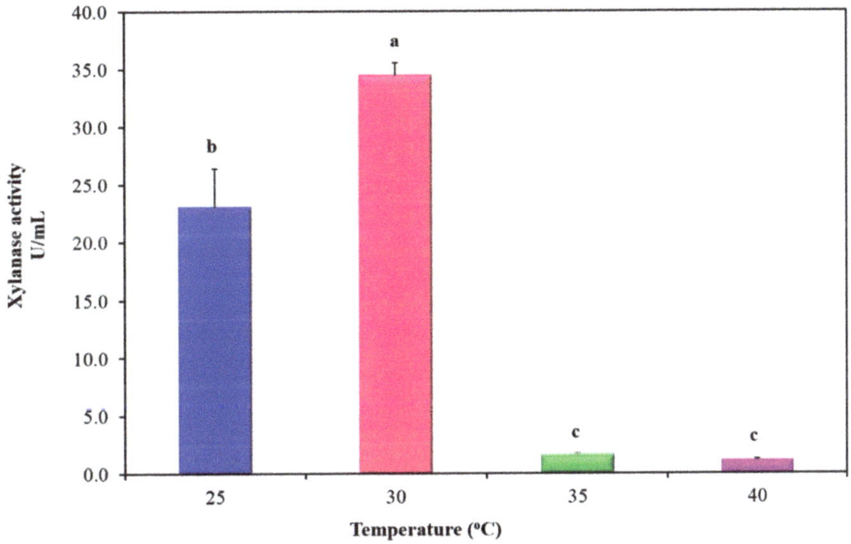


**Figure 4.** Influence of incubation temperature on the production of xylanase from *P. chrysogenum* A3 DSM105774 using minimal medium complemented with WS (0.5% ($w/v$)) for 6 days. Values are mean of 4 readings ± standard errors. Symbols (a, b, and c) displayed on histogram bars were used to distinguish the significance in the xylanase levels imposed by different inorganic nitrogen sources. The histogram bars bearing the same symbol letter indicated non-significant difference. The histogram bars bearing different symbol letters indicated a significant difference at $p$-value $\leq 0.05$.

Considering the data derived from OFAT experiments and preceding published reports, the influence of the following key factors: incubation time (days), WS concentration (%($w/v$)), initial pH of the production medium, yeast extract concentration (%($w/v$)), and ammonium sulphate concentration (%($w/v$)) on the process outcome was tested in the next experimental trials of PBD.

### *3.3. Screening of Key Factors Influencing Xylanase Production Using PBD*

The perceivable variations spanning from 4.47 to 15.45 U/mL in xylanase activity (Table 1a), produced by *P. chrysogenum* strain A3 DSM105774, among the twelve empirical runs, evidenced the indispensable need to carry out the optimization plan. Multiple linear regression analysis (Table 1b) evidenced three independent factors: WS concentration (%($w/v$)) (X1), pH of the production medium (X3) and yeast extract concentration (%($w/v$)) (X5) demonstrated improvements ($p \leq 0.05$) in the levels of xylanase produced by *P. chrysogenum* strain A3 DSM105774.

The model aptness was evaluated by the co-efficient $R^2$ that was computed to be 0.866. This $R^2$-value referred that 86% of the unevenness in the output might be explained by the model. The regression model had multiple co-efficient R-value of 0.93. At most, the best correlation between the predicted and experimental values could be achieved when the

R-value is close to 1.0. As a rule of thumb, the regression model with high significance could be inferred from its high F-value (23.3) and low $p$-value ($2.8 \times 10^{-7}$).

The regression coefficients had been calculated with the aid of coded to settle the full polynomial Equation (3) in order to describe the impact of the independent factors on xylanase production by *P. chrysogenum* strain A3 DSM105774.

$$Y = 10.69 + 3.03X1 + 0.04X2 - 0.074X3 - 0.209\ X4 - 1.25X5 \quad (3)$$

The regression analysis of PDB data inferred that the independent variables (incubation time and ammonium sulphate) did not elicit effects ($p > 0.05$) on the production of xylanase by *P. chrysogenum* strain A3 DSM105774. Consequently, they were settled at their initial levels in the following optimization step. Whilst the independent variables (WS concentration (X1), pH of production medium (X3), and concentration of yeast extract (X5) that exhibited magnitudes ($p \leq 0.05$) on xylanase production by *P. chrysogenum* strain A3 DSM105774 were assessed in the subsequent step of the optimization plan using BBD.

### 3.4. Pinpointing the Optimal Values of the Key Factors Switching the Production of Xylanase by BBD

The data of BBD trials for maximizing the production of xylanase by *P. chrysogenum* strain A3 DSM105774 are displayed in Table 2a. The tested three independent variables, derived from the PBD experiment, were WS concentration (%($w/v$)) (X1), pH of the production medium (X3) and yeast extract concentration (%($w/v$)) (X5). The three tried independent factors caused impacts ($p < 0.05$) on xylanase production by *P. chrysogenum* strain A3 DSM105774 through the next formulas of interactions (Table 2b); linear (X1 and X3), quadratic (X1.X1 and X3.X3), and cross form (X1.X3).

Judging of model aptness had been executed through computing co-efficient $R^2$. The regression model had co-efficient $R^2$ value of 0.96. This $R^2$ value deduced that 96% of the unevenness in the outcome might be justified by the model. Moreover, the regression model had a multiple correlation co-efficient $R$-value of 0.98. Normally, the best correlation between the predicted and experimental values could be realized when the $R$- value is close to 1.0. Typically, the high significance of the regression model is attributed to the high F-value and the low $p$-value. Here, the regression model demonstrated F-value (55.86) and $p$-value ($3.25 \times 10^{-12}$).

The second-order polynomial Equation (4) had been established with the aid of coded values to depict whole likely types of interactions of the tried independent factors triggering a significant consequence on xylanase production by *P. chrysogenum* strain A3 DSM105774.

$$Y = 38.45 + 15.92X1 - 12.97X3 - 0.74X5 - 7.59X1.X1 - 11.35X3.X3 - 2.88\ X5.X5 - 8.11X1.X3 + 0.52X1.X5 + 1.82X3.X5 \quad (4)$$

In an attempt to realize the optimized settings for xylanase levels by *P. chrysogenum* strain A3 DSM105774, the characteristics of the response shape was studied by portraying the 3D surface plots. The notion of 3D surface plots relies on investigating the influence of two independent factors simultaneously on output at utmost level of the third independent factor. The predicted stationary points exhibited saddle appearance; as elucidated in Figures 5–7. Hence, the ridge and canonical analyses were executed in order to pinpoint the optimum levels of the independent factors and the maximal output parallel.

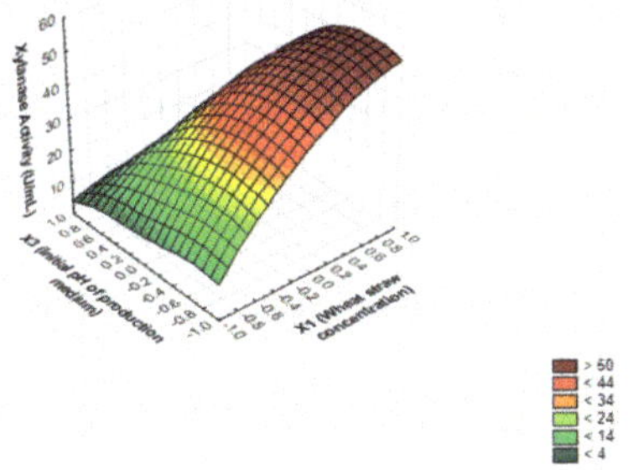

**Figure 5.** The 3D surface plot depicting the impact of WS concentration (%($w/v$)) and initial pH of the production medium at fixed value of yeast extract concentration (%($w/v$)) on the xylanase production from *P. chrysogenum* A3 DSM105774 in submerged state fermentation.

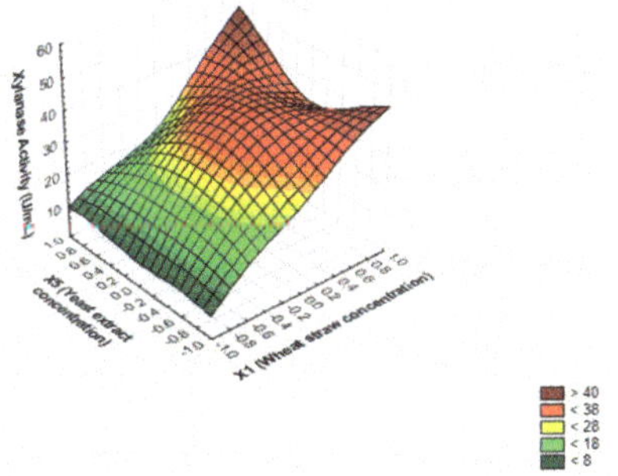

**Figure 6.** The 3D surface plot depicting the impact of WS concentration (%($w/v$)) and yeast extract concentration (%($w/v$)) at a fixed value of initial pH of the production medium on the xylanase production from *P. chrysogenum* A3 DSM105774 in submerged state fermentation.

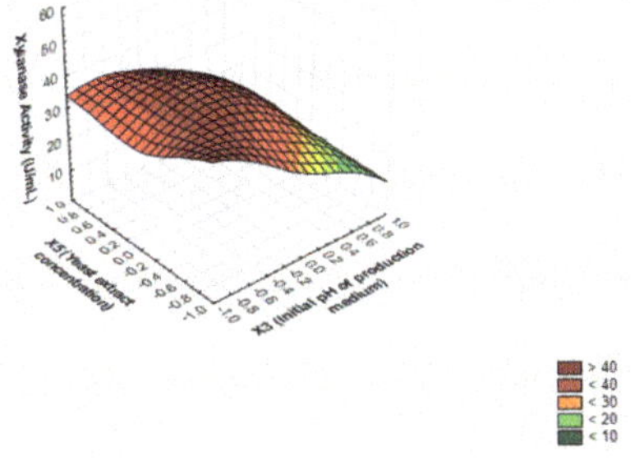

**Figure 7.** The 3D surface plot depicting the influence of yeast extract concentration (%($w/v$)) and initial pH of the production medium at a fixed value of WS concentration (%($w/v$)) on the production of xylanase from *P. chrysogenum* A3 DSM105774 in submerged state fermentation.

### 3.5. Canonical and Ridge Analyses

The canonical analysis does ascertain the nature of the predicted stationary point; minimum, maximum, or saddle. Eigenvectors in a second-order matrix and eigenvalues could elucidate the form of the response. The surface curvatures could be inferred by eigenvectors [45], whereas the magnitude and the signs of the eigenvalues are an actual signal of the surface form. The notion of the eigenvalues and the mathematical terms were previously clarified [45]. The positive and the negative eigenvalues could indicate responses with upward and downward curvatures, respectively (first rule of Myers). Whilst, the weight of the eigenvalue in its largest value marks off the response curvature in the related direction (second rule of Myers). The present findings prove the eigenvalues: [$\lambda_1 = -0.0924$, $\lambda_3 = -0.1675$ and $\lambda_5 = -0.4669$] for RSM model of xylanase production by *P. chrysogenum* strain A3 DSM105774. The negative eigenvalues of the RSM model revealed the upward curvature of the response in the direction of X5 (yeast extract concentration), with the largest absolute eigenvalue. However, the predicted stationary point for the xylanase activity by *P. chrysogenum* strain A3 DSM105774 was 59.73 U/mL with the predictor set: X1 = 1.679625, X3 = −1.199660, and X3 = −0.358546). Obviously, the predicted stationary point is localized outside the domain. Therefore, ridge analysis is mandatory to obtain further extrapolations for the maximum stationary point. Mostly, the ridge analysis figures the estimated ridge of optimum response from a predictor set at a radius d with steeply growing radii beginning from the origin [46]. Ridge analysis deduced that an increase in the response could happen without realizing the stationary point (threshold level) by moving along the rising ridge. For *P. chrysogenum* strain A3 DSM105774, the predicted rise in the response (xylanase) falling at a distance beyond 1.3 along the rising ridge is unrealistic as all the predictor combination sets would locate outside the explored domain (Table 3). The predictor set: X1 = 1.046, X3 = −0.759, X5 = −0.137 in terms of coded values at a distance of 1.3 inferred the highest, reliable, and estimated response of 56.7 U/mL. By applying the real values of the above-mentioned predictor combination set (3.8% (*w*/*v*), 5.1, and 0.098% (*w*/*v*) for X1 [WS concentration (%(*w*/*v*)], X3 (initial pH of the production medium), and X5 [yeast extract concentration(%(*w*/*v*))], respectively), the laboratory stationary point was 53.7 U/mL; representing 94.7% of the model validation). By the termination of the optimization strategy, the xylanase production by *P. chrysogenum* strain A3 DSM105774 was enhanced 8.95 fold when compared to its level at the beginning of the optimization plan.

**Table 3.** The estimated ridge of xylanase activity produced by *P. chrysogenum* strain A3 DSM105774 through the steepest ascent path of ridge analysis.

| Distance (d) | Independent Factor * | | | Xylanase Activity (U/mL) |
|---|---|---|---|---|
| | X1 | X3 | X5 | |
| 0.0 | 0.000 | 0.000 | 0.000 | 38.46 |
| 0.1 | 0.078 | −0.062 | −0.004 | 40.44 |
| 0.3 | 0.237 | −0.184 | −1.093 | 44.16 |
| 0.5 | 0.397 | −0.184 | −0.030 | 47.46 |
| 0.7 | 0.559 | −0.184 | −0.049 | 50.37 |
| 0.9 | 0.722 | −0.533 | −0.072 | 52.86 |
| 1.1 | 0.884 | −0.647 | −0.101 | 54.99 |
| 1.3 ** | 1.046 | −0.759 | −0.137 | 56.7 |
| 1.5 | 1.208 | −0.871 | −0.179 | 58.05 |
| 1.7 | 1.368 | −0.982 | −0.230 | 58.98 |
| 1.9 | 1.527 | −1.093 | −0.290 | 59.55 |
| 0.0 | 0.000 | 0.000 | 0.000 | 38.46 |
| 0.1 | 0.078 | −0.062 | −0.004 | 40.44 |

* Independent factors in terms of coded values. ** The predictor combination set exhibiting the highest yield of xylanase inferred from the steepest ascent path of ridge analysis. X1: WS concentration (%(*w*/*v*)), X3: initial pH of the production medium, and X5: yeast extract concentration (%(*w*/*v*)).

## 4. Discussion

The world xylanase mart is projected to sparkle in the coming years as a result of the continuous and increased obligation of xylanase in several industries. Hence, the continual searching for novel members of live xylanases, produced concomitantly with the most likely lowest expenditures and highest yield, is being addressed to cover the need of worldwide enzyme markets. Meanwhile, the management of the WS is becoming a vital goal to concurrently reduce the possible environmental hazards of this huge recalcitrant lignocellulose byproduct and beneficiate it into a value-added product.

In the view of the aforementioned, the current work attempted the valorization of WS into the value-added product xylanase by the locally novel *P. chrysogenum* A3 DSM105774 for the first time ever.

Primarily, the industrialization phase of the xylanase is confined to two prominent aspects: the satisfactory the yield of the elected producer strain and the cost-effectiveness of the bioprocess. In this regard, the indispensable need of searching for a cheap production medium reinforcing xylanase production via *P. chrysogenum* A3 DSM105774 is being obliged. Meanwhile, the whole bioprocess was optimized properly concerning the chemical and the physical key determinants to achieve this goal. The concept of classical methodology for attaining an optimized fermentation process relies on the one factor at a time (OFAT) where one independent factor is studied, and the other factors are fixed at steady levels [47]. This traditional process has some drawbacks mainly time-consuming, lack of precision, discrepancy of results, and unraveled effect of variables interactions. As a consequence, the shortages included in the OFAT approach were covered by employing the statistical sequential PBD and RSM designs. The statistical, sequential, and experimental designs are being broadly employed for the optimization of fermentation factors in the production of enzymes [23,48], antibiotics biodegradation [49], production of microbial pigments [50], production of extracellular pecticoligosaccharides [51], biosynthesis of chitooligosaccharides [52], biodegradation of chicken feather [53], etc. RSM covers a small number of experimental trials with a large number of factors. Additionally, RSM does solve multivariable equations to delimit the optimum levels of key factors regulating the output (response).

These designs were anticipated here in order to determine optimum levels of key factors regulating xylanase yield by *P. chrysogenum* A3 DSM105774. The present finding does indicate that WS can induce considerable levels of xylanase from *P. chrysogenum* A3 DSM105774. The literature of review shows a discrepancy in the appropriate and exploited lignocellulose byproduct which promotes the production of xylanase by an array of fungal strains. For example, wheat bran, persuaded the ultimate xylanase production by *P.chrysogenum* PCL501 [26] and *P. oxalicum* [30]. Whilst, corn cob and sugarcane bagasse did provoke the maximal xylanase production by *P. purpurogenum* [35] and *P. chrysogenum* F-15 [54], respectively. As a rule of thumb, the incorporation of zero-cost substrates (e.g., lignocellulose byproduct) in the production media will definitely decrease the expenditure of the bioprocess.

Among the physical independent variables, the incubation temperature is a major key determinant, controlling the fermentation processes [55]. Mostly, the optimal incubation temperature prompting the maximal xylanase production by the mesophilic fungi was reported to localize at 25–30 °C. The maximal xylanase production by *P. citrinum* xym2 [56], *P. chrysogenum* F-15 strain [54], and *P. glabrum* [57] was realized at 30, 20, and 25 °C, respectively. On the contrary, the maximum xylanase production by *P. purpurogenum* was accomplished at 40 °C [35]. At most, applying mild incubation temperatures (25–30 °C) in enzyme bioprocessing, would not only prevent the protein misfolding and its undesirable consequences [58] but also contribute greatly to energy saving, accompanying by lower bioprocesses' expenditures. Nevertheless, applying mild incubation temperatures for production in bioprocesses might increase the likelihood of exposure to bacterial contamination. This would in turn address the indispensable need for applying strict safety precaution protocols during the midstream processing to avoid bacterial contamination and its undesired consequences.

It is worth mentioning that the type of nitrogen source incorporated in the production medium is one of the key factors regulating the productivity of xylanases from the producers [33]. Unlike *P. chrysogenum* A3 DSM105774 with persuaded xylanase levels in the presence of ammonium sulphate, enhanced xylanase levels by *P. citrinum* xym2 [56] and *Penicillium* sp. AKB-24 [59] were achieved in the presence of di-ammonium hydrogen phosphate and urea (dosing of 0.08% and 0.12%), respectively. Likewise, *P. chrysogenum* A3 DSM105774 with prompted xylanase levels in the presence of yeast extract, significant positive consequences on the xylanase levels from *P. citrinum* isolate HZN13 [60] and *Penicillium* sp.WX-Z1 [61] were imposed by yeast extract.

The incubation time is a key determinant in the schedule of industrialized bioprocessing. Normally, the lengthening of a bioprocess period will increase the total bioprocess expenditure. This in turn focused on the obliged determination of the optimum mandatory period for xylanase synthesis by the fungal strain being studied. Enzyme bioprocessing directed by fungal members is somehow a lengthy process with a time spanning from four to fifteen days when compared to those directed by bacterial members. The optimal time encouraging xylanase production in combination with the maximum levels of xylanase by some reported fungal candidates was four days with 123.1 U/mL from *Penicillium* sp. [62], 5 days with 1.35 Units /mL from *P. chrysogenum* PCL501 [26], six days with 14.50 U/mL from *P. oxalicum* ZH-30 [30], eight days with 0.834 U/mL from *T. lanuginosus* A3-1 DSM 105773 [23], and twelve days with 1.42 U/mL from *Trametes versicolor* [63]. The compulsory time for xylanase production from *P. chrysogenum* A3 DSM105774 is localized within the abovementioned time span as five days.

The pH of the production medium could impose a significant influence on the magnitude value of the xylanase produced by the employed fungal strain. Mostly, the optimum pH prompting xylanase production from acidophiles and neutrophilies fungal candidates are localized in the strong acidic to a slightly neutral range of pH (3.0–7.7). The review of literature stated that the optimal initial pH of the production medium encouraging high xylanase values were pH 6.0 by *P. chrysogenum* F-15 [54], pH 7.72 by *P. oxalicum* ZH-30 [30], pH 5.5 by *P. glabrum* [57], pH 4.0 by *P.citrinum* isolate HZN13 [60], and pH 3.0 by *P. purpurogenum* [35]. *P. chrysogenum* strain A3 DSM105774 did exhibit almost a quite similar trend of an initial pH of the production medium of 5.1, encouraging high xylanase levels. The divergence in the initial pH of the production medium encouraging fungal xylanase production might be likely ascribed to the mechanism exploring the synthesis of xylanase by these disparate fungal strains. Additionally, the preferential initial pH of the production medium for fungal xylanase production might be a consequence of the pH of their natural habitats.

Bioprocess yield is a key element in the schedule of enzyme manufacturing to ensure bioprocess with satisfactory profits. A remarkable fold enhancement (8.95) in the level of xylanase produced by *P. chrysogenum* A3 DSM105774 under study was noticed by the end of the optimization plan. On the other hand, the level of xylanase from *T. lanuginosus* A3-1 DSM105773 was remarkably furthered by a fold improvement of 5.0 [23] and the xylanase yield by *P. purpurogenum* with heightened circumstances was 65.72% greater equaled to traditional conditions [35]. This fold of enhancement would confer not only the adequacy and precision of the applied statistical designs but also the indispensable need to perform the optimization step on the bench scale prior to the transfer to the industrialization step.

Comprehensively, there is an obvious discrepancy among different fungal xylanase producers regarding several issues; of most the importance the ultimate xylanase yield achieved by the employed fungal strain, duration time required for the completion of the bioprocess with the highest probable xylanase yield, type and concentration of each constituent in the production medium, and the type of the lignocellulose byproduct substrate provoking the highest possible xylanase yield. The underlying reasons behind all these discrepancy issues could be outlined in the following arguments [23]: (a) the potential of the producer fungal strain secreting the target enzyme and the co- helper hydrolytic enzyme cascades, (b) the type of substrate in terms of the lignocellulose byproduct, (c) the accessibility of the lignocellulose byproduct substrate to the producer fungal strain, (d) the chemical composition of the lignocellulose byproduct substrate, (e) the approach of the applied fermentation strategy either solid-state or submerged state fermentation, (f) the cultural conditions (e.g., the incubation temperature, the agitation speed, initial pH of the production medium, etc.), (g) the conditions of enzyme assay, (h) the definition of enzyme units, (i) the length of the fermentation process, (j) and the non-carbon/non-nitrogen constituents of the production medium acting as enzyme co-factors.

## 5. Conclusions

In conclusion, the locally isolated *Penicillium chrysogenum* strain A3 DSM105774 was successfully employed in the valorization of the accumulated lignocellulose byproduct WS to a value-added product xylanase with a fold enhancement of 8.95 compared to its level prior to the optimization strategy. Prospective studies will mainly focus on the feasibility of scaling up the current bioprocess using the pilot scale.

**Supplementary Materials:** The following are available online at https://www.mdpi.com/article/10.3390/jof7090696/s1, Figure S1: Representative screening of xylanase-producing fungi on beechwood xylan-containing agar plate showing a clear hydrolytic zone around the fungal colony against red background after flooding the plate with 0.1%($w/v$) Congo red.

**Author Contributions:** A.I.K.: supervision, conceptualization, and revising the manuscript. A.M.E.: supervision, methodology, conceptualization, data curation, writing—reviewing and editing the manuscript. A.A.M.: visualization, investigation, validation, data curation and writing—original draft preparation. H.S.M.: project administration, formal analysis, and software. All authors have read and agreed to the published version of the manuscript.

**Funding:** This research received no external funding.

**Conflicts of Interest:** The authors declare no conflict of interest.

## References

1. Anwar, Z.; Gulfraz, M.; Irshad, M. Agro-industrial lignocellulosic biomass a key to unlock the future bio-energy: A brief review. *J. Radiat. Res. Appl. Sci.* **2014**, *7*, 163–173. [CrossRef]
2. Gubitz, G.M.; Mansfield, S.D.; Bohm, D.; Saddler, J.N. Effect of endoglucanases and hemicelluloses in magnetic and floatation deinking of xerographic and laser printed papers. *J. Biotechnol.* **1998**, *65*, 209–215. [CrossRef]
3. McKendry, P. Energy production from biomass (part 1): Overview of biomass. *Bioresour. Technol.* **2002**, *83*, 37–46. [CrossRef]
4. Yasin, M.; Bhutto, A.W.; Bazmi, A.A.; Karim, S. Efficient utilization of rice-wheat straw to produce value added composite products. *Int. J. Chem. Env. Eng.* **2010**, *1*, 136–143.
5. Adapa, P.K.; Tabil, L.G.; Schoenau, G.J.; Canam, T.; Dumonceaux, T. Quantitative analysis of lignocellulosic components of non-treated and steam exploded barley, canola, oat and wheat straw using fourier transform infrared spectroscopy. *J. Agric. Sci. Technol.* **2011**, *B1*, 177–188.
6. WASDE. *World Agricultural Supply and Demand Estimates, 2010*; World Agricultural Outlook Board; United States Department of Agriculture: North Bend, WA, USA, 2010.
7. Drankham, K.; Carter, J.; Madl, R.; Klopfenstein, C.; Padula, F.; Lu, Y.; Warren, T.; Schmitz, N.; Takemoto, D. Antitumor activity of wheats with high orthophenolic content. *Nutr. Cancer* **2003**, *47*, 188–194. [CrossRef]
8. Mojsov, K. Application of solid state fermentation for cellulase enzyme production using *Trichoderma viride*. *Perspect. Innov. Econ. Bus.* **2010**, *5*, 108–110. [CrossRef]
9. Murray, T.D.; Bruehl, G.W. Composition of wheat straw infested with *Cephalosporium gramineum* and implications for its decomposition in soil. *Phytopathology* **1983**, *73*, 1046–1048. [CrossRef]
10. Goodall, C. *Biochar: Ten Technologies to Save the Planet*; Greystone Books: Vancouver, BC, Canada, 2010.
11. CWC and Domtar Inc. *Wheat Straw as a Fiber Source: Recycling Technology Assistance Partnership (RTAP)*; A Program of the Clean Washington Center; Clean Washington Center: Seattle, WA, USA, 1997.
12. Dupont, L.; Bounanda, J.; Domonceau, J.; Aplincourt, M. Metal ions binding onto a lignocellulosic substrate extracted from wheat bran: A NICA-donnon approach. *J. Colloid Interface Sci.* **2003**, *263*, 35–41. [CrossRef]
13. Al-Kindi, S.; Abed, R.M. Effect of Biostimulation using sewage sludge, soybean meal, and wheat straw on oil degradation and bacterial community composition in a contaminated desert soil. *Front. Microbiol.* **2016**, *7*, 240. [CrossRef] [PubMed]
14. Sharma, B.; Agrawal, R.; Singhania, R.R.; Satlewal, A.; Mathur, A.; Tuli, D.; Adsul, M. Untreated wheat straw: Potential source for diverse cellulolytic enzyme secretion by *Penicillium janthinellum* EMS-UV-8 mutant. *Biores. Tech.* **2015**, *196*, 518–524. [CrossRef]
15. Gao, H.; Chu, X.; Wang, Y.; Zhou, F.; Zhao, K.; Mu, Z.; Liu, Q. Media optimization for laccase production by *Trichoderma harzianum* ZF-2 using response surface methodology. *J. Microbiol. Biotechnol.* **2013**, *23*, 1757–1764. [CrossRef]

16. Balkan, B.; Ertan, F. Production of alpha amylase from *Penicillium chrysogenum* under solid-state fermentation by using some agricultural by-products. *Food Technol. Biotechnol.* **2007**, *45*, 439–442.
17. Yang, S.Q.; Yan, Q.J.; Jiang, Z.Q.; Li, L.T.; Tian, H.M.; Wang, Y.Z. High-level of xylanase production by the thermophilic *Paecilomyces themophila* J18 on wheat straw in solid-state fermentation. *Biores. Technol.* **2006**, *97*, 1794–1800. [CrossRef] [PubMed]
18. Michelin, M.; Maria de Lourdes, T.M.; Ruzene, D.S.; Silva, D.P.; Vicente, A.A.; Jorge, J.A.; Terenzi, H.F.; Teixeira, J.A. Xylanase and β-xylosidase Production by *Aspergillus ochraceus*: New perspectives for the application of wheat straw autohydrolysis liquor. *Appl. Biochem. Biotechnol.* **2012**, *166*, 336–347. [CrossRef]
19. Azzouz, Z.; Bettache, A.; Boucherba, N.; Prieto, A.; Martinez, M.J.; Benallaoua, S.; de Eugenio, L.I. Optimization of β-1,4-endoxylanase production by an *Aspergillus niger* strain growing on wheat straw and application in xylooligosaccharides production. *Molecules* **2021**, *26*, 2527. [CrossRef] [PubMed]
20. Ninawe, S.; Kuhad, R.C. Bleaching of wheat straw-rich soda pulp with xylanase from a thermoalkalophilic *Streptomyces cyaneus* SN32. *Biores. Technol.* **2006**, *97*, 2291–2295. [CrossRef] [PubMed]
21. Ali, U.F.; El-Dein, H.S.S. Production and partial purification of cellulase complex by *Aspergillus niger* and *A. nidulans* grown on water hyacinth blend. *J. Appl. Sci. Res.* **2008**, *4*, 875–891.
22. Sonia, K.G.; Chadha, B.S.; Saini, H.S. Sorghum straw for xylanase hyper-production by *Thermomyces lianuginosus* (D2W3) under solid-state fermentation. *Biores. Technol.* **2005**, *96*, 1561–1569. [CrossRef]
23. Matrawy, A.A.; Khalil, A.I.; Marey, H.S.; Embaby, A.M. Biovalorization of the raw agro-industrial waste rice husk through directed production of xylanase by *Thermomyces lanuginosus* strain A3-1 DSM 105773: A statistical sequential model. *Biomass Conv. Bioref.* **2020**. [CrossRef]
24. Rajoka, M.I.; Huma, T.; Khalid, A.M.; Latif, F. Kinetics of enhanced substrate consumption and endo-β-xylanase production by a mutant derivative of *Humicola lanuginosa* in solid-state fermentation. *World J. Microbiol. Biotechnol.* **2005**, *21*, 869–876. [CrossRef]
25. Sarkar, N.; Aikat, K. Cellulose and xylanase production from rice straw by a locally isolated fungus *Aspergillus fumigatus* NITDGPKA3 under solid state fermentation-statistical optimization by response surface methodology. *J. Technol. Innov. Renew. Energy* **2012**, *1*, 54–62.
26. Okafor, U.A.; Emezue, T.N.; Okochi, V.I.; Onyegeme-Okerenta, B.M.; NwodoChinedu, S. Xylanase production by *Penicillium chrysogenum* (PCL501) fermented on cellulosic wastes. *Afr. J. Biochem. Res.* **2007**, *1*, 48–53.
27. Yang, S.; Wang, Y.; Jiang, Z.; Hua, C. Crystallization and preliminary X-ray analysis of a 1,3–1,4-β-glucanase from *Paecilomyces thermophile*. *Acta Crystallogr. Sect. F Struct. Biol. Cryst. Commun.* **2008**, *64*, 754–756. [CrossRef]
28. Biswas, R.; Sahai, V.; Mishra, S.; Bisaria, V.S. Bioprocess strategies for enhanced production of xylanase by *Melanocarpus albomyces* IITD3A on agro-residual extract. *J. Biosci. Bioeng.* **2010**, *110*, 702–708. [CrossRef]
29. Goyal, M.; Kalra, E.L.; Sarren, V.K.; Soni, G. Xylanase production with xylan rich lignocellulosic wastes by a local soil isolate of *Trichoderma viride*. *Braz. J. Microbiol.* **2008**, *39*, 535–541. [CrossRef] [PubMed]
30. Yin, L.; Fengjie, C.; Zhiqiang, L.; Yingying, X.; Hui, Z. Improvement of xylanase production by *Penicillium oxalicum* ZH-30 using response surface methodology. *Enzym. Microb. Technol.* **2007**, *40*, 1381–1388.
31. Zahari, N.I.; Shah, U.K.M.; Asa'ari, A.Z.M.; Mohamad, R. Selection of potential fungi for production of cellulase-poor xylanase from rice straw. *Bioresources* **2016**, *11*, 1162–1175. [CrossRef]
32. Brown, J.A.; Collin, S.A.; Wood, T.M. Development of a medium for high cellulase, xylanase and fl-glucosidase production by a mutant strain (NTG III/6) of the cellulolytic fungus *Penicillium pinophilum*. *Enzym. Microbl. Technol.* **1987**, *9*, 355–360. [CrossRef]
33. Sun, X.; Zhang, R.; Zhang, Y. Production of lignocellulolytic enzymes by *Trametes gallica* and detection of polysaccharide hydrolase and laccase activities in polyacrylamide gels. *J. Basic Microbiol.* **2004**, *44*, 220–231. [CrossRef]
34. Oliveira, L.; Porto, A.; Tambourgi, E. Production of Xylanase and Protease by *Penicillium janthinellum* CRC 87M-115 from Different Agricultural Wastes. *Bioresour. Technol.* **2006**, *97*, 862–867. [CrossRef]
35. Sunkar, B.; Kannoju, B.; Bhukya, B. Optimized production of xylanase by *Penicillium purpurogenum* and ultrasound impact on enzyme kinetics for the production of monomeric sugars from pretreated corn cobs. *Front. Microbiol.* **2020**, *11*, 772. [CrossRef]
36. Camassola, M.; Dillon, A.J.J. Production of cellulases and hemicellulases by *Penicillium echinulatum* grown on pretreated sugar cane bagasse and wheat bran in solid-state fermentation. *Appl. Microbiol.* **2007**, *103*, 2196–2204. [CrossRef] [PubMed]
37. Querido, A.; Coelho, J.; Araújo, E.; Chaves-Alves, V. Partial purification and characterization of xylanase produced by *Penicillium expansum*. *Braz. Arch. Biol. Technol.* **2006**, *49*, 475–480. [CrossRef]
38. Bakri, Y.; Jacques, P.; Thonart, P. Xylanase production by *Penicillium canescens* 10-10c in Solid-state fermentation. *Appl. Biochem. Biotechnol.* **2003**, *105–108*, 737–748. [CrossRef]
39. Fusawat, P.; Rakariyatham, N. Potential of cellulase and xylanase production by fungal strains using corn husks as substrate. *Asia Pac. J. Sci. Technol.* **2014**, *19*, 229–234.
40. Adeleke, M.A.; Akatah, H.A.; Hassan, A.O.; Adebimpe, W.O. Microbial load and multiple drug resistance of pathogenic bacteria isolated from feaces and body surfaces of cockroaches in an urban area of southwestern Nigeria. *J. Microbiol. Biotechnol. Food Sci.* **2012**, *1*, 1448.
41. Smit, E.; Leeflang, P.; Glandorf, B.; Van Elsas, J.D.; Wernars, K. Analysis of fungal diversity in the wheat rhizosphere by sequencing of cloned PCR-amplified genes encoding *18S rRNA* and temperature gradient gel electrophoresis. *Appl. Environ. Microbiol.* **1999**, *65*, 2614–2621. [CrossRef]
42. Miller, G.L. Use of dinitrosalicylic acid reagent for determination of reducing sugar. *Anal. Chem.* **1959**, *31*, 426–428. [CrossRef]

43. Plackett, R.L.; Burman, J.P. The design of optimum multifactorial experiments. *Biometrika* **1946**, *33*, 305–325. [CrossRef]
44. Box, G.E.; Behnken, D.W. Some new three level designs for the study of quantitative variables. *Technometrics* **1960**, *2*, 455–475. [CrossRef]
45. Myers, R.H. *Response Surface Methodology*; Edwards Brothers: Ann Arbor, MI, USA, 1976.
46. Draper, N.R. Ridge analysis of response surfaces. *Technometrics* **1963**, *5*, 469–479.
47. Khucharoenphaisan, K.; Tokuyama, S.; Kitpreechavanich, V. Statistical optimization of activity and stability of β-xylanase produced by newly isolated *Thermomyces lanuginosus* THKU-49 using central composite design. *Afr. J. Biotechnol.* **2008**, *7*, 3599–3602.
48. Embaby, A.M.; Masoud, A.A.; Marey, H.S.; Shaban, N.Z.; Ghonaim, T.M. Raw agro-industrial orange peel waste as a low cost effective inducer for alkaline polygalacturonase production from *Bacillus licheniformis* SHG10. *SpringerPlus* **2014**, *3*, 327. [CrossRef] [PubMed]
49. Anan, A.; Ghanem, K.M.; Embaby, A.M.; Hussein, A.; El-Naggar, M.Y. Statistically optimized ceftriaxone sodium biotransformation through *Achromobacter xylosoxidans* strain Cef6: Unusual insight for bioremediation. *J. Basic Microbiol.* **2018**, *58*, 120–130. [CrossRef]
50. Embaby, A.M.; Hussein, M.N.; Hussein, A. Monascus orange and red pigments production by *Monascus purpureus* ATCC16436 through co-solid state fermentation of corn cob and glycerol: An eco-friendly environmental low cost approach. *PLoS ONE* **2018**, *13*, e0207755. [CrossRef] [PubMed]
51. Embaby, A.M.; Melika, R.R.; Hussein, A.; El-Kamel, A.; Marey, H.S. A Novel non-cumbersome approach towards biosynthesis of pectic-oligosaccharides by non-aflatoxigenic *Aspergillus* sp. Section Flavi Strain EGY1 DSM 101520 through citrus pectin fermentation. *PLoS ONE* **2016**, *11*, e0167981. [CrossRef]
52. Embaby, A.M.; Melika, R.R.; Hussein, A.; El-Kamel, A.H.; Marey, H.S. Biosynthesis of chitosan-oligosaccharides (COS) by non-aflatoxigenic *Aspergillus* sp. strain EGY1 DSM 101520: A robust biotechnological approach. *Process. Biochem.* **2018**, *64*, 16–30. [CrossRef]
53. Embaby, A.M.; Marey, H.S.; Hussein, A. A statistical-mathematical model to optimize chicken feather waste bioconversion via *Bacillus licheniformis* SHG10: A low cost effective and ecologically safe approach. *J. Bioprocess Biotech.* **2015**, *5*, 1.
54. Terrone, C.; Freitas, C.; Fanchini Terrasan, C.R.; Almeida, A.L.; Carmona, E. Agroindustrial biomass for xylanase production by *Penicillium chrysogenum*: Purification, biochemical properties, and hydrolysis of hemicelluloses. *Electron. J. Biotechnol.* **2018**, *33*, 39–45. [CrossRef]
55. Krishna, C. Solid-state fermentation systems—An overview. *Crit. Rev. Biotechnol.* **2005**, *25*, 1–30. [CrossRef] [PubMed]
56. Saha, S.P.; Ghosh, S. Optimization of xylanase production by *Penicillium citrinum* xym2 and application in saccharification of agro-residues. *Biocatal. Agric. Biotechnol.* **2014**, *3*, 188–196. [CrossRef]
57. Knob, A.; Beitel, S.M.; Fortkamp, D.; Terrasan, C.R.F.; Almeida, A.F.D. Production, purification, and characterization of a major *Penicillium glabrum* xylanase using brewer's spent grain as substrate. *BioMed Res. Int.* **2013**, *2013*, 728–735. [CrossRef]
58. Mogk, A.; Mayer, M.P.; Deuerling, E. Mechanisms of protein folding: Molecular chaperones and their application in biotechnology. *Chembiochem* **2002**, *3*, 807–814. [CrossRef]
59. Kumar, A.; Gautam, A.; Dutt, D. Co-Cultivation of *Penicillium* sp. AKB-24 and *Aspergillus nidulans* AKB-25 as a cost-effective method to produce cellulases for the hydrolysis of pearl millet stover. *Fermentation* **2016**, *2*, 12. [CrossRef]
60. Bagewadi, Z.K.; Mulla, S.I.; Shouche, Y.; Ninnekar, H.Z. Xylanase production from *Penicillium citrinum* isolate HZN13 using response surface methodology and characterization of immobilized xylanase on glutaraldehyde-activated calcium-alginate beads. *3 Biotech* **2016**, *6*, 164. [CrossRef]
61. Cui, F.; Zhao, L. Optimization of xylanase production from *Penicillium* sp. WX-Z1 by a two-step statistical strategy: Plackett-Burman and Box-Behnken experimental design. *Int. J. Mol. Sci.* **2012**, *13*, 10630–10646. [CrossRef]
62. Sridevi, A.; Golla, N.; Suvarnalatha, P. Production of xylanase by *Penicillium* sp. and its biobleaching efficiency in paper and pulp industry. *Int. J. Pharm. Sci. Res.* **2019**, *10*, 1307–1311.
63. Khalil., A.I.; Krakowiak, A.; Russel, S. Production of extracellular cellulase and xylanase by the ligninolytic white-rot fungus *Trametes versicolor* grown on agricultural wastes. *Ann. Agric. Sci.* **2002**, *47*, 161–173.

Journal of *Fungi*

MDPI

*Article*

# Integration of Solid State and Submerged Fermentations for the Valorization of Organic Municipal Solid Waste

**Gheorghe-Adrian Martău [1,2], Peter Unger [3], Roland Schneider [3], Joachim Venus [3], Dan Cristian Vodnar [1,2,*] and José Pablo López-Gómez [3,*]**

1 Institute of Life Sciences, University of Agricultural Sciences and Veterinary Medicine, Calea Mănăştur 3-5, 400372 Cluj-Napoca, Romania; adrian.martau@usamvcluj.ro
2 Faculty of Food Science and Technology, University of Agricultural Sciences and Veterinary Medicine, Calea Mănăştur 3-5, 400372 Cluj-Napoca, Romania
3 Leibniz Institute for Agricultural Engineering and Bioeconomy, Max-Eyth-Allee 100, 14469 Potsdam, Germany; punger@atb-potsdam.de (P.U.); rschneider@atb-potsdam.de (R.S.); jvenus@atb-potsdam.de (J.V.)
* Correspondence: dan.vodnar@usamvcluj.ro (D.C.V.); plopezgomez@atb-potsdam.de (J.P.L.-G.); Tel.: +40-747-341881 (D.C.V.); +49-177-3940305 (J.P.L.-G.)

**Abstract:** Solid state fermentation (SsF) is recognized as a suitable process for the production of enzymes using organic residues as substrates. However, only a few studies have integrated an evaluation of the feasibility of applying enzymes produced by SsF into subsequent hydrolyses followed by the production of target compounds, e.g., lactic acid (LA), through submerged-liquid fermentations (SmF). In this study, wheat bran (WB) was used as the substrate for the production of enzymes via SsF by *Aspergillus awamori* DSM No. 63272. Following optimization, cellulase and glucoamylase activities were 73.63 ± 5.47 $FPU/g_{ds}$ and 107.10 ± 2.63 $U/g_{db}$ after 7 days and 5 days of fermentation, respectively. Enzymes were then used for the hydrolysis of the organic fraction of municipal solid waste (OFMSW). During hydrolysis, glucose increased considerably with a final value of 19.77 ± 1.56 g/L. Subsequently, hydrolysates were fermented in SmF by *Bacillus coagulans* A166 increasing the LA concentration by 15.59 g/L. The data reported in this study provides an example of how SsF and SmF technologies can be combined for the valorization of WB and OFMSW.

**Keywords:** solid state fermentation; enzyme production; wheat bran; organic fraction of municipal solid waste; hydrolysis; *Aspergillus awamori*; *Bacillus coagulans*; lactic acid

**Citation:** Martău, G.-A.; Unger, P.; Schneider, R.; Venus, J.; Vodnar, D.C.; López-Gómez, J.P. Integration of Solid State and Submerged Fermentations for the Valorization of Organic Municipal Solid Waste. *J. Fungi* **2021**, *7*, 766. https://doi.org/10.3390/jof7090766

Academic Editor: Baojun Xu

Received: 2 September 2021
Accepted: 15 September 2021
Published: 16 September 2021



## 1. Introduction

The massive generation of solid waste has made their disposal an important challenge to overcome. Global municipal solid waste (MSW) production is expected to increase up to 70% by 2050 from 2010 million tons (Mt) in 2016 [1]. In addition to MSW, substantial global industrialization has also contributed to an unprecedented increase in the generation of industrial solid waste. In 2017, the rate of industrial waste generation was approximately 18 times higher than MSW [2]. A large portion of industrial wastes and MSW are organic residues, which can be used as substrates in fermentations.

The organic fraction of municipal solid wastes (OFMSW) represents an ample category of bio-waste (with an annual production of approximately 140 Mt just in the European Union (EU)) from households, restaurants, diverse businesses, yards and garden wastes, etc. [3,4]. The need for sustainable technologies to manage such abundant residues as the OFMSW and industrial organic wastes must be a priority. Until recently, OFMSW had been landfilled with adverse effects on the environment and wasting the potential of such residue for biotechnological applications.

Even though the composition of OFMSW may vary, these biowastes are typically rich in carbohydrates, proteins and lipids, making them a good substrate candidate for fermentation processes [5]. Most developed countries have efficient OFMSW and industrial organic waste sorting systems and facilities, which has allowed them to use such residues in for biogas production [6,7]. However, biogas value is low and thus, several works have been presented for the valorization of OFMSW through the production of other chemicals, for example, bio-oils, volatile fatty acids subsequently used for biogas production, or polyhydroxyalkanoates [8–11], hydrogen [12], ethanol, enzymes, organic acid, biopolymers and bioplastics [13–17]. Nonetheless, the polymeric carbohydrates (mainly starch, cellulose and hemicellulose) present in the OFMSW need to be hydrolyzed in order to release simple sugars for the fermentation, which increases costs.

Enzymatic hydrolysis requires mild process conditions, is highly specific and does not generate undesired side products, making it attractive for biowastes hydrolysis. However, the manufacturing of enzymes is expensive, leading to high costs of commercially available enzyme mixtures, raising the production costs of hydrolysis processes [5,18–20]. Solid state fermentation (SsF) is a green process that draws attention to enzyme production because it has a low cost, high yield and optimal use of agro-industrial by-products [21]. In SsF, microorganisms grow on a solid substrate with enough water to support microbial metabolism [21,22]. Using SsF as a production method of enzymes could offer some apparent economic and technical advantages over conventional submerged-liquid fermentation (SmF). These include a high product concentration and simple fermentation tools and low necessities for aeration and agitation during enzyme production [23]. Furthermore, SsF can directly use agricultural wastes, thus helping to prevent the negative environmental impact caused by their accumulation. In this context, numerous reports have shown that wheat bran (WB), the most abundant by-product produced from the wheat processing industry, to be a suitable substrate for the production of hydrolytic enzymes by SsF [20,24–26].

In recent years, the interest in cellulase and glucoamylase has increased due to the numerous potential applications for these enzymes. Glucoamylase can hydrolyze stepwise single glucose units from the non-reducing ends of amylose and amylopectin from starch or related polymers [27]. It has been extensively used in starch processing and bioconversion of organic waste [28], and many authors have exploited the abundance and availability of agricultural residues to synthesize amyloglucosidase in fermentations. Glucoamylase has been produced from WB, paddy husk, rice bran, wheat flour, cornflower, coconut seed flour, coconut oil cake, tea waste and other starch-containing wastes. *Aspergillus* spp. has been intensively studied for the possibility of using agricultural waste, such as rice bran, for glucoamylase production via SsF [29].

Based on the enzyme production from solid organic waste via SsF and their subsequent use in hydrolyses, a novel process has been described [30]. In such process, SsF, hydrolysis and SmF are used consecutively to produce specific compounds and valorize the organic solid wastes, as shown in Figure 1.

The present study aimed to combine SsF, hydrolysis and SmF to valorise the residues WB and OFMSW. This was achieved by (1) using a fungus specialized in enzyme production (*Aspergillus awamori*) via SsF and WB as a substrate (this fungus produces cellulase, glucoamylase and has the highest capacity to produce feruloyl esterase and amylase compared to other *Aspergillus* sp.); (2) the fermented substrate rich in enzymes obtained was used in the hydrolysis of OFMSW to increase sugars; (3) sugars rich hydrolysates were used as the substrate in SmF by a *Bacillus* specialized in L–lactic acid production. Lactic acid (LA) is an important building block with various applications in industry and that has had a renewed interest because its potential application in the polymer and cosmetic market.

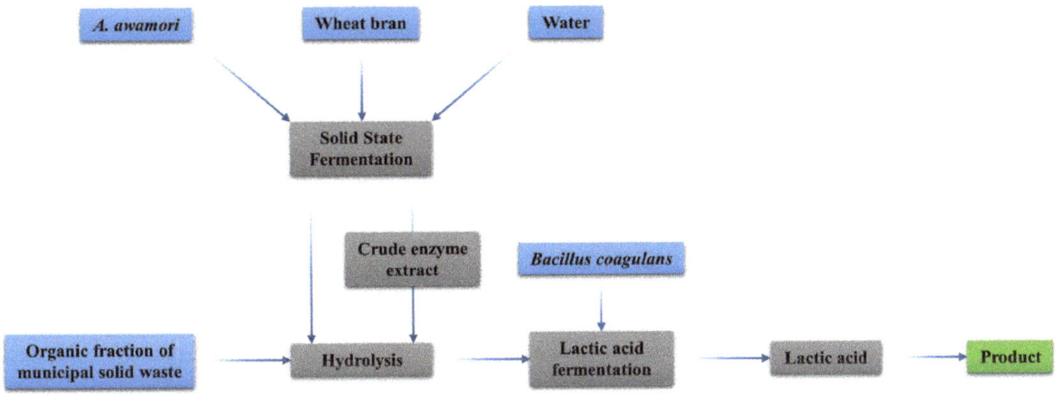


**Figure 1.** Novel bioprocess for producing enzymes from wheat bran to obtain lactic acid from the organic fraction of municipal solid waste.

## 2. Materials and Methods

### 2.1. Substrates

WB was provided by a farm (SC ALBATROS SRL) in Romania. The OFMSW was kindly delivered from the IMECAL SA company (L'Alcúdia, Valencia, Spain) and was obtained from an MSW treatment plant located in Valencia, Spain. To avoid undesirable fermentation with microorganisms already existing in the substrates, all the substrates were autoclaved for 15 min at 121 °C prior fermentations. The flask's content for SsF was mixed with a sterile spatula, and in the bioreactor WB and OFMSW mix, homogenization was achieved using the bioreactor mixing system [31]. Before the homogenization of the substrates, OFMSW was screened to remove inert constituents such as stones, plastic, glass, etc. Additionally, sterilized deionized water was used for all experiments. Culture media components for microorganisms' activation and inoculations and other reagents were of analytical grade, obtained from VWR International (Radnor, Pennsylvania, PA, USA) and agar (Agar plant for cell culture) obtained from AppliChem (Omaha, NE, USA).

### 2.2. Microorganisms and Culture Conditions

The experiments were conducted using *Aspergillus awamori* DSM No. 63272 and WB as a substrate for enzyme production. *Bacillus coagulans* A166 and OFMSW substrate for hydrolysis process and LA fermentation. All microorganisms were obtained from the Leibniz Institute for Agricultural Engineering and Bioeconomy (ATB) in Potsdam, Germany.

The propagation for *A. awamori* occurred in commercial potato dextrose agar (PDA) containing infusion from potatoes, 20 g/L glucose and 15 g/L agar. On the solidified agar, 100 µL pre-inoculum with spore suspension spread evenly with a glass Drigalsky spatula on the agar surface. For inoculation, 0.1% Tween 80 solution and sterile glass beads were used to remove spores from the agar plates. The plates inoculated were incubated at 30 °C for 5 days until the entire plates formed a uniform mass of black spores.

Pre-cultures for LA fermentations were passed in 250 mL shake flasks containing de Man, Rogosa and Sharpe (MRS) medium (glucose, 20.00 g/L; casein peptone, tryptic digest, 10.00 g/L; meat extract, 10.00 g/L; yeast extract, 5.00 g/L; Na-acetate, 5.00 g/L; $K_2HPO_4$, 2.00 g/L; $(NH_4)_3$ citrate, 2.00 g/L; Tween 80, 1.00 g/L; $MgSO_4 \times 7 H_2O$, 0.20 g/L; $MnSO_4 \times H_2O$, 0.05 g/L) a specific liquid medium suggested for use in the cultivation of *Lactobacillus* spp., and dolomite EVERZIT Dol 0.5–2.5 mm (Evers, Germany) [32]. After inoculation, the flasks were incubated at 150 rpm and 52 °C for 12–16 h [7]. The medium used for *Bacillus coagulans* was MRS broth with a final pH of 6.

### 2.3. Solid State Fermentation, Hydrolyses and Lactic Acid Fermentation

#### 2.3.1. Solid State Fermentation

WB was used for enzyme production via SsF, with an initial moisture content of 11.978 ± 0.05%, was adjusted to 80% using deionized sterile water. A 400 µL solution ($10^7$ spores/mL) of *A. awamori* was used to inoculate 10 g of WB in 250 mL shake flasks. The flasks content was mixed with a sterile spatula and incubated at 30 °C for 4 days. A whole flask was taken for each day of fermentation for sample analysis and evaluation of the SsF. The flask content was mixed and homogenized with a spatula, and then the required amount of fermented substrate was taken. Spores' suspension concentration was measured using a microscope (Zeiss, Oberkochen, Germany) and a Thoma cell counting chamber. Cellulase activity, glucoamylase activity, total reducing sugar test and pH value were monitored. In addition, the samples taken for the total reducing sugar test and pH were inactivated in a water bath for 10 min at 95 °C and kept frozen at −18 °C until further analysis [33].

#### 2.3.2. Hydrolysis Processes

A hydrolysis optimization was made in 300 mL flasks at 50 °C and 150 rpm of shaker without the pH adjusted for 68 h. Five samples were prepared with concentrations of fermented WB rich in enzymes (WBE) and OFMSW at ratios 1:9, 2:8, 3:7, 4:6 and 5:5 (mass/mass). The best ratio obtained previously was used for hydrolysis in a 1.5 L bioreactor (Sartorius Stedim Biotech AG, Göttingen, Germany) with a working mass of 1 kg. The temperature for hydrolysis in the bioreactor was 50 °C with 150 rpm of stirring and a pH 5, controlled by the addition of NaOH (20% *w*:*w*). The results obtained during the hydrolysis with WBE were compared to those obtained using a commercial enzymatic (CE) cocktail (Cellic CTec2; Novozymes A/S, Basgsværd, Denmark), a cellulose complex for the degradation of cellulose into fermentable sugars. The CE mixture was added at 1% (10 mL of CE for one Kg of the substrate). The hydrolysis lasted for 68 h, and samples were taken at times 0, 24, 48 and 68 h to quantify the formation of sugars. In addition, the experiments were made in triplicate, and the samples taken for the total reducing sugar test and pH were inactivated in a water bath for 10 min at 95 °C, after which kept in the freezer at −18 °C until further analysis.

#### 2.3.3. Lactic Acid Fermentation

LA fermentations were carried out successively after hydrolyses in the same 1.5 L BIOSTAT bioreactors. The fermentation parameters were adjusted at 52 °C, 200 rpm for 25 h and a pH of 6.0 was controlled by the addition of NaOH (20% *w*/*w*). Samples were taken every two hours to quantify glucose, fructose, disaccharide, xylose, arabinose, lactic and acetic acids. Cell viability was measured by spread plate method on MRS agar. The experiments were made in duplicate, and the samples taken for the sugars and acids quantification were inactivated in a water bath for 10 min at 95 °C and kept in the freezer at −18 °C until further analysis.

### 2.4. Enzymatic Extract

After appropriate culturing periods, the enzymes produced via SsF were extracted. The crude enzymatic extract was prepared by adding 5 g of the homogenized fermented sample (the whole flask homogenization was carried out with a sterile spatula for 5 min) and 40 mL of deionized water in a 250 mL flask. The flask was then placed on a rotary shaker (150 rpm, 30 min at room temperature). To collect the enzymatic extract, the samples were centrifuged at 4800× *g* for 15 min at 4 °C. The clarified supernatant was filtered using Whatman No. 1 filter paper to quantify total reducing sugars, cellulase activity, glucoamylase activity and pH [30].

#### 2.4.1. Total Reducing Sugar Test

The total reducing sugar concentration was measured using the DNS method proposed by Miller in 1959 [34]. This method provides a simple and fast way to handle several samples and was used for the optimization process. The determination is centered on the color reaction between reducing sugars and 3,5-dinitrosalicylic acid. The reaction yield was measured as absorbance of the sample at 540 nm using an LLG–uniSPEC 2 spectrophotometer. The total reducing sugar throughout the studies was expressed as grams of sugar per kilogram of fermented substrate ($g/kg_{fs}$).

#### 2.4.2. Enzymes Activity

Cellulase activity was measured using the protocol described by Ghose (1994) and recommended by IUPAC, using filter paper Whatman No. 1 as a substrate [35]. For this determination, a Whatman No. 1 filter paper strip (1 × 6 cm) with a weight of ≈50 mg was used as a substrate. The paper strip was then rolled, placed in a glass tube, and 1 mL of 0.05 M sodium citrate buffer, pH 4.8, was added, immersing and covering the paper, following, 0.5 mL of enzyme solution was added to the same tube. After this step, the tube was vortexed and incubated in a water bath at 50 °C for 60 min. At the end of the incubation, 3 mL of DNS reagent were added, the tubes were also vortexed and incubated in a water bath for 10 min at 95 °C. All the tubes were then placed in a cold water bath for 5 min, and 9 mL of deionized water was added to each tube. Finally, the tubes were vortexed, and the absorbance was measured at 540 nm against reagent blank using an LLG–uniSPEC 2 spectrophotometer. The cellulase activity was estimated, which would have released exactly 2.0 mg of glucose utilizing a plot of glucose liberated against cellulase concentration, was carried out. One $FPU/g_{ds}$ represents the enzyme unit per gram of initial dry solid substrate [24].

Glucoamylase activity was measured using the protocol described by Melikoglu and et al. in 2013 [30]. The glucoamylase activity of the enzyme solutions was estimated by measuring the amount of glucose released per minute using 1 mL potato starch (6% *w*/*v*) as substrate. Glucoamylase activity gave the difference between time zero and 10 min. Throughout the studies, the glucoamylase activity was expressed as $U/g_{db}$ dry basis. One unit (U) was estimated as the enzyme amount required to produce 1 mg of glucose/minute under the assay conditions.

### 2.5. *pH Measurements*

The pH measurement from SsF experiments was determined with a digital pH-meter (WTW, Weilheim, Germany) at room temperature using the enzymatic extract obtained.

### 2.6. *Analytical Assays*

During the hydrolyses and fermentations, quantification of glucose, fructose, disaccharide, xylose, arabinose, LA and acetic acids were measured via HPLC (Ultimate 3000 from the company DIONEX (Sunnyvale, CA, USA)). Equipment with the following parameters: a column Eurokat H (300 × 8 mm × 10 μm), company KNAUER; the mobile phase was 0.01 N sulfuric acid with a 0.8 mL/minute rate and a pressure of 65 bar connected with a detector RI–101 (SHODEX, Tokyo, Japan); the injection volume was 10 μL, and the auto-sampler was WPS–3000TSL analytical [33]. All the experiments, hydrolysis and fermentation were made in duplicate and triplicate, and the result are exposed as average ± standard deviation.

## 3. Results and Discussion

### 3.1. *Solid State Fermentation, Substrate Optimization and Enzymes Production*

In general, the initial culture and the fermentation substrate are well known to influence the production of enzymes. Therefore, in our experiments, WB and WB supplemented with glucose (WB+G) (moisture content was adjusted with glucose solution at a concentration of 30 g/L) were evaluated as substrates for SsF with *A. awamori* to produce enzymes with potential application in enzymatic hydrolyses. During 96 h of fermentation, WB and WB+G were able to support the fungi growth. The SsF control parameters showed a desirable behavior, maintaining the moisture content and a small reduction of the pH values. The initial pH value went from 6.032 ± 0.195 to 5.585 ± 0.119 and from 6.256 ± 0.106 to 5.103 ± 0.151 for WB and WB+G, respectively. Total reducing sugars and enzymatic activity were also monitored. It is remarkable that, at the end of the fermentation, after 96 h, the reducing sugars had similar values of 4.45 ± 0.54 $g/kg_{fs}$ for the simple WB substrate and 4.855 ± 0.837 $g/kg_{fs}$ for WB+G. In addition, an experiment containing the addition of yeast extract was performed. However, the results were worse than WB+G and the same as WB without nutrient supplementation.

Figure 2 shows the reducing sugars production and enzyme activities for the SsF of WB and WB+G over 96 h. The fermented WB+G substrate showed the highest cellulase activity, 73.35 ± 4.51 $FPU/g_{ds}$, after 96 h of fermentation. It was followed by fermentation without glucose addition 57.51 ± 3.10 $FPU/g_{ds}$, which showed a high activity after 96 h of fermentation. Additionally, cellulase production depended on the nature of the carbon source and other vital nutrients existing already in the substrate. Other studies evaluating WB as a substrate for SsF found similar cellulase activities using different fungi, with longer fermentation time (up to 120 h). However, these authors supplemented the media with $NH_4SO_4$, $KH_2PO_4$ and yeast extract to increase the nitrogen, carbon or mineral content [24,36,37]. In the literature, filter paper cellulase activities alternated from 1.7 $U/g_{ds}$ [38,39] to 437.5 $U/g_{ds}$ [40]. It is remarkable that, in this paper, we found similar values without supplements, and we set out to produce a fermented material easy to use and adaptable on an industrial scale for hydrolysis with large volumes. In our case, 96 h of fermentation are not enough for maximum cellulase activity, especially for WB. These results indicate that WB can be used by *A. awamori* to grow and produce enzymes without requiring nutrients supplementation.

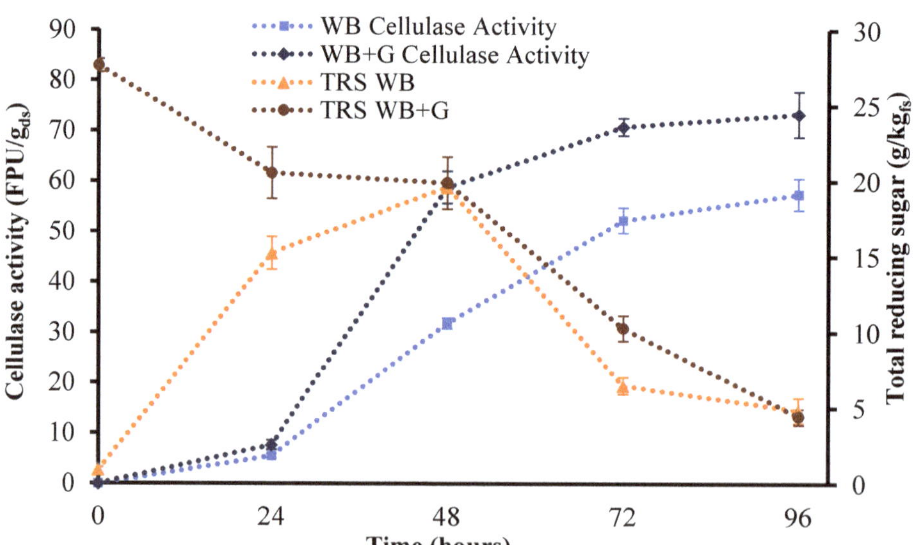


**Figure 2.** Cellulase activity and total reducing sugar (TRS) for wheat bran fermented by *A. awamori* during 96 h. Square and rhombus represent cellulase activity ($FPU/g_{ds}$); ··■··WB shows wheat bran substrate with sterile deionized water; ··♦··WB+G shows wheat bran substrate with the addition of glucose. Triangle and circle represent total reducing sugar; ··▲··WB shows wheat bran substrate with sterile deionized water; ··•··WB+G shows wheat bran substrate with the addition of glucose.

### 3.2. Effect of Fermentation Time on Enzymes Activity

Based on the previously obtained results, WB without nutrient supplementation was selected for the following experiments. During SsF, medium pH, nutrient concentration, temperature, moisture content and physical structure of the raw material continuously change. All these parameters affect microbial growth and are connected directly with enzyme production. Various reports cover different fermentation times for enzyme production via SsF, between 3 and 12 days [24,30,41,42]. Figure 3 shows total reducing sugars, cellulase and glucoamylase activity during 10 days of WB fermentation. In this article, the maximum value for cellulase activity was attained at 7 days of fermentation. In the first 5 days, cellulase activity increased rapidly, reaching 66.96 ± 4.41 FPU/$g_{ds}$, and after that, cellulase activity maintained a slow and constant growth peaking at day 7 with a value of 73.63 ± 5.47 FPU/$g_{ds}$. In addition, cellulase and glucoamylase activities showed a constant and fast evolution until the total reducing sugars were below 2.06 ± 0.35 g/$kg_{fs}$, after 5 days of fermentation. Glucoamylase activity peaked at day 5 with a value of 107.10 ± 2.63 U/$g_{db}$, followed by a constant decrease. These results are in line with those obtained by Kaushik et al. [41], who determined when *Aspergillus lentulus* showed the highest xylanase production via SsF using WB as the substrate and yeast extract as the nitrogen source. The fermentation profile showed a gradual increase in enzyme activity from 24 to 72 h, reaching its maximum value after 96 h and diminishing after 120 h until 144 h of fermentation [41]. Raghuwanshi et al. discussed bioprocessing of enhanced cellulase production using a mutant of *Trichoderma asperellum* RCK2011 via SsF and its application in hydrolysis of cellulose, released cellulases within 48 h of incubation. While the strain exhibited maximum FPase and CMCase after 4 days of fermentation, β–glucosidase production reached the maximum value only after 7 days [43]. A previous study reported that during depletion of the carbon sources in the fermentation medium, the organisms use these hydrolytic enzymes to produce sugars necessary for their own metabolic growth [44]. Various fungi have been reported to produce maximum cellulase at different time intervals; therefore, a direct comparison of time-dependent changes is difficult to perform [45]. These favorable results could be caused by WB, which has provided ample amounts of nutrients in the substrate, such as glucose and starch needed for fungal multiplication and enzyme production.

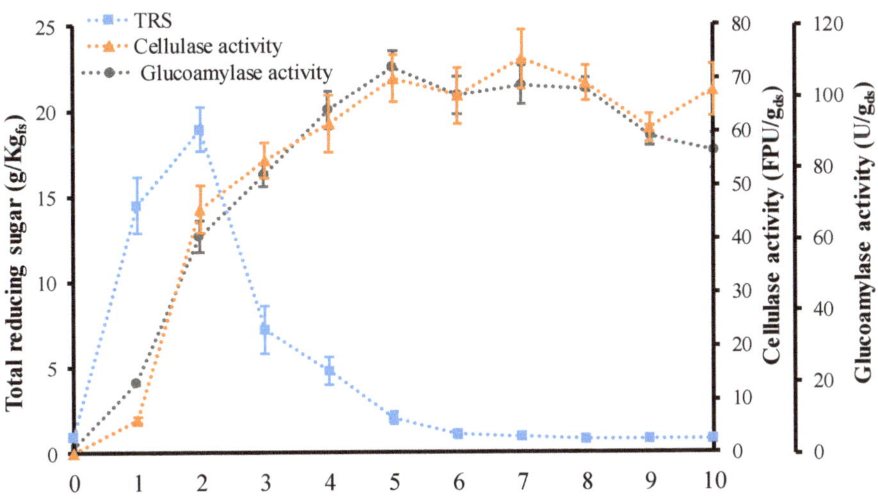


**Figure 3.** Total reducing sugar (TRS), cellulase and glucoamylase activity during 10 days of fermentation. Square represents total reducing sugars. Triangle represents cellulase activity, and the circle shows glucoamylase activity.

### 3.3. Hydrolysis Optimization

For the first time in 2008, Melikoglu detailed an integrated new system, using a part of the substrate for the production of hydrolytic enzymes via SsF, and continued with the hydrolysis of the remaining substrate to obtain in final a nutrient-rich hydrolyzate [46]. The nutrient-rich hydrolysate is able to be transformed into a specific product by suitable successive fermentations. Since 2008, many articles have been published analyzing the sequential application of SsF and SmF. It has been well-known that although enzymatic extracts can be used for the hydrolysis, direct addition of fermented solids can make the process more practical and economically feasible [47]. In our case, WBE solids were directly added to OFMSW at different ratios (mass/mass) to evaluate the hydrolysis potential of the WBE. Table 1 illustrates the results obtained during the optimization hydrolysis process. The maximum value of total reducing sugars resulted in a ratio of 4:6 (WBE:OFMSW), $39.53 \pm 0.42$ g/kg$_{fs}$ after 68 h. An equal ratio of 5:5 obtained a lower result by 12.6% after 68 h than 4:6 ratio. A ratio of 3:7, after 68 h, has a lower result with 3.1% than 4:6 ratio. In addition, a ratio of 2:8 has a lower result at the end of hydrolyzed with 11.7% compared with the best result. In the end, a ratio of 1:9 obtains just 72.1% of total reducing sugar comparing with the best result. These results are quite interesting, indicate that using a 5:5 ratio is not enough new substrate for the enzymes (enzymes remaining available in the substrate). Adding a lower ratio than 4:9, the enzymes were available in the substrate, and hydrolysis of OFMSW occurred but not entirely. However, obtaining a complete hydrolysis and, at the same time, using the enzymes entirely is very difficult to achieve. In our case, using a smaller amount of WBE from an economic point of view is recommended to capitalize on the enzymes obtained from SsF fully. In addition, the optimization hydrolysis results can be influenced by the substrate on which the fungi grew, in our case WB. It is believed that some of the starch could have been located in a lower region of WB where the fungi could not penetrate; also, the growth of the fungi could be uneven throughout the substrate, simply due to the lack of mixing [30].

**Table 1.** Optimization hydrolysis process of OFMSW with different concentrations of WBE.

| Substrate | | Total Reducing Sugar (g/kg$_{fs}$) | | | |
|---|---|---|---|---|---|
| **WBE (g)** | **OFMSW (g)** | **0 h** | **24 h** | **48 h** | **68 h** |
| 10 | 90 | $4.06 \pm 0.23$ | $20.99 \pm 0.49$ | $24.81 \pm 0.33$ | $28.50 \pm 0.49$ |
| 20 | 80 | $5.76 \pm 0.26$ | $24.72 \pm 0.41$ | $28.85 \pm 0.38$ | $34.92 \pm 0.43$ |
| 30 | 70 | $6.28 \pm 0.27$ | $26.26 \pm 0.35$ | $31.96 \pm 0.46$ | $38.32 \pm 0.41$ |
| 40 | 60 | $9.44 \pm 0.33$ | $26.91 \pm 0.46$ | $33.18 \pm 0.34$ | $39.53 \pm 0.42$ |
| 50 | 50 | $9.32 \pm 0.30$ | $27.55 \pm 0.45$ | $33.02 \pm 0.32$ | $34.49 \pm 0.47$ |

The total reducing sugar was expressed as grams of sugar per kilogram of fermented substrate (g/kg$_{fs}$). The experiments were made in triplicate, and the values represent the average and standard deviation of them. WBE–fermented wheat bran rich in enzymes; OFMSW–organic fraction municipal solid waste.

### 3.4. Hydrolysate Characterization and Lactic Acid Production

Table 1 shows the results for the total reducing sugar obtained in hydrolysis optimization, suggesting that OFMSW hydrolyzate could be a suitable substrate for microbial fermentations. Enzymes obtained by SsF were compared with a CE cocktail in two separate hydrolyses. According to the results obtained in the previous section, WBE was added to OFMSW at a ratio of 1:9 (mass/mass). Figure 4 illustrates the profiles for the hydrolysis and successive fermentation of OFMSW, monitoring glucose production/consumption and LA production during the process. Glucose was initially presented in both hydrolyses with concentrations of $6.64 \pm 0.19$ g/L and $3.68 \pm 0.19$ for the WBE and the CE, respectively. The higher concentration of glucose for the WBE experiment is the result from the previous SsF of the WB, which liberated sugars. Thus, an addition of 10% WBE increased glucose in the initial hydrolysis substrate. For all hydrolyses, glucose increases considerably during the first 24 h, followed by a slower increase until 68 h with a final value of $19.77 \pm 1.56$ g/L in the experiment with WBE and $29.00 \pm 0.65$ g/L in the experiment with CE.

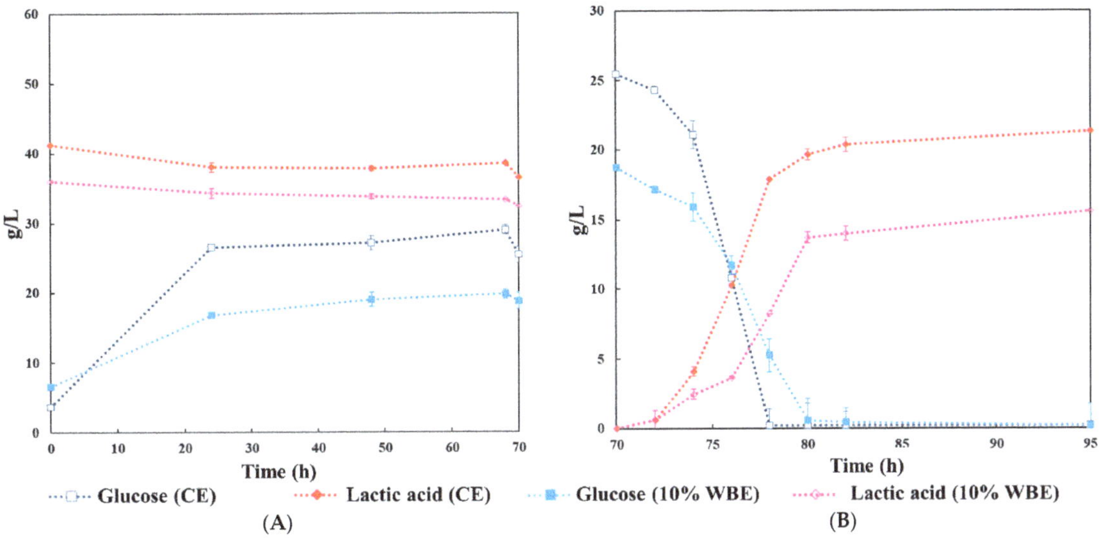


**Figure 4.** Hydrolysis and subsequent fermentation of OFMSW. (**A**) The graph illustrations the variation in the concentration of glucose and lactic acid during hydrolysis. (**B**) Glucose consumption and lactic acid production during lactic acid fermentation. CE–commercial enzymes; WBE–wheat bran rich in enzyme via SsF.

LA and acetic acids were present in both hydrolysates performed. The presence of these acids was detected from the start of the hydrolysis, showing that they were produced by naturally occurring organisms present in the OFMSW before sterilization [3,48]. Racemic LA was initially present in the hydrolysates with concentrations of 35.98 ± 0.98 g/L and 41.22 ± 1.39 g/L in the trials with WBE and CE, respectively. Probst et al., (2015) showed a natural production of an isomeric racemic mixture of LA in kitchen food waste and OFMSW [49]. Although a racemic mixture of LA can produce organic solvents, for example, ethyl lactate, high optical purity is essential for other applications [31,49]. Notably, for polylactic acid production, the ratio D– to L–lactic acid has a significant effect on the final product properties such as its thermal stability, crystallinity, biodegradation and commercially, a combination with a higher quantity of L–lactic acid is used and preferred [50,51].

During LA fermentation in 12 h, glucose was consumed almost entirely by *B. coagulans*. After 25 h of LA fermentation, LA increased by 15.59 g/L in the WBE batch and 21.32 g/L in the CE batch (Figure 4 B). This result indicates that only 73.12% LA was obtained in the WBE batch compared to the CE batch.

Table 2 shows variation in the OFMSW hydrolysate and LA fermentation composition using WBE, and Table 3 shows using CE. Glucose and xylose are the most abundant monosaccharides in lignocellulosic biomass, taking up 60–70 and 30–40% of their hydrolysates, respectively. By reason of the nature of the substrate, differences of this type between batches are difficult to avoid. The *B. coagulans* was able to consume xylose and glucose at the same time. However, in contrast with the case of glucose, the decrease of xylose ceased without being completely depleted [52]. Similar amounts of xylose were diminished in both batches, WBE batch until 7.29 ± 0.32 g/L, and CE batch was diminished until 7.46 ± 0.78 g/L. Contrasting lignocellulosic substrates, such as hardwood and softwoods, with a higher hemicellulose content, the xylose content of OFMSW hydrolysates was lower than the content of glucose [3]. Oppositely, at the start of the hydrolysis, xylose content was higher compared with glucose. An explanation may be most likely because part of the available glucose was fermented by LA bacteria from organisms present in the OFMSW.

**Table 2.** The average concentration of sugars, lactic and acetic acid obtained by hydrolysis and fermentation from OFMSW using WBE (g/L).

| | | Glucose | Fructose | Disaccharide | Xylose | Arabinose | Lactic Acid | Acetic Acid |
|---|---|---|---|---|---|---|---|---|
| Hydrolysis | Initial | 6.64 ± 0.19 | 9.58 ± 0.91 | 3.99 ± 0.74 | 10.18 ± 0.42 | N.D. | 35.98 ± 0.98 | 5.45 ± 0.11 |
| | Final | 19.77 ± 1.56 | 9.62 ± 0.36 | 2.65 ± 0.58 | 9.87 ± 0.13 | 0.62 ± 0.12 | 33.30 ± 0.36 | 5.38 ± 0.51 |
| LA fermentation | Initial | 18.77 ± 1.35 | 8.85 ± 0.31 | 1.88 ± 0.45 | 9.29 ± 0.45 | 0.54 ± 0.16 | 32.38 ± 0.85 | 5.30 ± 0.25 |
| | Final | N.D. | 7.06 ± 0.56 | 1.68 ± 0.61 | 7.29 ± 0.32 | N.D. | 47.97 ± 0.37 | 4.98 ± 0.15 |

The experiments were made in duplicate, and the values represent the average and standard deviation of them. OFMSW—organic fraction of municipal solid waste; WBE—Wheat bran rich in enzymes; N.D.—Not detected.

**Table 3.** The average concentration of sugars, lactic and acetic acid obtained by hydrolysis and fermentation from OFMSW using CE (g/L).

| | | Glucose | Fructose | Disaccharide | Xylose | Arabinose | Lactic Acid | Acetic Acid |
|---|---|---|---|---|---|---|---|---|
| Hydrolysis | Initial | 3.68 ± 0.19 | 10.08 ± 1.41 | 4.73 ± 0.30 | 11.14 ± 0.48 | N.D. | 41.22 ± 1.39 | 6.16 ± 0.18 |
| | Final | 29.00 ± 0.65 | 10.05 ± 0.57 | 2.63 ± 0.90 | 10.79 ± 0.20 | 0.51 ± 0.15 | 38.52 ± 0.60 | 6.03 ± 0.03 |
| LA fermentation | Initial | 25.46 ± 1.17 | 9.15 ± 0.30 | 2.26 ± 0.39 | 9.87 ± 0.45 | 0.50 ± 0.17 | 36.44 ± 0.52 | 5.88 ± 0.09 |
| | Final | N.D. | 5.28 ± 0.98 | 1.87 ± 0.75 | 7.46 ± 0.78 | N.D. | 57.76 ± 4.41 | 5.39 ± 0.28 |

The experiments were made in duplicate, and the values represent the average and standard deviation of them. OFMSW—organic fraction of municipal solid waste; CE—commercial enzymes; N.D.—Not detected.

Comparing the results from both tables, adding 10% WBE decreased the initial value for acetic acid, LA, xylose, disaccharide and fructose, indicating that the SsF did not experience contamination by LA bacteria [53]. In addition, disaccharides showed a constant decrease in both, hydrolysis and fermentation, reaching 2.65 ± 0.58 g/L for WBE hydrolysis and 1.68 ± 0.61 g/L at the end of fermentation, 2.63 ± 0.90 g/L for hydrolysis with CE, and 1.87 ± 0.75 g/L at the end of fermentation. However, previous articles reported by López-Gómez et al. show no significant variations in the concentration of disaccharides by the end of the fermentation [3,7]. Moreover, Aulitto et al. in 2017 reported that some strains of *B. coagulans* are struggling with the consumption of disaccharides with μmax values 50% lower than those obtained from glucose. This could explain why the concentration of disaccharides did not show a significant change throughout the SmF [54].

### 3.5. Mass Balance

Considering a ratio of 1:9, WBE to OFMSW, and the current SsF bioreactors available, scaling up the whole process can prove to be difficult. In terms of sugars, in order to obtain 1 kg of TRS, considering the value obtained in the optimization section, 35.09 kg of the mixture WBE and OFMSW (1:9) would be needed, more specifically 3.51 kg WBE and 31.58 kg OFMSW.

The results obtained in the previous section indicate that enzymes produced by SsF, mixed in a 1:9 ratio with OFMSW, have a 68.17% glucose release compared to the experiments in which CE was used. In addition, to obtain 1 kg of glucose, a 50.58 kg mix of WBE and OFSMW (ratio of 1:9) would be necessary. Furthermore, considering the same ratio, WBE:OFMSW, 64.14 kg of the mix would be required to produce 1 kg of LA. The amount of substrate required is reduced when using CE. For example, it would be possible to obtain 1 kg of glucose from 34.48 kg of OFMSW and 1 kg of LA from 46.90 kg of OFMSW. Naturally, this higher efficiency of CE can be offset by their cost.

WB is a good substrate for enzyme production, making it widely used and economically suitable. Its physical and chemical characteristics make WB an especially good substrate for SsF process. Thus, the combination of WB and OFMSW in the process takes advantage of their individual characteristics in order to have a process thar provides good enzymatic levels and at the same time exploits the carbohydrate potential in the OFMSW. Future work should target the optimization enzyme production in SsF with tests of different temperatures, moisture contents or even with different mixtures of nutrients (glucose +

yeast extract + minerals, etc.) to enhance the hydrolysis step. Additionally, studies using SsF bioreactors which could allow for larger volumes of WB and mixing would definitely provide useful results for the scale up of the system.

Further work could be carried out to investigate WB and OFMSW as important substrates for the fermentation process targeting other valuable compounds. Specially, fermentation products using thermophilic and extremophilic organisms could benefit from such SsF-SmF configuration. An essential advantage of the whole process is the working temperatures for enzyme production (*A. awamori* DSM No. 63272 using 30 °C), the hydrolysis (enzyme hydrolysis using 50 °C), followed by a LA fermentation (*B. coagulans* A166 using 52 °C). This difference in temperature means that the fungi are not activated in hydrolysis, and LA fermentation could start in the same reactor; without the need to sterilize the materials [47]. A challenge may be the autoclaving of substrates, recommended especially for OFMSW, in order to stop spontaneous fermentation with microorganisms already existing in the substrate, which can significantly depreciate its carbohydrate content. A potential alternative to circumvent this issue would be reducing to the maximum storage times or the addition of growth inhibitors during storage. Additionally, studies could be carried using extremophiles during the SmF which could reduce the chances of indigenous bacteria from the waste to grow.

## 4. Conclusions

In this study, the results obtained indicate that WB can be used by *A. awamori* to grow and produce enzymes without requiring nutrients supplementation. The SsF of WB was evaluated to obtain enzymes such as cellulase and glucoamylase. Cellulase and glucoamylase activities were $73.63 \pm 5.47$ FPU/$g_{ds}$ and $107.10 \pm 2.63$ U/$g_{db}$ after 7 days and 5 days of fermentation, respectively. Continuing with the hydrolysis of OFMSW, fermented WB solids containing the crude enzymes and mixed with OFMSW in a 1:9 ratio (WBE:OFMSW), showed a 68.17% glucose release compared to the experiments in which CE were used. During the SmF, a 73.12% of LA was obtained in the WBE batch compared to the CE batch. Further work, targeting the optimization of the SsF step, for a better provision of enzymes for the hydrolysis, would enhance these results. Previous reports have studied the conversion of the OFMSW into LA but only employing CE for the hydrolysis step. Nonetheless, utilizing wastes (i.e., WB) for the production of enzymes, instead of simple sugars, as is the case for CE, fits very well in the circular bioeconomy context. Furthermore, the integration of SsF and SmF for the conversion of OFMSW in a process in which only residues are needed as substrates, zero-waste are generated and that utilizes enzymes produced on-site, should represent a more economically attractive process.

**Author Contributions:** Conceptualization, G.-A.M. and J.P.L.-G.; methodology, G.-A.M., J.P.L.-G., P.U., R.S. and J.V.; software, G.-A.M. and J.P.L.-G.; validation, J.P.L.-G., P.U., R.S., J.V. and D.C.V.; formal analysis, G.-A.M. and J.P.L.-G.; investigation, G.-A.M., P.U. and R.S.; resources, J.V.; data curation, J.P.L.-G. and D.C.V.; writing—original draft preparation, G.-A.M.; writing—review and editing, J.P.L.-G.; visualization, G.-A.M., J.P.L.-G., P.U., R.S., J.V. and D.C.V.; supervision, J.P.L.-G. and D.C.V.; funding acquisition, D.C.V. All authors have read and agreed to the published version of the manuscript.

**Funding:** This research was funded by MCI-UEFISCDI, grant number TE 184, Project PN–III–P1-1.1-TE-2019-1748 and under the frame of European Social Found, Human Capital Operational Programme 2014–2020, project no. POCU/380/6/13/125171 and the publication was supported by funds from CASEE Fund for Incentives, project No: CASEE fund 2021-2.

**Institutional Review Board Statement:** Not applicable.

**Informed Consent Statement:** Not applicable.

**Data Availability Statement:** The data presented in this study are available within the article. Other data that support the findings of this study are available upon request from the corresponding authors.

**Conflicts of Interest:** The authors declare no conflict of interest. The funders had no role in the design of the study; in the collection, analyses or interpretation of data; in the writing of the manuscript or in the decision to publish the results.

## References


1. Khandelwal, H.; Dhar, H.; Thalla, A.K.; Kumar, S. Application of life cycle assessment in municipal solid waste management: A worldwide critical review. *J. Clean. Prod.* **2019**, *209*, 630–654. [CrossRef]
2. Kaza, S.; Yao, L.C.; Bhada-Tata, P.; Van Woerden, F. *What a Waste 2.0: A Global Snapshot of Solid Waste Management to 2050*; World Bank: Washington, DC, USA, 2018. [CrossRef]
3. López-Gómez, J.P.; Latorre-Sanchez, M.; Unger, P.; Schneider, R.; Lozano, C.C.; Venus, J. Assessing the organic fraction of municipal solid wastes for the production of lactic acid. *Biochem. Eng. J.* **2019**, *150*, 107251. [CrossRef]
4. Szabo, K.; Emőke Teleky, B.; Ranga, F.; Simon, E.; Lelia Pop, O.; Babalau-Fuss, V.; Kapsalis, N.; Cristian Vodnar, D. Bioaccessibility of microencapsulated carotenoids, recovered from tomato processing industrial by-products, using in vitro digestion model. *LWT* **2021**, *152*, 112285. [CrossRef]
5. Martau, G.A.; Calinoiu, L.F.; Vodnar, D.C. Bio-vanillin: Towards a sustainable industrial production. *Trends Food Sci. Technol.* **2021**, *109*, 579–592. [CrossRef]
6. Mmereki, D.; Baldwin, A.; Li, B. A comparative analysis of solid waste management in developed, developing and lesser developed countries. *Environ. Technol. Rev.* **2016**, *5*, 120–141. [CrossRef]
7. López-Gómez, J.P.; Alexandri, M.; Schneider, R.; Latorre-Sanchez, M.; Lozano, C.C.; Venus, J. Organic fraction of municipal solid waste for the production of L-lactic acid with high optical purity. *J. Clean. Prod.* **2020**, *247*, 119165. [CrossRef]
8. Papa, G.; Pepe Sciarria, T.; Carrara, A.; Scaglia, B.; D'Imporzano, G.; Adani, F. Implementing polyhydroxyalkanoates production to anaerobic digestion of organic fraction of municipal solid waste to diversify products and increase total energy recovery. *Bioresour. Technol.* **2020**, *318*, 124270. [CrossRef] [PubMed]
9. Korkakaki, E.; Mulders, M.; Veeken, A.; Rozendal, R.; van Loosdrecht, M.C.; Kleerebezem, R. PHA production from the organic fraction of municipal solid waste (OFMSW): Overcoming the inhibitory matrix. *Water Res.* **2016**, *96*, 74–83. [CrossRef] [PubMed]
10. Allegue, L.D.; Puyol, D.; Melero, J.A. Novel approach for the treatment of the organic fraction of municipal solid waste: Coupling thermal hydrolysis with anaerobic digestion and photo-fermentation. *Sci. Total Environ.* **2020**, *714*, 136845. [CrossRef] [PubMed]
11. Mitrea, L.; Calinoiu, L.F.; Precup, G.; Bindea, M.; Rusu, B.; Trif, M.; Stefanescu, B.E.; Pop, I.D.; Vodnar, D.C. Isolated Microorganisms for Bioconversion of Biodiesel-Derived Glycerol into 1,3-Propanediol. *Bull. Univ. Agric. Sci. Vet. Med. Cluj-Napoca-Food Sci. Technol.* **2017**, *74*, 43–49. [CrossRef]
12. Han, S.K.; Shin, H.S. Biohydrogen production by anaerobic fermentation of food waste. *Int. J. Hydrogen Energy* **2004**, *29*, 569–577. [CrossRef]
13. Sanders, J.; Scott, E.; Weusthuis, R.; Mooibroek, H. Bio-refinery as the bio-inspired process to bulk chemicals. *Macromol. Biosci.* **2007**, *7*, 105–117. [CrossRef]
14. Sakai, K.; Ezaki, Y. Open L-lactic acid fermentation of food refuse using thermophilic Bacillus coagulans and fluorescence in situ hybridization analysis of microflora. *J. Biosci. Bioeng.* **2006**, *101*, 457–463. [CrossRef]
15. Ohkouchi, Y.; Inoue, Y. Impact of chemical components of organic wastes on L(+)-lactic acid production. *Bioresour. Technol.* **2007**, *98*, 546–553. [CrossRef]
16. Uçkun Kiran, E.; Trzcinski, A.P.; Ng, W.J.; Liu, Y. Bioconversion of food waste to energy: A review. *Fuel* **2014**, *134*, 389–399. [CrossRef]
17. Mitrea, L.; Calinoiu, L.F.; Martau, G.A.; Szabo, K.; Teleky, B.E.; Muresan, V.; Rusu, A.V.; Socol, C.T.; Vodnar, D.C. Poly(vinyl alcohol)-Based Biofilms Plasticized with Polyols and Colored with Pigments Extracted from Tomato By-Products. *Polymers* **2020**, *12*, 532. [CrossRef]
18. Arte, E.; Huang, X.; Nordlund, E.; Katina, K. Biochemical characterization and technofunctional properties of bioprocessed wheat bran protein isolates. *Food Chem.* **2019**, *289*, 103–111. [CrossRef]
19. Chen, Y.; Kong, Q.; Chi, C.; Shan, S.; Guan, B. Biotransformation of aflatoxin B1 and aflatoxin G1 in peanut meal by anaerobic solid fermentation of *Streptococcus thermophilus* and *Lactobacillus delbrueckii* subsp. *bulgaricus*. *Int. J. Food Microbiol.* **2015**, *211*, 1–5. [CrossRef] [PubMed]
20. Costa, R.D.S.; de Almeida, S.S.; Cavalcanti, E.D.C.; Freire, D.M.G.; Moura-Nunes, N.; Monteiro, M.; Perrone, D. Enzymes produced by solid state fermentation of agro-industrial by-products release ferulic acid in bioprocessed whole-wheat breads. *Food Res. Int.* **2021**, *140*, 109843. [CrossRef] [PubMed]
21. Dhillon, G.S.; Oberoi, H.S.; Kaur, S.; Bansal, S.; Brar, S.K. Value-addition of agricultural wastes for augmented cellulase and xylanase production through solid-state tray fermentation employing mixed-culture of fungi. *Ind. Crop. Prod.* **2011**, *34*, 1160–1167. [CrossRef]
22. Rosales, E.; Pazos, M.; Ángeles Sanromán, M. Solid-state fermentation for food applications. In *Current Developments in Biotechnology and Bioengineering*; Pandey, A., Larroche, C., Soccol, C.R., Eds.; Elsevier: Amsterdam, The Netherlands, 2018; pp. 319–355. [CrossRef]
23. Camassola, M.; Dillon, A.J. Cellulases and xylanases production by *Penicillium echinulatum* grown on sugar cane bagasse in solid-state fermentation. *Appl. Biochem. Biotechnol.* **2010**, *162*, 1889–1900. [CrossRef] [PubMed]

24. Teigiserova, D.A.; Bourgine, J.; Thomsen, M. Closing the loop of cereal waste and residues with sustainable technologies: An overview of enzyme production via fungal solid-state fermentation. *Sustain. Prod. Consum.* **2021**, *27*, 845–857. [CrossRef]
25. Outeirino, D.; Costa-Trigo, I.; Oliveira, R.P.D.; Guerra, N.P.; Dominguez, J.M. A novel approach to the biorefinery of brewery spent grain. *Process. Biochem.* **2019**, *85*, 135–142. [CrossRef]
26. Kornbrust, B.A.; Forman, T.; Matveeva, I. *Applications of Enzymes in Breadmaking*; Woodhead Publishing: Cambridge, UK, 2012; pp. 470–498. [CrossRef]
27. Anto, H.; Trivedi, U.B.; Patel, K.C. Glucoamylase production by solid-state fermentation using rice flake manufacturing waste products as substrate. *Bioresour. Technol.* **2006**, *97*, 1161–1166. [CrossRef]
28. Wang, X.Q.; Wang, Q.H.; Liu, Y.Y.; Ma, H.Z. On-site production of crude glucoamylase for kitchen waste hydrolysis. *Waste Manag. Res.* **2010**, *28*, 539–544. [CrossRef]
29. Osho, M.B.; Solomon, T. Use of composite agro-substrates for amyloglucosidase synthesis and characterization by *Aspergillus niger* otf and *Aspergillus flavus* clor1 using solid state fermentation. *J. Microbiol. Biotechnol. Food Sci.* **2020**, *9*, 879–883. [CrossRef]
30. Melikoglu, M.; Lin, C.S.K.; Webb, C. Stepwise optimisation of enzyme production in solid state fermentation of waste bread pieces. *Food Bioprod. Process.* **2013**, *91*, 638–646. [CrossRef]
31. López-Gómez, J.P.; Perez-Rivero, C.; Venus, J. Valorisation of solid biowastes: The lactic acid alternative. *Process Biochem.* **2020**, *99*, 222–235. [CrossRef]
32. López-Gómez, J.P.; Alexandri, M.; Schneider, R.; Venus, J. A review on the current developments in continuous lactic acid fermentations and case studies utilising inexpensive raw materials. *Process Biochem.* **2019**, *79*, 1–10. [CrossRef]
33. López-Gómez, J.P.; Unger, P.; Schneider, R.; Venus, J. From Upstream to Purification: Production of Lactic Acid from the Organic Fraction of Municipal Solid Waste. *Waste Biomass Valorization* **2020**, *11*, 5247–5254. [CrossRef]
34. Miller, G.L. Use of Dinitrosalicylic Acid Reagent for Determination of Reducing Sugar. *Anal. Chem.* **1959**, *31*, 426–428. [CrossRef]
35. Yu, X.; Liu, Y.; Cui, Y.; Cheng, Q.; Zhang, Z.; Lu, J.H.; Meng, Q.; Teng, L.; Ren, X. Measurement of filter paper activities of cellulase with microplate-based assay. *Saudi. J. Biol. Sci.* **2016**, *23*, S93–S98. [CrossRef] [PubMed]
36. Behera, S.S.; Ray, R.C. Solid state fermentation for production of microbial cellulases: Recent advances and improvement strategies. *Int. J. Biol. Macromol.* **2016**, *86*, 656–669. [CrossRef]
37. Tiwari, R.; Nain, P.K.S.; Singh, S.; Adak, A.; Saritha, M.; Rana, S.; Sharma, A.; Nain, L. Cold active holocellulase cocktail from *Aspergillus niger* SH3: Process optimization for production and biomass hydrolysis. *J. Taiwan Inst. Chem. Eng.* **2015**, *56*, 57–66. [CrossRef]
38. Ma, F.Y.; Wang, J.J.; Zeng, Y.L.; Yu, H.B.; Yang, Y.; Zhang, X.Y. Influence of the co-fungal treatment with two white rot fungi on the lignocellulosic degradation and thermogravimetry of corn stover. *Process. Biochem.* **2011**, *46*, 1767–1773. [CrossRef]
39. Taherzadeh-Ghahfarokhi, M.; Panahi, R.; Mokhtarani, B. Optimizing the combination of conventional carbonaceous additives of culture media to produce lignocellulose-degrading enzymes by *Trichoderma reesei* in solid state fermentation of agricultural residues. *Renew. Energy* **2019**, *131*, 946–955. [CrossRef]
40. Pandey, A.K.; Edgard, G.; Negi, S. Optimization of concomitant production of cellulase and xylanase from *Rhizopus oryzae* SN5 through EVOP-factorial design technique and application in Sorghum Stover based bioethanol production. *Renew. Energy* **2016**, *98*, 51–56. [CrossRef]
41. Kaushik, P.; Mishra, A.; Malik, A. Dual application of agricultural residues for xylanase production and dye removal through solid state fermentation. *Int. Biodeterior. Biodegrad.* **2014**, *96*, 1–8. [CrossRef]
42. Thomas, L.; Parameswaran, B.; Pandey, A. Hydrolysis of pretreated rice straw by an enzyme cocktail comprising acidic xylanase from *Aspergillus* sp for bioethanol production. *Renew. Energy* **2016**, *98*, 9–15. [CrossRef]
43. Raghuwanshi, S.; Deswal, D.; Karp, M.; Kuhad, R.C. Bioprocessing of enhanced cellulase production from a mutant of *Trichoderma asperellum* RCK2011 and its application in hydrolysis of cellulose. *Fuel* **2014**, *124*, 183–189. [CrossRef]
44. Oberoi, H.S.; Babbar, N.; Dhaliwal, S.S.; Kaur, S.; Vadlani, P.V.; Bhargav, V.K.; Patil, R.T. Enhanced Oil Recovery by Pre-treatment of Mustard Seeds Using Crude Enzyme Extract Obtained from Mixed-Culture Solid-State Fermentation of Kinnow (*Citrus reticulata*) Waste and Wheat Bran. *Food Bioprocess. Technol.* **2010**, *5*, 759–767. [CrossRef]
45. Singhania, R.R.; Sukumaran, R.K.; Patel, A.K.; Larroche, C.; Pandey, A. Advancement and comparative profiles in the production technologies using solid-state and submerged fermentation for microbial cellulases. *Enzym. Microb. Technol.* **2010**, *46*, 541–549. [CrossRef]
46. Melikoglu, M. *Production of Sustainable Alternatives to Petrochemicals and Fuels Using Waste Bread as a Raw Material*; The University of Manchester: Manchester, UK, 2008.
47. López-Gómez, J.P.; Venus, J. Potential Role of Sequential Solid-State and Submerged-Liquid Fermentations in a Circular Bioeconomy. *Fermentation* **2021**, *7*, 76. [CrossRef]
48. Li, Y.B.; Park, S.Y.; Zhu, J.Y. Solid-state anaerobic digestion for methane production from organic waste. *Renew. Sust. Energ. Rev.* **2011**, *15*, 821–826. [CrossRef]
49. Probst, M.; Walde, J.; Pumpel, T.; Wagner, A.O.; Insam, H. A closed loop for municipal organic solid waste by lactic acid fermentation. *Bioresour. Technol.* **2015**, *175*, 142–151. [CrossRef] [PubMed]
50. Jem, K.J.; van der Pol, J.F.; de Vos, S. Microbial lactic acid, its polymer poly(lactic acid), and their industrial applications. In *Plastics from Bacteria: Natural Functions and Applications*; Chen, G.G.-Q., Ed.; Springer: Berlin/Heidelberg, Germany, 2010; pp. 323–346. [CrossRef]

51. Klotz, S.; Kaufmann, N.; Kuenz, A.; Prusse, U. Biotechnological production of enantiomerically pure d-lactic acid. *Appl. Microbiol. Biotechnol.* **2016**, *100*, 9423–9437. [CrossRef]
52. Kwak, S.; Jin, Y.S. Production of fuels and chemicals from xylose by engineered *Saccharomyces cerevisiae*: A review and perspective. *Microb. Cell Fact.* **2017**, *16*, 82. [CrossRef] [PubMed]
53. Teleky, B.E.; Martau, A.G.; Ranga, F.; Chetan, F.; Vodnar, D.C. Exploitation of Lactic Acid Bacteria and Baker's Yeast as Single or Multiple Starter Cultures of Wheat Flour Dough Enriched with Soy Flour. *Biomolecules* **2020**, *10*, 778. [CrossRef]
54. Aulitto, M.; Fusco, S.; Bartolucci, S.; Franzen, C.J.; Contursi, P. *Bacillus coagulans* MA-13: A promising thermophilic and cellulolytic strain for the production of lactic acid from lignocellulosic hydrolysate. *Biotechnol. Biofuels* **2017**, *10*, 210. [CrossRef]

Journal of
*Fungi*

MDPI

*Article*

# Waste Rose Flower and Lavender Straw Biomass—An Innovative Lignocellulose Feedstock for Mycelium Bio-Materials Development Using Newly Isolated *Ganoderma resinaceum* GA1M

**Galena Angelova [1], Mariya Brazkova [1,*], Petya Stefanova [1], Denica Blazheva [2], Veselin Vladev [3], Nadejda Petkova [4], Anton Slavov [4], Petko Denev [5], Daniela Karashanova [6], Roumiana Zaharieva [7], Atanas Enev [8] and Albert Krastanov [1]**

1 Department of Biotechnology, University of Food Technology, 26 Maritsa Blvd., 4002 Plovdiv, Bulgaria; g_angelova@uft-plovdiv.bg (G.A.); petyastefanova@uft-plovdiv.bg (P.S.); a_krastanov@uft-plovdiv.bg (A.K.)
2 Department of Microbiology, University of Food Technology, 26 Maritsa Blvd., 4002 Plovdiv, Bulgaria; d_blazheva@uft-plovdiv.bg
3 Department of Mathematics, Physics and Information Technologies, Faculty of Economics, University of Food Technologies, 26 Maritsa Blvd., 4002 Plovdiv, Bulgaria; v.p.vladev@abv.bg
4 Department of Organic and Inorganic Chemistry, University of Food Technologies, 26 Maritsa Blvd., 4002 Plovdiv, Bulgaria; petkovanadejda@abv.bg (N.P.); antons@uni-plovdiv.net (A.S.)
5 Laboratory of Biologically Active Substances, Institute of Organic Chemistry with Centre of Phytochemistry, Bulgarian Academy of Sciences, 139 Ruski Blvd., 4000 Plovdiv, Bulgaria; petkodenev@yahoo.com
6 Institute of Optical Materials and Technologies, Bulgarian Academy of Sciences, Acad. Georgy Bonchev Str., 1113 Sofia, Bulgaria; dkarashanova@yahoo.com
7 Department of Building Materials and Insulation, Faculty of Structural Engineering, University of Architecture, Civil Engineering and Geodesy, 1046 Sofia, Bulgaria; zaharieva_fce@uacg.bg
8 Byomic Ltd., 1000 Sofia, Bulgaria; atanas.enev@biomyc.eu
* Correspondence: mbrazkova@uft-plovdiv.bg

**Citation:** Angelova, G.; Brazkova, M.; Stefanova, P.; Blazheva, D.; Vladev, V.; Petkova, N.; Slavov, A.; Denev, P.; Karashanova, D.; Zaharieva, R.; et al. Waste Rose Flower and Lavender Straw Biomass—An Innovative Lignocellulose Feedstock for Mycelium Bio-Materials Development Using Newly Isolated *Ganoderma resinaceum* GA1M. *J. Fungi* **2021**, *7*, 866. https://doi.org/10.3390/jof7100866

Academic Editor: Baojun Xu

Received: 10 September 2021
Accepted: 12 October 2021
Published: 15 October 2021



**Abstract:** In this study, for the first time, the potential of rose flowers and lavender straw waste biomass was studied as feeding lignocellulose substrates for the cultivation of newly isolated in Bulgaria *Ganoderma resinaceum* GA1M with the objective of obtaining mycelium-based bio-composites. The chemical characterization and Fourier Transform Infrared (FTIR) spectroscopy established that the proximate composition of steam distilled lavender straw (SDLS) and hexane extracted rose flowers (HERF) was a serious prerequisite supporting the self-growth of mycelium bio-materials with improved antibacterial and aromatic properties. The basic physico-mechanical properties of the developed bio-composites were determined. The apparent density of the mycelium HERF-based bio-composites (462 $kg/m^3$) was higher than that of the SDLS-based bio-composite (347 $kg/m^3$) and both were much denser than expanded polystyren (EPS), lighter than medium-density fiber board (MDF) and oriented strand board (OSB) and similar to hempcrete. The preliminary testing of their compressive behavior revealed that the compressive resistance of SDLS-based bio-composite was 718 kPa, while for HERF-based bio-composite it was 1029 kPa and both values are similar to the compressive strength of hempcrete with similar apparent density. Water absorbance analysis showed, that both mycelium HERF- and SDLS-based bio-composites were hydrophilic and further investigations are needed to limit the hydrophilicity of the lignocellulose fibers, to tune the density and to improve compressive resistance.

**Keywords:** hexane extracted rose flowers; steam distilled lavender straw; mycelium bio-composites; *Ganoderma resinaceum*; apparent density; water absorbance; compressive resistance

## 1. Introduction

Roses and lavender are plant species, which have been highly appreciated for centuries, and because of their diverse biologically active substance content, they have an industrial impact on essential oil production worldwide [1,2]. Bulgaria and Turkey have long traditions in rose oil production and dominate the world market, followed by Iran, Morocco, France, Italy, India and China [1]. Bulgarian production satisfies 40% of the global demand for rose oil. In recent years a drastic increase in lavender cultivation and lavender oil yield is being observed in Bulgaria, which defines its leading position as a lavender oil producer, leaving France behind.

The most common methods for the production of essential oils are water-steam distillation, steam distillation and solvent (non-polar solvents and liquefied gases) extraction of the raw plant materials. Due to the relatively low concentration of essential oils in the crops, large amounts of plant material are treated and, respectively, a huge quantity of solid residual biomass is generated. More than 20,000 t of steam distilled lavender straw (SDLS), about 29,000 t of steam distilled rose flowers (SDRF) and hexane extracted rose flowers (HERF) are generated each year only in Bulgaria [1].

Generally, rose flower and lavender straw by-products are considered waste material and still, most of them are incinerated for energy recovery or discarded near the distilleries, which poses an environmental hazard. This waste biomass is referred to as lignocellulose waste, because of its high content of lignocellulose and hemicellulose compounds, pectic carbohydrates and polyphenols [1,3]. The valuable biological content of HERF and SDLS can be used in innovative "self-growing" technologies for obtaining value-added materials with sustainable social and economic impact.

In recent years, various lignocellulose wastes with agricultural and forestry origin are studied and used as substrates for the production of sustainable, renewable, biodegradable and eco-friendly natural-based bio-materials for indoor and outdoor applications [4]. The use of biological agents for binding the lignocellulose fibers and turning them into 100% natural and bio-recyclable bio-materials is the most environmentally friendly mechanism and it revolutionizes the traditional ideas for materials production. "Materials that are grown are better than manufactured" is an innovative promising approach in materials science, aiming to achieve unique self-grown material, which has attracted considerable research efforts in recent years [5–11].

Currently, numerous research teams explore the effect of various combinations between plant lignocellulose substrates and higher fungi to develop functional mycelium-based bio-composite materials with the potential to replace synthetic materials. Many studies have reported the ability of mushrooms, belonging to the *Trametes*, *Ganoderma* and *Pleurotus* genera, to be cultivated on a range of lignocellulose materials, including various types of cotton seed hulls, corn cobs, peanut shells, cotton from the textile industry, coffee pulp, paper, leaves, coconut powder [6,8,9,12–21] as well as low-quality organic waste streams like saw dust and straw and turning them into mycelium bio-composites [22]. The structural and macroscopic characteristics of the mycelium-based bio-composites are strongly dependent on the fungal species, the type and chemical composition of the substrates and both cultivation conditions and post-cultivation processing define their non-structural, semi-structural or specific applications [4,6,7,11,12,18,22–26]. Due to its fibrous structure and polymeric composition, the mycelium can serve as a matrix and the lignocellulose fibers as reinforcement for the newly obtained mycelium bio-material [6,7,13]. The fungal mycelium possesses the advantage of strength and durability performance and together with its non-toxicity, fire-resistance and hydrophobic capabilities show great potential as a binding element in mycelium bio-composites development [8,13,27].

The mycelium-based bio-composites are completely natural and can be composted after the end of their cycle of use, which would help the transition to a circular economy, keeping the value of materials and resources in the economy as long as possible and generating minimal waste [6,8,22]. They also have the potential to contribute to the new "green" economy by replacing many oil-based products and converting lignocellulose waste into value-added biodegradable products, which don't damage the ecosystem cycles. The lower water, energy and $CO_2$ requirements, together with the low environmental impact in comparison with the conventional industries, are the key advantages for the application of self-grown bio-composites [7,27,28].

The analysis of reference literature revealed that there are many studies showing the potential of waste rose flower and lavender straw biomass as raw materials for aroma and polysaccharides extraction, for recovery of biologically active substances and their application in the food industry and medicine [1,29]. Lavender distilled straw separately or in combination with other lignocellulose substrates was studied by Ratiarisoa et al. [30] and was successfully recycled as a bio-aggregate for building materials using metakaolin as a binder.

At present, there are no reports on the usage of rose and lavender industrial processing by-products as feeding substrates for mushroom growth aiming at the obtaining of mycelium-based bio-materials, although their chemical composition suggests they are fully compatible with fungal growth. Lavender and lavandin distilled straw was successfully used as a feeding substrate for *Pycnoporus cinnabarinus* cultivation by Lesage-Meessen et al. [29], but for lignocellulose-acting enzyme synthesis.

The focus of this study is to explore the chemical characteristics of hexane extracted rose flowers and steam distilled lavender straw, which are a significant waste of the essential oil-processing industry and to reveal their realistic potential to be used as feedstock, promoting vegetative fungal growth. The ability of a newly isolated in Bulgaria fungal strain to grow on HERF and SDLS and convert them into mycelium-based bio-materials was studied. The assessment of the basic physical and mechanical properties of these mycelium-based bio-materials opens possibilities for further in-depth investigations and the production of mycelium HERF- and SDLS-based bio-materials for various applications.

## 2. Materials and Methods

### 2.1. Materials

#### 2.1.1. Substrates

The hexane extracted rose flowers (*Rosa damascena* Mill.) and the steam distilled lavender straw (*Lavandula angustifolia* Mill.) were provided by Galen-N (2020 crop, Zelenikovo distillery, Brezovo region, Bulgaria). The HERF was stored at −18 °C and prior use was dried at 40 °C (Figure 1a) The SDLS was collected immediately after the steam distillation processes from the discarding area near the distillery, air-dried at 40 °C and stored at room temperature (Figure 1b). Before use, the SDLS was milled and sieved (particle size 1–5 mm).

(a) (b)

**Figure 1.** Hexane extracted rose flowers (**a**) and steam distilled lavender straw (**b**).

2.1.2. Fungal Isolate

The fruiting body of the newly isolated basidiomycete was collected from a forest near Maritza River, Bulgaria in May 2019. In order to isolate a pure culture, the fruiting body was first rinsed very well with tap water and cut into 20 to 30 mm pieces. They were then surface sterilized with 70% ethanol for 20 sec, followed by 10 min in 0.1% $Ca(ClO)_2$ and finally rinsed in sterile distilled water. The samples were further sliced to 2.5 by 2.5 mm pieces with sterile scalpel and aseptically transferred on Rose Bengal Chloramphenicol Agar (RBCA) (HIMEDIA, India). The plates were incubated in darkness at 28 °C for 14 days and were visually monitored daily. The fungal culture was isolated and purified by several transfers of growing mycelium on fresh medium. The pure fungal culture was maintained at 4 °C on Mushroom Complete medium (MCM), containing g/L: glucose—20, $KH_2PO_4$—0.5, $K_2HPO_4$—1.0, $MgSO_4$—0.5, peptone—2.0, yeast extract—2.0, Agar—2.0, pH 4.8–5.2.

*2.2. Substrate Preparation Procedure*

2.2.1. Preparation for Cultivation

150 g of feeding substrate (SDLS or HERF) was moisturized to final humidity of 65% with a solution containing (g/L): $MgSO_4$—0.5, $KH_2PO_4$—0.5, $K_2HPO_4$—1.0, peptone—2.0, yeast extract—2.0. After the addition of 0.1% $CaCO_3$ the substrate was mixed well and transferred to autoclavable mushroom growing bags ($SacO_2$, Belgium) (100 × 350 mm) and sterilized at 121 °C for 45 min to render the substrate inert. The substrate was left to cool down for 12 h and was used as feeding substrates for solid-state cultivation.

2.2.2. Preparation for Chemical Analysis

SDLS or HERF (30 g), milled and sieved (fraction with average particle size 1.5–1.6 mm was used), was treated with preheated 150 mL 70% ethanol for 1 h at 60 °C with constant stirring (120 rpm), then left for 24 h at room temperature (20 °C). The biomass was filtered through cloth (250 mesh), and the remaining residue was treated with additional 100 mL 70% ethanol at the same conditions. After the second filtration, the extracts were combined and used for analyses.

*2.3. Chemical Characterization of the Substrate*

2.3.1. Proximate Composition Determination

The total dietary fibers were determined using K-TDFR–100A (Megazyme, Wicklow, Ireland), according to AOAC method 991.43 "Total, soluble and insoluble dietary fibers in foods" (First action 1991) and AACC method 32–07.01 "Determination of soluble, insoluble and total dietary fibers in foods and food products" (Final approval 10–16–91). The uronic acid, cellulose, lignin and non-cellulosic polysaccharides in the SDLS and HERF were determined according to Pla et al. [31]. Inorganic matters (ash) were determined after ashing of a 5 g sample at 605 °C.

2.3.2. Determination of Total and Individual Polyphenols

The total polyphenol content was determined according to Singleton and Rossi [32] using Folin-Ciocalteu's reagent (Sigma-Aldrich, Schaffhausen, Switzerland). Gallic acid (Sigma-Aldrich, Steinheim, Germany) was employed for calibration and the results were expressed as gallic acid equivalents (GAE) per gram dry matter of the extracts. HPLC analyses of the phenolic components were performed on an Agilent 1220 HPLC system (Agilent Technology, Santa Clara, CA, USA), equipped with a binary pump and UV-Vis detector. A wavelength of 280 nm was used. The separation of phenolic compounds was performed using an Agilent TC–C18 column (5 μm, 4.6 mm × 250 mm) at 25 °C. The mobile phases consisted of 0.5% acetic acid (A) and 100% acetonitrile (B) at a flow rate of 0.8 mL/min. The gradient conditions started with 14% B, increased linearly to 25% B between the 6th and the 30th min, then to 50% B at the 40th min. The standard compounds (gallic acid, 3,4-dihydroxy benzoic acid, chlorogenic acid, caffeic acid, p-coumaric acid, ferulic acid, ellagic acid, catechin, epicatechin, rutin, naringin, myricetin, quercetin, naringenin and kaempferol) were purchased from Sigma-Aldrich (Steinheim, Germany).

The individual volatile and non-volatile compounds in the rose and lavender ethanol extracts were determined according to the following procedures:

(1) Non-volatile substances—0.2 mL ethanol extract was lyophilized and 50 μL pyridine and 50 μL N,O-Bis-(trimethylsilyl)-trifluoroacetamide (BSTFA) were added. The sample was incubated at 70 °C for 40 min. For analysis 1.0 μL from the solution was injected on a gas chromatograph Agilent GC 7890 with mas-selective detector Agilent MD 5975 and column HP-5ms (30 m with diameter 0.32 mm and 0.25 μm thicknesses). The following temperature regimen was used—initial temperature 100 °C (hold for 2 min) then increase to 180 °C with 15 °C/min (hold for 1 min) and increase of the temperature to 300 °C with 5 °C/min (hold for 10 min); injector and detector temperatures—250 °C, helium was used s carrier gas at 1.0 mL/min. The scanning range of the mass-selective detector was m/z = 50–550 in split-split mode (10:1).

(2) Volatile substances—1.0 mL ethanol extract was extracted with 1.0 mL dichloromethane (triple). The combined organic layers were dried under vacuum at 30 °C. To the dry residue 100 μL dichloromethane was added. For analysis 1.0 μL from the solution was injected on a gas chromatograph Agilent GC 7890 with mas-selective detector Agilent MD 5975 and column HP-5ms. The following temperature regimen was used—the initial temperature was 40 °C and then increase to 300 °C with 5 °C/min (hold for 10 min); injector and detector temperatures—250 °C, helium was used as carrier gas at 1.0 mL/min. The scanning range of the mass-selective detector was m/z = 40–400 in splitless mode. The individual compounds were identified by comparing the retention times and the relative index (RI) with those of standard substances and mas-spectral data from libraries of The Golm Metabolome Database (http://csbdb.mpimp-golm.mpg.de/csbdb/gmd/gmd.html, accessed on 5 November 2020) and NIST'08 (National Institute of Standards and Technology, Gaithersburg, MD, USA).

### *2.4. Molecular Identification of the Basidiomycete Isolate by ITS1-5.8S-ITS2 rRNA Gene Sequence Analysis*

Prior to DNA extraction, the basidiomycete isolate was cultivated for 7 days on MCM agar plates. The fungal mycelium was scraped out with a sterile spatula (100–300 mg) and transferred to a 2 mL microtube. The DNA extraction was conducted by a modified CTAB method, according to Stefanova et al. [33]. The quality and concentration of the DNA extracts were determined by spectrophotometric measurements using Shimadzu UV-VIS spectrophotometer (Shimadzu Corporation, Japan). The ITS-5.8S-ITS2 region was amplified by forward primer ITS 4 (5′-TCCTCCGCTTATTGATATGC-3′) and reverse primer ITS 5 (5′-GGAAGTAAAAGTGCTAACAAGG-3′) (Metabion, Martinsried, Germany) [34]. The PCR reaction mix contained 1 μL of DNA (50 ng), 0.5 μM of each primer and 8 μL of Red-Taq DNA Polymerase Master Mix (Canvax Biotech, S.L., Córdoba, Spain) in total volume of 20 μL. The amplification was carried out in a PCR 2720 Thermal Cycler (Applied Biosystems, Waltham, MA, USA) using the following program: initial denaturation at 95 °C for 10 min, followed by 35 cycles of denaturing at 94 °C for 1 min, annealing at 52 °C for 1 min, extension at 72 °C for 1 min, and final extension at 72 °C for 7 min. The PCR product was visualized in 1% agarose gel stained with SafeView (NBS Biologicals, Huntingdon, UK) at 100 V for 50 min using VWR Mini Electrophoresis System (VWR, Darmstadt, Germany) and MiniBis Pro (DNR Bio-Imaging Systems, Israel) for gel visualization. The PCR product was cut out from the gel and purified with Clean-Easy™ Agarose Purification Kit (Canvax Biotech, S.L., Córdoba, Spain). The sequencing of the PCR product was performed by Microsynth Seqlab (Göttingen, Germany). The resulting sequence was analyzed using BLAST algorithm [35] and compared to the nucleotide sequences in the GenBank database (www.ncbi.nlm.nih.gov, accessed on 21 April 2021). The ITS1-5.8S-

ITS2 rRNA gene sequence of *Ganoderma resinaceum* GA1M was deposited in the GenBank database and an accession number was assigned.

### 2.5. Solid-State Cultivation for Mycelium Bio-Composite Obtaining

#### 2.5.1. Inoculum Preparation

The vegetative inoculum was prepared to form a 7-day culture of *G. resinaceum* grown on MCM slant-agar. Each 500 mL Erlenmeyer flask, containing 100 mL MCM broth, was inoculated with vegetative biomass from a single MCM-slant culture and incubated at 28 °C and 220 rpm for 7 days for pellets formation. The pellets were harvested by sterile filtration and were used for inoculation of the SDLS and HERF feeding substrates. The dry weight of the pellets was determined on moisture analyzer MAC 50/NH (RADWAG, Radom, Poland).

#### 2.5.2. Inoculation of the Substrates and Mycelium Growth

The pellets (10% $w/w$) were mixed with the substrates aseptically and left to grow in the mushroom growing bags in darkness at 25 °C for 7 days until the mycelium completely enveloped the substrate's particles. Every day the mixtures were visually monitored and mixed further to ensure visible uniform mycelium growth and full coverage of the particles of the substrates and the formation of "pre-grown" substrate.

The desired sample geometry was achieved as the ready "pre-grown" substrates were aseptically transferred into 3D-printed molds (55 × 55 × 55 mm), made of poly-lactic acid (PLA), consisting of two parts to ensure easy removal of the ready composites. They were manually filled tightly by layering to produce compact and dense samples and then covered with transparent perforated foil to create a microenvironment with high humidity (about 98%), necessary for fungal respiration rate, and incubated at 25 °C, 95% humidity in Nüve climatic chamber (Nüve, Turkey) (Figure 2a). The samples reached the required shape and size after 7 days. Then they were de-molded (Figure 2b) and the material was put back in the incubation chamber to grow for another 6 days in order to develop full external homogeneous growth on the sides and to form an external mycelium skin.

(a) (b)

**Figure 2.** Cultivation of *G. resinaceum* for obtaining mycelium-based bio-composites: in moulds (**a**) and de-moulded (**b**).

The samples were dried in a drying oven SLW 32 (POL-EKO-APARATURA, Poland) at 60 °C for 8 h in order to remove the moisture and inactivate the mycelium. After drying out, the samples of the two types of mycelium bio-composites (HERF-based and SDLS-based) had the approximate size of 40 × 40 × 40 mm and were used for further tests.

### 2.6. Scanning Electron Microscopy (SEM)

The visualization of the surface and core morphology of the dried mycelium-based bio-composites was performed by digitized scanning electron microscope (SEM) Philips 515 (Philips, Netherland) in secondary electrons imaging (SEI) mode. All analyses were performed at an accelerating voltage of 8 kV and different magnifications between 200× and 5000×. A preliminary preparation of the samples was applied, including fixation of pieces of the corresponding composites samples on standard SEM stubs and subsequent metallization of their surfaces by deposition of an Au film.

### 2.7. Fourier Transform Infrared Spectroscopy (FTIR)

The FTIR spectra of the feeding substrates SDLS and HERF and the obtained SDLS and HERF-based mycelium bio-composites previously ground and homogenized (2 mg) were collected on a Fourier transform infrared (FTIR) spectrophotometer VERTEX 70v (Brucker, Germany) in KBr pellets (Honeywell/Fluka FTIR grade > 99%). The spectra were recorded in the 4000–400 $cm^{-1}$ range at 132 scans with a spectral resolution of 2 $cm^{-1}$. Three samples of each type were measured to ensure reproducibility.

### 2.8. Basic Physical and Mechanical Characterization

The apparent density was determined according to ISO 29470:2020. The short and long-term water absorption was measured according to ISO 29767:2019 and ISO 16535:2019, respectively. EN 826:2013 was used for compression behavior determination. Three samples of approximate size 40 × 40 × 40 mm were used for each test of the mycelium HERF- and SDLS-based bio-composites. The average value and the standard deviation were calculated. Despite the structural differences between the core and the near-surface zones of the samples, the testing was performed on intact samples, in order to take into consideration the specific properties of bio-composites due to the mycelium "coating" of the samples. Thus, the results reflected the overall behavior, but were highly influenced by the surface/volume ratio. In case the bio-composites are to be used in applications where the edges need to be trimmed or the surface layer removed, the tests must be carried out on the relevant representative samples.

The parallel piped samples were dried out in a ventilated oven at +60 °C until constant mass md (precision of 0.01 g) was reached. The dimensions were measured by a caliper with a precision of 0.01 mm. Each dimension (length, width, height) was measured three times—at the edges and in the middle and the average value was reported. The apparent density of each sample was calculated by the formula:

$$\rho_a = \frac{m_d}{a.b.h} \quad (1)$$

where, $\rho_a$ is the apparent dry density, in kg/m$^3$; $m_d$ is the mass in dry condition, in kg and $a$, $b$, $h$ are the dimensions (length, width and height) of the samples in dry condition, in m.

The capillary absorption was measured by a partial immersion (1 cm of the height of samples) in water at +20 ± 3 °C. The samples were placed on a grid, allowing free water circulation below the sample surface. The water level was maintained constant for 24 h. After removal from the water the surface was wiped off with a moist cloth and the mass of each sample was then measured. The calculations were made according to Formula (2):

$$W_c = \frac{m_{w,c} - m_d}{a.b} \quad (2)$$

where $W_c$ is the capillary absorption per unit area for 24 h, in kg/m$^2$; $m_d$ is the mass in dry condition, in kg; $m_{w,c}$ is the mass in wet condition after 24 h of partial immersion in water, in kg and $a$, $b$ are the width and length of the samples, in m.

Water absorption was measured after short-term (1 day) and long-term (28 days) immersion in water. The samples were placed between 2 grids and covered with 1 cm of water at +20 ± 3 °C. The bottom grid allowed free circulation of water, while the upper grid served as a support for additional weights on the samples in order to avoid their emergence. After removal from the water the surface was wiped off with a moist cloth and the mass of each sample was then measured.

The water absorption was calculated as follows:

$$W^{m}{}_{a,1d} = \frac{m_{w,1} - m_d}{m_d}.100 \quad (3)$$

$$W^{m}{}_{a,28d} = \frac{m_{w,28} - m_d}{m_d}.100 \quad (4)$$

$$W^{v}{}_{a,1d} = \frac{(m_{w,1} - m_d)/\rho_w}{a.b.h}.100 \quad (5)$$

$$W^{v}{}_{a,28d} = \frac{(m_{w,28} - m_d)/\rho_w}{a.b.h}.100 \quad (6)$$

where $W^{m}{}_{a,1d}$ is the short term water absorption after 1 day of full immersion in water, expressed per unit mass, in % wt; $W^{v}{}_{a,1d}$ is the short term water absorption after 1 day of full immersion in water, expressed per unit volume, in % vol.; $W^{m}{}_{a,28d}$ is the long-term water absorption after 28 days of full immersion in water, expressed per unit mass, in % wt.; $W^{v}{}_{a,28d}$ is the long term water absorption after 28 days of full immersion in water, expressed per unit volume, in % vol.; $m_d$ is the mass in dry condition, in kg; $m_{w,1d}$ is the mass in wet condition after 1 day of full immersion in water, in kg; $m_{w,28d}$ is the mass in wet condition after 28 days of full immersion in water, in kg; $a$, $b$, $h$ are the dimensions (length, width and height) of the samples in dry condition, in m and $\rho_w$ is the density of water at +20 °C, i.e., 997 kg/m$^3$.

The compressive stress at 10% relative deformation was calculated as follows:

$$\sigma_{10} = \frac{F_{10}}{a.b} \quad (7)$$

where, $\sigma_{10}$ is the compressive stress at 10% deformation, in MPa; $F_{10}$ is the load at which the test specimen reaches the required longitudinal deformation of 10%, in N and $a$, $b$ are the width and length of the samples, in mm.

### 2.9. Statistical Analysis

All the experiments were conducted in triplicate and the values were expressed as mean ± SD. Statistical significance was detected by analysis of variance (ANOVA, Tukey's test; value of $p < 0.05$ indicated statistical difference).

## 3. Results and Discussion

### 3.1. Chemical Characterization of the HERF and SDLS and Assessment of Their Potential as Feedstock for Mycelium Growth and New Bio-Materials Development

The selection of HERF and SDLS as substrates in this study was mainly based on the fact that they are generated globally and in Bulgaria in significant amounts as essential oil industry waste by-products. The solid rose and lavender processed biomass are underutilized and mostly thrown away by distilleries. These circumstances are a prerequisite for the sustainable valorization of rose and lavender by-products and their transformation into new mycelium bio-composites with added value. Moreover, the industrial processing of essential oil plants (steam and steam-water distillation, extraction with organic solvents or supercritical gases) leads naturally to sterilization, which could simplify the procedure for mushroom growth substrate preparation.

Knowing the chemical composition of these waste materials is of critical importance not only to the stimulation of the vegetative growth and development of the basidiomycete strain but to the formation of a homogenous mycelium bio-composite with proper integrity and innovative aromatic and bactericidal properties.

The proximate composition of the HERF and SDLS revealed that there were significant differences in the analyzed compound content (Table 1). The concentration of lignocellulose carbohydrates in both studied substrates was higher than that of non-cellulosic polysaccharides. Higher content of cellulose and lignin was determined in SDLS—38.16 and 24.48 g/100 g DW, respectively. In HERF the concentrations of cellulose and lignin were about 24% lower and were 29.13 and 18.57 g/100 g DW, respectively. This is dependent on the type of raw material (rose and lavender) used by the distilleries. Lavender is a crop that is much easier to harvest mechanically and the whole stem is mowed, while only the valuable part of the roses—the flower, is harvested. This explains the higher content of cellulose and lignin in SDLS. It is well known that the inner part of the rose flower is a good source of pectic polysaccharides [36], which was reflected in the higher uronic acid concentration in the HERF (8.95 g/100 g DW) in comparison with the 3.54 g/100 g DW for SDLS. The difference in the concentration of non-cellulosic polysaccharides was small—14.57 g/100 g DW in HERF and 13.79 g/100 g DW in SDLS. The lignin content of SDLS (24.48%) and HERF (18.57%) was closer to that of softwood (18–26%) and grass (17–24%) than the one of hardwoods (27–33%) [37].

**Table 1.** Proximate composition of hexane extracted rose flowers (HERF) and steam distilled lavender straw (SDLS).

<table>
<tr><th></th><th>HERF, g/100 g DW</th><th>SUM</th><th>SDLS, g/100 g DW</th><th>SUM</th></tr>
<tr><td>Fibers (total)</td><td>$75.14 \pm 0.19$ [b]</td><td>-</td><td>$81.54 \pm 0.18$ [a]</td><td>-</td></tr>
<tr><td>Uronic acids</td><td>$8.95 \pm 0.17$ [a]</td><td>-</td><td>$3.54 \pm 0.12$ [b]</td><td>-</td></tr>
<tr><td>Ash</td><td>$3.10 \pm 0.16$ [b]</td><td>-</td><td>$6.59 \pm 0.11$ [a]</td><td>-</td></tr>
<tr><td>Non-cellulosic polysaccharides</td><td>$14.57 \pm 0.2$ [a]</td><td rowspan="3">$62.27 \pm 0.21$ [b]</td><td>$13.79 \pm 0.15$ [a]</td><td rowspan="3">$76.43 \pm 0.21$ [a]</td></tr>
<tr><td>Cellulose</td><td>$29.13 \pm 0.17$ [b]</td><td>$38.16 \pm 0.21$ [a]</td></tr>
<tr><td>Lignin</td><td>$18.57 \pm 0.16$ [b]</td><td>$24.48 \pm 0.14$ [a]</td></tr>
</table>

Data were expressed as mean ± SD ($n$ = 3); [a,b]—different letters indicate significantly different values in the rows (Student's $T$ test; value of $p < 0.05$ indicated statistical significance).

The variation of the type and initial concentration of polysaccharides in both studied substrates is an important factor affecting the mycelium fungal growth, the colonization rate of the substrates and the physico-mechanical properties of the obtained mycelium-based composite.

The content of lignocellulose compounds, together with the non-cellulosic polysaccharides and uronic acid could act as an essential carbon source for the vegetative development of the studied basidiomycete strain. Non-cellulosic polysaccharides and uronic acid would probably stimulate the initial fungal biomass production and later partial de-polymerization of lignin and cellulose would occur. According to Singh and Singh [38], basidiomycetes have their own mechanism for penetration of lignocellulose materials starting through nutrient-rich parenchyma cells, which are composed of easily metabolized components. Further, due to the thin hyphae formation and synthesis of a specific ligninolytic enzyme system, they can bore through the cell wall and support full colonization of the lignocellulose substrates.

The lignin and cellulose content and lignocellulose fibers reinforce mycelium-based bio-composite materials because of the crystallinity of cellulose and the protecting role of lignin in the lignocellulose matrix [39]. A large number of various chemical groups in the lignin macromolecules is a prerequisite for modifications and new functional groups occurrence by enzymatic or biological methods to improve adhesion and to support aggregation of lignocellulose particles [40]. The presence of lignin also helps to improve the stiffness of the bio-composites and protecting the ready bio-materials against microbiological attack [39].

The potential substrates were further analyzed for the presence and quantity of polyphenol compounds (total and individual) and the results are presented in Table 2.

**Table 2.** Total polyphenol content, individual flavonoids and phenolic acids in hexane extracted rose flowers (HERF) and steam distilled lavender straw (SDLS).

| | HERF, g/100 g DW | SDLS, g/100 g DW |
|---|---|---|
| Total polyphenols | 5.70 ± 0.20 [b] | 1.14 ± 0.07 [a] |
| Quercetin | 0.30 ± 0.002 [b] | 0.03 ± 0.001 [a] |
| Quercetin-3-β-glucoside | 0.48 ± 0.002 [b] | 0.04 ± 0.001 [a] |
| Rutin | 0.98 ± 0.002 | nd |
| Myricetin | 0.09 ± 0.001 [b] | 0.028 ± 0.001 [a] |
| Kaempferol | 0.04 ± 0.001 [b] | 0.01 ± 0.001 [a] |
| Catechin | 0.33 ± 0.002 [b] | 0.33 ± 0.001 [a] |
| Epicatechin | 0.23 ± 0.001 [b] | 0.25 ± 0.001 [a] |
| Neochlorogenic acid | 0.13 ± 0.002 [a] | 0.16 ± 0.001 [b] |
| 3,4-dihydroxy benzoic acid | 0.20 ± 0.001 [a] | 0.51 ± 0.001 [b] |
| p-Coumaric acid | 0.05 ± 0.001 | nd |
| Ferulic acid | 0.07 ± 0.001 [a] | 0.16 ± 0.001 [b] |
| Gallic acid | 0.12 ± 0.002 [a] | 0.29 ± 0.002 [b] |
| Rosmarinic acid | 0.05 ± 0.001 [a] | 0.13 ± 0.001 [b] |
| Cinnamic acid | 0.02 ± 0.001 | nd |

nd—not determined; Data are expressed as mean ± SD ($n$ = 3); [a,b]—different letters indicate significantly different values in the raws (Student's $T$ test; value of $p < 0.05$ indicated statistical significance).

Both rose and lavender essential oil industry by-products were found to be rich in polyphenols [3,41]. The fivefold difference in the total polyphenol content: 5.70 and 1.14 g/100 g DW for HERF and SDLS, respectively, is mainly due to the differences in industrial processing. The fresh rose flowers were extracted with a non-polar solvent (hexane), which resulted in the preservation of a large part of the polyphenols in the residual biomass, while the lavender was steam-distilled leading to partial extraction of phenol compounds. The highest amounts of individual polyphenols in HERF were found for rutin and quercetin-3-β-glucoside (glycosides, which are poorly soluble in hexane), while in SDLS 3,4-dihydroxy benzoic acid and gallic acids (0.51 and 0.29 g/100 g DW, respectively) were predominately identified.

The presence of polyphenols in the rose and lavender by-products could influence positively the mycelium growth because they might be used as a carbon source on account of the enzyme-oxidizing system of the basidiomycete isolate. The polyphenol content also suggests bacteriostatic and bactericidal effects of the substrates, which could support the initial sterilization procedure, as well as serve as a natural bio-protective barrier against bacterial contamination in the process of bio-composite production and application.

The process of essential oil production (either by extraction, or distillation) is never exhausting and part of the aroma compounds remain in the solid or liquid by-products. These aroma substances, mainly terpenes, often possess antimicrobial properties and could also serve as bio-preservative agents. Their presence in the plant matrix could contribute to the final pleasant aroma of the mycelium-based bio-composites. Phenethyl alcohol, β-citronellol, geraniol, β-cubebene, β-linalool and nerol were found in higher amounts in the HERF and β-linalool, dihydro-linalyl acetate, lavandulol, α and β-caryophyllene, lavandulyl acetate and α-terpineol were distinctive for the SDLS by-products (Table 3).

**Table 3.** Polar volatile metabolites in 70% ethanol extracts of hexane extracted rose flowers (HERF) and steam distilled lavender straw (SDLS), % of TIC (total ion current).

| Compound | RI | HERF | SDLS |
|---|---|---|---|
| α-Pinene | 940 | 0.87 ± 0.08 [a] | 0.25 ± 0.06 [b] |
| β-Pinene | 980 | 0.61 ± 0.05 [a] | 1.54 ± 0.08 [b] |
| β-Myrcene | 991 | 0.34 ± 0.07 [a] | 1.19 ± 0.09 [b] |
| p-Cymene | 1019 | - | 0.54 ± 0.07 |
| Limonene | 1025 | - | 3.55 ± 0.15 |
| Eucalyptol | 1031 | - | 3.18 ± 0.16 |
| cis-beta-Ocimene | 1040 | - | 5.41 ± 0.21 |
| trans-beta-Ocimene | 1050 | - | 3.37 ± 0.19 |
| γ-Terpinene | 1062 | 0.91 ± 0.08 [a] | 0.38 ± 0.06 [b] |
| cis-Linalool oxide | 1073 | - | 0.19 ± 0.05 |
| trans-Linalool oxide | 1078 | - | 0.29 ± 0.05 |
| Terpinene | 1087 | 2.00 ± 0.10 | - |
| β-Linalool | 1097 | 3.84 ± 0.10 [a] | 18.91 ± 0.15 [b] |
| Phenethyl alcohol | 1110 | 21.07 ± 0.17 | - |
| cis-Rose oxide | 1112 | 0.45 ± 0.05 | - |
| trans-Rose oxide | 1127 | 0.25 ± 0.04 | - |
| Camphor | 1146 | - | 0.48 ± 0.07 |
| Borneol | 1169 | - | 0.58 ± 0.10 |
| Lavandulol | 1171 | - | 6.12 ± 0.21 |
| Terpin-4-ol | 1178 | 0.81 ± 0.07 [a] | 3.10 ± 0.11 [b] |
| α-Terpineol | 1189 | 0.51 ± 0.05 [a] | 3.13 ± 0.09 [b] |
| β-Citronellol | 1228 | 11.26 ± 0.18 | - |
| Nerol | 1230 | 3.78 ± 0.15 | - |
| Geraniol | 1255 | 8.24 ± 0.21 [a] | 0.28 ± 0.10 [b] |
| Linalyl acetate, dihydro- | 1275 | - | 18.14 ± 0.16 |
| (±)-Lavandulyl acetate | 1290 | - | 4.93 ± 0.13 |
| Citronellyl acetate | 1354 | 0.19 ± 0.05 | - |
| Eugenol | 1356 | 2.84 ± 0.09 | - |
| Neryl acetate | 1364 | 2.09 ± 0.08 [a] | 0.95 ± 0.08 [b] |
| Geranyl acetate | 1383 | 0.63 ± 0.07 [a] | 2.94 ± 0.11 [b] |
| β-Bourbonene | 1384 | 3.16 ± 0.18 [a] | 0.20 ± 0.09 [b] |
| β-Cubebene | 1389 | 5.90 ± 0.15 | - |
| β-Elemene | 1390 | 0.51 ± 0.04 | - |
| Methyl eugenol | 1401 | 0.44 ± 0.05 | - |
| β-Caryophyllene | 1419 | 1.58 ± 0.12 [a] | 7.20 ± 0.18 [b] |
| α-Humulene (α-Caryophyllene) | 1454 | 0.34 ± 0.05 [a] | 5.06 ± 0.12 [b] |
| Germacrene D | 1479 | 0.39 ± 0.06 [a] | 2.76 ± 0.09 [b] |

**Table 3.** *Cont.*

| Compound | RI | HERF | SDLS |
|---|---|---|---|
| α-Farnesene | 1508 | 0.56 ± 0.05 [a] | 0.27 ± 0.04 [b] |
| β-Bisabolene | 1510 | 0.18 ± 0.04 [a] | 0.20 ± 0.03 [a] |
| trans-Nerolidol | 1564 | 2.60 ± 0.10 [a] | 0.27 ± 0.08 [b] |
| Spathulenol | 1575 | 1.63 ± 0.12 [a] | 0.19 ± 0.07 [b] |
| Caryophyllene oxide | 1580 | 0.32 ± 0.05 [a] | 0.30 ± 0.04 [a] |
| γ-Eudesmol | 1631 | 0.30 ± 0.04 [a] | 0.42 ± 0.05 [a] |
| β-Eudesmol | 1649 | 0.25 ± 0.05 [a] | 0.22 ± 0.03 [a] |
| α-Eudesmol | 1651 | 0.89 ± 0.03 [a] | 0.34 ± 0.05 [b] |
| Farnesol | 1714 | 0.34 ± 0.04 [a] | 0.55 ± 0.07 [a] |
| n-Nonadecane | 1901 | 4.29 ± 0.18 [a] | 0.17 ± 0.04 [b] |
| n-Eicosane | 2000 | 3.83 ± 0.15 [a] | 0.32 ± 0.08 [b] |
| n-Heneicosane | 2100 | 0.27 ± 0.06 [a] | 0.34 ± 0.08 [a] |
| n-Docosane | 2200 | 0.75 ± 0.08 | - |
| n-Tricosane | 2300 | 4.46 ± 0.21 | - |
| n-Tetracosane | 2400 | 1.60 ± 0.12 | - |
| n-Pentacosane | 2500 | 1.41 ± 0.12 | - |
| n-Hexacosane | 2600 | 1.41 ± 0.11 | - |

RI—retention index; Data are expressed as mean ± SD ($n$ = 3); [a,b]—different letters indicate significantly different values in the raws (Student's $T$ test; value of $p < 0.05$ indicated statistical significance).

The chemical characteristics of SDLS and HERF showed potential to stimulate mycelium growth of the basidiomycete isolate and to support self-growth of mycelium bio-materials with improved aroma and bactericidal properties.

The use of HERF and SDLS as a feedstock for new mycelium-based bio-composite production could be accepted as an innovation because at present there are no reports on the usage of these by-products for mycelium-based bio-composite formation. The chemical composition of the SDLS and HERF is a serious prerequisite for the development of mycelium-based bio-composites with improved antibacterial and aromatic properties.

### *3.2. Solid State Cultivation of Ganoderma Resinaceum GA1M on HERF and SDLS for Mycelium-Based Bio-Composite Formation*

The molecular identification of the new regionally isolated basidiomycete was performed by amplification of the ITS1-5.8S-ITS2 region and the obtained PCR product was subjected to sequence analysis. The resulting sequence was analyzed using BLAST algorithm and compared to the nucleotide sequences in the GenBank database (www.ncbi.nlm.nih.gov, accessed on 10 September 2021). The strain was identified with high percent confidence (99.67%) as *Ganoderma resinaceum*. The ITS1-5.8S-ITS2 rRNA gene sequence *of G. resinaceum* GA1M was deposited in the GenBank under accession number MW996753.

*G. resinaceum* belongs to the white-rot fungi (*Basidiomycota, Polyporales*) and is most commonly found as a saprophyte on wood trunks, but also as a parasite on dying trees, mainly in forests in the temperate zone. In Bulgaria, this fungus grows most often on deciduous trees and in rare cases on conifers.

Our interest in *G. resinaceum* for the present experimental research was based on preliminary results from the investigation of the growth rate of this strain's mycelium on different lignocellulose materials (the data are not published). It exhibited relatively fast growth, which was suitable for the purposes of this study.

In the development of new bio-composites, the whole process was based mainly on the solid-state cultivation of *G. resinaceum* on HERF and SDLS and comprised of several stages. The strain utilization of the studied substrates started with relatively rapid growth on both substrates resulting in the formation of a dense well-developed mycelium network and almost full colonization was observed. Mycelium bio-composites with a homogeneous structure and proper integrity were obtained after 20 days. The first sign of fungal growth was detected on the surface of HERF (Figure 3a) 48 h after inoculation while the visual growth on SDLS (Figure 3b) started after 72 h. Although *G. resinaceum* showed a slight delay in growth on lavender straw waste during the first 72 h, compared to rose flowers, fully homogenous "pre-growth" substrates were produced after 7 days of cultivation for both substrates.

(a) (b)

**Figure 3.** "Pre-growth" substrates—hexane extracted rose flowers (**a**) and steam distilled lavender straw (**b**). The cultivation of "pre-growth" substrates went on into special molds for another 7 days, where the fungal strain continued to increase its mycelial mass, to synthesize enzymes degrading the substrate polysaccharides and to produce metabolites, which additionally support the binding of the lignocellulose fibers of the substrate, envelope them and contribute to the final shape and structure of the biomaterial (Figure 2a).

The visual observation showed that the strain grew very well on the top surface of both studied substrates, which were in contact with the air, and formed velvety aerial mycelium, resembling skin. At the bottom and on the sides, which were in contact with the mold, the mycelium was less dense than that at the top surface (Figure 2b). When the cultivation period was completed, the substrate surfaces were coated with a white mycelial layer with small brownish spots, and had a relatively regular three-dimensional shape (Figure 4a,d). The brownish spots are commonly reported and accepted as a result of the de-polymerization of polysaccharides and the melanin production in the early stages of lignocellulose substrates degradation [4,27]. The moisture content of the substrate and the relative humidity of the air during the vegetative growth stage had a direct effect on the mycelium growth given that they influence both nutrient availability and fungal degradation. When the mycelium reached optimal growth and fully covered and bound the fibers of the substrates, the samples were dried at 60 °C for 8 h to cease fungal growth and remove the moisture. During this process, the mycelium adhered tightly to the fibers of the substrate and as a result, mycelium HERF/SDLS-based bio-composites with specific visual characteristics were obtained.

(a) (b) (c) (d) (e) (f)

**Figure 4.** Mycelium steam distilled lavender straw (SDLS)-based bio-composites grown for 20 days before (**a**) and after drying at 60 °C for 8 h (**b**), Scanning Electron Microscopy (SEM) image of mycelium surface layer (**c**); mycelium hexane extracted rose flowers (HERF)-based bio-composites grown for 20 days before (**d**) and after drying at 60 °C for 8 h (**e**), SEM image of mycelium surface layer (**f**).

As presented in Figure 4b,e, both materials had acquired a yellowish color of the mycelium layer after drying compared to the color before. According to Apples et al. [22], the change in color during the process of drying or thermal pressing is common and is probably due to Maillard reactions between the sugars and proteins present in the fungal cell walls and the plant materials or it is caused by low water content in the area.

Visually both mycelium-based materials appeared to have randomly oriented fibers tightly covered by fungal mycelium. The SDLS-based bio-composite showed bigger conformational stability and less deformation after drying compared to the HERF-based one. This could be explained by the type of used substrates and their cellulose and lignin content. These two polysaccharides are mainly responsible for the conformational stability of the mycelium-based bio-composites [39]. The better dimensional stability of the SDLS-based bio-composites was probably a result of the use of the whole lavender straw, while only the flower was used form of the roses. The significant rigidity and sturdiness of the lavender straw supported the conformational stability unlike the fragility of rose flowers. The dimensional stability of rose flower-based bio-composite could be improved by blending it with more rigid substrates. Both composites had surface texture, which could be described as satiny and velvety to the touch, but the HERF-based mycelium bio-composite surface was slightly wavier and rougher.

The SEM images of the surface of both mycelium bio-composites after drying are shown in Figure 4c,f. They depict a dense and adequately developed mycelium layer, forming a fibrous macro-porous structure. As can be seen in the images, the filaments appeared to flatten. The same findings were reported by Haneef et al. [6]. According to the authors, the thermal treatment for 7 h at 60 °C stopped the mycelium growth and the hyphae were not supported by internal hydrostatic pressure.

The visual observation of the cross-section of both mycelium composites showed that the fungal colonization was denser when closer to the air-contacted sides of the materials than in their center (Figure 5). The cross-section of the mycelium SDLS-based bio-composite exhibited more porosity (Figure 5a) and the hyphae had fully colonized the space between the lavender straw fibers at the outer part of the material, but they had not completely penetrated the center. Less clearly visible mycelium growth was observed during the visual inspection of the cross-section of the mycelium bio-composites made of HERF. The fungal hyphae and fibers looked glued together and the structure of the HERF-based bio-composite was denser and tougher (Figure 5b). As oxygen stimulates the mycelium growth, the difficulties in air penetration to the center of the material is one of the possible reasons for weak mycelium growth in the core, together with fiber structure and heat production by the mycelium during the degradation processes [9,19,22]. The SEM images confirmed the less porous internal part of the mycelium HERF-based bio-composite compared to the SDLS-based one (Figure 5c,d). According to Lelivelt et al. [42], if the thickness of the material is increased, there will be a point at which the center becomes too hot or too anaerobic to allow any growth at all. As proposed by other authors, a longer cultivation time could improve the colonization of the core of the composite [21,43].

**Figure 5.** Visual observation of cross-section of mycelium (**a**) steam distilled lavender straw (SDLS)- and (**b**) hexane extracted rose flowers (HERF)-based bio-composites; Scanning Electron Microscopy (SEM) images of cross-section of (**c**) SDLS- and (**d**) HERF-based bio-composites.

The fungal growth rate, the full coverage of the composite with aerial mycelium and the dense hypha network in the center of the materials influence their physical and mechanical characteristics. They, in turn, determine the application of the developed bio-materials.

### 3.3. FTIR Spectroscopy

FTIR spectroscopy was used to characterize the functional groups of both substrates (HERF and SDLS) and the obtained mycelium-based bio-composites. The FTIR spectra of the mycelium HERF/SDLS-based bio-composites were compared to the initial HERF and SDLS substrates (Figures 6 and 7). FTIR bands were assigned in detail taking note of literature data for other mycelium-based bio-composites produced on the base of different combinations between substrates and basidiomycetes (Table 4). In general, the infrared absorption spectra of the mycelium bio-composites were associated with the biomolecules that compose them, e.g., lipids (3000–2800 cm$^{-1}$, ~1737 cm$^{-1}$ ester bonds), proteins (amide I at 1700–1600 cm$^{-1}$, amide II and III at 1575–1300 cm$^{-1}$), nucleic acids (1255–1245 cm$^{-1}$), and polysaccharides (1200–900 cm$^{-1}$), which was in accordance with the bands previously reported in the literature [6,17,40].

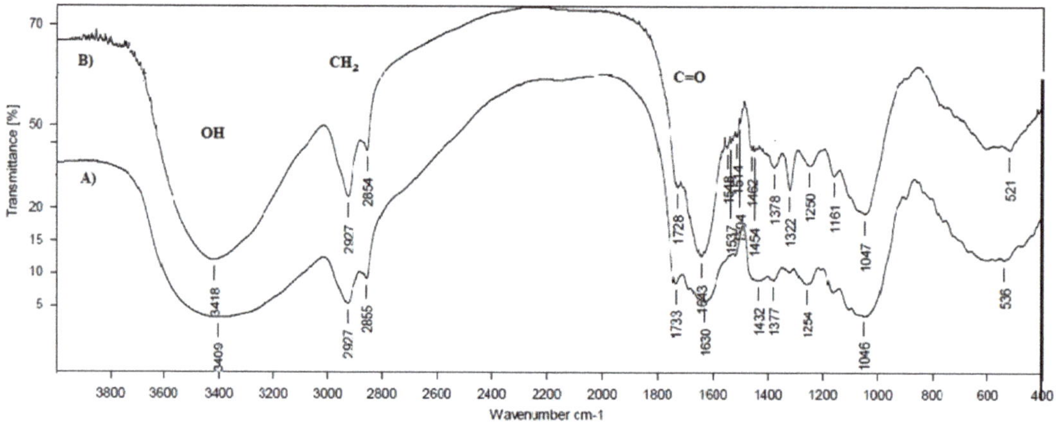


**Figure 6.** FTIR spectra of steam distilled lavender straw (SDLS) substrate (**A**) and mycelium SDLS-based bio-composite (**B**).

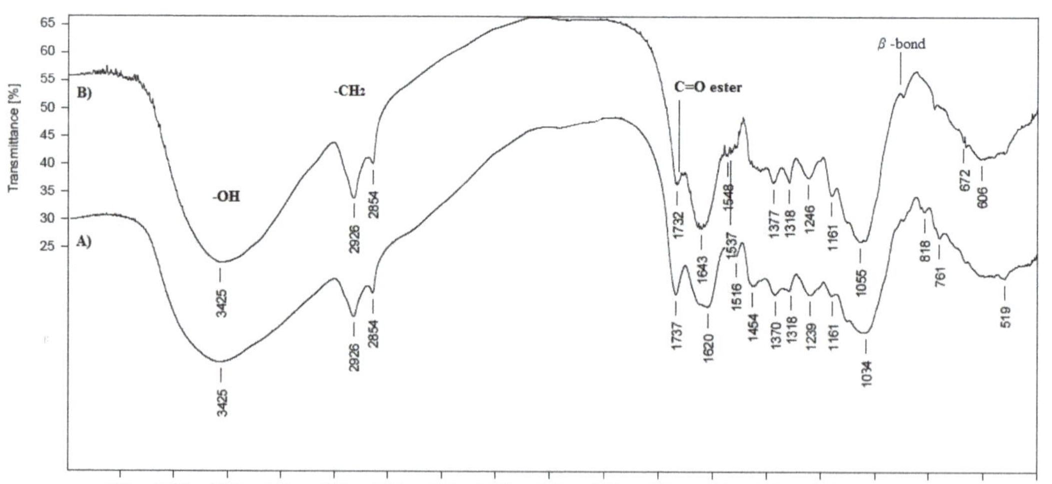


**Figure 7.** FTIR spectra of hexane extracted rose flowers (HERF) substrate (**A**) and mycelium HERF-based bio-composite (**B**).

**Table 4.** Bands assignment for the FTIR characterization of hexane extracted rose flowers (HERF), steam distilled lavender straw (SDLS) and mycelium HERF/SDLS-based bio-composite.

| Assignment | Mycelium Component (Main Contribution) | HERF | SDLS | Mycelium HERF-Based Bio-Composite | Mycelium SDLS-Based Bio-Composite |
|---|---|---|---|---|---|
| | | Vibration Frequency, $cm^{-1}$ | | | |
| O-H stretching vibration of intra and inter hydrogen bond | Polysaccharides | 3429 | 3409 | 3425 | 3418 |
| C-H stretches in methyl and methylene groups $CH_2$, asymmetric stretching | Lipids, Polysaccharides | 2966 | 2927 | 2926 | 2927 |
| $CH_2$ symmetric stretching | Lipids, Polysaccharides | 2854 | 2855 | 2854 | 2854 |
| C=O stretching in xylans (hemicellulose), pectins; ester bonds | Lipids, Polysaccharides | 1737 | 1733 | 1732 | 1728 |
| (amide I in β -sheets secondary structures) Absorbed O-H associated with lignin or cellulose; a characteristic band for cis- HRC = CR'H bonds, | Proteins; Lipids, Polysaccharides | 1652; 1620 | 1630 | 1665 1643 | 1643 |
| Amide II | Proteins | - | - | 1548 | 1548 |
| a characteristic band of the type C=C; C=C stretching of aromatic ring (syringyl) in lignin C=C stretcing of aromatic ring (guaiacyl) in lignin C=C stretching of aromatic ring (syringyl) in lignin | Polysaccharides | 1516 | 1516 | 1537;1548 | 1537;1548 |
| νC-Hs($CH_2$) in pyranise ring, | Polysaccharides | 1454 | 1432 | - | 1454;1462 |
| C-H bending in chitin, cellulose and hemicellulose | Polysaccharides | 1370 | 1377 | 1377 | 1378 |
| $CH_2$ wagging in cellulose | Polysaccharides | 1318 | 1318 | - | - |
| $PO_2$- asymmetric stretching, C-O stretching in lignin and xylan, Nucleic acids | Nucleic acids Polysaccharides | 1239 | 1246 | 1254 | 1250 |
| 1C-O-C vibration in cellulose and hemicellulose | Polysaccharides | 1150 | 1161 | 1152 | 1161 |
| | | - | 1114 | - | 1110 |
| C-O valence vibration from C3-O3H | Polysaccharides | 1057 | 1055 | 1046 | 1047 |
| C-O stretching in cellulose | Polysaccharides | 1053 | 1055 | 1055 | 1047 |

Table 4. Cont.

| Assignment | Mycelium Component (Main Contribution) | HERF | SDLS | Mycelium HERF-Based Bio-Composite | Mycelium SDLS-Based Bio-Composite |
|---|---|---|---|---|---|
| | | Vibration Frequency, $cm^{-1}$ | | | |
| C-C stretching | Polysaccharides | 1034 | 1055 | 1046 | 1047 |
| β(COH), β(CH) of C-1, νs(COC) in glycosidic linkage, ring modes | Polysaccharides | 928 | 916 | 916 | 919 |
| Anomer C-group, Glucan β-anomer C-H bending, C-H deformation in cellulose, ν(CC), β(CCH) | Polysaccharides | 899 | 878 | 855 | 842 |
| $CH_2$-rocking | Polysaccharides | 761 | - | 769 | - |

The ratio of the band intensity of the absorption associated with the C-H bending mode of chitin (~1374 $cm^{-1}$) to the one of the C-C stretching of polysaccharides (~1043 $cm^{-1}$) remained constant and unchanged. The IR bands in the region of 1000–1200 $cm^{-1}$ were related to stretching vibrations of C-O-C and C-O, respectively. The FTIR bands at 847 $cm^{-1}$ characterized the β-1,4 bonds vibrations. In our case, the presence of β-bonds in the mycelium HERF- and SDLS-based bio-composites was clearly observed (Table 4 and Figures 6 and 7). Bruscato et al. [17] also reported that the bands between 915 and 1110 $cm^{-1}$ were assigned to the D-Glc*p* unit and β-configuration of the sugar units in the polysaccharide at 890 $cm^{-1}$. In our study, the β-configuration of the sugar units in the polysaccharide matrix was preserved and these bonds did not disappear (Table 4).

The degradation of lavender straw and rose flower waste by *G. resinaceum* led to a small decrease in the intensities of carbohydrates at 1733 $cm^{-1}$, 1158 $cm^{-1}$, 897 $cm^{-1}$ and a similar observation was reported for the flax composites produced by *T. versicolor* [40].

Relative increases in protein and lipid bands were detected in both mycelium HERF- and SDLS-based bio-composites, compared to the initial substrates. A large band (18 $cm^{-1}$, from 1686 to 1668 $cm^{-1}$ to lower wavenumbers) was ascribed to the amide I of β-turns [6]. In addition, the bands at 1548 $cm^{-1}$, typical for amide II, were found only in mycelium grown on HERF and SDLS and were completely absent in the used waste substrates (Table 4). Bands due to the interaction of the polysaccharides with the mycelium at 1551 $cm^{-1}$ (lignin bands) and 1318 $cm^{-1}$ (cellulose bands) were not found in the mycelium HERF/SDLS-based bio-composites (Table 4). Contrary to the reports of Elsacker et al. [40], in our study, the decrease in carbohydrates (especially cellulose) was most pronounced at 1317 $cm^{-1}$ (strong vibration) as a result of *G. resinaceum* growth. However, the FTIR spectra also revealed a small band at 1377-1378 $cm^{-1}$ (weak vibrations), assigned to chitin. The ratio of bands intensities between the mycelium HERF/SDLS-based bio-composites and the HERF and SDLS substrates (1510 $cm^{-1}$/1510 $cm^{-1}$) was 3,6 for SDLS and 1.67 for HERF. This ratio revealed a higher intensity of lignin band for the SDLS-based bio-composite than the lavender straw. In comparison, Elsacker et al. [40] reported, that the depolymerisation of lignin by *T. versicolor* occurred to a greater extent in flax (0.97) than in hemp (0.57).

Haneef et al. [6] compared the FTIR spectra of the mycelium species, independent of the feeding substrates, and found that *G. lucidum* showed a higher contribution of lipids, whereas *Pleurotus ostreatus* showed relatively more intense bands that were due to polysaccharides. Similar to FTIR bands obtained from *Pycnoporus sanguineus*, *P. albidus* and *Lentinus velutinus* growth on wheat bran [17], the bands of the bio-composites from SDLS and HERF could not be observed, probably because they were consumed due to overlap with bands of cellulose, hemicellulose or lignin.

Comparing the data from the current research with reference literature data, it is clear that the chemical nature of the substrates is also responsible for distinct changes in the FTIR spectra of the mycelium-based bio-material. In conclusion, a decrease in cellulose bands and decomposition of lignin were observed and amide II bands appeared, which confirmed the use of HERF and SDLS as substrates for the *G. resinaceum* growth in both mycelium bio-composites.

### 3.4. Basic Physical and Mechanical Characterization of the Mycelium HERF/SDLS-Based Bio-Composites

Taking into consideration the physical structure and possible application of the mycelium-based bio-composites, standardized methods for thermal insulating products for building applications were used for the determination of some basic physical and mechanical properties. The results are summarized in Table 5. The apparent density of the materials is considered as one of the main parameters, allowing the prediction of other significant characteristics (porosity, water absorption, strength, thermal conductivity, etc.) and it is a good indicator of the mechanical properties [11,20,44]. For mycelium-based bio-composites, it has also been established that density is an important factor regarding the competitiveness of these materials and their applications [23]. Our analyses revealed that the mycelium HERF-based bio-composite had higher density (462 kg/m$^3$) as compared to the SDLS-based bio-composite (347 kg/m$^3$). These results reflect the porosity of the mycelium bio-composites, the differences in the chemical composition and particles sizes and distribution of the used waste rose flowers and lavender straw together with the mycelium growth.

**Table 5.** Physical and mechanical properties of the mycelium hexane extracted rose flowers (HERF)/ steam distilled lavender straw (SDLS)-based bio-composites.

| Mycelium Bio-Composite | Apparent Density $\rho_a$ | Capillary Absorption Wc | Water Absorption | | | | Compressive Resistance at 10% Deformation $\sigma_{10}$ |
|---|---|---|---|---|---|---|---|
| | | | 1 Day | | 28 Days | | |
| | | | $W^v_{a,1d}$ | $W^m_{a,1d}$ | $W^v_{a,28d}$ | $W^m_{a,28d}$ | |
| | kg/m$^3$ | kg/m$^2$ | %vol. | %wt. | %vol. | %wt. | kPa |
| HERF | 462 ± 10.2 | 3.4 ± 0.38 | 20.3 ± 3.45 | 43.9 ± 7.5 | 58.2 ± 7.2 | 126.0 ± 15.7 | 1029 ± 51 |
| SDLS | 347 ± 3.7 | 6.5 ± 0.20 | 39.8 ± 5.72 | 114.6 ± 14.5 | 85.0 ± 3.5 | 245.0 ± 10.1 | 718 ± 22 |

The standard deviation is determined with triplicate samples (mean ±standard deviation).

Based on literature data about the density of natural lignocellulose substrates varying from 1.2 to 1.5 g/cm$^3$ [44], porosity is considered to be a significant factor that could affect the densities of mycelium-based composites [23]. The cross-section of the SDLS-based mycelium composites (Figure 5a) showed the presence of more macro-pores allowing a bigger amount of hyphae inside the material. The structure of the HERF-based bio-composites was more compact and tougher, especially in the core of samples, thus leading to a higher apparent density of the mycelium HERF-based bio-composites. (Figure 5b). Our findings of the higher apparent density of the mycelium HERF-based bio-composites in comparison with the SDLS-based bio-composite is in contrast with a previous study [6] according to which higher amount of branched hyphae led to a more compact structure and higher density of the developed mycelium-based bio-composites. Our findings are in conformity with those of Tacer-Caba et al. [20], according to which fewer branched hyphae of *P. ostreatus* grown on oat husk and rapeseed cake resulted in a denser material. The degradation of lignocellulose substrates and their substitution with fungal biomass caused a decrease in the bio-composite density because the reported density of pure mycelium is quite low 0.03–0.05 g/cm$^3$ [13].

The density of the mycelium HERF/SDLS-based bio-composites was similar to non-pressed mycelium-based bio-composites obtained in different studies using various lignocellulose substrates, which vary from 59 to 590 kg/m$^3$ [7–9,17,22]. The mycelium HERF/SDLS-based bio-composites were denser than expanded polystyrene (EPS) (22–50 kg/m$^3$) [20,23] and lighter in comparison with wood-based composites such as medium-density fiber board (MDF) (500–1000 kg/m$^3$) and oriented strand board (OSB) (550–700 kg/m$^3$) [22]. The density of the HERF/SDLS-based mycelium bio-composites was also similar to hempcrete, which is between 400–500 kg/m$^3$ and 200–250 kg/m$^3$, obtained by on-site pouring and spray method, respectively [45].

Depending on the intended use of the mycelium HERF/SDLS bio-composites—for non-structural or semi-structural applications (i.e., thermal insulation or particleboards), the density might be further decreased by mixing the HERF and SDLS with other lignocellulose substrates with low density and an increase in the fungal biomass in the core of the bio-composites or it could be increased by cold or thermal pressing.

Water absorption is another very important parameter that influences quality, durability and the application range of a mycelium-based bio-composite [40]. The capillary absorption, short-term water absorption and long-term water absorption of the HERF/SDLS-based mycelium bio-composites were evaluated.

Water absorption is the easiest way to characterize the open porosity of materials, especially the absorption through capillarity. As can be seen in Table 5, the capillary absorption of the HERF-based mycelium bio-composites was about two times lower (3.4 kg/m$^2$) than the SDLS-based mycelium bio-composite (6.5 kg/m$^2$), which confirmed that the HERF-based mycelium bio-composites had a smaller amount of coarse pores than the SDLS-based ones. The obtained results were higher than those reported by Elsacker et al. [40] for a mycelium chopped hemp, flax- and straw-based bio-composite varying between 2 and 3.8 kg/m$^2$. The results on water absorption after one day of full immersion, which can be taken as representative for the finer porosity, confirmed the same—the water absorption of the mycelium HERF-based bio-composites was two times lower (ca. 20% vol.), as compared to that of the SDLS-based bio-composites (ca. 40% vol.). The long-term water absorption of the studied bio-composites reached 58%vol. (126% wt.) for the HERF-based bio-composite and 85%vol. (245% wt.) for the SDLS-based bio-composites. This correlates with reference literature data for bio-composites, where values up to 350% wt. were presented [4,24,46]. Apples et al. [22] also reported that the water absorption of different mycelium lignocellulose-based bio-composites ranged from 43% wt. to 508% wt. after only 8 days of immersion in water. The water absorption of the studied bio-composites exceeded considerably the water absorption of some hemp hurds composites, reported to be between 6.3% wt. and 25.8% wt. [47] and was much higher than the long-term water absorption of conventional thermal insulating material—between 0.2% vol. for extruded polystyrene (XPS) to 2.8% vol. for EPS, after 2 months of immersion [48]. Thus, it seems the present mycelium bio-composites would not be suitable for applications involving frost attack, because in that case the long-term water absorption shall be limited to 2%. However, the water absorption of the bio-composites could be significantly modified, because it is due not only to the open (to water) porosity, but also to the water suction by the used HERF and SDLS. Further testing shall clarify the impact of water absorption on the bio-composites swelling and the mycelium changes. The behavior of mycelium-based bio-composites during water immersion strongly depends on the hydrophobic properties of the fungal mycelium [20,23] and the hydrophilicity of lignocellulose fibers [46]. The hydrophobicity of basidiomycetes is mainly a result of the content of a low-weight protein, called hydrophobin. This protein was also reported to affect the process of adhesion of lignocellulose particles [6,20,23]. To limit the hydrophilicity of the lignocellulose fibers and reduce the water absorbance of the mycelium HERF/SDLS-based bio-composites, further investigations are needed in order to obtain well-developed mycelium, fully enveloped substrates particles together with an intact outer mycelium layer that covers the entire bio-composite. Another option would be to cover the bio composites with different bio-coatings (chitosan, carrageenan and xanthan), which was reported to considerably decrease the water absorbance [24].

In the preliminary testing of compressive behavior, no fracture was observed, because the bio-composites were quite deformable. Therefore, the behavior at compression is characterized by the compressive stress, corresponding to 10% of longitudinal strain, i.e., compressive stress at 10% relative deformation. In order to create smooth and parallel loading surfaces, a preliminary treatment by a thin layer of cement paste was applied and the testing was performed after 48 h. The compressive resistance of the lighter mycelium SDLS-based bio-composite was 718 kPa, while it was 1029 kPa for the relatively denser mycelium HERF-based bio-composite. Both values are quite higher than the range of compressive strength of different mycelium-based bio-composites, reported by Amstivslaski et al. [49], which was between 29 and 567 kPa. It could be concluded that our results also surpassed significantly the typical values for some conventional thermal insulating materials, such as several standard categories of EPS, which do not exceed 120 kPa for short-term loading [50]. However, it shall be mentioned that there is a significant impact of the density on the compressive behavior of thermal insulating materials and thus various data can be found. Because of this Nava et al. [24] stated, that the compressive strength of mycelium-based bio-composites was lower than most of the EPS categories and our results for compressive resistance were closer to the compressive strength of hempcrete, which has a similar apparent density [45].

The experimental data on compressive behavior of the mycelium HERF/SDLS-based bio-composite are probably due to the successful combination between the used fibrous substrates and the basidiomycete strain *G. resinaceum*. It was observed that the mycelium of *Ganoderma* spp. demonstrated quite robust consistency as compared to other basidiomycetes [24].

Due to the smaller size of the samples than the required by EN 12667:2001, the coefficient of thermal conductivity (lambda) was not measured. A rough estimation on the lambda range was carried out based on the apparent density of the studied bio-composites, the available literature data on mycelium-based bio-composites and data on other similar thermal insulating products.

It was difficult to assess lambda only based on apparent density, because the densities were closer to that of inorganic thermal insulation materials (such as cellular concrete, porous ceramic, etc.), while the bio-composites are an organic material, with quite a big volume of open porosity (such as EPS). The thermal insulation with the closest profile is hempcrete, but its binder is inorganic (lime, cement, magnesia binder, etc.). Therefore, the developed mycelial bio-composites are quite specific, combining a fibrous component (waste lavender straw and rose flower biomass) and an organic binder (mycelium). Considering only the density values, the maximum thermal conductivity coefficient, based on cellular concrete and hempcrete data, can be expected to be 0.12–0.17 W/m.K for the HERF-based composites and 0.09–0.11 W/m.K for the SDLS bio-composites [24]. The study of Jones et al. [11] showed that when the apparent density of mycelium composites varied between 59 and 552 kg/m$^3$, the values of lambda varied between 0.04 and 0.18 W/m.K. If thermal insulation application is targeted, the density of the bio-composites should be significantly reduced and lambda might be quite low as well. For density in the range of 95–135 kg/m$^3$ the reported data for lambda was between 0.04 and 0.06 W/m.K, i.e., very similar to that of conventional thermal insulating materials [40].

## 4. Conclusions

This is the first study reporting that waste hexane extracted rose flowers and steam distilled lavender straws showed realistic potential to be utilized as a feedstock for new mycelium-based bio-composites development by a new locally isolated and molecularly identified *Ganoderma resinaceum* GA1M. The visual observation and SEM images of the surface and cross-section of the mycelium HERF/SDLS-based bio-composites both showed adequately developed aerial mycelium layer, fibrous internal macroporous structure and proper integrity. Basic physical and mechanical characterization of the obtained mycelium HERF/SDLS-based bio-composites revealed promising potential and after additional optimization they could be suggested for non-structural or semi-structural applications and interior design.

**Author Contributions:** Conceptualization, G.A., A.E. and A.K.; methodology, G.A. and A.E.; software, M.B.; validation, G.A., M.B. and P.S.; formal analysis, V.V., N.P., A.S., P.D., D.K. and R.Z.; investigation, M.B., P.S., V.V., N.P., A.S., P.D., D.K. and R.Z.; resources, G.A.; writing—original draft preparation, G.A.; writing—review and editing, D.B. and M.B.; visualization, M.B.; supervision, A.K.; project administration, G.A.; funding acquisition, A.K. All authors have read and agreed to the published version of the manuscript.

**Funding:** This research was funded by THE NATIONAL SCIENCE FUND OF BULGARIA under contract No. KП-06-H37/4 from 06.12.2019, "Novel mycelium based bio-composites—a new alternative for environmental sustainability".

**Institutional Review Board Statement:** Not applicable.

**Informed Consent Statement:** Not applicable.

**Data Availability Statement:** The raw/processed data required to reproduce these findings cannot be shared at this time as the data also forms part of an ongoing study.

**Acknowledgments:** The authors wish to thank Byomic Ltd. for the generous support and valuable advice throughout the experimental process.

**Conflicts of Interest:** The authors declare no conflict of interest.

## References

1. Slavov, A.; Vasileva, I.; Stefanov, L.; Stoyanova, A. Valorization of wastes from the rose oil industry. *Rev. Environ. Sci. Bio/Technology* **2017**, *16*, 309–325. [CrossRef]
2. Lesage-Meessen, L.; Bou, M.; Sigoillot, J.-C.; Faulds, C.B.; Lomascolo, A. Essential oils and distilled straws of lavender and lavandin: A review of current use and potential application in white biotechnology. *Appl. Microbiol. Biotechnol.* **2015**, *99*, 3375–3385. [CrossRef]
3. Slavov, A.; Denev, P.; Panchev, I.; Shikov, V.; Nenov, N.; Yantcheva, N.; Vasileva, I. Combined recovery of polysaccharides and polyphenols from Rosa damascena wastes. *Ind. Crop. Prod.* **2017**, *100*, 85–94. [CrossRef]
4. Sun, W.; Tajvidi, M.; Hunt, C.; McIntyre, G.; Gardner, D.J. Fully Bio-Based Hybrid Composites Made of Wood, Fungal Mycelium and Cellulose Nanofibrils. *Sci. Rep.* **2019**, *9*, 3766. [CrossRef]
5. Zolotovsky, K. Guided Growth: Design and Computation of Biologically Active Materials. Ph.D. Thesis, Massachusetts Institute of Technology, Cambridge, MA, USA, 2017.
6. Haneef, M.; Ceseracciu, L.; Canale, C.; Bayer, I.S.; Heredia-Guerrero, J.A.; Athanassiou, A. Advanced Materials from Fungal Mycelium: Fabrication and Tuning of Physical Properties. *Sci. Rep.* **2017**, *7*, srep41292. [CrossRef]
7. Jones, M.; Huynh, T.; Dekiwadia, C.; Daver, F.; John, S. Mycelium Composites: A Review of Engineering Characteristics and Growth Kinetics. *J. Bionanoscience* **2017**, *11*, 241–257. [CrossRef]
8. Jones, M.; Bhat, T.; Huynh, T.; Kandare, E.; Yuen, K.K.R.; Wang, C.-H.; John, S. Waste-derived low-cost mycelium composite construction materials with improved fire safety. *Fire Mater.* **2018**, *42*, 816–825. [CrossRef]
9. Jones, M.; Huynh, T.; John, S. Inherent species characteristic influence and growth performance assessment for mycelium composite applications. *Adv. Mater. Lett.* **2018**, *9*, 71–80. [CrossRef]
10. Jones, M.P.; Lawrie, A.C.; Huynh, T.T.; Morrison, P.D.; Mautner, A.; Bismarck, A.; John, S. Agricultural by-product suitability for the production of chitinous composites and nanofibers utilising *Trametes versicolor* and *Polyporus brumalis* mycelial growth. *Process. Biochem.* **2019**, *80*, 95–102. [CrossRef]
11. Jones, M.; Mautner, A.; Luenco, S.; Bismarck, A.; John, S. Engineered mycelium composite construction materials from fungal biorefineries: A critical review. *Mater. Des.* **2020**, *187*, 108397. [CrossRef]
12. Manan, S.; Ullah, M.W.; Ul-Islam, M.; Atta, O.M.; Yang, G. Synthesis and applications of fungal mycelium-based advanced functional materials. *J. Bioresour. Bioprod.* **2021**, *6*, 1–10. [CrossRef]
13. Islam, M.R.; Tudryn, G.; Bucinell, R.; Schadler, L.; Picu, R.C. Morphology and mechanics of fungal mycelium. *Sci. Rep.* **2017**, *7*, 13070. [CrossRef]
14. Grimm, D.; Wösten, H.A.B. Mushroom cultivation in the circular economy. *Appl. Microbiol. Biotechnol.* **2018**, *102*, 7795–7803. [CrossRef]
15. Perez, R.; Luccioni, M.; Kamakaka, R.; Clamons, S.; Gaut, N.; Stirling, F.; Adamala, K.P.; Silver, P.A.; Endy, D. Enabling community-based metrology for wood-degrading fungi. *Fungal Biol. Biotechnol.* **2020**, *7*, 1–15. [CrossRef]
16. Higgins, C.; Margot, H.; Warnquist, S.; Obeysekare, E.; Mehta, K. Mushroom cultivation in the developing world: A comparison of cultivation technologies. In Proceedings of the 2017 IEEE Global Humanitarian Technology Conference (GHTC), IEEE, San Jose, CA, USA, 19–22 October 2017; pp. 1–7.
17. Bruscato, C.; Malvessi, E.; Brandalise, R.N.; Camassola, M. High performance of macrofungi in the production of mycelium-based biofoams using sawdust—Sustainable technology for waste reduction. *J. Clean. Prod.* **2019**, *234*, 225–232. [CrossRef]

18. Liu, R.; Long, L.; Sheng, Y.; Xu, J.; Qiu, H.; Li, X.; Wang, Y.; Wu, H. Preparation of a kind of novel sustainable mycelium/cotton stalk composites and effects of pressing temperature on the properties. *Ind. Crop. Prod.* **2019**, *141*, 111732. [CrossRef]
19. Alves, R.M.E.; Alves, M.L.; Campos, M.J. Morphology and Thermal Behaviour of New Mycelium-Based Composites with Different Types of Substrates. In Proceedings of the 2nd International Conference on Experimental and Computational Mechanics in Engineering, Banda Aceh, India, 13–14 October 2020; Gabler: Wiesbaden, Germany; pp. 189–197.
20. Tacer-Caba, Z.; Varis, J.J.; Lankinen, P.; Mikkonen, K.S. Comparison of novel fungal mycelia strains and sustainable growth substrates to produce humidity-resistant biocomposites. *Mater. Des.* **2020**, *192*, 108728. [CrossRef]
21. Santos, I.S.; Nascimento, B.L.; Marino, R.H.; Sussuchi, E.M.; Matos, M.P.; Griza, S. Influence of drying heat treatments on the mechanical behavior and physico-chemical properties of mycelial biocomposite. *Compos. Part B Eng.* **2021**, *217*, 108870. [CrossRef]
22. Appels, F.V.; Camere, S.; Montalti, M.; Karana, E.; Jansen, K.; Dijksterhuis, J.; Krijgsheld, P.; Wösten, H.A. Fabrication factors influencing mechanical, moisture- and water-related properties of mycelium-based composites. *Mater. Des.* **2019**, *161*, 64–71. [CrossRef]
23. Girometta, C.; Picco, A.M.; Baiguera, R.M.; Dondi, D.; Babbini, S.; Cartabia, M.; Pellegrini, M.; Savino, E. Physico-Mechanical and Thermodynamic Properties of Mycelium-Based Biocomposites: A Review. *Sustainability* **2019**, *11*, 281. [CrossRef]
24. Nava, J.A.L.; González, J.M.; Chacón, X.R.; Luna, J.A.N. Assessment of Edible Fungi and Films Bio-Based Material Simulating Expanded Polystyrene. *Mater. Manuf. Process.* **2015**, *31*, 1085–1090. [CrossRef]
25. Jiang, L.; Walczyk, D.; McIntyre, G.; Bucinell, R.; Tudryn, G. Manufacturing of biocomposite sandwich structures using mycelium-bound cores and preforms. *J. Manuf. Process.* **2017**, *28*, 50–59. [CrossRef]
26. Läkk, H.; Krijgsheld, P.; Montalti, M.; Wösten, H. Fungal based biocomposite for habitat structures on the Moon and Mars. In Proceedings of the 69th International Astronautical Congress (IAC), Bremen, Germany, 1–5 October 2018.
27. Ridzqo, I.F.; Susanto, D.; Panjaitan, T.H.; Putra, N. Sustainable Material: Development Experiment of Bamboo Composite through Biologically Binding Mechanism. *IOP Conf. Series: Mater. Sci. Eng.* **2020**, *713*, 012010. [CrossRef]
28. Attias, N.; Danai, O.; Abitbol, T.; Tarazi, E.; Ezov, N.; Pereman, I.; Grobman, Y.J. Mycelium bio-composites in industrial design and architecture: Comparative review and experimental analysis. *J. Clean. Prod.* **2020**, *246*, 119037. [CrossRef]
29. Lesage-Meessen, L.; Bou, M.; Ginies, C.; Chevret, D.; Navarro, D.; Drula, E.; Bonnin, E.; Del Río, J.C.; Odinot, E.; Bisotto, A.; et al. Lavender- and lavandin-distilled straws: An untapped feedstock with great potential for the production of high-added value compounds and fungal enzymes. *Biotechnol. Biofuels* **2018**, *11*, 217. [CrossRef] [PubMed]
30. Ratiarisoa, R.V.; Magniont, C.; Ginestet, S.; Oms, C.; Escadeillas, G. Assessment of distilled lavender stalks as bioaggregate for building materials: Hygrothermal properties, mechanical performance and chemical interactions with mineral pozzolanic binder. *Constr. Build. Mater.* **2016**, *124*, 801–815. [CrossRef]
31. Pla, M.F.D.E.; González, P.; Sette, P.; Portillo, F.; Rojas, A.M.; Gerschenson, L.N. Effect of processing on physico-chemical characteristics of dietary fibre concentrates obtained from peach *(Prunus persica* L.) peel and pulp. *Food Res. Int.* **2012**, *49*, 184–192. [CrossRef]
32. Singleton, V.L.; Rossi, J.A. Colorimetry of total phenolics with phosphomolybdic-phosphotungstic acid reagents. *Am. J. Enol. Vitic.* **1965**, *16*, 144–158.
33. Stefanova, P.; Taseva, M.; Georgieva, T.; Gotcheva, V.; Angelov, A. A Modified CTAB Method for DNA Extraction from Soybean and Meat Products. *Biotechnol. Biotechnol. Equip.* **2013**, *27*, 3803–3810. [CrossRef]
34. Toju, H.; Tanabe, A.; Yamamoto, S.; Sato, H. High-Coverage ITS Primers for the DNA-Based Identification of *Ascomycetes* and *Basidiomycetes* in Environmental Samples. *PLoS ONE* **2012**, *7*, e40863. [CrossRef]
35. Altschul, S.F.; Gish, W.; Miller, W.; Myers, E.W.; Lipman, D.J. Basic local alignment search tool. *J. Mol. Biol.* **1990**, *215*, 403–410. [CrossRef]
36. Slavov, A.; Panchev, I.; Kovacheva, D.; Vasileva, I. Physico-chemical characterization of water-soluble pectic extracts from *Rosa damascena*, *Calendula officinalis* and *Matricaria chamomilla* wastes. *Food Hydrocoll.* **2016**, *61*, 469–476. [CrossRef]
37. Calvo-Flores, F.G.; Dobado, J.A.; Isac-García, J.; Martín-Martínez, F.J. *Lignin and Lignans as Renewable Raw Materials*; Wiley: Hoboken, NJ, USA, 2015.
38. Singh, T.; Singh, A.P. White and Brown Rot Fungi as Decomposers of Lignocellulosic Materials and Their Role in Waste and Pollution Control. *Advances in Endophytic Fungal Research* **2016**, 233–247. [CrossRef]
39. Yang, Z.; Zhang, F.; Still, B.; White, M.; Amstislavski, P. Physical and Mechanical Properties of Fungal Mycelium-Based Biofoam. *J. Mater. Civ. Eng.* **2017**, *29*, 04017030. [CrossRef]
40. Elsacker, E.; Vandelook, S.; Brancart, J.; Peeters, E.; De Laet, L. Mechanical, physical and chemical characterisation of myceli-um-based composites with different types of lignocellulosic substrates. *PLoS ONE* **2019**, *14*, e0213954. [CrossRef]
41. Slavov, A.; Vasileva, I.; Denev, P.; Dinkova, R.; Teneva, D.; Ognyanov, M.; Georgiev, Y. Polyphenol-rich extracts from essential oil industry wastes. *Bulg. Chem. Commun.* **2021**, *52*, 78–83.
42. Lelivelt, R.J.J.; Lindner, G.; Teuffel, P.; Lamers, H. The production process and compressive strength of Mycelium-based materials. In Proceedings of the First International Conference on Bio-based Building Materials, Clermont-Ferrand, France, 22–25 June 2015; pp. 1–6.
43. Feng, N.-J.; Zhai, H.; Lai, Y.-Z. On the chemical aspects of the biodelignification of wheat straw with Pycnoporus sanguineus and its combined effects with the presence of Candida tropicalis. *Ind. Crop. Prod.* **2016**, *91*, 315–322. [CrossRef]

44. Pelletier, M.; Holt, G.; Wanjura, J.; Lara, A.; Tapia-Carillo, A.; McIntyre, G.; Bayer, E. An evaluation study of pressure-compressed acoustic absorbers grown on agricultural by-products. *Ind. Crop. Prod.* **2017**, *95*, 342–347. [CrossRef]
45. Demir, I.; Doğan, C. Physical and mechanical properties of hempcrete. *Open Waste Manag. J.* **2020**, *13*, 26–34. [CrossRef]
46. Ziegler, A.R.; Bajwa, S.G.; Holt, G.A.; McIntyre, G. Evaluation of physico-mechanical properties of mycelium reinforced green biocomposites made from cellulosic fibers. *Appl. Eng. Agric.* **2016**, *32*, 931–938. [CrossRef]
47. Stevulova, N.; Cigasova, J.; Purcz, P.; Schwarzova, I.; Kacik, F.; Geffert, A. Water absorption behavior of hemp hurds composites. *Mater.* **2015**, *8*, 2243–2257. [CrossRef]
48. Pakkala, T.A.; Lahdensivu, J. Long-term water absorption tests for frost insulation materials taking into account frost attack. *Case Stud. Constr. Mater.* **2014**, *1*, 40–45. [CrossRef]
49. Amstislavski, P.; Yang, Z.; White, M.D. United States Patent Application Pubblication. WO 2017/132523 Al, 2017.
50. Available online: https://www.specifiedby.com/styrene-packaging-insulation-ltd/stylite-eps-geofill-void-formers/geofill-compressive-strength-factsheet_4eeaa0ea.pdf (accessed on 11 May 2021).

Journal of *Fungi*

MDPI

*Article*

# The Bicarbonate Transporter (MoAE4) Localized on Both Cytomembrane and Tonoplast Promotes Pathogenesis in *Magnaporthe oryzae*

**Yuejia Dang [1,2,3], Yi Wei [1,2,3], Penghui Zhang [3], Xinchun Liu [3], Xinrui Li [3], Shaowei Wang [3], Hao Liang [3] and Shi-Hong Zhang [1,2,3,*]**

1 College of Plant Protection, Shenyang Agricultural University, Shenyang 110866, China; dangyj@syau.edu.cn (Y.D.); wyziyu@syau.edu.cn (Y.W.)
2 Center for Extreme-Environmental Microorganisms, Shenyang Agricultural University, Shenyang 110866, China
3 College of Plant Sciences, Jilin University, Changchun 130062, China; zhangph18@mails.jlu.edu.cn (P.Z.); lxc18@mails.jlu.edu.cn (X.L.); lixr19@mails.jlu.edu.cn (X.L.); wangsw15@mails.jlu.edu.cn (S.W.); lianghao8218@mails.jlu.edu.cn (H.L.)
* Correspondence: zhangsh89@syau.edu.cn

**Citation:** Dang, Y.; Wei, Y.; Zhang, P.; Liu, X.; Li, X.; Wang, S.; Liang, H.; Zhang, S.-H. The Bicarbonate Transporter (MoAE4) Localized on Both Cytomembrane and Tonoplast Promotes Pathogenesis in *Magnaporthe oryzae*. *J. Fungi* **2021**, *7*, 955. https://doi.org/10.3390/jof7110955

Academic Editor: Baojun Xu

Received: 21 October 2021
Accepted: 8 November 2021
Published: 11 November 2021

**Publisher's Note:** MDPI stays neutral with regard to jurisdictional claims in published maps and institutional affiliations.



**Abstract:** Bicarbonate ($HCO_3^-$) transporter family including the anion exchanger (AE) group is involved in multiple physiological processes through regulating acid-base homeostasis. $HCO_3^-$ transporters have been extensively studied in mammals, but fungal homologues of AE are poorly understood. Here, we characterized the AE group member (MoAE4) in *Magnaporthe oryzae*. MoAE4 exhibits more sequence and structure homologies with the reported AE4 and BOR1 proteins. In addition to the common sublocalization on cytomembrane, MoAE4 also localizes on tonoplast. Yeast complementation verified that MoAE4 rescues boron sensitivity and endows $NaHCO_3$ tolerance in the *BOR1* deleted yeast. *MoAE4* gene is bicarbonate induced in *M. oryzae*; and loss of *MoAE4* (Δ*MoAE4*) resulted in mycelial growth inhibited by $NaHCO_3$. Lucigenin fluorescence quenching assay confirmed that Δ*MoAE4* accumulated less $HCO_3^-$ in vacuole and more $HCO_3^-$ in cytosol, revealing a real role of MoAE4 in bicarbonate transport. Δ*MoAE4* was defective in conidiation, appressorium formation, and pathogenicity. More $H_2O_2$ was detected to be accumulated in Δ*MoAE4* mycelia and infected rice cells. Summarily, our data delineate a cytomembrane and tonoplast located $HCO_3^-$ transporter, which is required for development and pathogenicity in *M. oryzae*, and revealing a potential drug target for blast disease control.

**Keywords:** anion exchange protein 4 (AE4); $HCO_3^-$ transporter; tonoplast; pathogenicity; *Magnaporthe oryzae*

## 1. Introduction

The bicarbonate anion ($HCO_3^-$)-transporter family, also known as the SLC4 (solute carrier 4) transporter family, functions to transport $HCO_3^-$ across the plasma membrane and in the maintenance of intracellular pH value. $HCO_3^-$ transporter proteins have been extensively studied in mammalians and invertebrates. In mammals there are 14 genes which encode proteins with bicarbonate transport activity [1]. According to the physiological activity, bicarbonate transporters can be classed into three major groups: $Cl^-/HCO_3^-$ exchangers (AEs), $Na^+/HCO_3^-$ cotransporters (NBCs), and $Na^+$ dependent $Cl^-/HCO_3^-$ exchangers (NDCBEs). The $Cl^-/HCO_3^-$ exchangers AE1-3 are about 53–56% identical to one another at the amino-acid level. The electrogenic $Na^+/HCO_3^-$ cotransporters NBCe1 and NBCe2 are about 28–34% identical to the AEs. The electroneutral $Na^+/HCO_3^-$ transporters NBCn1, NDCBE, and NBCn2 are about 30–34% identical to the AEs, and about 39–50% identical to the electrogenic NBCs. Therefore, the deduced amino acid sequences of $HCO_3^-$ transporter proteins show a high degree of similarity to anion exchangers [2].

$HCO_3^-$ transporters share many common features in membrane topology, glycosylation, and inhibition by stilbene disulfonate inhibitors; but they are different in some ways such as the nature of transport activity and the subsidiary ions carried [1,2]. All function by an electroneutral mechanism, exchanging $Cl^-$ for $HCO_3^-$ across the plasma membrane, driven by the respective gradients of the transport substrates [1].

The three identified anion exchangers (AE1-3) mediate the electroneutral exchange of one monovalent anion for another across the plasma membrane. The well characterized anion exchanger 1 (AE1), the first bicarbonate transporter cloned and sequenced $Cl^-/HCO_3^-$ exchanger, is the erythrocyte band 3 glycoprotein that contains a membrane domain responsible for transport function. The crystal structure of the AE1 anion exchanger domain reveals a transport mechanism with which to understand the many mutations in the protein that lead to diseases [3,4].

The AE4 (SLC4A9) was originally reported to facilitate $Cl^-/HCO_3^-$ exchange [5,6]. Based on the strong phylogenetic clustering of AE4 with reported $Na^+/HCO_3^-$ co-transporters, AE4 was regarded as a $Na^+/HCO_3^-$ co-transporter, not a $Cl^-/HCO_3^-$ exchanger [2]. Native AE4 activity in mouse salivary gland acinar cells supports $Na^+$-dependent $Cl^-/HCO_3^-$ exchange that is comparable with that obtained upon heterologous expression of mouse AE4 and human AE4 in CHO-K1 cells. Particularly, AE4 mediates $Cl^-/HCO_3^-$ exchange activity in the presence of $K^+$ as well as $Cs^+$, $Li^+$, and $Rb^+$ [7–9].

The AE4 gene is conserved in a variety of species. Homologous sequences and crystal structures of $HCO_3^-$ transporters have been identified not only in mammals, but also in fungi and plants [10]. The budding yeast *Saccharomyces cerevisiae* genome harbors the YNL275w gene (ScAE4), which showed some sequence identity to band 3 [11]. The YNL275w is transcribed at an extremely low level and not induced in response to nitrogen starvation. In addition, the YNL275w disruption mutant did not show any phenotype alteration under normal growth conditions. However, YNL275w homologue BOR1 is involved in tolerance to boric acid and the maintenance of the protoplasmic boron concentration [12,13], and BOR1 regulates a saturable uphill boron efflux, with characteristics consistent with a bicarbonate-independent exchange of extracellular $H^+$ for intracellular $H_3BO_3$ [14]. Similarly, *Arabidopsis* YNL275w (*BOR1*) also supports plant boron tolerance [15]. Recently, several SLC family members such as SbtA (AN4904), SbtB (AN0218), and SB (AN2730) were characterized in *Aspergillus nidulans* [16]. SbtB functions as a BOR1 homologue; but SB appears to be a $HCO_3^-$ transporter.

The rice blast fungus *M. oryzae* is the causal agent of blast disease worldwide. Host infection is initiated by developed conidia, which occurs outside plant cells and involves conidium germination, tube elongation, appressorium maturation, and differentiation [17,18]. After penetration, successful development of invasive hyphae determines the severity of blast [19,20]. However, during invasive hyphae growth and development in the host plant, *M. oryzae* undergoes various harsher obstacles involving not only plant-derived passive and active resistance such as the accumulation of reactive oxygen species (ROS), antimicrobial compounds, and pathogenesis-related proteins for instance [21–23], but also *in planta* nitrogen starvation, high- $HCO_3^-$, and low-oxygen stresses [24–26]. To colonize the host successfully, *M. oryzae* must ensure a basic standard to survive these adverse environments. In this research, we biologically analyzed the *M. oryzae* AE4 homologue gene through creating deletion mutant and complementary strains. Biologic and molecular data reveal that MoAE4 is a cytomembrane and tonoplast localized $HCO_3^-$ transporter. Importantly, MoAE4 required for pathogenicity provides a new target for blast disease control.

## 2. Materials and Methods

### 2.1. Sequence Alignment Assays

The MoAE4 (MGG_15203) gene and amino acid sequences were acquired from the NCBI database (https://www.ncbi.nlm.nih.gov/, accessed on 25 March 2021). The protein tertiary and transmembrane structures were predicted using I-TASSER (https://zhanglab.ccmb.med.umich.edu/I-TASSER/, accessed on 25 March 2021), TMHMM Server v. 2.0

(http://www.cbs.dtu.dk/services/TMHMM/, accessed on 16 May 2021), and Softberry (http://www.softberry.com/berry.phtml, accessed on 16 May 2021). In addition, the amino acid sequence was aligned using the DNAMAN program, and the phylogenic tree was drawn using MEGA7.0.9 software.

### 2.2. Fungal Strains and Culture Conditions

*M. oryzae* strain JJ88 was used as the wild type. It was isolated and purified from *Oryza sativa* cultivar Jijing88, a variety that is widely planted in Jilin Province, China. All the fungal strains were cultured on complete media (CM) agar plates and maintained on paper filters at −20 °C (CM [10 g/L glucose, 2 g/L peptone, 1 g/L yeast extract, 1 g/L casamino acids, 0.1% (*V*/*V*) trace elements, 0.1% (*V*/*V*) vitamin supplement, 0.5 g/L $MgSO_4$, 6 g/L $NaNO_3$, 0.5 g/L KCl, and 1.5 g/L $KH_2PO_4$, pH 6.5]). For conidiation, the strains were inoculated on oatmeal–tomato agar medium (OMA) at 24 °C for 7 days in the dark [27]. The strains were grown continually for 3 days while illuminated under fluorescent lights after the aerial hyphae of the strains had been removed by washes with sterile distilled water.

*S. cerevisiae* BY4741 and the *ScBor1* deletion mutant strains (Invitrogen, Beijing, China) were used for functional complementation test. The yeast *S. cerevisiae* transformation was performed by the lithium acetate procedure. For yeast gene expression, YPB-ADHpt promoter and terminator regions of ADH1 gene in YPB1 was used [28]. All yeast strains were cultured according to Li et al. [29]. Δ*ScBor1* of *S. cerevisiae* was transferred in MoAE4 and MoACT, respectively. The mutants of Δ*ScBor1*, Δ*ScBor1*/*MoAE4*, Δ*ScBor1*/*MoACT*, and wild type were inoculated on to the Solid YPD medium Plates with 100 mM $H_3BO_3$ and 50 mM $NaHCO_3$, respectively.

### 2.3. Assays for the Subcellular Localization of MoAE4

The localization of *MoAE4* in the wild type strain was observed by tagging it with the *Bgl* II-*Spe*I sites of green fluorescent protein (GFP) of vector pCAMBIA1303 at its C-terminus. We generated transgenic strains expressing GFP-tagged *MoAE4* fusion gene in the wild type of *M. oryzae* (pCAMBIA1303-MoAE4:: GFP). Fluorescent microscopic observation was carried out by using hyphae (6d) and conidia (6d). To visualize the cytoplasmic membrane and vacuolar membrane, vegetative hyphae and conidia were treated with 2 µg/mL FM4-64 (AAT Bioquest, Sunnyvale, CA, USA) solution for 30–60 min before observed [30] under laser scanning confocal microscope (Olympus fluoview FV3000, Olympus, Tokyo, Japan).

### 2.4. Targeted Gene Deletion and Complementation

To generate the *MoAE4* replacement construct pXEH20, the upstream (1155 bp) and downstream (1195 bp) fragments of *MoAE4* were amplified using primers MoAE4-L-S/MoAE4-L-A and MoAE4-R-S/MoAE4-R-A, respectively. The resulting PCR products were cloned into the *Spe*I-*Kpn*I and *Xba*I-*Hind*III sites of vector pXEH2.0. The knockout vector was introduced into *Agrobacterium tumefaciens* strain AGL-1 and then transformed into the wild type *M. oryzae* using the *A. tumefaciens*-mediated transformation (ATMT) method as previously described [31]. Transformants were selected and cultured in 200 µg/mL hygromycin. The transformants were identified using PCR with primers HYG-S/HYG-A, MoAE4-LHYG-S/MoAE4-LHYG-A, and MoAE4-G-S/MoAE4-G-A.

The entire *MoAE4* sequence was amplified using a PCR technique with MoAE4-C-S/MoAE4-C-A and inserted into the hygromycin resistant vector pCAMBIA1303 for complementation into the mutant strain. The reconstructed pCAMBIA1303-*MoAE4* was transformed into the Δ*MoAE4* mutant strain and designated Δ*MoAE4*/*MoAE4*. The complemented strain was also confirmed by PCR with HYG-S/HYG-A and MoAE4-G-S/MoAE4-G-A.

To further verify the gene deletion and complementation, the expression of the wild type, Δ*MoAE4* mutant, and Δ*MoAE4*/*MoAE4* strains was amplified using qRT-PCR with qRT-MoAE4-S/qRT-MoAE4-A and Actin-S/Actin-A, and the strains were identified. The primers for gene deletion and complementation are listed in Table S2.

### 2.5. Quantitative Real-Time PCR (qRT-PCR)

The total RNA was isolated from mycelia that had been harvested from 5-day-old CM media using the TRIzol reagent (Invitrogen, Carlsbad, CA, USA). First strand cDNA was synthesized using an oligo (dT) primer from total RNA, which had been treated with DNase I. Subsequently, qRT-PCR was performed using an ABI7500 System (Applied Biosystems, Foster City, CA, USA) and SYBR Premix Ex Taq (TaKaRa, Dalian, China). The relative mRNA levels were calculated using the $2^{-\Delta\Delta Cq}$ ($C_q = C_{qgene} - C_{qactin}$) method. The *M. oryzae* actin gene (MGG_03982.6) was utilized as a reference gene for normalization. Each sample was tested in three replicates in each experiment. The primer sequences used for qRT-PCR are shown in Table S2.

### 2.6. Assays for Conidial Production, Growth, and Development

The strains (wild type, Δ*MoAE4*, and Δ*MoAE4*/*MoAE4*) were cultured on PDA media to understand the effect on *MoAE4* conidial production, and the conidia were cultured on OMA media as previously described. A volume of 200 μL of a $1 \times 10^5$/mL conidial solution was placed on OMA medium. After 3 days of cultivation at 28 °C, sterile water was added to remove the hyphae, and a piece of the culture medium was cut with a blade and placed on a glass slide. It was then placed in a moisturizing box and incubated at 28 °C. The piece was observed under a Nikon Eclipse 80i microscope at 6, 12, 24, and 48 h after it had been cut. The strains were then stained with lactophenol cotton blue to observe the conidiophore stalks and hyphae under a light microscope [32]. Additionally, the conidia were collected with 2 mL of sterile water after 3 days of culture on OMA media and counted with a hemocytometer. Each strain was repeated three times, and the experiment was conducted in triplicate.

Conidia of the wild type, Δ*MoAE4*, and Δ*MoAE4*/*MoAE4* were cultured on OMA media and collected to observe the germination of conidia and formation of appressoria. The conidial suspension was adjusted to $1 \times 10^5$/mL and added drop wise to a hydrophobic cover slips under a microscope at 1, 2, 3, 4, 5, and 6 h. Each strain was repeated three times, and the experiment was conducted in triplicate.

### 2.7. Rice Sheath Penetration and Plant Infection Assays

To determine the pathogenicity of *MoAE4*, the wild type, Δ*MoAE4*, and Δ*MoAE4*/*MoAE4* strains were inoculated on OMA media to collect the conidia as previously described. The fourth leaf stage of rice seedlings (*Oryza sativa* cv. Lijiangxintuanheigu) was assayed for infection following the spraying of 2 mL of a conidial suspension ($5 \times 10^4$ conidia/mL in 0.2% gelatin). The inoculated plants were placed in the dark in a dew chamber for 24 h at 28 °C and then transferred to a growth chamber with a photoperiod of 16 h for 7 days.

Conidial suspensions (100 μL, $5 \times 10^4$ conidia/mL) were injected into seedling leaf sheaths using a 1 mL syringe, and the inoculated plants were placed in a moist chamber as described previously. The formation of lesions and necrosis around the inoculation sites was examined when the injection-wounded leaves unfolded at different time points after the injection. The mean infectious hyphal (IH) growth rates and movement to the adjacent cells were determined from 100 germinated conidia per treatment at 12, 24, and 48 h post inoculation (hpi) and repeated in triplicate as previously described. The leaf sheaths were trimmed at the time points indicated and observed using a Nikon Eclipse 80i microscope. This experiment was performed with three independent replicates, and the representative results from one of these experiments are presented.

### 2.8. Assays for $NaHCO_3$ Treatment

To illustrate the effect of different concentrations of $NaHCO_3$ on the expression of *MoAE4* gene, wild type strains of *M. oryzae* were cultivated on PDA with 0, 12.5, 25, 37.5, 50, 62.5, and 75 mM $NaHCO_3$ at 28 °C for 7 days.

To investigate the effects of sodium bicarbonate stress on the wild type, Δ*MoAE4*, and Δ*MoAE4*/*MoAE4* strains, each strain was cultured on PDA with $NaHCO_3$ at final concentrations of 0, 12.5, 25, 37.5, 50, 62.5, and 75 mM at 28 °C for 7 days, and the diameters of fungal strains were photographed using a digital camera (EOS 800D, Canon, Inc., Tokyo, Japan) and measured after inoculation. Each assay was repeated three times independently for each strain, and the experiment was performed in triplicate. Further, the wild type of mycelium treated under different concentrations of $NaHCO_3$ was collected for expression patterns of AE4.

### 2.9. Assays for $HCO_3^-$ Transport and Intracellular pH Measurements

To determine the relationship between *MoAE4* and $HCO_3^-$ transport, confocal microscope was performed with lucigenin (bis-N-methylacridinium nitrate) (MCE, Shanghai, China), a compound that is used as a chemiluminescent probe to indicate the presence of superoxide anion radicals in cells in alkaline conditions 28 [33,34]. The microscopy enabled the detection of direction of $HCO_3^-$ transport in the conidia and hyphae. First, conidia and hyphae from the strains (wild type, Δ*MoAE4*, and Δ*MoAE4*/*MoAE4*) were treated at 28 °C with 0.4 M mannitol in a solution of 50 mM $NaHCO_3$ for 2 h, then washed and added 0.4 M mannitol to continue recovery at 28 °C for 2 h. A solution of only 0.4 M mannitol served as the control. The strains were incubated with 10 mM lucigenin and observed with a 470 nm fluorescence microscope.

The intracellular pH was measured using the dual-excitation ratio method with the pH sensitive dye 20,70-bis-(2-carboxyethyl)-5- (and-6)-carboxyfluorescein acetoxymethyl ester (BCECF-AM) (Mock, Sigma, Shanghai, China) to detect the wild type, Δ*MoAE4*, and Δ*MoAE4*/*MoAE4* strains treated under 50 mM $NaHCO_3$ as previously described. The pH sensitive dye was excited at 460 and 488 nm using a digital fluorescence microscopy system, and the fluorescence emitted at 520 nm was detected. The ratios of background-corrected emission intensities (I488/I460) were transformed into intracellular pH [35,36].

### 2.10. $H_2O_2$ Treatment and Endogenous $H_2O_2$ Measurements

To investigate the effects of exogenous oxidative stress on the wild type, Δ*MoAE4*, and Δ*MoAE4*/*MoAE4* strains, each strain was cultured on CM agar that contained 2.5 mM or 5 mM $H_2O_2$ for 7 days at 28 °C.

The $H_2O_2$ content was determined as previously described for plants [37]. Hydrogen peroxide ($H_2O_2$) was extracted by homogenizing 3 g of mycelia from the wild type, Δ*MoAE4*, and Δ*MoAE4*/*MoAE4* strains in 6 mL of cold acetone. The homogenate was then centrifuged at 3500× *g* for 5 min at room temperature, and the resulting supernatant was designated as the sample extract. Next, 0.1 mL of titanium reagent (5% [*w*/*v*] titanic sulfate in concentrated $H_2SO_4$) was added to 1 mL of the sample extract, followed by the addition of 0.2 mL of strong aqueous ammonia to precipitate the peroxide-titanium complex. The precipitated sample was centrifuged at 3000× *g* for 10 min at room temperature; the supernatant was discarded, and the precipitate was then solubilized in 5 mL of 2 M $H_2SO_4$. The absorbance of the samples was determined at 415 nm against a blank of 2 M $H_2SO_4$. The $H_2O_2$ concentration in the samples was determined by comparing the absorbance against a standard curve of a 0–5 mM titanium-$H_2O_2$ complex that was prepared according to Cui et al. [38].

The production of $H_2O_2$ was monitored by staining with 3,3′-diaminobenzidine (DAB) as an assay [39]. The hyphae of the wild type, Δ*MoAE4*, and Δ*MoAE4*/*MoAE4* strains were cultured in CM media for 5 days and then incubated in the dark in a 1 mg/mL solution of DAB at room temperature for 8 h. The samples were washed with sterile water and observed under a Nikon light microscope. This experiment was performed in triplicate and repeated three times for each strain. Similarly, leaf sheath cells of rice infected by wild type, mutant, and complementation strains were stained DAB at 36 hpi.

The conidia of the wild type, Δ*MoAE4*, and Δ*MoAE4*/*MoAE4* strains were extracted with DMSO [40].

### 2.11. Statistical Analysis

All the experiments were performed at least three times. The mean ± SD of the strain diameter, germination rate, and relative expression were determined using SPSS Statistics 22 (IBM, Inc., Armonk, NY, USA). Error bars represent the standard deviation. * indicates a statistically significant difference at $p < 0.05$. ** indicates a highly significant difference at $p < 0.01$. *** indicates a highly significant difference at $p < 0.001$. **** indicates a highly significant difference at $p < 0.0001$.

## 3. Results

### 3.1. The Bicarbonate Transporter AE4 Homologue in M. oryzae

Homologous sequences of AE4 proteins have been reported in a variety of species. Based on the conserved amino acid sequences of several reported AE4 proteins, a single homolog of AE4 (MGG_15203) was searched in the *M. oryzae* genome (http://fungalgenomics.ca/wiki/Fungal_Genomes, accessed on 25 March 2021). The *M. oryzae AE4*, termed as *MoAE4*, with a length of 2169 bp open reading frame, encodes a protein of 701 amino acids. The protein sequence alignments delineated the 10 transmembrane-spanning domains in the MoAE4 gene (Figure S1A,B).

The phylogenic tree indicated that *MoAE4* was closely related to the fungal group (Figure 1A), sharing 75.6% identity with *A. nidulans* SbtA gene, and 57.5% identity with *S. cerevisiae* YNL275w gene. The transmembrane-spanning domains and tertiary (3D) structures of MoAE4 were predicted with the web-based TMHMM Server v.2.0 (http://www.cbs.dtu.dk/services/TMHMM/, accessed on 16 May 2021) and I-TASSER (http://zhanglab.ccmb.med.umich.edu/I-TASSER/, accessed on 25 March 2021). Both the N-terminal and C-terminal ends are membrane inside (Figures 1B and S1B), which are responsible for transport activity and transmembrane domain anchoring [41]. The two regions TM1 and TM8, together with the N-terminal and C-terminal ends, form the so-called gate domain and core domain. The three conserved ligand sites ($P^{129}$, $A^{132}$, and $F^{487}$) and the active site $N^{218}$ harbors in the C-terminal end (core domain) (Figures 1C,D and S1C).

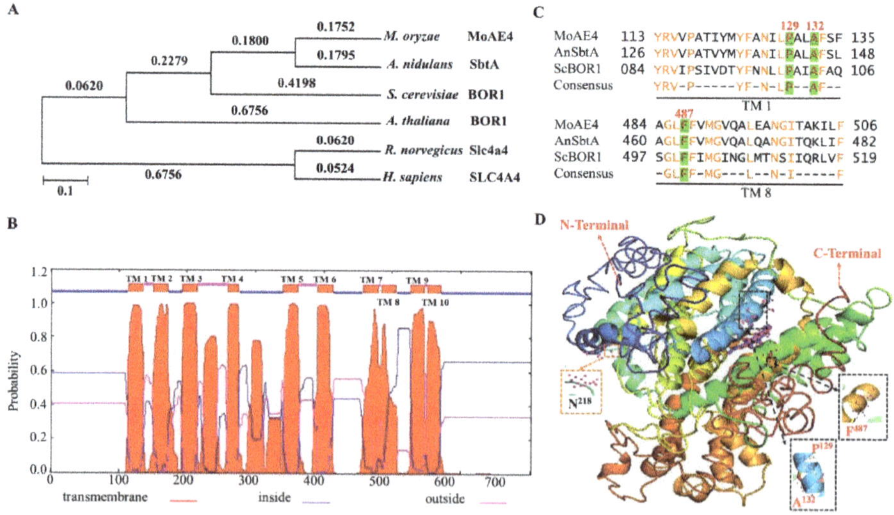


**Figure 1.** Structure of the MoAE4 and phylogenetic analysis. (**A**) Phylogenetic tree. It was constructed with reported anion exchange protein 4 homologs from *M. oryzae*, *A. nidulans*, *S. cerevisiae*, *A. thaliana*, *R. norvegicus*, *H. sapiens*, indicating that MoAE4 has a relatively close relationship with the fungal group. (**B**) TMHMM posterior probabilities for WEBSEQUENCE (on-line analysis). A schematic diagram of the MoAE4 protein transmembrane domain of *M. oryzae* is shown indicating the 10 transmembrane domains and marked the amino acid inside and outside the membrane. (**C**) Sequence alignment. The

transmembrane domain 1 and 8 of MoAE4 was compared with closely related to the fungal group *A. nidulans* SbtA gene and *S. cerevisiae* BOR1 gene. (**D**) Tertiary structure (3-D). The structure was predicted using the web-based server I-TASSER that shows the ligand binding site residues of transmembrane domain $P^{129}$, $A^{132}$, and $F^{487}$. The active site residue is $N^{218}$.

According to the 3D structures (Figure 1D and Figure S1C), the conformations with 10 TM helices of MoAE4 were characterized by two inverted repeats that are intertwined to form both core and gate domains, which appears to be similar with that of AE1 or BOR1 [3,10]. The homologous protein sequences and the typical domain patterns reveal MoAE4 is a member of the $HCO_3^-$ transporter AE4 group.

### 3.2. MoAE4 Localizes on Cytomembrane and Tonoplast and Functions in Yeast

Most $HCO_3^-$ transporters function on cytoplasm membrane [16]. To test where MoAE4 occurred, we generated transgenic strains expressing GFP-tagged *MoAE4* fusion gene in the wild type of *M. oryzae* (Figure S2A). Fluorescent microscopic observation was carried out by using hyphae (6d) and conidia. The lipophilic dye (FM™ 4-64 Dye, AAT Bioquest, USA) was used for observing the cytoplasmic membrane and vacuolar membrane. A strong green fluorescence signal of the MoAE4-GFP protein co-localized with FM4-64 red fluorescence was detected on the cytoplasmic membrane in young hyphae and conidia (Figure 2A,B); interestingly, a strong co-localized yellow fluorescence signal was also detected on the vacuole membrane in hyphae (Figure 2A,B). By comparison, in the wild type or untransformed strains, the background green fluorescence was too weak to be detected. This result suggests that MoAE4 was targeted to cytomembrane and tonoplast.

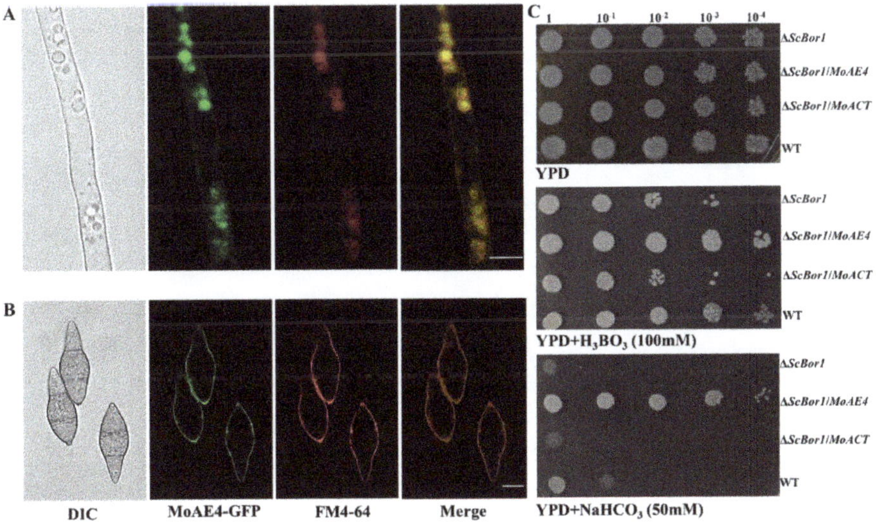


**Figure 2.** MoAE4 subcellular localization and functions in yeast. (**A**) Subcellular localization in hyphae (6d). Green fluorescence signals of the MoAE4–GFP protein of 6–day–old hyphae was examined by confocal microscopy and co–localized with FM4–64 on cytomembrane and tonoplast. Scale bar = 10 μm. (**B**) Subcellular localization in conidia. A strong green fluorescence signal of the MoAE4–GFP protein co–localized with FM4–64 red fluorescence was detected on the cytoplasmic membrane in conidia. Scale bar = 10 μm. (**C**) Functional complementation of MoAE4 for ScBor1 in *S. cerevisiae*. In total, 10 μL droplets containing the indicated concentration of yeast cells were inoculated on to the Solid YPD medium Plates (100 mM $H_3BO_3$ and 50 mM $NaHCO_3$ added, respectively). The *MoAE4* gene could functionally reverse the defect of the Δ*ScBor1* mutant in boric acid and $NaHCO_3$ tolerance. Representative plates were photographed 3 days post-inoculation.

Yeast bicarbonate transporters are boric acid tolerant [12–14]. Based on the *ScBor1/AE4* deletion mutant (Δ*ScBor1*), the complementary yeast strains were created by using *MoAE4*.

As a result, the *MoAE4* gene could functionally reverse the defect of the Δ*ScBor1* mutant in boric acid tolerance (Figure 2C), suggesting *MoAE4* functions as yeast Bor1. Different from the yeast BOR1, MoAE4 also endowed the $NaHCO_3$ tolerance in the mutant strains (Figure 2C), implying MoAE4 may be a $HCO_3^-$ transporter.

### 3.3. MoAE4 Transports Cytosolic $HCO_3^-$ to Vacuole and Cell outside

To identify the function of MoAE4 in response to $NaHCO_3$, *M. oryzae* was cultivated under $NaHCO_3$ stress conditions. By using the knockout mutant strain of *MoAE4* (Δ*MoAE4*) and the complemented strain (Δ*MoAE4*/*MoAE4*) (Figure S2B–D), the growth of the tested strains was assessed. When cultivated on complete media (CM) plates at 25 °C without $NaHCO_3$, both the Δ*MoAE4* and Δ*MoAE4*/*MoAE4* strains grew at a rate similar to that of the wild type, and their colony morphologies exhibited little difference (Figure 3A,B). When subjected with $NaHCO_3$ of different concentrations, all strains including the wild type were repressed in mycelial growth under $NaHCO_3$ stress. Impressively, Δ*MoAE4* almost stopped growing at 25 mM of $NaHCO_3$, but the wild type and complementary strains were capable of growing even at the concentration of 37.5 mM (Figure 3A,B), suggesting a role of Δ*MoAE4* in tolerance to $NaHCO_3$. The expression patterns in response to $NaHCO_3$ treatment also illustrated this point (Figure 3C).

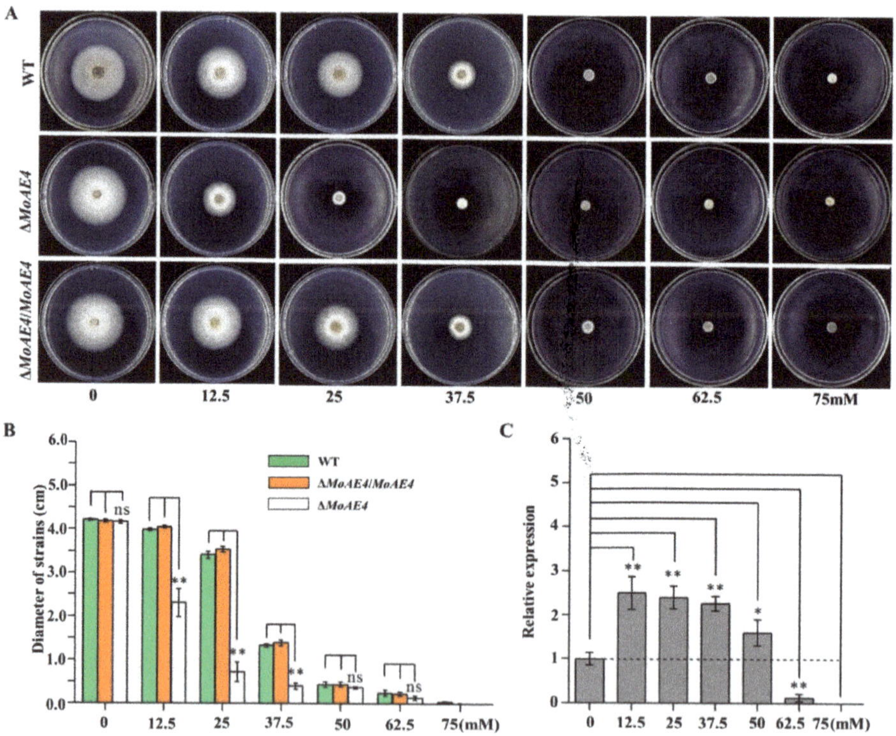


**Figure 3.** $NaHCO_3$ stress assay of the wild type and created strains and expression patterns of MoAE4. (**A**) $NaHCO_3$ stress assay. Δ*MoAE4* strains were more sensitive to $NaHCO_3$ stress than the wild-type strains. The strains were cultured in PDA media at 28 °C under 0–75 mM different concentrations of $NaHCO_3$ and representative colonies were photographed 7 days post-inoculation. (**B**) The colonies diameter of the wild type, the Δ*MoAE4* and Δ*MoAE4*/*MoAE4* mutant strains following treatments under different concentrations of $NaHCO_3$. (**C**) Transcription abundance of MoAE4 under different concentrations of $NaHCO_3$. Data represent the means ± standard deviation (SD) from three independent experiments in which triplicate plates were examined for each strain in each experiment. ns $p > 0.05$. * $p < 0.05$. ** $p < 0.01$.

To determine the $HCO_3^-$ transport activity of *MoAE4*, lucigenin (bis-N-methylacridinium nitrate), a chemiluminescent probe, was used as to detect the existence of anion radicals in cells under alkaline conditions [33,34]. After treated with $NaHCO_3$ solution (50 mM), the tested strains incubated with 10 mM lucigenin were observed with a 470 nm fluorescence microscope. As expected, lucigenin probed $HCO_3^-$ with green fluorescence signals was accumulated in vacuoles of the wild type and Δ*MoAE4*/*MoAE4* strains; however, in Δ*MoAE4*, green fluorescence signals were only detected in cytoplasm (Figure 4A,B). Accordingly, intracellular pH value increased significantly in Δ*MoAE4* (Figures 4C,D and S3). These results indicated that *MoAE4* functions as a *bona fide* $HCO_3^-$ transporter.

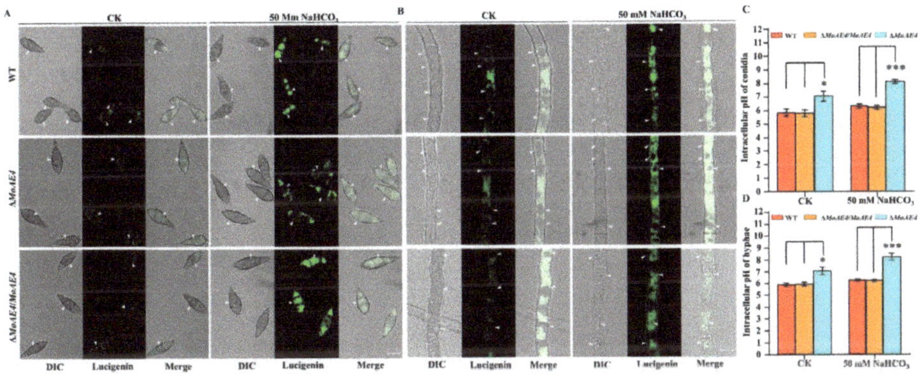


**Figure 4.** The fluorescence observations and intracellular pH value of the wild type, the Δ*MoAE4* and Δ*MoAE4*/*MoAE4* mutant strains. (**A**) The green fluorescence signals in hyphae. The light (**left**), fluorescence confocal (**middle**), and merge (**right**) microscope observations of strains hypha. Bar = 10 μm. (**B**) The green fluorescence signals in conidia. The light (**left**), fluorescence confocal (**middle**), and merge (**right**) microscope observations of strains connidia. Bar = 10 μm. (**C**) The intracellular pH value of conidia. Δ*MoAE4* strains were obviously higher than the wild-type strains under 50 mM sodium bicarbonate. (**D**) The intracellular pH value of the strains in hyphae. Under 50 mM sodium bicarbonate conditions, the intracellular pH of Δ*MoAE4* strains remained high in hyphae. The data represent means ± standard deviations (SD) of three experiments. * $p < 0.05$. *** $p < 0.001$.

*3.4. MoAE4 Is Important for Conidiation and Appressorium Development*

Conidiation and appressorium formation were analyzed among the Δ*MoAE4*, Δ*MoAE4*/*MoAE4*, and wild type strains. The sparse conidiophores with less conidia were observed in the deletion mutant Δ*MoAE4*, however, both the Δ*MoAE4*/*MoAE4* and wild type produced thick conidiophores and more conidia (Figures 5A–C and S4).

The conidium germination rate of all strains including the wild type was similar at 4–6 h, although Δ*MoAE4* appeared to be a little slow in conidium germination at 1–4 h (Figures 5D and S4B,C). In terms of appressorium formation, Δ*MoAE4*/*MoAE4* had the formation rate similar as the wild type did; but Δ*MoAE4* was severely affected (Figure 5E). As conidia and appressoria are essential factors for disease cycle and infection, *MoAE4* is proposed to be involved in pathogenesis.

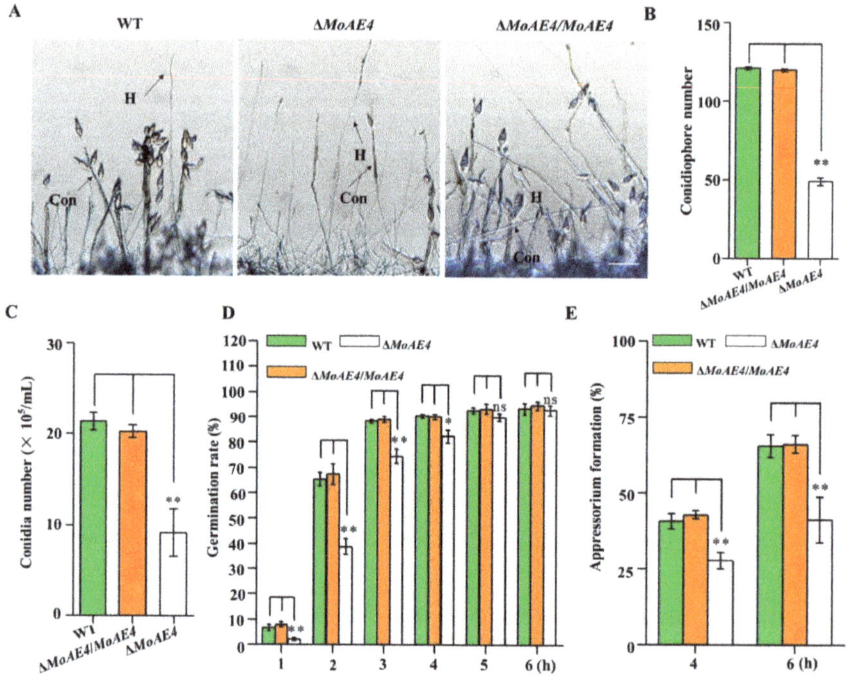


**Figure 5.** Conidium and appressorium development analysis of the wild type and created strains. (**A**) Conidiophores stained with lactophenol cotton blue. The conidiophores of the wild type, the Δ*MoAE4* and Δ*MoAE4*/*MoAE4* strains induced for 48 h were stained with lactophenol cotton blue, and observed and counted under a light microscope at room temperature. The hyphae are stained blue, whereas the conidiophore stalks are in gray. Bar = 50 μm. Con, conidiophores; H, hyphae. (**B**) Statistical analysis of the conidiophores number of the wild type, the Δ*MoAE4* and Δ*MoAE4*/*MoAE4* mutant strains. (**C**) Statistical analysis of conidial production in the strains. The conidia were harvested from the 3-day-old mycelium grown on OMA media, and counted using a hemocytometer for all the three strains. (**D**) Conidial germination rate. Conidial germination was measured on a hydrophobic cover slips and was calculated under the microscope at 1, 2, 3, 4, 5, and 6 h post inoculation. (**E**) Appressorial formation rate. Appressorial formation was measured on a hydrophobic cover slips and was calculated under the microscope at 4 and 6 h per inoculation. The analysis was performed using an independent samples *t*-test. ns $p > 0.05$. * $p < 0.05$. ** $p < 0.01$. Error bars indicate the mean ± SD from three independent experiments.

### 3.5. Requirement of MoAE4 for Pathogenicity in M. oryzae

In order to characterize the function of MoAE4 in pathogenic development, pathogenicity assays were carried out using conidia collected from Δ*MoAE4*, Δ*MoAE4*/*MoAE4*, and the wild type. When intact susceptible rice seedlings were spraying-inoculated, at 7 days post inoculation (dpi), some acute expansive disease lesions were observed in rice leaves by the wild type and Δ*MoAE4*/*MoAE4*; but no lesions were formed in rice leaves by the Δ*MoAE4* (Figure 6). Similarly, when drop-inoculation was assayed, only the wild type and complementary strains still showed pathogenicity.

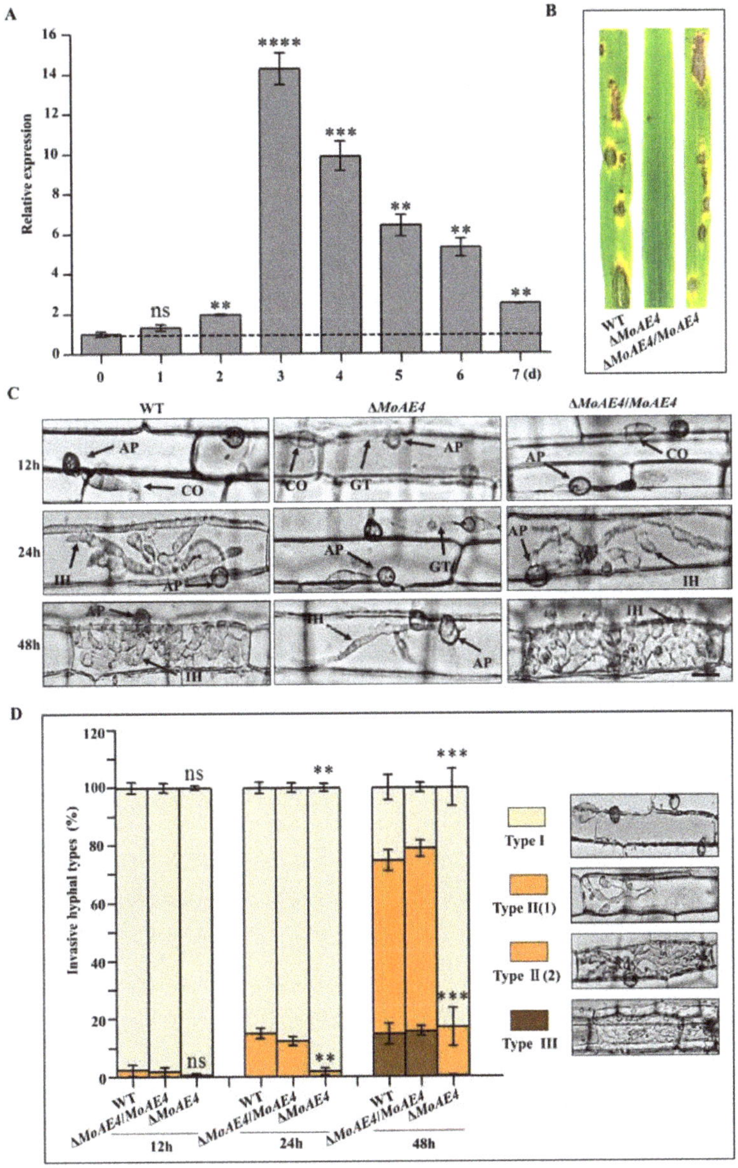


**Figure 6.** Pathogenesis analysis of the wild type and created strains. (**A**) Transcription abundance of *MoAE4* during disease development. (**B**) Spray-inoculation assay. (**C**) Rice leaf sheath infection assay. Scale bar = 10 μm. IH, infectious hyphae; CO, Conidium; GT, Germination tube; AP, appressorium. (**D**) The infection rate was calculated according to the number of type I to type III events. The infection status of more than 100 germinated conidia per leaf sheath was scored at 12, 24, and 48 h post inoculation. Type I, conidia with mature appressoria; Type II, primary hyphae formed, infectious hyphae extended and branched in one cell; Type III, infectious hyphae crossing to neighboring cells. Values represent the averages of five measurements ± standard deviation. The statistical analysis was performed using a one-way ANOVA with Tukey's multiple comparison test. The averages were taken from the quadruplicate analysis. Values are based on three biological samples and error bars indicate SD, ns $p > 0.05$. ** $p < 0.01$; *** $p < 0.001$. **** $p < 0.0001$.

Leaf sheath infection assays were performed to examine the infection effects of the *MoAE4* in rice host. At 12 hpi, most mature (black) appressoria have been formed in the wild type and Δ*MoAE4*/*MoAE4*, but less in Δ*MoAE4*. At 24 hpi, invasive hyphae of the wild type and Δ*MoAE4*/*MoAE4* commenced to branch in rice cells, but the primary infectious hyphae were just formed in Δ*MoAE4*. At 48 hpi, the majority of invasive hyphae of the wild-type and Δ*MoAE4*/*MoAE4* branched and started entering neighboring cells, but the Δ*MoAE4* strains did not due to the defects in appressorium formation (Figure 6C,D).

To decipher the exact action of *MoAE4* during pathogenic development, we defined the three types of infection hyphae according to their developmental morphologies. Then we quantified the proportion of the three types of infection hyphae based on 100 germinated conidia in the inoculated leaf sheath (Figure 6D). As a result, at 48 hpi more than 60% of inoculated conidia from Δ*MoAE4*/*MoAE4* and wild type formed branched infectious hyphae in one cell, of which about 20% extended to neighboring cells of rice (type II and III), suggesting the requirement of *MoAE4* in pathogenesis.

### *3.6. MoAE4 Is Important for $H_2O_2$ Tolerance and Clearance Inside or Outside Cells*

To address the relationship between *MoAE4* and endogenous $H_2O_2$, the wild type and mutant strains were cultured on CM agar supplemented with 2.5 or 5 mM $H_2O_2$ at 28 °C for 5 days. As a result, Δ*MoAE4* was markedly inhibited in mycelial growth under $H_2O_2$ stress, indicating a role of *MoAE4* in oxidative stress tolerance (Figure 7A,B). As rice plant accumulates more $H_2O_2$ during pathogen-rice interaction, and *MoAE4* expression increases with pathogenic development of *M. oryzae*, we speculate that MoAE4 is responsible for the clearance of host-derived $H_2O_2$ during infection. To test this, DAB staining was used to identify the endogenous ROS accumulated in the cells of rice leaf sheath infected by *M. oryzae* at 36 hpi (Figure 7C). In the leaf sheaths inoculated with the Δ*MoAE4* strains, more than 60% of the infected cells investigated were stained dark brown; in contrast, less than 20% of the infected cells were stained light brown or colorless as Δ*MoAE4*/*MoAE4* and wild type (Figure 7C,D), displaying loss of $H_2O_2$ scavenging function in Δ*MoAE4*. Additionally, endogenous $H_2O_2$ was measured in *M. oryzae* more $H_2O_2$ accumulated in Δ*MoAE4* than in the wild type and Δ*MoAE4*/*MoAE4* (Figure 7D). These results reveal that *MoAE4* is responsible for regulating $H_2O_2$ levels exogenous, endogenous, or plant-derived.

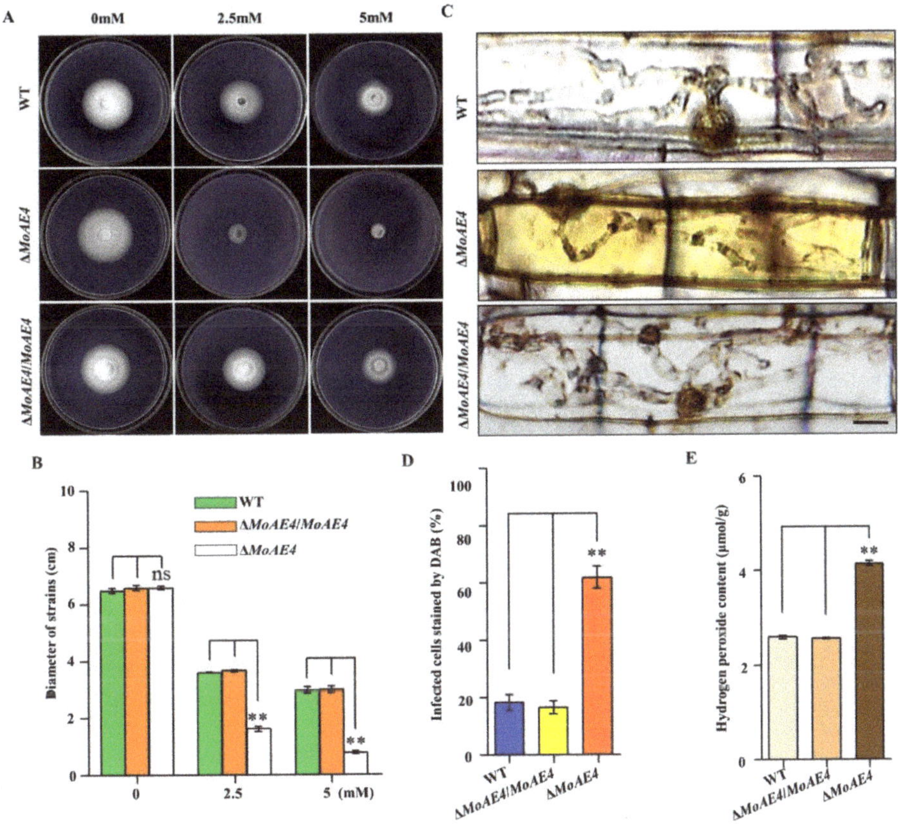


**Figure 7.** Comparison of oxidative stress, DAB staining, and endogenous $H_2O_2$ among the wild type, the Δ*MoAE4*, and Δ*MoAE4*/*MoAE4* strains. (**A**) $H_2O_2$ stress assay. The strains were cultured in CM media for 7 days at 28 °C with 2.5 or 5 mM $H_2O_2$. (**B**) The colonies diameter of the wild type, Δ*MoAE4*, and Δ*MoAE4*/*MoAE4* following treatment with 2.5 or 5 mM $H_2O_2$. (**C**) DAB staining of leaf sheath cells of rice infected by wild type, mutant and complementation strains at 36 hpi. Scale bar = 10 μm. (**D**) Statistical analysis of DAB staining of leaf sheath cells infected by different strains. (**E**) Endogenous $H_2O_2$ assay. The strains of hyphae of Endogenous $H_2O_2$ were determined as described in Experimental Procedures. The above experiments were performed in triplicate and repeated three independent times for each strain. Error bars represent the ± SD of three independently repeated samples, ns $p > 0.05$. ** $p < 0.01$.

## 4. Discussion

The family of bicarbonate transport proteins are involved in a wide-range of physiological processes in humans and mammals [1,2,42]. Mutation or dysregulation of these transporters results in physiological diseases in humans [43]. Therefore, bicarbonate transporters have attracted medical attention and have been extensively studied. In fungi, such as the unicellular organism yeast, $CO_2$ can diffuse directly out of yeast cells, so the anion exchanger/carbonic anhydrase system and $HCO_3^-$ transporter are proposed to be dispensable [32]. Indeed, the AE1 homologue, YNL275w has been verified as an anion transporter just serving for boron detoxification or tolerance in *S. cerevisiae* [12–14]. Bicarbonate gradients modulate growth and colony morphology in *A. nidulans* [44]. Bioinformatically, other filamentous fungi harbor homologous of mammal $HCO_3^-$ transporters. In *A. nidulans*, there are at least five members of SLC family, but the most homologous *SbtB* still showed boron transporter, not $HCO_3^-$ transporter [16]. Considering the exclusive activity of BOR1 in fungi and plants, YNL275w and homologues should be grouped into the secondary bicarbonate transporter family specific for boron. The *bona fide* $HCO_3^-$ transport function in fungal YNL275w homologues is actually unknown. In this research, we demonstrated that the MoAE4, as a YNL275w homologue, plays a role in $HCO_3^-$ transport. Importantly, MoAE4 located to cytomembrane and tonoplast promotes conidiation, appressorium formation, and pathogenesis in *M. oryzae*.

In general, AEs localize to cytoplasmic membrane [10,35]. GFP-tagged Bor1p were detected to localize preferentially to the vacuole and that cells lacking Bor1p have fragmented vacuoles [45]. Recently, kidney anion exchanger 1(kAE1) has been detected on vacuole [46]. These findings suggest Bor1p functions on both cytomembrane and tonoplast. Additionally, in our study, MoAE4 was localized on cytomembrane and tonoplast (Figure 2A; Table S1). In hyphae, MoAE4 protein was concentrated on tonoplast; but in conidia, a strong fluorescence signal was on cytomembrane, suggesting the different subcellular patterns of MoAE4. This may reflect the specific function of MoAE4 in different developmental stages of *M. oryzae*.

In addition to the homology with BOR1 in sequences and 3-D structures (Figure 1), MoAE4 rescued the defect of Δ*ScBOR1* in boron tolerance, suggesting an authentic BOR1 homologue. However, MoAE4 also endowed the $NaHCO_3$ tolerance for the mutant strains, for even the wild type failed to grow normally under $NaHCO_3$ stress (Figure 2C). Particularly, loss of *MoAE4* resulted in the severe mycelial growth inhibition in Δ*MoAE4* compared with the wild type and complementary strains (Figure 3A,B), implying the $HCO_3^-$ transport activity in MoAE4. Based on the lucigenin fluorescence quenching assay, a great amount of $HCO_3^-$ was detected in the cytosol of Δ*MoAE4*, but not in vacuoles of Δ*MoAE4*; on the contrary, $HCO_3^-$ was only detected in vacuole in the wild type and Δ*MoAE4*/*MoAE4* (Figure 4), revealing the MoAE4-associated bicarbonate resistant mechanism, by which MoAE4 removes redundant $HCO_3^-$ from cytosol to vacuole and outside cells.

As a pathogenic fungus, conidiation and appressorium formation are key processes for disease cycle and infection [17,18]. MoAE4 loss resulted in the impaired conidiophore formation and then decreased conidial and appressorial productivity (Figure 5). We speculate that this may be related to the excessively accumulated $H_2O_2$ in the MoAE4 deletion mutant (Figure 7D). After all, Δ*MoAE4* became sensitive under $H_2O_2$ stress (Figure 7B,C). At this point, the reduced pathogenicity in Δ*MoAE4* could also be partially explained because Δ*MoAE4* was exposed to a high hydrogen peroxide stress both in vivo and in vitro (Figure 7A,E and Figure S6).

$CO_2$, as a labile molecule, is the oxidation waste product of mitochondrial respiration. In humans, redundant $CO_2$ must be released, or the equilibrium with $HCO_3^- + H^+$ will be disturbed. The ability of $HCO_3^-$ to undergo pH-dependent conversions is central to its physiological role [1]. $CO_2$ enters the cytoplasm through the membrane and is rapidly hydrated forming carbonic acid ($H_2CO_3$). This acid is dissociated into $H^+$ and $HCO_3^-$ by intracellular carbonic anhydrases [7]. Under normal medium culture conditions, pathogenic fungi such as yeast may not require the anion exchanger/carbonic anhydrase system to help the release of metabolic $CO_2$ [11]. However, the interaction system between host plant and pathogen forms a whole multicellular organism, which should face a challenge in moving membrane impermeant bicarbonate from inside the cell where it is produced to the environment for disposal. In addition, during interactions between plant and pathogen, respiration from both pathogen and plant must be enhanced to produce available energy [24,47–49]. Therefore, we propose a pathogenic model mediated by MoAE4/MoCA (carbonic anhydrase) system (Figure 8). Under cultivation conditions, the metabolic $CO_2$ can be released freely from *M. oryzae*, and both *MoAE4* and *MoCA* genes are at low levels of expression because of the equilibrium of ($CO_2 + H_2O \rightleftharpoons HCO_3^- + H^+$) (Figure 8A). In the process of invasive hyphae growth, the infected plant cell is a relatively high concentration of $CO_2$ and low concentration of $O_2$ microenvironment; and accordingly, the diffusion of fungal $CO_2$ to the outside (cytosol of plant cell) is hindered. The upregulated

MoCA (Figure S5) will increase the concentration of $HCO_3^-$, which leads to MoAE4 being upregulated, then MoAE4 transports $HCO_3^-$ to the vacuole or to plant cells (Figure 8B).

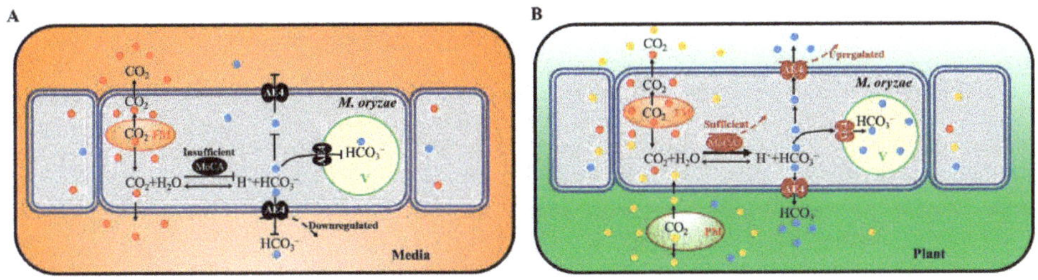


**Figure 8.** A pathogenic model mediated by MoAE4/MoCA (carbonic anhydrase) system. (**A**) *M. oryzae* is under cultivation conditions. (**B**) *M. oryzae* infects the host cells. FM, Fungal Mitochondrion; PM, Plant mitochondrion; V, vacuole; MoCA, *M. oryzae* carbonic anhydrase.

In the model, MoAE4 maintains the homeostasis of intracellular $CO_2$-$HCO_3^-$ system, which further ensures the intracellular acid-base balance in cells [50]. According to the importance of the acid-base equilibrium in multiple physiological activities [51], we imply $HCO_3^-$ regulated by MoAE4 is a pathogenic signal for blast disease development. Actually, in our recent research, a low dose of sodium carbonate can induce the expression of a large number of disease-related genes in wild type, not in the *MoAE4* deletion mutant (unpublished data), revealing a potential drug target for blast disease control. To decipher the regulation mechanism, much work remains to be completed.

## 5. Conclusions

In the rice blast fungus, MoAE4 exhibits more sequence and structure homology with AE proteins. Additionally, MoAE4, localized on cytomembrane and tonoplast, possesses boron and $NaHCO_3$ tolerance in yeast. Lucigenin fluorescence quenching assay indicated that MoAE4 has $HCO_3^-$ transport activity. Meanwhile, Δ*MoAE4* contained more $H_2O_2$ than the wild type and complementary strains did, implying a role of MoAE4 in energy metabolism. Importantly, MoAE4 is involved in conidiation, appressorium formation, and pathogenicity in *M. oryzae*. Overall, MoAE4, as a cytomembrane and tonoplast $HCO_3^-$ transporter, promotes pathogenesis of *M. oryzae*. Based on these results, a pathogenic model mediated by MoAE4 is proposed.

**Supplementary Materials:** The following are available online at https://www.mdpi.com/article/10.3390/jof7110955/s1, Figure S1. Sequence alignment and prediction of MoAE4 structure. Figure S2. The construction strategies for MoAE4 deletion and complementation strains. Figure S3. The green fluorescence signals detection of intracellular pH in conidia. Figure S4. The loss of the *MoAE4* gene has a negative effect on sporulation, appressorial development. Figure S5. Transcription abundance of *MoCA* during disease development. Figure S6. Standard curve line and the endogenous $H_2O_2$ of strains hyphae. Table S1. Identifying sub-cellular location. Table S2. Primers used in this study.

**Author Contributions:** S.-H.Z.: conceptualization, writing—review and editing the manuscript, supervision, and funding acquisition the research; Y.D.: writing and editing original draft preparation, methodology, conceptualization, and data curation; Y.W.: data curation, investigation, validation, and project administration; P.Z., X.L. (Xinchun Liu), X.L. (Xinrui Li), S.W. and H.L.: formal analysis, and software. All authors have read and agreed to the published version of the manuscript.

**Funding:** This work was supported by the Natural Science Foundation of China (Grant Nos.31670141 to Y.W. and 31671972 to S.-H.Z.); the Local scientific research of Department of Education of Liaoning Province of China (Grant No.01032920021 to S.-H.Z.); Special talent introduction of Shenyang Agricultural University of China (Grant No. 880420019 to S.-H.Z.); Postdoctoral funding of Shenyang Agricultural University of China (Grant No. 770221003 to Y.D.).

**Conflicts of Interest:** The authors declare no conflict of interest.

## References

1. Cordat, E.; Casey, J.R. Bicarbonate Transport in Cell Physiology and Disease. *Biochem. J.* **2009**, *417*, 423–439. [CrossRef]
2. Romero, M.F.; Chen, A.P.; Parker, M.D.; Boron, W.F. The SLC4 Family of Bicarbonate ($HCO_3^-$) Transporters. *Mol. Asp. Med.* **2013**, *34*, 159–182. [CrossRef]
3. Shieh, Y.W.; Minguez, P.; Bork, P.; Auburger, J.J.; Guilbride, D.L.; Kramer, G.; Bukau, B. Operon Structure and Cotranslational Subunit Association Direct Protein Assembly in Bacteria. *Science* **2015**, *350*, 678–680. [CrossRef]
4. Hatae, H.; Inaka, K.; Okamura, R.; Furubayashi, N.; Kamo, M.; Kobayashi, T.; Abe, Y.; Iwata, S.; Hamasaki, N. Crystallization of Human Erythrocyte Band 3, the Anion Exchanger, at the International Space Station "KIBO". *Anal. Biochem.* **2018**, *559*, 91–93. [CrossRef] [PubMed]
5. Tsuganezawa, H.; Kobayashi, K.; Iyori, M.; Araki, T.; Koizumi, A.; Watanabe, S.I.; Kaneko, A.; Fukao, T.; Monkawa, T.; Yoshida, T.; et al. A New Member of the $HCO_3^-$ Transporter Superfamily Is an Apical Anion Exchanger of β-Intercalated Cells in the Kidney. *J. Biol. Chem.* **2001**, *276*, 8180–8189. [CrossRef] [PubMed]
6. Ko, S.B.H.; Luo, X.; Hager, H.; Rojek, A.; Choi, J.Y.; Licht, C.; Suzuki, M.; Muallem, S.; Nielsen, S.; Ishibashi, K. AE4 Is a DIDS-Sensitive $Cl^-/HCO_3^-$ Exchanger in the Basolateral Membrane of the Renal CCD and the SMG Duct. *Am. J. Physiol. Cell Physiol.* **2002**, *283*, 1206–1218. [CrossRef] [PubMed]
7. Peña-Münzenmayer, G.; Catalán, M.A.; Kondo, Y.; Jaramillo, Y.; Liu, F.; Shull, G.E.; Melvin, J.E. Ae4 (Slc4a9) Anion Exchanger Drives Cl- Uptake-Dependent Fluid Secretion by Mouse Submandibular Gland Acinar Cells. *J. Biol. Chem.* **2015**, *290*, 10677–10688. [CrossRef]
8. Peña-Münzenmayer, G.; George, A.T.; Shull, G.E.; Melvin, J.E.; Catalán, M.A. Ae4 (Slc4a9) Is an Electroneutral Monovalent Cation-Dependent $Cl^-/HCO_3^-$ Exchanger. *J. Gen. Physiol.* **2016**, *147*, 423–436. [CrossRef]
9. Vera-Sigüenza, E.; Catalán, M.A.; Peña-Münzenmayer, G.; Melvin, J.E.; Sneyd, J. A Mathematical Model Supports a Key Role for Ae4 (Slc4a9) in Salivary Gland Secretion. *Bull. Math. Biol.* **2018**, *80*, 255–282. [CrossRef]
10. Coudray, N.; Seyler, S.L.; Lasala, R.; Zhang, Z.; Clark, K.M.; Dumont, M.E.; Rohou, A.; Beckstein, O.; Stokes, D.L. Structure of the SLC4 Transporter Bor1p in an Inward-Facing Conformation. *Protein Sci.* **2017**, *26*, 130–145. [CrossRef] [PubMed]
11. Zhao, R.; Reithmeier, R.A.F. Expression and Characterization of the Anion Transporter Homologue YNL275w in *Saccharomyces cerevisiae*. *Am. J. Physiol. Cell Physiol.* **2001**, *281*, 33–45. [CrossRef]
12. Nozawa, A.; Takano, J.; Kobayashi, M.; Von Wirén, N.; Fujiwara, T. Roles of BOR1, DUR3, and FPS1 in Boron Transport and Tolerance in *Saccharomyces cerevisiae*. *FEMS Microbiol. Lett.* **2006**, *262*, 216–222. [CrossRef] [PubMed]
13. Takano, J.; Kobayashi, M.; Noda, Y.; Fujiwara, T. Saccharomyces Cerevisiae Bor1p Is a Boron Exporter and a Key Determinant of Boron Tolerance. *FEMS Microbiol. Lett.* **2007**, *267*, 230–235. [CrossRef] [PubMed]
14. Jennings, M.L.; Howren, T.R.; Cui, J.; Winters, M.; Hannigan, R. Transport and Regulatory Characteristics of the Yeast Bicarbonate Transporter Homolog Bor1p. *Am. J. Physiol. Cell Physiol.* **2007**, *293*, 468–476. [CrossRef]
15. Takano, J.; Noguchi, K.; Yasumori, M.; Kobayashi, M.; Gajdos, Z.; Miwa, K.; Hayashi, H.; Yoneyama, T.; Fujiwara, T. *Arabidopsis* Boron Transporter for Xylem Loading. *Nature* **2002**, *420*, 337–340. [CrossRef] [PubMed]
16. Villarino, M.; Etxebeste, O.; Mendizabal, G.; Garzia, A.; Ugalde, U.; Espeso, E.A. Boron Tolerance in *Aspergillus nidulans* Is Sustained by the SltA Pathway through the SLC-Family Transporters SbtA and SbtB. *Genes* **2017**, *8*, 188. [CrossRef]
17. Howard, R.J.; Ferrari, M.A.; Roach, D.H.; Money, N.P. Penetration of Hard Substrates by a Fungus Employing Enormous Turgor Pressures. *Proc. Natl. Acad. Sci. USA* **1991**, *88*, 11281–11284. [CrossRef]
18. Talbot, N.J. On the Trail of a Cereal Killer: Exploring the Biology of *Magnaporthe grisea*. *Annu. Rev. Microbiol.* **2003**, *57*, 177–202. [CrossRef]
19. Heath, M.C.; Valent, B.; Howard, R.J.; Chumley, F.G. Interactions of Two Strains of *Magnaporthe grisea* with Rice, Goosegrass, and Weeping Lovegrass. *Can. J. Bot.* **1990**, *68*, 1627–1637. [CrossRef]
20. Kankanala, P.; Czymmek, K.; Valent, B. Roles for Rice Membrane Dynamics and Plasmodesmata during Biotrophic Invasion by the Blast Fungus. *Plant Cell* **2007**, *19*, 706–724. [CrossRef]
21. Hammond-Kosack, K.E.; Parker, J.E. Deciphering Plant-Pathogen Communication: Fresh Perspectives for Molecular Resistance Breeding. *Curr. Opin. Biotechnol.* **2003**, *14*, 177–193. [CrossRef]
22. Nürnberger, T.; Brunner, F.; Kemmerling, B.; Piater, L. Innate Immunity in Plants and Animals: Striking Similarities and Obvious Differences. *Immunol. Rev.* **2004**, *198*, 249–266. [CrossRef]
23. Egan, M.J.; Wang, Z.Y.; Jones, M.A.; Smirnoff, N.; Talbot, N.J. Generation of Reactive Oxygen Species by Fungal NADPH Oxidases Is Required for Rice Blast Disease. *Proc. Natl. Acad. Sci. USA* **2007**, *104*, 11772–11777. [CrossRef]
24. Grahl, N.; Puttikamonkul, S.; Macdonald, J.M.; Gamcsik, M.P.; Ngo, L.Y.; Hohl, T.M.; Cramer, R.A. In Vivo Hypoxia and a Fungal Alcohol Dehydrogenase Influence the Pathogenesis of Invasive Pulmonary Aspergillosis. *PLoS Pathog.* **2011**, *7*, e1002145. [CrossRef]
25. Choi, J.; Chung, H.; Lee, G.W.; Koh, S.K.; Chae, S.K.; Lee, Y.H. Genome-Wide Analysis of Hypoxia-Responsive Genes in the Rice Blast Fungus, *Magnaporthe oryzae*. *PLoS ONE* **2015**, *10*, e0134939. [CrossRef]
26. Fernandez, J.; Marroquin-Guzman, M.; Wilson, R.A. Mechanisms of Nutrient Acquisition and Utilization during Fungal Infections of Leaves. *Annu. Rev. Phytopathol.* **2014**, *52*, 155–174. [CrossRef] [PubMed]

27. Cui, X.; Wei, Y.; Wang, Y.H.; Li, J.; Wong, F.L.; Zheng, Y.J.; Yan, H.; Liu, S.S.; Liu, J.L.; Jia, B.L.; et al. Proteins Interacting with Mitochondrial ATP-Dependent Lon Protease (MAP1) in *Magnaporthe oryzae* Are Involved in Rice Blast Disease. *Mol. Plant Pathol.* **2015**, *16*, 847–859. [CrossRef] [PubMed]
28. Bertram, G.; Swoboda, R.K.; Gooday, G.W.; Gow, N.A.R.; Brown, A.J.P. Structure and Regulation of the Candida Albicans ADH1 Gene Encoding an Immunogenic Alcohol Dehydrogenase. *Yeast* **1996**, *12*, 115–127. [CrossRef]
29. Li, Z.; Pei, X.; Zhang, Z.; Wei, Y.; Song, Y.; Chen, L.; Liu, S.; Zhang, S.H. The Unique GH5 Cellulase Member in the Extreme Halotolerant Fungus *Aspergillus glaucus* CCHA Is an Endoglucanase with Multiple Tolerance to Salt, Alkali and Heat: Prospects for Straw Degradation Applications. *Extremophiles* **2018**, *22*, 675–685. [CrossRef] [PubMed]
30. Zheng, W.; Zhou, J.; He, Y.; Xie, Q.; Chen, A.; Zheng, H.; Shi, L.; Zhao, X.; Zhang, C.; Huang, Q.; et al. Retromer Is Essential for Autophagy-Dependent Plant Infection by the Rice Blast Fungus. *PLoS Genet.* **2015**, *11*, e1005704. [CrossRef]
31. Dang, Y.; Wei, Y.; Wang, Y.; Liu, S.; Julia, C.; Zhang, S.H. Cleavage of PrePL by Lon Promotes Growth and Pathogenesis in *Magnaporthe oryzae*. *Environ. Microbiol.* **2020**, *23*, 4881–4895. [CrossRef] [PubMed]
32. Zhou, Z.; Li, G.; Lin, C.; He, C. Conidiophore Stalk-Less1 Encodes a Putative Zinc-Finger Protein Involved in the Early Stage of Conidiation and Mycelial Infection in *Magnaporthe oryzae*. *Mol. Plant-Microbe Interact.* **2009**, *22*, 402–410. [CrossRef]
33. Qi, D.; Wang, D.; Zhang, C.; Tang, X.; He, J.; Zhao, Y.; Deng, W.; Deng, X. Vaspin Protects against LPS-Induced ARDS by Inhibiting Inflammation, Apoptosis and Reactive Oxygen Species Generation in Pulmonary Endothelial Cells via the Akt/GSK-3β Pathway. *Int. J. Mol. Med.* **2017**, *40*, 1803–1817. [CrossRef] [PubMed]
34. Oosthuizen, M.M.J.; Greyling, D. Hydroxyl Radical Generation: The Effect of Bicarbonate, Dioxygen and Buffer Concentration on PH-Dependent Chemiluminescence. *Redox Rep.* **2001**, *6*, 105–116. [CrossRef]
35. Schwiening, C.J.; Boron, W.F. Regulation of Intracellular PH in Pyramidal Neurones from the Rat Hippocampus by Na(+)-dependent Cl(-)-$HCO_3^-$ Exchange. *J. Physiol.* **1994**, *475*, 59–67. [CrossRef]
36. Yang, L.; Zhu, L.; Xu, Y.; Zhang, H.; Ye, W.; Mao, J.; Chen, L.; Wang, L. Uncoupling of $K^+$ and $Cl^-$ Transport across the Cell Membrane in the Process of Regulatory Volume Decrease. *Biochem. Pharmacol.* **2012**, *84*, 292–302. [CrossRef] [PubMed]
37. Brennan, T.; Frenkel, C. Involvement of Hydrogen Peroxide in the Regulation of Senescence in Pear. *Plant Physiol.* **2020**, *59*, 411–416. [CrossRef]
38. Cui, X.; Wei, Y.; Xie, X.L.; Chen, L.N.; Zhang, S.H. Mitochondrial and Peroxisomal Lon Proteases Play Opposing Roles in Reproduction and Growth but Co-Function in the Normal Development, Stress Resistance and Longevity of *Thermomyces lanuginosus*. *Fungal Genet. Biol.* **2017**, *103*, 42–54. [CrossRef]
39. Liu, X.; Qian, B.; Gao, C.; Huang, S.; Cai, Y.; Zhang, H.; Zheng, X.; Wang, P.; Zhang, Z. The Putative Protein Phosphatase MoYvh1 Functions Upstream of MoPdeH to Regulate the Development and Pathogenicity in *Magnaporthe oryzae*. *Mol. Plant-Microbe Interact.* **2016**, *29*, 496–507. [CrossRef]
40. Rakotonirainy, M.S.; Héraud, C.; Lavédrine, B. Detection of Viable Fungal Spores Contaminant on Documents and Rapid Control of the Effectiveness of an Ethylene Oxide Disinfection Using ATP Assay. *Luminescence* **2003**, *18*, 113–121. [CrossRef]
41. van den Akker, E.; Satchwell, T.J.; Williamson, R.C.; Toye, A.M. Band 3 Multiprotein Complexes in the Red Cell Membrane; of Mice and Men. *Blood Cells Mol. Dis.* **2010**, *45*, 1–8. [CrossRef] [PubMed]
42. Carignan, C.C.; Punshon, T.; Karagas, M.R. HHS Public Access. *Physiol. Behav.* **2016**, *176*, 100–106. [CrossRef]
43. Alka, K.; Casey, J.R. Bicarbonate Transport in Health and Disease. *IUBMB Life* **2014**, *66*, 596–615. [CrossRef]
44. Rodríguez-Urra, A.B.; Jimenez, C.; Dueñas, M.; Ugalde, U. Bicarbonate Gradients Modulate Growth and Colony Morphology in *Aspergillus nidulans*. *FEMS Microbiol. Lett.* **2009**, *300*, 216–221. [CrossRef] [PubMed]
45. Decker, B.L.; Wickner, W.T. Enolase Activates Homotypic Vacuole Fusion and Protein Transport to the Vacuole in Yeast. *J. Biol. Chem.* **2006**, *281*, 14523–14528. [CrossRef]
46. Sarder, H.A.M.; Li, X.; Funaya, C.; Cordat, E.; Schmitt, M.J.; Becker, B. Saccharomyces Cerevisiae: First Steps to a Suitable Model System To Study the Function and Intracellular Transport of Human Kidney Anion Exchanger 1. *mSphere* **2020**, *5*, e00802-19. [CrossRef] [PubMed]
47. Rios, J.A.; de Ávila Rodrigues, F.; Debona, D.; Silva, L.C. Photosynthetic Gas Exchange in Leaves of Wheat Plants Supplied with Silicon and Infected with *Pyricularia oryzae*. *Acta Physiol. Plant.* **2014**, *36*, 371–379. [CrossRef]
48. Domiciano, G.P.; Cacique, I.S.; Freitas, C.C.; Filippi, M.C.C.; DaMatta, F.M.; Do Vale, F.X.R.; Rodrigues, F.Á. Alterations in Gas Exchange and Oxidative Metabolism in Rice Leaves Infected by *Pyricularia oryzae* Are Attenuated by Silicon. *Phytopathology* **2015**, *105*, 738–747. [CrossRef]
49. Torres, M.A.; Dangl, J.L. Functions of the Respiratory Burst Oxidase in Biotic Interactions, Abiotic Stress and Development. *Curr. Opin. Plant Biol.* **2005**, *8*, 397–403. [CrossRef]
50. Massey, M.K.; Reiterman, M.J.; Mourad, J.; Luckie, D.B. Is CFTR an Exchanger? Regulation of $HCO_3^-$ Transport and Extracellular PH by CFTR. *Biochem. Biophys. Rep.* **2021**, *25*, 100863. [CrossRef]
51. Chen, Y.; Cann, M.J.; Litvin, T.N.; Iourgenko, V.; Sinclair, M.L.; Levin, L.R.; Buck, J. Soluble Adenylyl Cyclase as an Evolutionarily Conserved Bicarbonate Sensor. *Science* **2000**, *289*, 625–628. [CrossRef] [PubMed]

Journal of *Fungi*

MDPI

*Article*

# *Tr*-milRNA1 Contributes to Lignocellulase Secretion under Heat Stress by Regulating the Lectin-Type Cargo Receptor Gene *Trvip36* in *Trichoderma guizhouence* NJAU 4742

**Tuo Li [1], Jinding Liu [2], Qin Wang [1], Yang Liu [1], Ting Li [1], Dongyang Liu [1,*] and Qirong Shen [1]**

1 Jiangsu Provincial Key Lab of Solid Organic Waste Utilization, Jiangsu Collaborative Innovation Center of Solid Organic Wastes, Educational Ministry Engineering Center of Resource-Saving Fertilizers, Nanjing Agricultural University, Nanjing 210095, China; 2018203039@njau.edu.cn (T.L.); 2020103111@stu.njau.edu.cn (Q.W.); 2021203046@stu.njau.edu.cn (Y.L.); 2020103119@stu.njau.edu.cn (T.L.); shenqirong@njau.edu.cn (Q.S.)

2 College of Information Management, Nanjing Agricultural University, Nanjing 210095, China; liujd@njau.edu.cn

* Correspondence: liudongyang@njau.edu.cn

**Abstract:** Background: MicroRNA plays an important role in multifarious biological processes by regulating their corresponding target genes. However, the biological function and regulatory mechanism of fungal microRNA-like RNAs (milRNAs) remain poorly understood. Methods: In this study, combined with deep sequencing and bioinformatics analysis, milRNAs and their targets from *Trichoderma guizhouence* NJAU 4742 were isolated and identified under solid-state fermentation (SSF) by using rice straw as the sole carbon source at 28 °C and 37 °C, respectively. Results: A critical milRNA, TGA1_S04_31828 (*Tr*-milRNA1), was highly expressed under heat stress (37 °C) and adaptively regulated lignocellulase secretion. Overexpression of *Tr*-milRNA1 (OE-*Tr*-milRNA1) did not affect vegetative growth, but significantly increased lignocellulose utilization under heat stress. Based on the bioinformatics analysis and qPCR validation, a target of *Tr*-milRNA1 was identified as *Trvip*36, a lectin-type cargo receptor. The expression of *Tr*-milRNA1 and *Trvip36* showed a divergent trend under SSF when the temperature was increased from 28 °C to 37 °C. In addition, the expression of *Trvip*36 was suppressed significantly in *Tr*-milRNA1 overexpression strain (OE-*Tr*-milRNA1). Compared with the wild type, deletion of *Trvip*36 (Δ*Trvip*36) significantly improved the secretion of lignocellulases by reducing the retention of lignocellulases in the ER under heat stress. Conclusions: *Tr*-milRNA1 from NJAU 4742 improved lignocellulose utilization under heat stress by regulating the expression of the corresponding target gene *Trvip*36. These findings might open avenues for exploring the mechanism of lignocellulase secretion in filamentous fungi.

**Keywords:** lignocellulase secretion; *Trichoderma guizhouence* NJAU 4742; heat stress; milRNA; posttranscriptional regulation

**Citation:** Li, T.; Liu, J.; Wang, Q.; Liu, Y.; Li, T.; Liu, D.; Shen, Q. *Tr*-milRNA1 Contributes to Lignocellulase Secretion under Heat Stress by Regulating the Lectin-Type Cargo Receptor Gene *Trvip36* in *Trichoderma guizhouence* NJAU 4742. *J. Fungi* **2021**, *7*, 997. https://doi.org/10.3390/jof7120997

Academic Editor: Baojun Xu

Received: 7 October 2021
Accepted: 20 November 2021
Published: 23 November 2021



## 1. Introduction

Small RNA (sRNA)-induced RNA interference (RNAi) is a broad biological process that can lead to sequence-specific degradation or translational repression of target mRNAs [1,2]. The common feature of RNAi pathways is that sRNAs bind to Argonaute (AGO) proteins and guide the RNA-induced silencing complex (RISC) to RNAs with complementary sequences [3,4]. RNAi is highly conserved in eukaryotes, including the majority of the fungal kingdom [5]. Fungal RNA interference was first discovered in *Neurospora crassa*. Introducing fragments of *albino-1* (*al-1*) or *albino-3* (*al-3*), which are required for carotenoid biosynthesis, reduced *al-1*, or *al-3* mRNA levels and resulted in an albino phenotype [6]. With the expansion of fungal genome information, RNAi pathway components have been found in most fungal species. Studies on *Schizosaccharomyces pombe* and other fungi have revealed various small RNA biogenesis pathways, suggesting that RNAi related pathways are utilized in all cellular processes to adapt to complex external environments [7].

sRNAs, defined by their length of 18–24 nucleotides, play significant roles in growth and development processes, pathogenicity and stress response [2,8,9]. Usually, miRNAs are a class of interior sRNAs found in eukaryotes, that function in many processes, such as development, biotic and abiotic reactions, and defense [10]. Previously, miRNA-like RNAs (milRNAs), with similar characteristics of miRNAs in animals and plants, were recognized and corroborated to be generated through at least four diverse pathways in *N. crassa* [11]. Interestingly, at least four distinct mechanisms have been discovered to produce milRNAs. Dicer-independent small interfering RNAs (disiRNAs) with a size of 21 or 22 nt were also identified in *Neurospora* [11]. Afterward, a large number of milRNAs were identified in fungi. Fifteen milRNAs could regulate mycelium growth and conidiogenesis processes in *Metarhizium anisopliae* [12], and 27 sRNAs were found to have miRNA-like precursor structures in *Botrytis cinerea* [13]. In addition, the presence and differential expression of *Trichoderma reesei* milRNAs when *T. reesei* was cultivated in basal medium supplemented with 3% Avicel (microcrystalline cellulose powder) or 2% glucose, implied that milRNA might function in *T. reesei* growth and cellulase induction [14]. Nevertheless, analogous studies have mainly focused on bioinformatics predictions, but the biogenesis and biological function of milRNAs are rarely investigated.

*Trichoderma guizhouense* NJAU 4742, a saprophytic filamentous fungi isolated from mature compost, has recently attracted wide attention due to its potential to promote plant growth [14,15]. Similar to other heterotrophs, NJAU 4742 usually relies on particular host organisms or substrates for their nutrition. During the colonization process in different habitats including soils and plant roots, NJAU 4742 always feeds on dead fungi or efficiently degrades different plant debris to obtain enough energy resources [16]. Thus, the synthesis and secretion of lignocellulases by NJAU 4742 is one of the critical parameters during the colonization process. Nevertheless, various factors affect its biological functions, especially when the topsoil temperature is too high. Our previous results showed that the secretion of lignocellulases by NJAU 4742 was significantly suppressed at 37 °C (unpublished results). Here, deep sequencing and molecular assays were used to reveal the mechanism of lignocellulase secretion in NJAU 4742 under posttranscriptional regulation mediated by *Tr*-milRNAs. The preliminary analysis of the results indicated that TGA1_S04_31828 (*Tr*-milRNA1) adaptively regulates a lectin-type cargo receptor (*Trvip36*), which might be related to the lignocellulase secretion under heat stress. VIP36 is an intracellular lectin cycling between the endoplasmic reticulum (ER) and the Golgi apparatus, and it is deemed to serve as a cargo receptor in the transport and classification of glycoproteins [17]. VIP36 has also been shown to recycle human α1-antitrypsin from the Golgi compartment back to the ER, and silencing VIP36 improved α1-antitrypsin production. This fact demonstrates that this cargo receptor has a protein retention function [18]. In *Aspergillus oryzae*, deletion of VIP36 improved heterologous protein secretion, and the results of ER-enriched cellular fractions revealed that VIP36 was involved in the retention of heterologous proteins in the ER [19]. As VIP36 could form a stable complex with the molecular chaperone BiP, it appeared to be involved in the quality control of secretory proteins [17]. Therefore, lectin-type cargo receptors might alter the secretion of cellulases, which would be potential suitable models for studying the relationship between cargo receptors and secretory proteins in filamentous fungi.

The objective of this study was to uncover the role of *Tr*-milRNA in lignocellulose utilization of NJAU4742 under heat stress. Further studies investigated the *Tr*-milRNA1 regulation mechanism and the involvement of lectin-type cargo receptors in the intracellular transfer of cellulases in NJAU 4742 under heat stress, which aids readers to have a deeper understanding of lignocellulose biodegradation by filamentous fungi.

## 2. Materials and Methods

### 2.1. Strains and Culture Conditions

*T. guizhouense* NJAU 4742 isolated from a compost sample was maintained in our laboratory, and its genome sequence was deposited in the NCBI database (Accession No. LVVK00000000.1). Mandels' salt solution without organic components (1.4 g $L^{-1}$ $(NH_4)_2SO_4$, 2.0 g $L^{-1}$ $KH_2PO_4$, 0.3 g $L^{-1}$ $CaCl_2$, 0.3 g $L^{-1}$ $MgSO_4$, 5 mg $L^{-1}$ $FeSO_4{\cdot}7H_2O$, 20 mg $L^{-1}$ $CoCl_2$, 1.6 mg $L^{-1}$ $MnSO_4$ and 1.4 mg $L^{-1}$ $ZnSO_4$) supplemented with 2% (*w*/*v*) rice straw was used for lignocellulase production under SSF [20]. Spores were obtained from incubated PDA plates followed by filtration with four layers of sterilized gauze, and the spore suspension was adjusted to $1.0 \times 10^7$ spores·$mL^{-1}$ by quantifying spores on a hemocytometer.

### 2.2. Small RNA cDNA Library Construction and High-Throughput Sequencing

Mycelial samples were harvested after 72 h by using rice straw as the sole carbon source at 28 °C (T28) and 37 °C (T37). Total RNA was extracted using the RNeasy® Plant Mini Kit (Qiagen, Hilden, Germany) and handled with DNase I (TaKaRa) by following the manufacturer's instructions. RNA concentration and purity were evaluated in an Agilent 2100 Bioanalyzer (Agilent, Santa Clara, CA, USA) to check RNA integrity. Sequencing libraries were generated by employing NEBNext® Multiplex Small RNA Library Prep Set for Illumina® (New England BioLabs Inc., Beverly, MA, USA) according to the manufacturer's instructions. The clustering of each index-coded sample was carried out on a cBot Cluster Generation System using TruSeq SR Cluster Kit v3-cBot-HS (Illumina, San Diego, CA, USA) according to the manufacturer's instructions. Finally, libraries were sequenced through an Illumina HiSeq 2500 platform.

### 2.3. Tr-milRNAs Sequence Analysis and Target Prediction

The original reads from sequencing data were filtered by removing poor quality reads, adaptor contamination reads and reads longer than 30 nt or shorter than 18 nt. The standard-compliant reads of small RNAs were aligned to the reference NJAU 4742 genome (https://bioinfo.njau.edu.cn/tgn4742/, accessed on 20 November 2021). The alignment analysis was performed on the CLC Genomics Workbench 12. The sequences that corresponded to known miRNAs were defined by matching to the miRNA database (miRBase 22.0) [21]. The unannotated sRNA sequences were aligned to the reference NJAU 4742 genome to identify precursor sequences for novel miRNAs. Novel miRNAs were predicted by miRDeep2 with a stem-loop structure [22]. The R package DEGseq software was used to analyze the differentially expressed miRNAs. TargetFinder and psRNATarget were used to predict the candidate target genes of *Tr*-milRNAs [23,24].

### 2.4. Relative Expression of Tr-milRNA1 and the Corresponding Target Genes

Samples of different treatments were collected at 0, 24, 36, 48, 60 and 72 h post inoculation (hpi). Total RNA was extracted using the RNeasy® Plant Mini Kit (Qiagen, Germany) according to the manufacturer's instructions. Expression of *Tr*-milRNA1 was determined by stem-loop qRT–PCR as previously described [25]. First strand cDNA was synthesized by miRNA First Strand cDNA Synthesis (Vazyme Biotech, Nanjing, China) with the stem-loop RT primer based on the manufacturer's instructions. PCR detection was performed by a *Tr*-milRNA1-specific forward primer and a universal reverse primer. The NJAU 4742 18S rRNA biogenesis gene (18S) was used as a control. qRT–PCR was performed using the miRNA Universal SYBR qPCR Master Mix (Vazyme Biotech, Nanjing, China) according to the manufacturer's instructions. For the determination of transcript levels of the corresponding target genes, cDNA synthesis was completed by the PrimeScript RT Reagent Kit (RR036A, Takara, Dalian, China) according to the manufacturer's instructions. qRT-PCR was performed using SYBR Premix Ex Taq II (RR820A, Takara, Dalian, China) and the CFX connect™ Real-Time system (Bio–Rad, Hercules, CA, USA). Transcription levels of the target genes were normalized by the $2^{-\Delta Ct}$ method, and translation elongation

Factor 1 alpha (*Tef*) was used as the housekeeping gene. All primers used in this study are listed in Table S1.

### 2.5. Deletion and Overexpression of Fungal Small RNAs

For targeted deletion of *Tr*-milRNA1, the 5′ and 3′ flanking regions of *Tr*-milRNA1 were amplified by PCR from NJAU 4742 genomic DNA. *HygB* was used as a resistance gene and the fragment was amplified by PCR from plasmid pcDNA1 (Vienna University of Technology). The three fragments were fused by CloneAmp HiFi PCR Premix (Takara, Japan) according to the manufacturer's instructions. The *Tr*-milRNA1 deletion mutant was generated via a gene replacement strategy using the polyethylene glycol (PEG)-mediated protoplast transformation procedure as described in Zhang et al. [26].

To overexpress *Tr*-milRNA1 in NJAU 4742, the 400 bp fragment surrounding primary *Tr*-milRNA1 (a fragment of approximately 200 bp upstream and downstream of primary *Tr*-milRNA1) was amplified by PCR from NJAU 4742 genomic DNA and then introduced into plasmid pcDNA1 by the ClonExpress-II One Step Cloning Kit (Vazyme Biotech, Nanjing, China). In pcDNA1: *Tr*-milRNA1 precursor construct was expressed under the control of the cDNA1 promoter (Figure S1b). Mutated *Tr*-milRNA1 (OE-Mut-*Tr*-milRNA1) expression construct was obtained by a Fast Site-Directed Mutagenesis Kit (Tiangen) according to the manufacturer's instructions. The sequence of constructs was confirmed by sequencing (Tongyong Biological Technology, Chuzhou, China). The vectors of OE-*Tr*-milRNA1, OE-Mut-*Tr*-milRNA1 and empty vector were separately transformed into NJAU 4742 by using the polyethylene glycol (PEG)-mediated protoplast transformation procedure [26]. The relative expression level of *Tr*-milRNA1 between wt, mutants OE-*Tr*-milRNA1, OE-Mut-*Tr*-milRNA1 and empty vector transformant (EV) was measured following the above method (relative expression of *Tr*-milRNA1 at different sampling times). A diagram shows the strategy of *Tr*-milRNA editing and mutant detection used in this study (Figure S1a,b,d,f,g), and the primers are given in Table S1.

### 2.6. Generation of the Target Gene Mutants

To obtain single knockout mutant homologous recombination fragments of *Trvip36*, the 5′ and 3′ flanking regions of the gene open-reading frame (ORF) were amplified by PCR from NJAU 4742 genomic DNA, and the two fragments were ligated with *Hygb* through CloneAmp HiFi PCR Premix (Clontech) according to the manufacturer's instructions. To generate a *Trvip36* overexpression mutant homologous recombination fragment, three fragments including the 5′ flanking regions of the gene ORF, the promoter fragment and the ORF of the *Trvip36* gene fragment were amplified by PCR from NJAU 4742 genomic DNA and then fused with *Hygb* by using CloneAmp HiFi PCR Premix (Clontech). *Trvip36* deletion and overexpression mutants were generated by using homologous recombination and a polyethylene glycol (PEG)-mediated protoplast transformation system [26]. A diagram shows the strategy of gene editing and mutant detection used in this study (Figure S1a,c,e), and the primers are given in Table S1.

### 2.7. Growth and Enzyme Activity Assays of NJAU 4742

Equally harvested biomass samples of different treatments were transferred to the medium with rice straw as the sole carbon source and incubated at 37 °C to determine the growth rate. For enzyme activity assays, 1 mL of fresh spore suspension ($1.0 \times 10^7$ spores·mL$^{-1}$) of different strains was inoculated for SSF. All samples of different treatments were collected on the 4th day, and three biological replicates were collected at each sampling point. Filter paper activity (FPA) and endoglucanase activity (EG) were measured according to the method described by Xue et al. [27] with filter paper (Whatman NO.1) and CMC-Na (Sigma, St. Louis, MO, USA) as the substrates. Xylanase activity (XYL) was assayed with oat spelts xylan (Sigma, St. Louis, MO, USA) as the substrate [28]. The reaction system was executed in 0.1 M acetate buffer (pH 4.8) at 50 °C for 10 min, after which the DNS method was used to measure the released reducing sugars. The cellobiohydrolase activity (CBH) was determined in 0.1 M acetate buffer at 50 °C for 30 min with pNPC (Sigma, St. Louis, MO, USA) as the substrate according to Liu et al. [29]. One enzyme activity unit was defined as the amount of enzyme required to liberate 1 µmol glucose or pNP per minute under the assayed conditions.

*2.8. Confocal Imaging of EGL-GFP and CBH-GFP in the ER and Golgi Apparatus between wt and ΔTrvip36*

The DNA fractions of the representative lignocellulase genes including endoglucanase (*egl*, A1A110863.1) and cellobiohydrolase (*cbh*, A1A102028.1) were amplified by DNA of NJAU 4742, and the eGFP fragment was amplified by using plasmid pEGFP-N1 (Clontech, Mountain View, CA, USA). The DNA fractions of different cellulase genes were ligated with the GFP fragment by the overlapping-PCR technique based on the instructions of CloneAmp HiFi PCR Premix (Clontech, Mountain View, CA, USA). Mutants of endoglucanase-GFP (*egl*-GFP) and cellobiohydrolase-GFP (*cbh*-GFP) were obtained through homologous recombination based on wt and Δ*Trvip36* (Figures S1h and S2). Subsequently, these mutants labeled with GFP in situ were cultured in medium with rice straw as the sole carbon source for three days, and the freshly cultured mycelium was incubated with ER-Tracker™ Red (Thermo Fisher Scientific, Cat. M7512, Ex/Em = 587 nm/615 nm) or the BODIPY™ TR Ceramide (Thermo Fisher Scientific, D7540, Ex/Em = 589 nm/616 nm) at a concentration of 100 nM at room temperature for 5 min in darkness. Subsequently, the cellulase-GFP and ER or Golgi apparatus stained hyphae were imaged under a confocal fluorescence microscope (TCS SP8, Leica, Germany), and the fluorescence intensity and colocalization analysis were performed by plot profile in ImageJ software according to Zhao et al. [30]. All fluorescence intensity values of EGL-GFP and CBH-GFP in the ER and Golgi apparatus were counted when the fluorescence intensity value of ER-Tracker or Golgi-Tracker was valid (higher than 0). Data were completed as at least three independent biological replicates. Statistical data were expressed as means ± standard errors (SE) from all valid pixels.

## 3. Results

*3.1. Identification and Quantification of milRNAs in NJAU 4742*

In order to determine whether milRNAs are involved in the regulation of cellulase secretion in NJAU 4742 under heat stress, two small RNA libraries were generated from NJAU 4742, grown using rice straw rice straw as the sole carbon source at 28 °C (T28) and 37 °C (T37), respectively. In total, 56 putative milRNAs were identified in NJAU 4742, and 47 and 46 *Tr*-milRNAs were identified at T28 and T37, respectively (Table 1 and Figure 1a). *Tr*-milRNAs were enriched at 22 nt and seemed to possess a strong preference for uracil (Figure 1b,c). Interestingly, 10 and 9 *Tr*-milRNAs were specifically expressed at T28 or T37, respectively. Thirty-seven *Tr*-milRNAs were coexpressed in T28 and T37.

After normalization, 20 *Tr*-milRNAs with transcripts per million (TPM) values higher than 10 were identified. TGA1_S20_236772 and TGA1_S20_238808 exhibited a high abundance at 28 °C. In contrast, TGA1_S04_31828 (*Tr*-milRNA1), TGA1_S02_9056, TGA1_S05_43489 (*Tr*-milRNA2), TGA1_S06_56271 (*Tr*-milRNA3), TGA1_S10_93851 (*Tr*-milRNA4), TGA1_S17_158248 (*Tr*-milRNA5), TGA1_S17_162471, TGA1_S17_175238, TGA1_S17_175348, TGA1_S18_185930 (*Tr*-milRNA6), TGA1_S19_204554, TGA1_S20_223477, TGA1_S20_238860, TGA1_S20_238862, TGA1_S22_248446, TGA1_S22_248536, TGA1_S22_253299, TGA1_S22_257134 and TGA1_S30_293034 showed a high abundance at 37 °C (Table 1 and Figure 1d). At a false discovery rate <0.05 and $\log_{10}$ fold change >1 or <−1, six significantly upregulated *Tr*-milRNAs including *Tr*-milRNA1, *Tr*-milRNA2, *Tr*-milRNA3, *Tr*-milRNA4, *Tr*-milRNA5 and *Tr*-milRNA6 were selected (Figure 1e). This result indicated that at least several *Tr*-milRNAs might be involved in the regulation of lignocellulose utilization at different temperatures.

**Table 1.** Identification and expression abundance of milRNAs in NJAU 4742 by miRNA deep sequencing.

| miRNA_ID | Mirdeep2_Score | RNAfold_Results | TPM (T28) | TPM (T37) | Mature_Sequence |
|---|---|---|---|---|---|
| TGA1_S20_236772 | 2,295,965.4 | yes | 983,492.79 | 976,779.71 | uuuuugagauacuccgcaacgac |
| TGA1_S20_238860 | 31,534.5 | yes | 11,201.07 | 16,331.53 | uacguaggacuuuaccgugacgu |
| TGA1_S06_56271 | 115.5 | yes | 12.30 | 112.44 | uugccgaguggcagaggacuggcau |
| TGA1_S17_175348 | 105 | yes | 23.71 | 47.53 | uggaaguugaaucgagaagcccu |
| TGA1_S17_158248 | 85.8 | yes | 6.96 | 89.12 | gcaucugauuuccacccuuggguu |
| TGA1_S19_204554 | 43.5 | yes | 10.77 | 28.55 | agcuuuuggcuuuccagaacccgu |
| TGA1_S19_204542 | 38 | yes | 11.73 | 11.75 | uggcuggacggcccagagggccu |
| TGA1_S30_293034 | 14.6 | yes | 2.86 | 35.64 | cgcaggcucgauuguugucu |
| TGA1_S14_132594 | 13.2 | yes | 0.18 | 4.88 | cgccaacggcccucgcccggcuac |
| TGA1_S19_188258 | 12.5 | yes | 2.18 | 2.91 | caucugcauguugucucugg |
| TGA1_S10_93851 | 11.9 | yes | 0.56 | 18.67 | gucuagcacucuacuuuggcau |
| TGA1_S17_156622 | 9.3 | yes | 0.30 | 6.57 | uuuuuacacagauaccaguaggu |
| TGA1_S20_237911 | 7.5 | yes | 0.77 | 5.62 | ccaccaggccgcucaagacua |
| TGA1_S03_23817 | 7.1 | yes | 0.68 | 5.62 | cacugaccugcuccucgcacag |
| TGA1_S17_161581 | 7.1 | yes | 0 | 1.69 | caucuagcuuggacggcagcg |
| TGA1_S11_99749 | 6.5 | yes | 0.30 | 5.62 | ucgucacccuugaggcggaa |
| TGA1_S02_7777 | 5.4 | yes | 0.86 | 3.66 | auugcgaugccuggucagcuacuc |
| TGA1_S22_248536 | 5.4 | yes | 0.38 | 12.40 | ugaggagucugaagauggagaggaa |
| TGA1_S03_27309 | 5.3 | yes | 0.60 | 7.80 | guggacagauuagcugacccgcggg |
| TGA1_S15_136221 | 4.5 | yes | 0.36 | 0 | accaacagcggacauugcuccaa |
| TGA1_S11_101486 | 4 | yes | 0.60 | 0 | ugcuccaaaugagaaucgaagu |
| TGA1_S11_106603 | 3.7 | yes | 0.76 | 0 | ugccugguacauguacgga |
| TGA1_S07_71495 | 3.5 | yes | 1.80 | 0.85 | ugaccaagaacuucgacgucuu |
| TGA1_S15_142991 | 3.3 | yes | 0 | 2.54 | uuuggaccgugucuggaacgcua |
| TGA1_S17_178322 | 3.3 | yes | 0 | 5.62 | ucauauucucagcacuuggaau |
| TGA1_S02_16357 | 3.2 | yes | 0 | 1.70 | agguaacgucugguggcaa |
| TGA1_S05_35063 | 2.9 | yes | 0 | 4.98 | gcaagaucaaaacucaaa |
| TGA1_S17_175238 | 2.8 | yes | 4.34 | 13.62 | uggaaguugaaucgagaagccc |
| TGA1_S09_88615 | 2.7 | yes | 0 | 1.70 | cguguaccagagcgucau |
| TGA1_S08_81282 | 2.5 | yes | 0.60 | 0 | uccgauagggauguucgggcu |
| TGA1_S20_238862 | 2.5 | yes | 4634.89 | 4921.04 | uaggacuuuaccgugacgucc |
| TGA1_S22_257134 | 2.5 | yes | 1.45 | 12.20 | cgguggauugaacgggacucuuggu |
| TGA1_S05_43489 | 2.4 | yes | 28.74 | 491.72 | cucgggagaaggcggccu |
| TGA1_S07_78907 | 2.3 | yes | 0.36 | 0 | gucuucuugaucucuuau |
| TGA1_S18_185930 | 2.3 | yes | 25.59 | 160.54 | acccggcuuggagaugug |
| TGA1_S20_209761 | 2.3 | yes | 0.38 | 0 | ugaugauucggcuaguucggacag |
| TGA1_S22_248446 | 2.3 | yes | 0.94 | 10.71 | cggcgaggcuguguuucagcga |
| TGA1_S20_223477 | 2.2 | yes | 445.56 | 655.43 | uacguaggacucuaccgugacau |
| TGA1_S20_238808 | 2.2 | yes | 44.36 | 38.21 | guucgaggguugaaauga |
| TGA1_S06_55136 | 2 | yes | 0 | 1.70 | auguaugccuccugagauua |
| TGA1_S16_151294 | 2 | yes | 0 | 1.70 | acuucauccauagauaucgcaa |
| TGA1_S17_162471 | 2 | yes | 3.95 | 13.62 | uggaaguugaaucgagaagccc |
| TGA1_S19_199524 | 1.9 | yes | 0.18 | 0 | uaccucucauucacugcagug |
| TGA1_S02_9056 | 1.8 | yes | 1.77 | 26.56 | ggaaacaagguuguucugacuau |
| TGA1_S13_112804 | 1.8 | yes | 0.60 | 0 | uuauggugguacggcagagagu |
| TGA1_S04_31828 | 1.7 | yes | 0.30 | 23.35 | gguucgacucccggcuugu |
| TGA1_S14_132761 | 1.6 | yes | 7.96 | 6.95 | aacgugcaauugcuaccaa |
| TGA1_S20_229688 | 1.6 | yes | 2.62 | 5.62 | ugagggaccggauucgcca |
| TGA1_S22_253299 | 1.6 | yes | 15.54 | 38.89 | auccaaaagcucggcuuu |
| TGA1_S23_271212 | 1.5 | yes | 0 | 5.62 | agcgaggacauuuaugac |
| TGA1_S05_33550 | 1.4 | yes | 2.47 | 27.03 | accuagagaacgaugguucccauu |
| TGA1_S07_67387 | 1.3 | yes | 0.71 | 0 | aagcagauugcgagggggucauuug |
| TGA1_S22_265703 | 1.1 | yes | 0.38 | 9.86 | gggcagucuguuggacuccggu |
| TGA1_S07_76476 | 1 | yes | 1.14 | 0.85 | cucggucuguuguggauugguc |
| TGA1_S19_189543 | 0.7 | yes | 1.14 | 0 | aaggacauuuuggaggag |
| TGA1_S07_65040 | 0.1 | yes | 1.50 | 5.73 | ucgaaguugugguuguagugguagu |

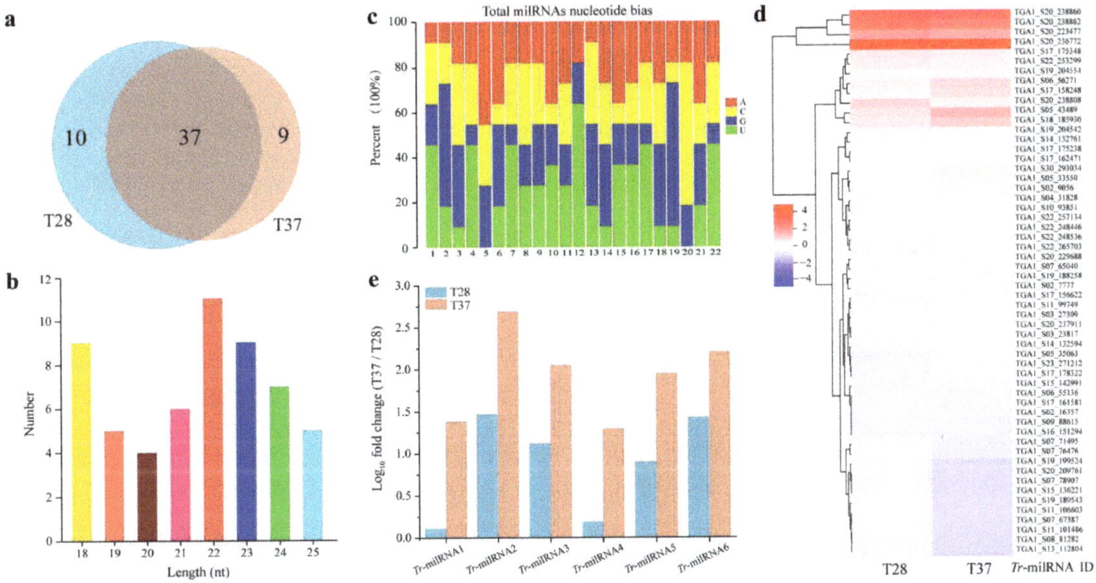


**Figure 1.** *Tr*-milRNAs are differentially expressed under SSF using rice straw as the sole carbon source at 28 °C (T28) and 37 °C (T37). (**a**) Different quantities of *Tr*-milRNAs were isolated from T28 (blue) and T37 (orange). Ten *Tr*-milRNAs and nine *Tr*-milRNAs are expressed specifically at T28 or T37, respectively. Thirty-seven *Tr*-milRNAs are coexpressed in T28 and T37; (**b**) length distribution of *Tr*-milRNAs; (**c**) nucleotide bias of *Tr*-milRNAs; (**d**) $Log_{10}$ (TPM) normalized values of all *Tr*-milRNAs were used to generate the heatmap. Red and blue colors indicate relatively high and low *Tr*-milRNA expression, respectively, and white equals median abundance; (**e**) normalized read numbers of *Tr*-milRNAs with transcripts per million (TPM) values higher than 10 in T28 or T37.

### 3.2. Identification of Candidate Target Genes Related to Tr-milRNAs of NJAU 4742

To investigate target genes of putative *Tr*-milRNAs, two rigorous target prediction software programs (TargetFinder and psRNATarget) were used to predict potential targets in the coding sequences, 3′UTR and 5′UTR of the NJAU 4742 genome. Gene ontology (GO) enrichment analysis indicated that candidate target genes were predictably involved in metabolic process, cellular process, single-organism process, catalytic activity, cell and membrane binding (Figure S3a). *Tr*-milRNA target genes also seem to be enriched through KEGG pathways including carbohydrate metabolism, catabolism, transport and translation (Figure S3b). To further ascertain whether the *Tr*-milRNA target genes were involved in the utilization of lignocellulose at different temperatures, putative target genes predictably involved in the secretion, synthesis and transportation of lignocellulases were selected. These findings might reveal that the differences in lignocellulose utilization at different temperatures are precisely regulated by *Tr*-milRNAs and their targets.

### 3.3. Tr-milRNA1 Exhibits a Critical Role in Lignocellulose Utilization at Different Temperatures

To explore the possible roles of *Tr*-milRNAs in lignocellulose utilization at different temperatures, overexpression transformants of *Tr*-milRNAs that were significantly upregulated at 37 °C were generated to screen their function under heat stress. The results showed that *Tr*-milRNA1 overexpression mutant (OE-*Tr*-milRNA1) could significantly increase lignocellulose utilization and also exhibited faster growth rate of mycelium using rice straw as the sole carbon source under heat stress than wt. Additionally, the pri-*Tr*-milRNA1 (primary transcript of *Tr*-milRNA1) deletion mutant (Δ*Tr*-milRNA1) exhibited normal growth, but showed a slight reduction in lignocellulose utilization capacity (Figure 2a). Various lignocellulosic enzyme activities were systematically assessed to determine the utilization efficiency of lignocellulose for filamentous fungi [31]. Therefore, different lignocellulases, including endoglucanase (EG), cellbiohydrolase (CBH), xylanase (XYL), and filter paper activity (FPA) were detected under SSF in different treatments. Noteworthy, EG and CBH activities increased significantly compared with wt under heat stress ($p < 0.05$). Correspondingly, various lignocellulosic enzyme activities of Δ*Tr*-milRNA1 decreased at different levels compared to wt (Figure 2b). These findings suggested that *Tr*-milRNA1 might be critical in regulating the utilization of lignocellulose at different temperatures.

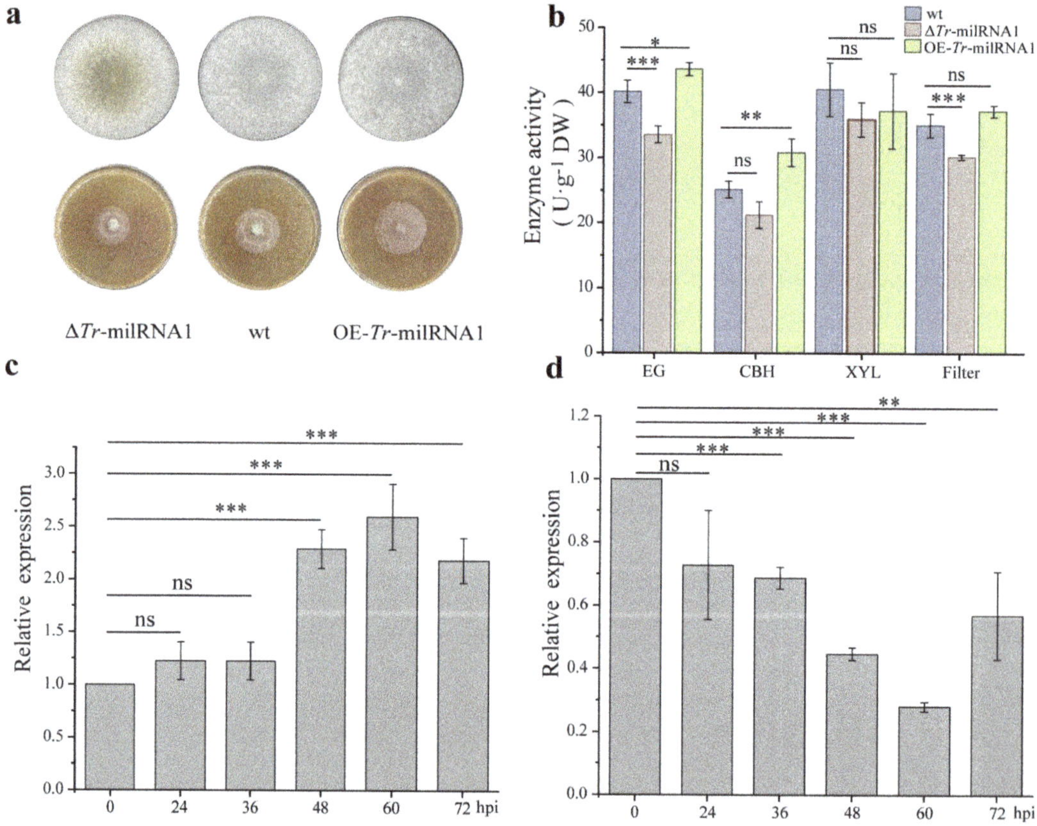


**Figure 2.** Overexpression of *Tr*-milRNA1 increased cellulase activities by silencing the target gene *Trvip36*. (**a**) Biomass comparison of wt and mutants on PDA at 28 °C (top row) and rice straw medium at 37 °C (bottom row); (**b**) the hydrolase activities, including FPA, EG, CBH and XYL of wt and mutants by using rice straw as the sole carbon source under heat stress; (**c**) relative expression levels of *Tr*-milRNA1 from 0 to 72 hpi at 37 °C. Relative expression levels of *Tr*-milRNA1 were normalized to gene 18S rRNA and calibrated to the levels of wt at 0 hpi (set as 1.0). (**d**) relative expression levels of target *Trvip36* from 0 to 72 hpi at 37 °C. Relative expression levels of *Trvip36* were normalized to the *Tef* gene and calibrated to the levels of wt at 0 hpi (set as 1). Data were calculated from three biological replicates, and error bars represent ± SDs. * $p < 0.05$, ** $p < 0.01$, *** $p < 0.001$. A $p$ value < 0.05 was regarded as statistically significant and ns refers to no significance.

### 3.4. Expression of *Trvip36* Could Be Regulated by *Tr*-milRNA1

Among the preselected target genes of *Tr*-milRNA1, we found that *Tr*-milRNA1 could base pair with several genes predictably related to lignocellulose utilization, including genes encoding carbohydrate hydrolases, transport proteins, and energy homeostasis related proteins. To determine the interaction between *Tr*-milRNA1 and the candidate target genes, we performed target verification tests. The relative expression of *Tr*-milRNA1 and predicted target genes were determined using stem-loop qRT–PCR and qRT–PCR to determine if the expression levels of *Tr*-milRNA1 and candidate target genes are correlated. During SSF, *Tr*-milRNA1 was upregulated at 36, 48, and 60 hpi and downregulated at 72 hpi (Figure 2c). Interestingly, we found that the transcript levels of the lectin-type cargo receptor gene *Trvip36* were downregulated at 0 to 60 hpi and then upregulated at 72 hpi (Figure 2d). However, the results for the other predicted target genes were illogical and contradictory. Thus, the negatively correlated expression of *Tr*-milRNA1 and *Trvip36* might indicate that *Trvip36* could be target of *Tr*-milRNA1.

### 3.5. *Tr-milRNA1 Regulates Target Gene Expression in a Sequence-Specific Manner*

*Tr*-milRNA1 had a 64 bp putative precursor, which might form a typical hairpin structure (Figure 3a). To confirm the regulatory mechanism of *Tr*-milRNA1, a mutated *Tr*-milRNA1 overexpression transformant (OE-Mut-*Tr*-milRNA1) was generated by site-specific mutagenesis (Figure 3b). Subsequently, the transcript level of *Trvip36* was quantified in wt, OE-*Tr*-milRNA1, OE-Mut-*Tr*-milRNA1, and EV at 60 hpi under SSF. Interestingly, the expression of *Trvip36* was significantly suppressed in OE-*Tr*-milRNA1, but not in OE-Mut-*Tr*-milRNA1 or EV (Figure 3c). These results suggested that *Tr*-milRNA1 might regulate the expression of *Trvip36* in a sequence-specific manner.

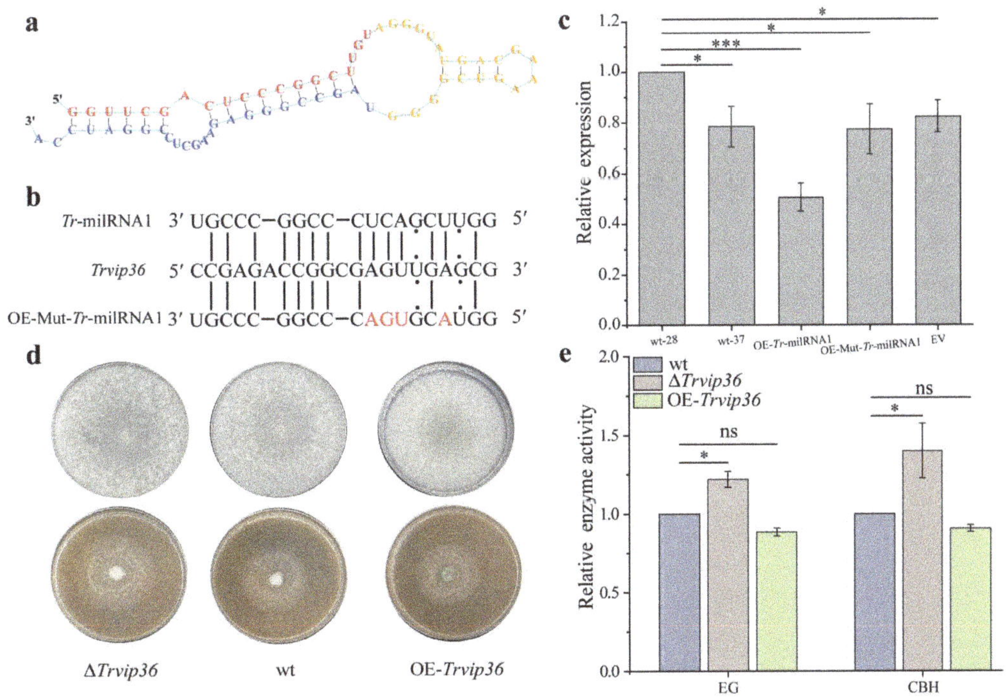


**Figure 3.** The function of *Tr*-milRNA1 in regulating the target gene *Trvip36*. (**a**) The precursor of *Tr*-milRNA1 could form a hairpin structure. The red line indicates the sequence of mature *Tr*-milRNA1; (**b**) alignment of *Tr*-milRNA1 and mutated *Tr*-milRNA1 (OE-Mut-*Tr*-milRNA1) with target gene (*Trvip36*) at the predicted binding sites; (**c**) relative expression of *Trvip36* in wt at 28 °C, and in wt, OE-*Tr*-milRNA1, OE-Mut-*Tr*-milRNA1 and EV at 37 °C. Relative expression of *Trvip36* was normalized to the *Tef* gene and calibrated to the levels of wt at 28 °C (set as 1.0). Data were calculated from three biological replicates, and error bars represent ± SDs. * $p < 0.05$, *** $p < 0.001$. A $p$ value < 0.05 was regarded as statistically significant and ns refers to no significance; (**d**) growth of wt and different mutants on PDA at 28 °C (top row) and rice straw medium at 37 °C (bottom row), respectively; (**e**) relative activities of EG and CBH between wt, Δ*Trvip36* and OE-*Trvip36* by using rice straw as the sole carbon source under heat stress.

### 3.6. Deletion of *Trvip36* Improved the Secretion of Lignocellulases of NJAU 4742 under Heat Stress

Deletion (Δ*Trvip36*) and overexpression (OE-*Trvip36*) strains of *Trvip36* were generated to determine the function of lignocellulose utilization under heat stress. Compared to wt, both mutants showed no difference in vegetative growth (Figure 3d). Nevertheless, Δ*Trvip36* and OE-*Trvip36* exhibited opposite growth trends, and Δ*Trvip36* showed better growth rate relative to wt by using rice straw as the sole carbon source under heat stress (Figure 3d). Previous results showed that the EG and CBH activities increased significantly in OE-*Tr*-milRNA1. Thus, these two enzyme activities were also measured in Δ*Trvip36* and OE-*Trvip36* to determine whether these changes were associated with the expression level of *Trvip36*. In correlation with *Trvip36*, EG and CBH activities increased significantly in Δ*Trvip36* and slightly decreased in OE-*Trvip36* (Figure 3e). These findings suggested that activation of *Trvip36* might suppress lignocellulase secretion, and we concluded that impaired expression of *Trvip36* regulated by *Tr*-milRNA1 contributes to the utilization of lignocellulose under heat stress.

### 3.7. Deletion of *Trvip36* Reduces the Retention of Lignocellulases in the ER of NJAU 4742 under Heat Stress

In mammalian cells, VIP36 dynamically localized to the cis-Golgi and ER-Golgi intermediate compartment, and modification of its N-linked carbohydrate together with other evidence suggested that VIP36 possessed a role in trafficking through later Golgi compartments [32,33]. VIPL, a VIP36-like protein, is widely expressed in animal cells and mainly localizes to the ER [34]. In filamentous fungi, VIP36 was also found to localize in the ER and Golgi compartments and a higher fraction of VIP36 co-localized with the Golgi marker (Golgi protein *Ao*Grh1 fused to EGFP) than with the ER marker (the ER membrane protein *Ao*ClxA fused to EGFP) [19]. We inferred that they might interfere with the normal intracellular transport of cellulases. Thus, strains expressing *egl*-GFP or *cbh*-GFP fusion proteins were generated in wt and Δ*Trvip*36 genetic backgrounds to track their intracellular localization. Cellulases tagged with eGFP were successfully expressed in hyphae of different mutants. To elucidate the aggregation level of cellulases in the ER and Golgi apparatus, the fluorescence intensities of EGL-GFP and CBH-GFP were detected in the ER and Golgi apparatus after staining with ER-Tracker™ Red and BODIPY™ TR Ceramide, respectively (Figure 4a,b,e,f). EGL-GFP and CBH-GFP intensities were analyzed in each pixel by ImageJ software. The average intensity of GFP fluorescence was calculated to measure the expression levels of EGL-GFP and CBH-GFP in the ER and Golgi apparatus. In wt, the average intensity of EGL-GFP and CBH-GFP was $20.13 \pm 1.70$ and $31.65 \pm 1.32$ in the ER, $50.29 \pm 1.60$ and $21.71 \pm 2.03$ in the Golgi apparatus, respectively. Compared to wt, Δ*Trvip36* exhibited lower average intensities of EGL-GFP ($15.03 \pm 0.99$) and CBH-GFP ($17.21 \pm 0.94$) fluorescence in the ER ($p < 0.01$). In contrast, higher average fluorescence intensities levels of EGL-GFP ($86.85 \pm 4.01$) and CBH-GFP ($69.09 \pm 3.18$) were distributed in the Golgi apparatus of Δ*Trvip36* ($p < 0.001$) (Figure 4c,d,g,h and Figure S4). These results suggest that VIP36 could interfere with the secretion of EGL and CBH by increasing the ER retention rate, which is consistent with the increased secretion of EGL and CBH in Δ*Trvip36*.

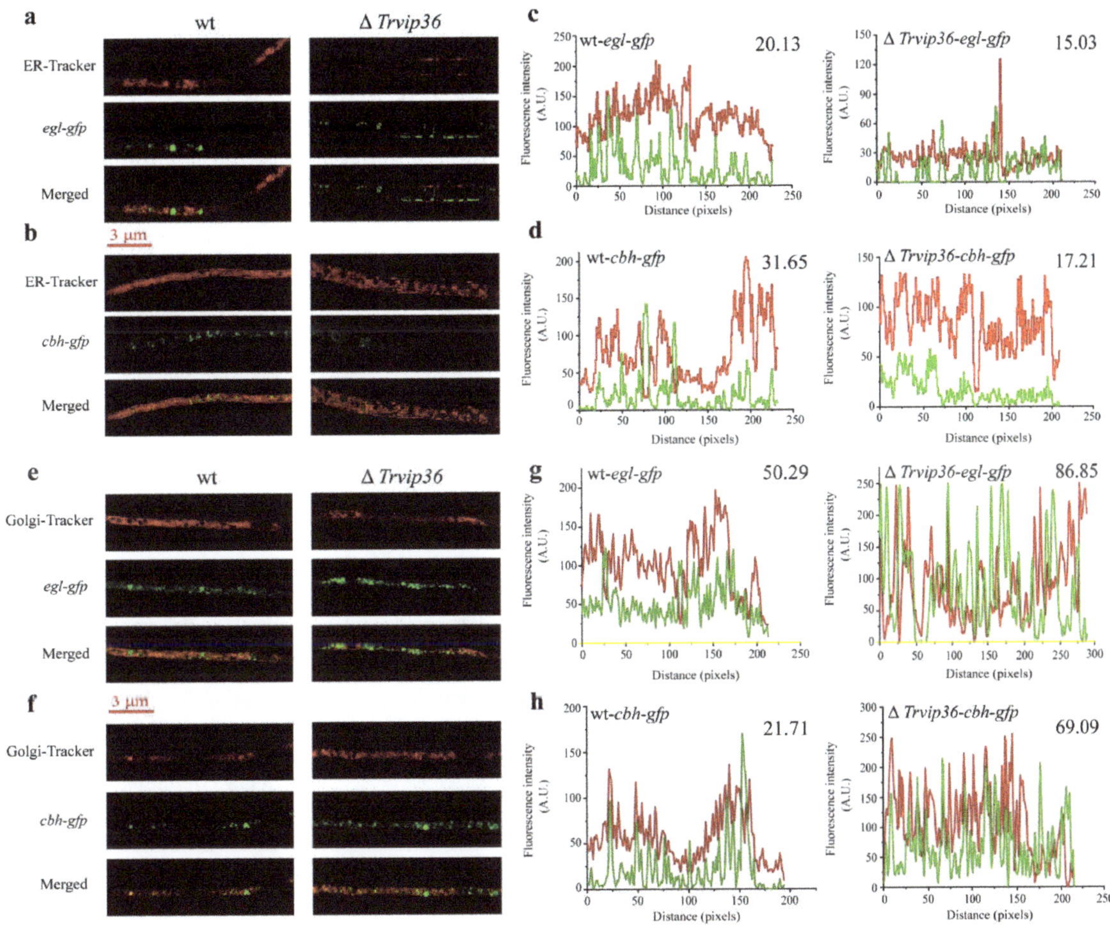


**Figure 4.** Localization analysis of EGL-GFP and CBH-GFP in wt and Δ*Trvip36* genetic backgrounds after staining hyphae with ER-Tracker™ and BODIPY™ TR Ceramide, as markers of ER and Golgi compartments. (**a**,**b**) Confocal images of hyphae from *egl*-GFP and *cbh*-GFP fusion strains after staining with ER-Tracker™ Red to track endoglucanase and cellobiohydrolase localization in the ER between wt and Δ*Trvip36*. Bar = 3 μm; (**c**,**d**) the fluorescence intensity of the two channels (ER-Tracker (red) and GFP (green)) detected in wt and Δ*Trvip36*; (**e**,**f**) confocal images of hyphae from the *egl*-GFP and *cbh*-GFP fusion strains after staining with BODIPY™ TR ceramide to track endoglucanase and cellobiohydrolase localization in the Golgi apparatus between wt and Δ*Trvip36*. Bar = 3 μm; (**g**,**h**) the fluorescence intensity of the two channels (Golgi-Tracker (red) and GFP (green)) detected in wt and Δ*Trvip36*.

## 4. Discussion

Normally, most soil fungi experience a constantly fluctuating environment. Throughout the evolution process, fungi have developed different modes of reproduction and the ability to adapt to complex environments [35]. Generally, fungi operate in narrower temperature ranges, and a decrease in temperature may cause fungal dormancy, while an increase in temperature would lead to the apoptosis of fungi [36]. In fact, a substantial increase in temperature causes attenuation and eventually leads to the death of many organism [37]. The changes associated with heat stress responded to a set of proteins that promoted the survival of the organism. A major portion of these proteins were termed heat-shock proteins (Hsps). Hsps play a major role in cell cycle progression, replication, transcription and post-translational processes, including protein folding, stability, transportation and degradation, and they were also reported to activate many key signal transducers in fungi [38,39]. Therefore, filamentous fungi could precisely regulate the expression of specific genes to maintain essential biological functions under high temperature environment.

In our previous study, we found that cellulase secretion in strain NJAU 4742 was largely limited under heat stress during the SSF process, which seriously inhibited its ability to obtain carbon source. In addition, through proteomic analysis, it was found that some differentially expressed genes responded to temperature changes to regulate cellulase secretion (unpublished data). In this study, we first reported that *Tr*-milRNA was involved in the regulation of cellulase secretion under heat stress. It is generally accepted that miRNAs from plants are integral components of plant responses to disadvantageous environmental conditions [40,41]. For example, in *Arabidopsis*, miR398 was shown to be quickly induced under heat stress, accompanied by the downregulation of its target genes in response to heat stress [42]. Similarly, miR156 could accelerate the sustained expression of heat stress responsive genes through SPL genes [43]. Here, *Tr*-milRNA1, from filamentous fungi NJAU 4742, was revealed to be related to cellulase secretion under heat stress during SSF by regulating its target gene *Trvip36*.

Although numerous milRNAs have been identified in fungi [44,45], no homologous miRNAs were found in NJAU 4742 by mapping the clean reads of *Tr*-milRNA1 to miRBase 22.0. A similar result was discovered in *T. reesei* and *F. oxysporum*, which indicated a large degree of species specificity of fungal milRNAs [46]. More interestingly, there have been few relevant studies on the regulation of cellulase secretion by milRNAs, especially under heat stress. Previously, differential expression of milRNAs in *T. reesei* was discovered when *T. reesei* was cultivated in Avicel medium or glucose medium, which implied that milRNA might be involved in fungal growth and cellulase secretion [14]. In *Ganoderma lucidum*, several milRNA target genes were also identified by high-throughput sequencing and degradome analysis, and were considered glycoside hydrolases involved in the biosynthesis of polysaccharides [47]. Thus, it seemed that milRNAs were related to the regulation of cellulose degradation but there was a lack of in-depth evidence. In plants and animals, miRNAs could play vital regulatory roles by targeting mRNAs for cleavage or translational repression [10], while the regulatory mechanism of milRNAs in fungi is largely unknown. In this study, *Tr*-milRNA1 negatively regulated the gene expression of lectin-type cargo receptor by suppressing the expression of *Trvip36*. There was evidence that *Tr*-milRNA1 regulated target gene expression in a sequence specific mode. During SSF, upregulation of *Tr*-milRNA1 increased the repression effect of *Trvip36*, which promoted cellulase secretion and exhibited better growth rate under heat stress. Deletion mutant of *Trvip36* also showed a significant increase in cellulosic enzyme activities, which might be due to the reduction in cellulase retention in the ER compared to wt. Consistent with this speculation, in *Aspergillus oryzae*, *AoVip*36 was shown to have a protein retention function in the ER and the deletion of *AoVip*36 could increase the secretion of heterologous proteins [19]. In mammals, glycoprotein $\alpha$1-antitrypsin was determined to be the retention target of VIP36, and it exhibited a significant promotion of intracellular transport upon silencing VIP36 [18]. VIP36 could also form a complex with BiP independently to increase its N-glycan binding activity, and overproduction of a lectin-deficient form of VIP36 could decrease the secretion of clusterin [17]. These results suggested that VIP36 could also negatively interfere with protein transport in an N-glycan-independent manner [48]. In addition, it was reported that VIP36 could be recycled between the Golgi and the ER and carry out the retrograde transport of glycoproteins [49]. Moreover, it is worth noting that heat and nutrient deprivation stress could lead to the generation of protein aggregates in the ER [50]. Avoiding the excessive aggregation of proteins in the ER could alleviate ER stress, which would be conducive to protein synthesis and secretion [51]. Thus, we speculated that deletion of *Trvip36* might also contribute to reducing the retrograde transport of lignocellulases and preventing the aggregation of cellulases in the ER to relieve ER stress thereby increasing the secretion of cellulases by using rice straw as the sole carbon source under heat stress. In addition, it was reported that deletion of VIP36 would impair the secretion of α-amylase [19,52], which might suggest that some unnecessary pathways would be decreased to avoid excessive energy consumption of filamentous fungi under stress (Figure 5). In fact, the main component of rice straw is lignocellulose [53], and producing more lignocellulases is the most important way to obtain energy and maintain vegetative growth, especially under heat stress. However, its internal regulatory mechanism between *Trvip36* and the secretion of α-amylase in NJAU 4742 still needs in-depth and detailed study in the future.

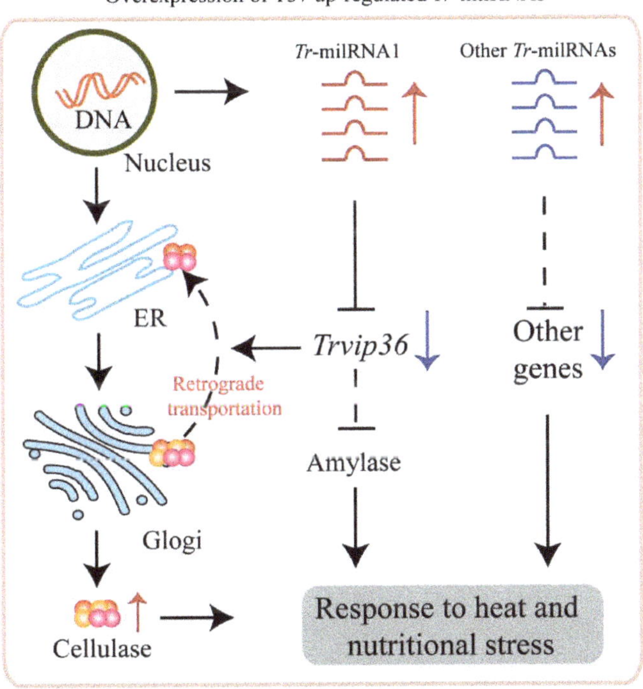


**Figure 5.** Proposed model for the regulation of NJAU 4742 by *Tr*-milRNAs during the lignocellulose utilization process under heat stress. *Tr*-milRNA1 is involved in regulating cellulase secretion by revoking the expression of *Trvip36* to decrease the cellulase retrograde transportation.

## 5. Conclusions

Overall, this study demonstrated that *Tr*-milRNA1 from NJAU 4742 played a critical role in cellulase secretion by regulating the endogenous target gene *Trvip36*. These results provide important evidence to determine the roles of milRNA and their corresponding target gene during the utilization of lignocellulose in filamentous fungi under heat stress.

**Supplementary Materials:** The following are available online at https://www.mdpi.com/article/10.3390/jof7120997/s1, Figure S1: Knockout or over-expression of *Tr*-milRNA or gene in NJAU 4742. (a) Schematic diagram for *Tr*-milRNA or gene disruption by double crossover recombination; (b) schematic diagram for *Tr*-milRNA over-expression; (c) schematic diagram for gene over-expression; (d) analysis of expression level of *Tr*-milRNA relative to 18S in wt, OE-*Tr*-milRNA1, OE-Mut-*Tr*-milRNA1, and empty vector transformant (EV) strains determined by qPCR. (e) analysis of expression level of *Trvip36* gene relative to *Tef* in wt and OE-*Trvip36* strain determined by qPCR. Data were calculated from three biological replicates. Error bars represent ± SDs. * $p < 0.05$, ** $p < 0.01$, *** $p < 0.001$. A *P*-value < 0.05 was regarded as statistically significant. The expression values are normalized to wt; (f,g) verification of Δ*Tr*-milRNA1 (f), Δ*Trvip36* (g), wt-*egl*-GFP, Δ*Trvip36*-*egl*-GFP, wt-*cbh*-GFP, and Δ*Trvip36*-*cbh*-GFP (h) mutants by PCR to verify homologous recombination. Figure S2:

Diagram of construction principle for prepaing the mutants of lignocellulases-eGFP fusion strains. Arm1 and Arm2 were used as two arms of homologous recombination and *Hygb* gene was used as the biomaker for screening. Figure S3: GO (a) and KEGG (b) enrichment of candidate *Tr*-milRNAs target genes. Figure S4: The average intensity of EGL-GFP and CBH-GFP fluorescence in the ER and Golgi apparatus of wt and Δ*Trvip36* after staining hyphae with ER-Tracker™ and BODIPY™ TR Ceramide, as markers of ER and Golgi compartments. Data were calculated from all valid pixels in the ER and Golgi apparatus, and error bars represent ± SEs. * $p < 0.05$, ** $p < 0.01$, *** $p < 0.001$. A $p$ value < 0.05 was regarded as statistically significant and ns refers to no significance. Table S1: All PCR primers used in this study.

**Author Contributions:** T.L. (Tuo Li) performed the majority of the experiments and wrote the manuscript; J.L. analyzed the sequencing data of *Tr*-milRNAs; Q.W. determined the lignocellulases activities; Y.L. constructed the mutant OE-*Tr*-milRNA1; T.L. (Ting Li) constructed the mutant Δ*Trvip36*; D.L. and Q.S. assisted in the drafting and revision of the manuscript. D.L. was the corresponding author. All authors have read and agreed to the published version of the manuscript.

**Funding:** This research was financially supported by National Natural Science Foundation of China (31972513), the National Key R & D Program of China (2018YFD0500201), and the Fundamental Research Funds for the Central Universities (KYZ201716).

**Institutional Review Board Statement:** Not applicable.

**Informed Consent Statement:** Not applicable.

**Data Availability Statement:** The data that support the findings of this study areavailable on request from the corresponding author.

**Acknowledgments:** We gratefully acknowledge all fundings for supporting this research work.

**Conflicts of Interest:** The authors declare no conflict of interest. The funders had no role in the design of the study; in the collection, analyses or interpretation of data; in the writing of the manuscript or in the decision to publish the results.

## References

1. Chang, S.S.; Zhang, Z.; Liu, Y. RNA Interference Pathways in Fungi: Mechanisms and Functions. *Annu. Rev. Microbiol.* **2012**, *66*, 305–323. [CrossRef]
2. Castel, S.E.; Martienssen, R.A. RNA interference in the nucleus: Roles for small RNAs in transcription, epigenetics and beyond. *Nat. Rev. Genet.* **2013**, *14*, 100–112. [CrossRef] [PubMed]
3. Holoch, D.; Moazed, D. RNA-mediated epigenetic regulation of gene expression. *Nat. Rev. Genet.* **2015**, *16*, 71–84. [CrossRef]
4. Zhang, T.; Zhao, Y.L.; Zhao, J.H.; Wang, S.; Jin, Y.; Chen, Z.Q.; Fang, Y.Y.; Hua, C.L.; Ding, S.W.; Guo, H.S. Cotton plants export microRNAs to inhibit virulence gene expression in a fungal pathogen. *Nat. Plants* **2016**, *2*, 16153. [CrossRef] [PubMed]
5. Li, L.; Chang, S.S.; Liu, Y. RNA interference Pathways in Filamentous Fungi. *Cell. Mol. Life Sci.* **2010**, *67*, 3849–3863. [CrossRef]
6. Romano, N.; Macino, G. Quelling: Transient inactivation of gene expression in *Neurospora crassa* by transformation with homologous sequences. *Mol. Microbiol.* **1992**, *6*, 3343–3353. [CrossRef]
7. Verdel, A.; Moazed, D. RNAi-directed assembly of heterochromatin in *Fission yeast*. *FEBS Lett.* **2005**, *579*, 5872–5878. [CrossRef]
8. Carthew, R.W.; Sontheimer, E.J. Origins and mechanisms of miRNAs and siRNAs. *Cell* **2009**, *136*, 642–655. [CrossRef]
9. Mallory, A.C.; Vaucheret, H. Functions of microRNAs and related small RNAs in plants. *Nat. Genet.* **2006**, *38*, S31–S36. [CrossRef] [PubMed]
10. Bartel, D.P. MicroRNAs: Genomics, Biogenesis, Mechanism, and Function. *Cell* **2004**, *116*, 281–297. [CrossRef]
11. Lee, H.C.; Li, L.; Gu, W.; Xue, Z.; Yi, L. Diverse Pathways Generate MicroRNA-like RNAs and Dicer-Independent Small Interfering RNAs in Fungi. *Mol. Cell* **2010**, *38*, 803–814. [CrossRef]
12. Zhou, Q.; Wang, Z.; Zhang, J.; Meng, H.; Huang, B. Genome-wide identification and profiling of microRNA-like RNAs from *Metarhizium anisopliae* during development. *Fungal Biol.* **2012**, *116*, 1156–1162. [CrossRef]
13. Weiberg, A.; Wang, M.; Lin, F.M.; Zhao, H.; Zhang, Z.; Kaloshian, I.; Huang, H.D.; Jin, H. Fungal Small RNAs Suppress Plant Immunity by Hijacking Host RNA Interference Pathways. *Science* **2013**, *342*, 118–123. [CrossRef]
14. Kang, K.; Zhong, J.; Jiang, L.; Liu, G.; Gou, C.Y.; Wu, Q.; Wang, Y.; Luo, J.; Gou, D. Identification of microRNA-Like RNAs in the filamentous fungus *Trichoderma reesei* by solexa sequencing. *PLoS ONE* **2013**, *8*, e76288. [CrossRef]
15. Jin, Y.; Zhao, J.H.; Zhao, P.; Zhang, T.; Wang, S.; Guo, H.S. A fungal milRNA mediates epigenetic repression of a virulence gene in *Verticillium dahliae*. *Philos. Trans. R. Soc. Lond. Ser. B Biol. Sci.* **2019**, *374*, 20180309. [CrossRef]
16. Druzhinina, I.S.; Seidl-Seiboth, V.; Herrera-Estrella, A.; Horwitz, B.A.; Kenerley, C.M.; Monte, E.; Mukherjee, P.K.; Zeilinger, S.; Grigoriev, I.V.; Kubicek, C.P. *Trichoderma*: The genomics of opportunistic success. *Nat. Rev. Microbiol.* **2011**, *9*, 749–759. [CrossRef] [PubMed]

17. Daisuke, N.; Osamu, S.; Norihito, K.; Naoki, M.; Kazuo, Y. Stable interaction of the cargo receptor VIP36 with molecular chaperone BiP. *Glycobiology* **2007**, *17*, 913–921.
18. Reiterer, V.; Nyfeler, B.; Hauri, H.P. Role of the lectin VIP36 in post-ER quality control of human alpha1-antitrypsin. *Traffic* **2010**, *11*, 1044–1055. [CrossRef]
19. Hoang, H.D.; Maruyama, J.I.; Kitamoto, K.; Brakhage, A. Modulating Endoplasmic Reticulum-Golgi Cargo Receptors for Improving Secretion of Carrier-Fused Heterologous Proteins in the Filamentous Fungus *Aspergillus oryzae*. *Appl. Environ. Microbiol.* **2015**, *81*, 533–543. [CrossRef]
20. Miao, J.; Wang, M.; Ma, L.; Li, T.; Huang, Q.; Liu, D.; Shen, Q. Effects of amino acids on the lignocellulose degradation by *Aspergillus fumigatus Z5*: Insights into performance, transcriptional, and proteomic profiles. *Biotechnol. Biofuels* **2019**, *12*, 1–19. [CrossRef] [PubMed]
21. Ana, K.; Sam, G.J. miRBase: Integrating microRNA annotation and deep-sequencing data. *Nucleic Acids Res.* **2011**, *39*, D152–D157.
22. Mackowiak, S.D. Identification of novel and known miRNAs in deep-sequencing data with miRDeep2. *Curr. Protoc. Bioinform.* **2011**, *36*, 12.10.1–12.10.15. [CrossRef]
23. Bo, X.; Wang, S. TargetFinder: A software for antisense oligonucleotide target site selection based on MAST and secondary structures of target mRNA. *Bioinformatics* **2005**, *21*, 1401–1402. [CrossRef]
24. Dai, X.; Zhao, P.X. psRNATarget: A plant small RNA target analysis server. *Nucleic Acids Res.* **2011**, *39*, W155–W159. [CrossRef]
25. Varkonyi-Gasic, E.; Wu, R.; Wood, M.; Walton, E.F.; Hellens, R.P. Protocol: A highly sensitive RT-PCR method for detection and quantification of microRNAs. *Plant Methods* **2007**, *3*, 12. [CrossRef]
26. Zhang, J.; Miao, Y.; Rahimi, M.J.; Zhu, H.; Steindorff, A.; Schiessler, S.; Cai, F.; Pang, G.; Chenthamara, K.; Xu, Y. Guttation capsules containing hydrogen peroxide: An evolutionarily conserved NADPH oxidase gains a role in wars between related fungi. *Environ. Microbiol.* **2019**, *21*, 2644–2658. [CrossRef]
27. Xue, D.; Lin, D.; Gong, C.; Peng, C.; Yao, S. Expression of a bifunctional cellulase with exoglucanase and endoglucanase activities to enhance the hydrolysis ability of cellulase from a marine *Aspergillus niger*. *Process Biochem.* **2017**, *52*, 115–122. [CrossRef]
28. Miao, Y.; Kong, Y.; Li, P.; Li, G.; Liu, D.; Shen, Q.; Zhang, R. Effect of CBM1 and linker region on enzymatic properties of a novel thermostable dimeric GH10 xylanase (Xyn10A) from filamentous fungus *Aspergillus fumigatus Z5*. *AMB Express* **2018**, *8*, 1–10. [CrossRef]
29. Liu, D.; Li, J.; Zhao, S.; Zhang, R.; Wang, M.; Miao, Y.; Shen, Y.; Shen, Q. Secretome diversity and quantitative analysis of cellulolytic *Aspergillus fumigatus Z5* in the presence of different carbon sources. *Biotechnol. Biofuels* **2013**, *6*, 1–16. [CrossRef]
30. Zhao, X.; Yong, F.; Yang, Y.; Yu, Q.; Ding, M. Elaiophylin, a novel autophagy inhibitor, exerts antitumor activity as a single agent in ovarian cancer cells. *Autophagy* **2015**, *11*, 1849–1863. [CrossRef] [PubMed]
31. Wi, S.G.; Cho, E.J.; Lee, D.S.; Lee, S.J.; Lee, Y.J.; Bae, H.J. Lignocellulose conversion for biofuel: A new pretreatment greatly improves downstream biocatalytic hydrolysis of various lignocellulosic materials. *Biotechnol. Biofuels* **2015**, *8*, 228. [CrossRef] [PubMed]
32. Hara-Kuge, S. Involvement of VIP36 in Intracellular Transport and Secretion of Glycoproteins in Polarized Madin-Darby Canine Kidney (MDCK) Cells. *J. Biol. Chem.* **2002**, *277*, 16332. [CrossRef] [PubMed]
33. Shimada, O.; Hara-Kuge, S.; Yamashita, K.; Tosaka-Shimada, H.; Yanchao, L.; Einan, L.; Atsumi, S.; Ishikawa, H. Localization of VIP36 in the post-Golgi secretory pathway also of rat parotid acinar cells. *J. Histochem. Cytochem. Off. J. Histochem. Soc.* **2003**, *51*, 1057–1063. [CrossRef] [PubMed]
34. Neve, E.; Svensson, K.; Fuxe, J.; Pettersson, R.F. VIPL, a VIP36-like membrane protein with a putative function in the export of glycoproteins from the endoplasmic reticulum. *Exp. Cell Res.* **2003**, *288*, 70–83. [CrossRef]
35. Lamoth, F.; Juvvadi, P.R.; Steinbach, W.J. Heat shock protein 90 (Hsp90) in fungal growth and pathogenesis. *Curr. Fungal Infect. Rep.* **2014**, *8*, 296–301. [CrossRef]
36. Lamoth, F.; Juvvadi, P.R.; Fortwendel, J.R.; Steinbach, W.J. Heat shock protein 90 is required for conidiation and cell wall integrity in Aspergillus fumigatus. *Eukaryot. Cell* **2012**, *11*, 1324–1332. [CrossRef]
37. Bhabhra, R.; Askew, D. Thermotolerance and virulence of *Aspergillus fumigatus*: Role of the fungal nucleolus. *Med. Mycol.* **2005**, *43*, S87–S93. [CrossRef]
38. Kregel, K.C. Invited Review: Heat shock proteins: Modifying factors in physiological stress responses and acquired thermotolerance. *J. Appl. Physiol.* **2002**, *192*, 2177–2186. [CrossRef]
39. Verghese, J.; Abrams, J.; Wang, Y.; Morano, K.A. Biology of the Heat Shock Response and Protein Chaperones: *Budding Yeast (Saccharomyces cerevisiae)* as a Model System. *Microbiol. Mol. Biol. Rev.* **2012**, *76*, 115–158. [CrossRef]
40. Katiyar-Agarwal, S.; Jin, H. Role of Small RNAs in Host-Microbe Interactions. *Annu. Rev. Phytopathol.* **2010**, *48*, 225. [CrossRef]
41. Li, S.; Castillo-Gonzalez, C.; Yu, B.; Zhang, X. The functions of plant small RNAs in development and in stress responses. *Plant J.* **2017**, *90*, 654. [CrossRef]
42. Guan, Q.; Lu, X.; Zeng, H.; Zhang, Y.; Zhu, J. Heat stress induction of miR398 triggers a regulatory loop that is critical for thermotolerance in *Arabidopsis*. *Plant J.* **2013**, *74*, 840–851. [CrossRef] [PubMed]
43. Cui, L.G.; Shan, J.X.; Shi, M.; Gao, J.P.; Lin, H.X. The miR156-SPL 9-DFR pathway coordinates the relationship between development and abiotic stress tolerance in plants. *Plant J.* **2014**, *80*, 1108–1117. [CrossRef] [PubMed]
44. Xu, M.; Guo, Y.; Tian, R.; Gao, C.; Guo, F.; Voegele, R.T.; Bao, J.; Li, C.; Jia, C.; Feng, H. Adaptive regulation of virulence genes by microRNA-like RNAs in *Valsa mali*. *New Phytol.* **2020**, *227*, 899–913. [CrossRef]

45. Feng, H.; Xu, M.; Gao, Y.; Liang, J.; Guo, F.; Guo, Y.; Huang, L. *Vm*-milR37 contributes to pathogenicity by regulating glutathione peroxidase gene VmGP in *Valsa mali*. *Mol. Plant Pathol.* **2021**, *22*, 243–254. [CrossRef]
46. Chen, R.; Jiang, N.; Jiang, Q.; Sun, X.; Wang, Y.; Zhang, H.; Hu, Z. Exploring microRNA-like small RNAs in the filamentous fungus *Fusarium oxysporum*. *PLoS ONE* **2014**, *9*, e104956.
47. Js, A.; Lw, A.; Yang, L.B.; Qq, B.; Bw, B.; Sl, A.; Chang, L.A. Identification of milRNAs and their target genes in *Ganoderma lucidum* by high-throughput sequencing and degradome analysis. *Fungal Genet. Biol.* **2020**, *136*, 103313.
48. O'Keefe, S. *Small Molecule Inhibitors Targeting Protein Biogenesis at the Endoplasmic Reticulum*; The University of Manchester: Manchester, UK, 2019.
49. Kamiya, Y.; Kato, K. Sugar Recognition by Intracellular Lectins That Determine the Fates of Glycoproteins. *Trends Glycosci. Glycotechnol.* **2006**, *18*, 231–244. [CrossRef]
50. Tyedmers, J.; Mogk, A.; Bukau, B. Cellular strategies for controlling protein aggregation. *Nat. Rev. Mol. Cell Biol.* **2010**, *11*, 777–788. [CrossRef] [PubMed]
51. Buchberger, A.; Bukau, B.; Sommer, T. Protein quality control in the cytosol and the endoplasmic reticulum: Brothers in arms. *Mol. Cell* **2010**, *40*, 238–252. [CrossRef]
52. Hara-Kuge, S.; Seko, A.; Shimada, O.; Tosaka-Shimada, H.; Yamashita, K. The binding of VIP36 and α-amylase in the secretory vesicles via high-mannose type glycans. *Glycobiology* **2004**, *14*, 739–744. [CrossRef] [PubMed]
53. Gu, Y.; Chen, X.; Liu, Z.; Zhou, X.; Zhang, Y. Effect of inoculum sources on the anaerobic digestion of rice straw. *Bioresour. Technol.* **2014**, *158*, 149–155. [CrossRef] [PubMed]

Journal of *Fungi*

MDPI

*Review*

# Fungal Laccases: The Forefront of Enzymes for Sustainability

**Martina Loi [1], Olga Glazunova [2], Tatyana Fedorova [2], Antonio F. Logrieco [1] and Giuseppina Mulè [1,*]**

1 Institute of Sciences of Food Production, National Research Council, Via Amendola 122/O, 70126 Bari, Italy; martina.loi@ispa.cnr.it (M.L.); antonio.logrieco@ispa.cnr.it (A.F.L.)
2 A.N. Bach Institute of Biochemistry, Research Center of Biotechnology of the Russian Academy of Sciences, 119071 Moscow, Russia; olga.a.glas@gmail.com (O.G.); fedorova_tv@mail.ru (T.F.)
* Correspondence: giuseppina.mule@ispa.cnr.it; Tel.: +39-08-0592-9329

**Abstract:** Enzymatic catalysis is one of the main pillars of sustainability for industrial production. Enzyme application allows minimization of the use of toxic solvents and to valorize the agro-industrial residues through reuse. In addition, they are safe and energy efficient. Nonetheless, their use in biotechnological processes is still hindered by the cost, stability, and low rate of recycling and reuse. Among the many industrial enzymes, fungal laccases (LCs) are perfect candidates to serve as a biotechnological tool as they are outstanding, versatile catalytic oxidants, only requiring molecular oxygen to function. LCs are able to degrade phenolic components of lignin, allowing them to efficiently reuse the lignocellulosic biomass for the production of enzymes, bioactive compounds, or clean energy, while minimizing the use of chemicals. Therefore, this review aims to give an overview of fungal LC, a promising green and sustainable enzyme, its mechanism of action, advantages, disadvantages, and solutions for its use as a tool to reduce the environmental and economic impact of industrial processes with a particular insight on the reuse of agro-wastes.

**Keywords:** enzymes; sustainability; fungal laccase; solid state fermentation; agro-wastes; immobilization; catalysis

**Citation:** Loi, M.; Glazunova, O.; Fedorova, T.; Logrieco, A.F.; Mulè, G. Fungal Laccases: The Forefront of Enzymes for Sustainability. *J. Fungi* **2021**, 7, 1048. https://doi.org/10.3390/jof7121048

Academic Editor: Baojun Xu

Received: 17 November 2021
Accepted: 4 December 2021
Published: 7 December 2021



## 1. Introduction

Enzyme biotechnology is a green and efficient process which has been applied to various industrial operations as an alternative to chemical catalysis with advantages in terms of sustainability and efficiency [1]. Indeed, enzymatic catalysis enables the tackling of most issues that are related to environmental sustainability, especially regarding the use of hazardous chemicals, and represents an important tool that may be applied at an industrial level.

The use of enzymes may significantly reduce the environmental impact of industrial operations. Nonetheless, enzyme production and formulation are energy- and labor-intensive processes, which may incur in environmental impacts as well. In addition, techno-economic analyses of enzyme production revealed that the technological readiness and process economics for enzyme scale-up and use in industrial processes are not always encouraging [1]. Sustainability refers to all actions devoted to satisfying the needs of present generations without jeopardizing the ability of future generations to meet their own needs. Hence, sustainability is a complex and holistic concept which brings together the economic, social, and environmental fields [2]. Sustainability in industry encompasses the reduction of pollution that is caused by chemicals; the hazards deriving from their use; the utilization of abundant, carbon-neutral biomasses or wastes as raw materials (i.e., circularity); the use of renewable energy sources; and the use of efficient conversion technologies [3].

Today the synthesis, conversion, removal, or remediation of chemicals is often performed by means of enzyme biotechnology in different sectors, such as pharmaceutical, food and beverage, feed, biofuel, textile, paper and pulp, and leather industries. Unsurprisingly, the global enzyme market has been increasing in recent years. It was valued USD 9.9

billion in 2019 [4] and it is expected to continue growing at a compound annual growth rate of 6.8–7.1%, drawn in by food, beverages, laundry, and detergents industries [4,5].

Approximately 60% of industrial enzymes are produced from fungi, 24% from bacteria, 4% from yeasts, and the remaining 10% from plants and animals [6]. Hydrolases account for 75% of the entire enzyme production worldwide [6].

Lignin-modifying enzymes from white-rot fungi are the forefront of enzyme technology for sustainability because they allow the degradation of the phenolic components of lignin and release the cellulose that is used in microbial fermentations to produce energy or valuable bioactive ingredients [7].

Owing to the enzymatic treatment, the shift from energy intensive processes, such as steam or chemical-based ones, to mild and environmentally friendly processes is feasible [8], while valorizing the plant material and applying circularity in food systems. In this regard, plant material deriving from food wastes still represents a poorly exploited resource which may enter new production paths. Indeed, the Food and Agriculture Organization estimated that in 2019 17% of total food that was available to consumers was discarded at household, retailer, and restaurant levels [9,10].

Biobased methods for the breakdown of lignocellulosic biomasses are essential to allow the complete exploitation of cellulose and its further valorization and are gaining massive popularity nowadays.

Laccase (LC, benzenediol: oxygen oxidoreductases, EC 1.10.3.2) is a multicopper oxidase which has a long history of application in de-lignification, lignocellulosic fiber modification, chemical synthesis, and many other industrial applications [11]. LCs have also shown great potential in the bioremediation of toxic pollutants such as dyes [12,13], polycyclic aromatic hydrocarbons [14], endocrine disruptors [15], and mycotoxins [16]. It is worth mentioning that this is of particular interest for the reuse of material which can be potentially contaminated and unsafe to use. Its versatility and environmentally friendly catalysis opens up countless possibilities for the valorization of such material in the era of sustainability.

Therefore, this review aims to give an overview of fungal LC, a promising green and sustainable enzyme, its mechanism of action, advantages, disadvantages, and solutions for its use as a tool to reduce the environmental and economic impact of industrial processes with a particular insight on the reuse and valorization of agro-wastes.

## 2. Laccase: A Case Study

### *2.1. Laccase: Overview, Sources*

Laccase (p-diphenol: dioxygen oxidoreductase EC 1.10.3.2) belongs to the group of multicopper oxidases (MCOs) that catalyze one-electron oxidation of a wide range of compounds with concomitant reduction of molecular oxygen to water. Electron donor substrates for laccases can be substituted monophenols, polyphenols, aromatic amines, aromatic thiols, as well as some other aromatic compounds and certain inorganic metal ions [17].

Laccase was first isolated over 100 years ago from the Chinese lacquer tree (*Rhus vernicifera*) [17]. A decade later, the first fungal laccase was described [18]. For a long time, laccase was considered as an enzyme that was present only in fungi or higher plants, but later laccases were also found in bacteria, insects, and algae [19]. To date, according to the BRENDA database (http://www.brenda-enzymes.org/ accessed on 24 June 2021), about 280 laccases have been isolated and characterized. However, the modern classification of laccases that is based on phylogenetic analysis considers only fungal (basidiomycetous and ascomycetous) laccases as "true" laccases (laccases *sensu stricto*) [20].

The issue of the biological (or physiological) role of laccases is still controversial. Laccases can perform a variety of functions in different organisms including both the processes of synthesis of new compounds from simple monomeric substrates and the destruction of polymeric substrates. However, it is believed that the main biological function of laccases is lignification in higher plants and delignification in fungi. Other functions that have been proposed for laccases are participating in morphogenesis, pathogenesis, fruiting body formation, sporulation, spore protection, humification, metal ion homeostasis, and others [21–23].

A typical fungal genome contains several nonallelic laccase genes which form a multigene family [24]. However, the biological role of various laccase izoenzymes (i.e., the products of different genes) is still unclear. Some izoenzymes could be secreted by a fungus under specific conditions while others may be obtained only by recombinant expression [25]. In addtition, post-translational modification of laccase proteins (i.e., glycosylation) leads to a wide spectrum of laccase isoforms which can be produced by fungi.

The active site of the laccase contains four copper ions which are classified according to their spectral characteristics as copper ions of the T1, T2, and T3 types. The T1 copper ions are characterized by strong absorption at 605–610 nm. In the oxidized state, it gives laccases a characteristic blue color. The T3 copper ions are characterized by a shoulder at 330 nm, while the T2 copper ions have no signal in the absorption spectra. In EPR spectra, T1 and T2 copper ions are characterized by hyperfine splitting, while a pair of T3 copper ions has no signal due to strong anti-ferromagnetic coupling [26]. The oxidation of the phenolic (and other electron donor) substrates takes place at the mononuclear T1 center which contains one T1 copper ion (Cu1). This copper ion serves as the primary electron acceptor. Molecular oxygen is reduced to water at the trinuclear T2/T3 centers which is formed by one T2 copper ion (Cu2) and two T3 copper ions (Cu3).

An important characteristic of laccases is the redox potential of the T1 copper center of the enzyme. The value of the redox potential determines whether a substrate can be oxidized by laccase. In some cases (mostly for the substrates with a redox potential that is higher than 700 mV), the efficiency of substrate oxidation by laccase directly depends on the difference between the redox potentials of the substrate and the T1 center of the enzyme [27,28]. However, the efficiency of catalysis is influenced not only by the redox potential difference, but also by the structure of the substrate-binding pocket of the laccase and the structure of the substrate [28–30].

Depending on the value of their redox potential, laccases are usually divided into enzymes with low, medium, and high redox potential [31]. The redox potential of fungal laccases is, as a rule, higher than that of enzymes of plant and bacterial origin [32]. The group of enzymes with a low redox potential (up to 460 mV vs. NHE) includes mainly plant and bacterial laccases, the group with medium redox potential (460–710 mV vs. NHE) includes laccases of *Ascomycete* and *Basidiomycete* fungi, whereas high redox potential (more than 710 mV vs. NHE) is characteristic mainly of laccases of wood-degrading white-rot fungi [33,34].

The redox potential of the T2/T3 center affects catalysis of laccases to a lesser extent [35]. The redox potential value of the T2 center was measured for laccases of *Trametes hirsuta* (400 mV) [36] and *R. vernicifera* (365 mV) [37]. The redox potential value of the T3 center is 756 mV for the laccase of *T. hirsuta* [38] and 434 mV for the laccase of *R. vernicifera* [37].

### 2.2. Physicochemical and Catalytic Properties, Enzyme Stability

Nowadays, the most studied groups are the fungal and bacterial laccases. Most laccases are glycoproteins with a molecular weight of 50–140 kDa [31–39]. Usually, laccases exist in a monomeric form but there are also laccases that form homo- and hetero-dimers. The isoelectric points of laccases are in a wide range of pH from 2.6 to 7.0. Fungal laccases are characterized by isoelectric points near 4.0, while plant laccases, as a rule, have neutral pI values [39,40].

The carbohydrate part of a laccase molecule can comprise of up to 50% of the total mass of the protein; plant laccases are characterized by a higher carbohydrate content (usually from 20 to 45%), while fungal laccases have a slightly lower carbohydrate content (usually from 10 to 25%) [41]. N-glycosylation is almost always found in fungal laccases. Branched mannose chains are bonded to asparagine residues via two N-acetylglucosamine residues at 3–10 sites [35]. Presumably, the role of carbohydrates is to stabilize copper centers, participate in the process of protein secretion, protect the molecule from proteolysis, and increase the thermal stability [42,43]. Moreover, glycosylation can affect the activity of laccases [44–47]. Differences in glycosylation can also be responsible for the presence of a large number of laccase isoforms differing in molecular weight and pI values [48–50].

For fungal laccases, it is typical to have an acidic pH optima. The optimum for 2,2′-azinobis(3-ethylbenzthiazoline-6-sulfonic acid) (ABTS) oxidation is generally below pH 4.0, while the oxidation optima for phenolic compounds such as 2,6-dimethoxyphenol, guaiacol, and syringaldazine are in the pH range from 4.0 to 7.0 [40].

Typically, the temperature optimum for laccase activity is in the range from 50 to 70 °C. Some laccases have a temperature optimum that is below 35 °C [40]. The thermal stability of laccases that are isolated from different sources varies greatly, even in laccases that are from different isolates of the same species [42]. The thermal stability of laccases depends on many physicochemical factors—the packaging of the protein globule, the hydrophobicity, the number of intramolecular hydrogen bonds and salt bridges, the distribution of charged amino acid residues on the protein surface, or an increased content of certain amino acid residues [42].

Laccases are capable of catalyzing the oxidation of a very wide range of substrates including various xenobiotic organic compounds such as aromatic amines, polycyclic aromatic hydrocarbons, synthetic dyes, antibiotics, etc. [19,25]. However, the typical substrates for laccases are various substituted phenols. The efficiency of catalysis of various compounds is also influenced by the number and nature of substituents in the phenolic ring [51]. Due to a higher redox potential, laccases of basidiomycetes have a broader substrate specificity in comparison with laccases of ascomycetes, plants, and bacteria [52].

Additionally, the substrate specificity of laccases can be extended by using mediators —low molecular weight compounds that act as electron shuttles. It is suggested that mediators can promote catalysis through two different mechanisms. First, in reactions with compounds whose direct oxidation by laccase is sterically hindered due to their big size or/and complex structure, mediators can enable substrate oxidation by acting as electron transfer agents. Second, in reactions with compounds whose direct oxidation by laccase is impossible due to their high redox potential, mediators can enable substrate oxidation by promoting alternative (non-enzymatic) oxidative pathways [53–56]. From the chemical perspective, laccase mediator should be a small-sized compound which could be transformed by laccase into a stable but still reactive oxidized form and then reduced to the initial form by the final substrate; hence recycling during the reaction. It should be noted that most of the mediators that are described in the literature are not true redox mediators because they undergo secondary chemical reactions (i.e., are consumed during the reaction cycle). Among the secondary reactions, the coupling reactions are of great relevance, especially in organic synthesis as they allow the production of novel compounds. Homomolecular coupling in the reaction consists of two oxidized substrates to form a dimer or an oligomer. Heteromolecular coupling involves an oxidized LC substrate and a non-laccase) to create new hybrid molecules [55]. Figure 1A shows the scheme of laccase oxidation of phenolic derivatives with and without the participation of a mediator. In addition, three examples of reaction couplings are reported (Figure 1B).

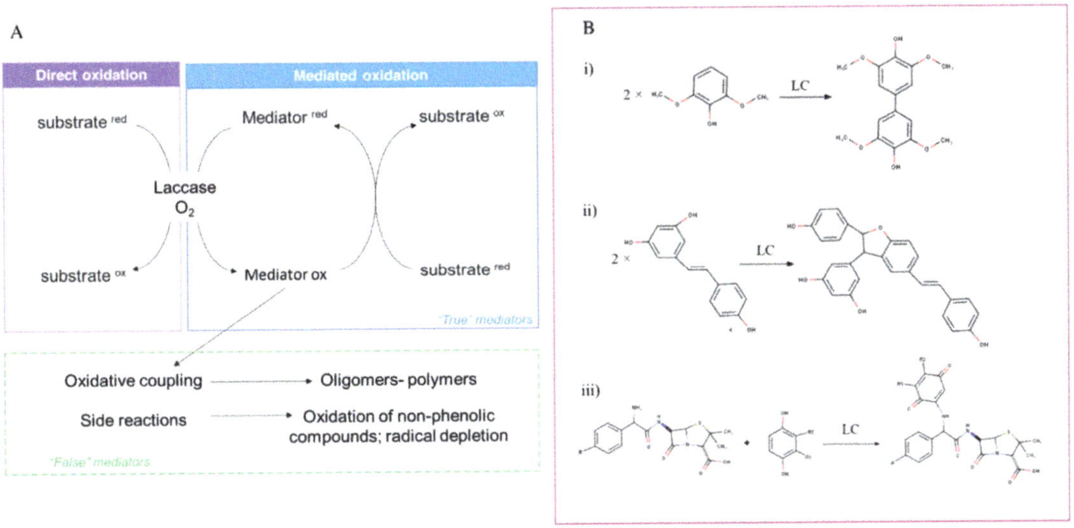


**Figure 1.** Oxidation of substrates by laccase. Panel (**A**) shows the direct and mediated oxidation mechanisms. The route of "false" mediators is also depicted. Panel (**B**) shows three examples of coupling reactions. In particular homomolecular coupling of 2,6-dimethoxyphenol (i) and resveratrol (ii), and the heteromolecular coupling of aminopenicillin and cathecol (iii).

An example of "false" mediators is represented by natural phenolic compounds which can undergo polymerization reactions. The synthetic mediators HBT and VA also undergo chemical degradation after a few cycles. True mediators include a very limited number of compounds, including inorganic ion complexes (potassium octocyanomolybdate and octocyanotungstate, FeII complexes) and organic compounds (ABTS, TEMPO) [55].

### 2.3. Laccase Structure and Mechanism of Catalysis

Usually, amino acid sequences of laccases, *sensu stricto*, contain 520–550 residues including signal peptide [32]; it is worth noting that the sequence identity of laccases of organisms that belong to different classes is 20–30%. Multiple alignment of amino acid sequences of more than 100 fungal and plant laccases allowed the identification of four conserved regions (L1–L4) distinguishing laccases from other proteins from the MCO family [57]. Figure 2 shows the sequence of the laccase LacA from *T. hirsuta* with the regions L1–L4 highlighted in purple. These regions contain 12 amino acid residues coordinating the copper ions in the active site of the enzyme.

```
AVGPVADLTITDAAVSPDGFSRQAVVVNGVTPGPLVAGNIGDRFQLNVIDNLTNHTMLKSTSIHWHGFFQ 70
      L1                                      L2
HGTNWADGPAFINQCPISPGHSFLYDFQVPDQAGTFWYHSHLSTQYCDGLRGPFVVYDPNDPHASRYDVD 140

NDDTVITLADWYHTAAKLGPRFPLGADATLINGKGRAPSDTTAELSVIKVTKGKRYRFRLVSLSCDPNHT 210

FSIDGHNLTIIEVDSVNSQPLEVDSIQIFAAQRYSFVLDANQAVDNYWIRANPNFGNVGFDGGINSAILR 280

YDGAPAVEPTTNQTTSVKPLNEVDLHPLVSTPVPGSPSSGGVDKAINMAFNFNGSNFFINGASFVPPTVP 350
                                                          L3
VLLQILSGAQTAQDLLPSGSVYVLPSNASIEISFPATAAAPGAPHPFHLHGHTFAVVRSAGSTVYNYDNP 420
                                       L4
IFRDVVSTGTPAAGDNVTIRFDTNNPGPWFLHCHIDFHLEGGFAVVMAEDTPDVKAVNPVPQAWSDLCPT 480

YDALDPNDQ
```

**Figure 2.** Amino acid sequence of *T. hirsuta* LacA laccase (gb | KP027478.1). The L1-L4 regions are shown in purple, the copper ion Cu1 ligands are shown in orange, Cu2—in green, and Cu3—in blue. Cysteine residues that are involved in the formation of disulfide bonds are shown in yellow.

The first three-dimensional structure was determined in 1998 for a laccase from the fungus *Coprinus cinereus* [58]. At present, the PDB database (www.rcsb.org, accessed on 24 June 2021) contains more than 180 structures of laccases of bacteria (17 species) and fungi (25 species). The 3D structures of fungal laccases are very similar to those of other MCOs, such as ascorbate oxidase, bilirubin oxidase, and ceruloplasmin. Three cupredoxin-like domains are sequentially connected to each other: domain I contains ~1–130 amino acid residues, domain II—~130 and 310 residues, and domain III—~310–500 residues (Figure 3). Each domain has the Greek key β barrel topology which is typical for all proteins of the MCO family [59,60]. As a rule, the structure is stabilized at least by two disulphide bridges connecting the first domain with the other two domains. The T1 center is located in domain III, while the T2/T3 center (TNC) is located between domains I and II. The distance between the T1 and T2/T3 centers is approximately 13 Å, but they are connected via a conservative His-Cys-His motif that is involved in the electron transfer from the T1 center to the TNC [23]. Bacterial laccases can have a three-domain structure, similar to the fungal laccases. However, in addition to three-domain laccases, there are also two-domain bacterial laccases which form trimers [59].

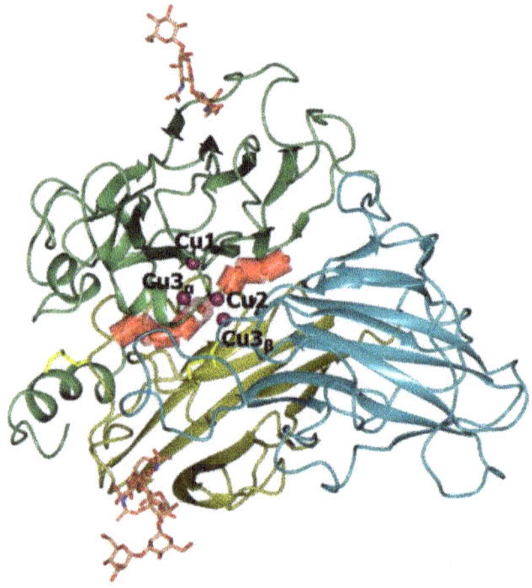


**Figure 3.** The overall structure of *Coriolopsis caperata* laccase (PDB 3JHV, [60]). The first domain is shown in gold, the second in green, and the third in blue color. Copper atoms are shown with purple spheres. Sugars are shown with stick models, atoms are colored by type (C—orange, O—red, N—blue). Disulfide bridges are shown in yellow.

In all laccases there is an identical structure of the T1 center where the oxidation of the electron donor substrate occurs. The copper ion in the T1 center is coordinated by the side chains of two histidine residues and a cysteine residue (Figure 4). Moreover, in the vicinity of the Cu1 copper ion (~5 Å), the side chains of another two amino acid residues are located (highlighted in gray in Figure 4). One of them, isoleucine, is strictly conservative among laccases. In fungal laccases, the second position is usually occupied by phenylalanine or leucine residues, whereas in bacterial laccases a methionine residue is located there [61]. There is a correlation between the redox potential of laccases and the nature of the non-conservative amino acid residue in the nearest surrounding of the copper ion in the T1 active center. High redox potential laccases have a Phe residue in this position, occasionally Leu, middle redox potential laccases predominantly have a Leu residue and sometimes Phe, and low redox potential laccases have a Met residue [31]. This observation was confirmed using site-specific mutagenesis [39,62–64]. In addition, the redox potential of the T1 center of laccases depends on the composition and structure of the loops around the Cu1 copper center [65–67]; laccases which have a higher number of large hydrophobic residues (e.g., Phe) in the vicinity of the Cu1 copper ion that cover it from the solvent are characterized by a higher redox potential [34].

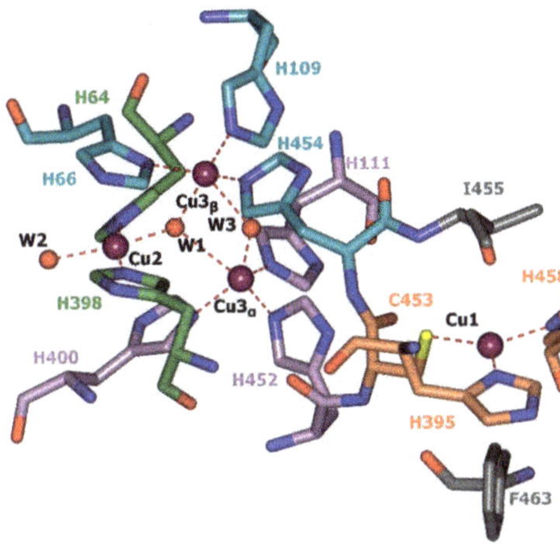


**Figure 4.** Active center of *T. hirsuta* laccase (PDB: 3FPX, [61]). Copper ions are shown in purple, oxygen atoms in red, and nitrogen atoms in blue. The carbon atoms of histidine residues coordinating copper ions Cu1, Cu2, $Cu3_\alpha$, and $Cu3_\beta$ are shown in coral, green, purple, and blue colors, respectively. Carbon atoms of non-coordinating amino acid residues from the nearest surrounding of the copper ion Cu1 are shown in gray. T2 and T3 water channels are shown with red cylinders and water molecules inside them.

For bacterial and fungal laccases, a number of structures of laccase complexes with electron donor substrates were determined [65–71] which allowed the highlighting of the loops that form the substrate-binding pocket. In three-domain laccases, the substrate-binding pocket for oxidizable phenolic compounds was found to be formed by the loops of domains I and II located near the T1 center (Figure 5). The substrate-binding pocket is formed mainly by hydrophobic amino acid residues (Figure 5, residues Phe162, Leu164, Phe265, Phe332, Phe337, Pro391); it also includes the His458 residue coordinating the Cu1 copper ion. Presumably, a polar residue at position Asp206 is also involved in substrate binding (Figure 5). It should be noted, however, that the resolution of the structures of the complexes of laccases with substrates was not high and the position of the substrate molecule was different for different laccases and different substrates, which indicates non-specific binding. Moreover, the amino acid composition of the loops of the substrate-binding pocket varies greatly among laccases from different organisms [34], which makes it difficult to predict the binding mode of the substrate. In addition, it was shown that the more flexible substrate-binding pocket loops can also facilitate the oxidation of bulky substrates with a rigid structure despite the redox potential value of the laccases [28].

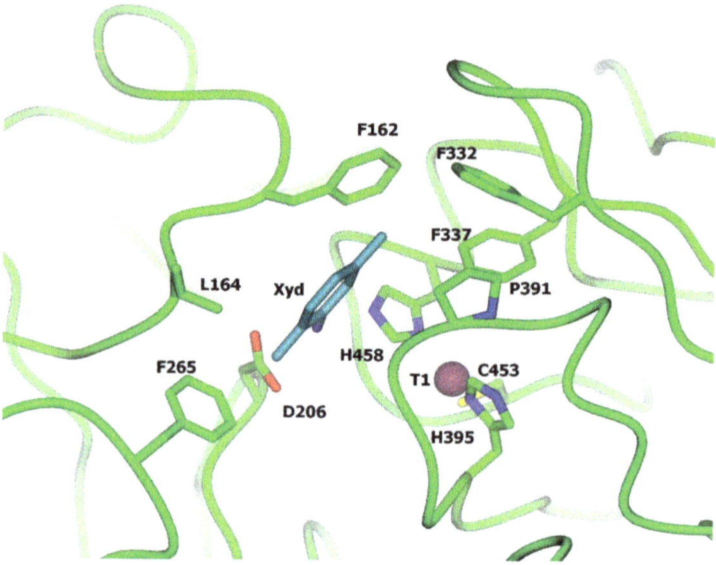


**Figure 5.** Substrate-binding pocket of *T. versicolor* laccase (structure of the laccase complex with 2,5-xylidine, PDB 1KYA [68].

As it was mentioned before, there are three copper ions in the TNC where the molecular oxygen is reduced by laccase. The T2 copper ion is coordinated by side chains of two histidine residues and T3 copper ions—by three histidine residues each (Figure 3). TNC can be reached by water and protons from the reaction medium through the T3 and T2 water channels. The mechanism of oxygen reduction to water by laccase was originally proposed by Solomon et al. on the basis of spectroscopic data and the three-dimensional structure of a laccase-related enzyme, ascorbate oxidase [72]; later, it was confirmed [73] and detailed [74] by the X-ray diffraction analysis. Modern views on the mechanism of oxygen reduction to water by laccases are summarized in the review [25] and updated with X-ray diffraction data in [74]. The oxygen reduction process consists of two stages of two-electron reduction processes. The catalytic cycle begins with a fully reduced enzyme (all copper ions are in the +1 oxidation state, Figure 6A) with two oxygen ligands (W2 and W3, Figure 4) that are bound to copper ions of TNC. The Cu2 copper ion forms a bond with the W2 oxygen ligand that is located outside the T2/T3 center. When a chloride or fluoride ion binds at this position, laccase inhibition takes place [75]. The W3 is located between the pair of Cu3 ions and can be substituted by molecular oxygen (Figure 6A, and 6B). Then, during the first two-electron reduction, molecular oxygen oxidizes both copper ions to the +2 oxidation state and becomes a deprotonated peroxide molecule (first two-electron reduction, Figure 6C). It can be protonated from the side of the T3 water channel. The second two-electron reduction process involves the electron transfer to peroxide from the Cu2 copper ion and from the Cu1 copper ion (via the Cu3 copper ion, Figure 6D). As a result, an O-O bond in the peroxide is cleaved and Cu1 and Cu2 ions become oxidized (+2). Subsequent protonation of oxygen ligands leads to the formation of a hydroxide ion and a water molecule (Figure 6E). The release of oxygen ligands from the TNC takes place due to the reduction of copper ions of the T2/T3 center by electrons that are transferred from the Cu1 copper ion (Figure 6F). The $OH^-$ ligand at position W1 replaces the W3 water molecule which leaves the active site. The $OH^-$ ligand can be protonated and become a water molecule. Thus, the system returns to its original state (Figure 6A).

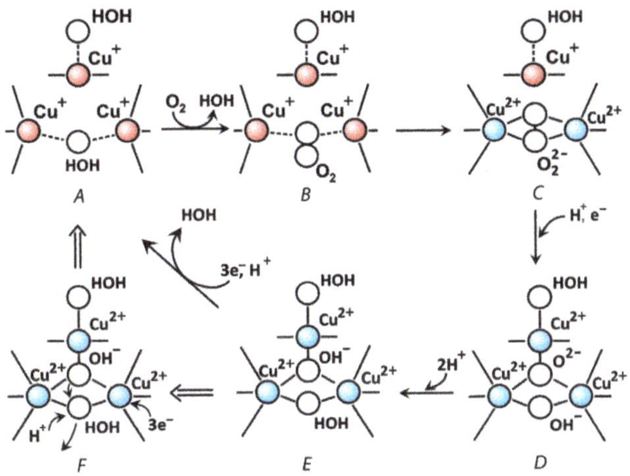

**Figure 6.** Mechanism of molecular oxygen reduction by laccases to water: $O_2 + 4e^- + 4H^+ \rightarrow 2H_2O$. Intermediate states are shown in Figure (**A–F**). Panel F shows the mechanism of the release of a water molecule from the TNC. Coordination and covalent bonds are shown by solid lines. Ion-dipole electrostatic interactions are shown with dotted lines.

### *2.4. Laccase Industrial Application*

LCs have become relevant industrial enzymes due to their broad range of activity towards phenolic and non-phenolic compounds. As recently reviewed by Moreno et al. [76], many industrial applications can be performed by LC, as well as in combination with redox mediators, especially for recalcitrant compounds. From the industrial perspective, LC mediators should be environmentally friendly and cheap. Currently, the most effective mediators are chemically synthesized compounds, however, they are usually hazardous and quite expensive to produce [77]. This leads to the development of alternative uses of naturally-occurring phenolic mediators originating from lignin biodegradation or from vascular plants either as extractives or lignin-forming monomers [78].

In textile and paper pulp, fungal LC is used for bio-bleaching, i.e., the degradation of synthetic dyes and inks, which are chemically heterogeneous chromogenic compounds. They can be toxic and genotoxic, and, once released into water bodies, they reduce the oxygen content and the photosynthetic activity of microorganisms [78]. Due to their synthetic origin and complex aromatic structures, they are recalcitrant to the conventional microbial degradation or decolorization. LC is able to degrade them but often in combination with the synthetic mediators hydroxybenzotriazole (HBT), ABTS and violuric acid (VA) [8,12,79].

In biorefinery, LC application has different purposes: (i) delignification, (ii) detoxification, i.e., reduction of the phenolic content, and (iii) reducing the adsorption of hydrolytic enzymes to lignin. Delignification is a mandatory step to make carbohydrates more accessible to carbohydratases in the subsequent saccharification process. Nonetheless, phenols can be released after enzymatic and chemical delignification as well as steam explosion. Phenols act as inhibitors of enzymes and fermentative yeasts, thus a detoxification process may be needed to boost both the saccharification and fermentation yields [76].

In food applications, LC is used to remove the phenols that are responsible for turbidity and color instability in wine, beer, and juices [80,81]. Alternatively, LC can catalyze the formation of crosslinks between proteins, also by oxidizing natural mediators, which act as crosslinking agents [82]. Crosslinks in food matrix generated by LC allow to modify the rheological, technological, and nutritional properties of vegetables, cereals, and milk-based products, thus creating novel foods with improved characteristics [83–86].

Bioremediation by LC encompasses the oxidation of phenols from industrial effluents, dyes, and toxic compounds, such as mycotoxins, polycyclic aromatic hydrocarbons, endocrine disruptors, and antibiotics [86–89]. Also in this case, redox mediators are often needed to ensure an effective level of detoxification.

Laccase use in organic synthesis has been increasing in recent years. Substrates such as substituted phenols, polyamines, anilines, aromatic and alkyl amines, and benzenethiol are the starting point for biocatalytic reactions for the syntheses of bioactive compounds for pharmaceutical and nutraceutical applications [90]. The synthesis of new compounds may include simple oxidation/amination/thiolation reactions and oxidative coupling reactions. Homomolecular coupling can be used to produce compounds that are endowed with new or improved biological activities. LC-mediated coupling was used on the simple phenols, such as 2,6-dimethoxyphenol to obtain compounds with increased antioxidant activity; on complex phenolic compounds, such as rutin, to improve their solubility in water; or resveratrol, to produce compounds with antiproliferative activity towards colon cancer cells [91] or new antibiotics (e.g., aminopenicillins, aminocephalosporins, aminocar-bacephems, and sulfonamide derivatives). Eventually, LC- mediated homomolecular coupling was also exploited to reduce the concentration of several water pollutants including estradiol, triclosan, and cumylphenol [92–94].

Grafting, e.g., coupling reactions involving a natural or synthetic polymer, can be used to functionalize polymers to confer new features. LC-mediated grafting has multiple applications, including environmental pollution control, modification of lignocellulosic material/foods/ingredients in textile industry, biosensors, food industry, pharmaceutical industry, and organic synthesis [95].

LCs, together with several other oxidoreductases, have found applications in the construction of enzymatic biosensors and bio-cells for biomedical, food safety, and environmental monitoring purposes. While in the classic biosensors, LC is immobilized directly on an electrode, in bio-cells the enzyme are displayed on the membrane of the whole microorganism that is grown on an electrode [96]. The advantages rely on the longer lifetime due to the possibility of the microorganisms duplicating and creating a biofilm on the electrode [97].

## 3. Re-Use of Agro-Wastes for Laccase Production

### 3.1. Laccase Production by Solid State Fermentation Using Agro-Wastes

Enzymes can be produced by two different approaches: solid state (SSF) and submerged fermentation (SmF) [6]. SSF is performed on a heterogeneous, porous, low moisture, solid material, where microorganisms are able to grow whether on or within the surface material. On the contrary, SmF is performed in a liquid medium in which optimized nutrients are dissolved.

SSF holds a great potential for LC production by filamentous fungi, as they are able to colonize the substrate and promote the hyphal growth on or in the particle. SSF mimics the natural fungal environment and naturally contains LC inducers, such as flavonoids and phenols. This is a great advantage because the use of inducers, such as metal ions (copper and, to a lesser extent, manganese), synthetic compounds such as ABTS, or ethanol in SmF is often needed to achieve satisfactory enzyme yield and poses both economic and environmental concerns. [98,99].

SSF allows the easy production of extracellular enzymes, has lower energy requirements, lower environmental impact, lower risk of contamination, easier downstream processing, and higher volumetric and enzyme yield than SmF [100].

White-rot fungi are among the best LC producers because they naturally use LC together with peroxidases to degrade lignin and access the plant polysaccharides which are then used as carbon source [101]. Lignin is the structural component of plants and it is responsible for their impermeability, resistance towards microbial attacks, and oxidative stress. Owing to its resistance, it represents a major obstacle to reconvert agro-wastes into valuable compounds [102]. Lignin is also recalcitrant to mild chemical and biological degradation techniques and it can be efficiently degraded only by microorganisms which possess a wide array of oxidative enzymes. White-rot fungi are key players in the global carbon cycling and among the main producers of LCs and peroxidases which enable them to degrade this recalcitrant biopolymer [96,103,104].

The carbon to nitrogen (C/N) ratio is a crucial parameter for fungal growth by SSF, as it influences the mycelial growth and the production of several compounds. The optimal C/N ratio has to be assessed for each fungal species, although higher C/N ratios (30–40) have been reported to promote the fastest mycelial growth (colonization rate), fruiting bodies, and polysaccharide production, while lower values (10–20) were reported to give the highest overall biomass and LC production [98,99]. To balance C/N ratio, ammonia or organic sources, such as yeast extract, can be added to significantly increase LC production [105].

The reactor type and airflow system are also important determinants for the fungal growth because they determine how the air, containing $O_2$, is provided; how the correct moisture is maintained; and how the metabolic heat and the $CO_2$ are dissipated [106].

Using agro-wastes for SSF allows the use of cost effective resources to reduce the environmental impact of their disposal and to boost LC yield thanks to the presence of inducers [93]. Spent grains, straw and residues of cereal harvesting, fruit peels, and general food waste are perfect examples of economic and sustainable ingredients for LC production by SSF; i.e., this material can be further exploited to produce energy. The most recent studies on LC production by solid state fermentation (SSF) on sustainable substrates are reported in Table 1.

Brewers' spent grains (BSG) are the solid by-product of beer production and they are usually disposed as cattle feed. Nonetheless, they can be used as material for LC production by SSF. Additionally, lignin degradation by fungi can drastically enhance polyphenol recovery from lignocellulosic biomass [100,107].

SSF with BSG was performed with *T. versicolor*, obtaining a maximum LC production of 560 U/L after 7 days and 3.4-fold increase in the extraction of total polyphenols [108]. Dhillon et al. used BSG that was supplemented with different inducers. Despite the fact that they obtained good results without inducers (up to 2343 IU/g dry weight in flasks and 2956 I U/g dry weight in trays), the addition of different compounds significantly increased LC production. The best one was phenol (10,108 IU/g dry weight in flasks and 13,506 IU/g dry weight in trays) [109].

**Table 1.** Laccase production by solid state fermentation (SSF) on sustainable substrates.

| SSF Substrate | Species | Substrate Composition and Growth Parameters | LC Activity | Reference |
|---|---|---|---|---|
| Brewers' spent grains (BSG) | *Trametes versicolor* | Lignin 8.53%, cellulose 16.1%, hemicellulose 20%, ash 5.3% DW.<br>Growth at 27 °C for 14 days | 560 U/L after 7 days | [108] |
| BSG added with LC inducers | *T. versicolor* | Lignin 12.4%, cellulose 13.8%, hemicellulose 30%, ash 2.6% DW.<br>Growth at 30 °C for 16 days | 13,506 IU/g using 10 mg/kg phenol as inducer after 12 days | [109] |
| Wheat straw (WS), bean stalk (BS), and red grass (RG) | *Lentinula edodes* | WS: Lignin 7.58%, cellulose 68.93%, hemicellulose 11.16% DW.<br>BS: Lignin 11.27%, cellulose 64.65%, hemicellulose 11.27% DW.<br>RG: Lignin 7.5%, cellulose 69.22%, hemicellulose 10.69% DW.<br>Growth at 26 °C for 40 days | 579 U/g DW after 25 days<br>258 U/g DW after 25 days<br>390 U/g DW after 35 days | [110] |
| Corn stalk (raw and steam exploded) | *T. versicolor* | Lignin 22.43%, cellulose 33.96%, hemicellulose 13.95% DW.<br>Growth at 30 °C for 7 days | 2600 U/g after 15 days | [111] |
| Rice straw (raw and ammonia-treated) | *Funalia trogii* | - | 172 U/g after 14 days | [112] |
| Wheat and rice straw and bran | *Tricholoma giganteum* | -<br>Growth at 30 °C for 20 days | 89,800 U/g after 16 days | [113] |
| Sorghum Bagasse | *Coriolus versicolor* | Lignin 25.14%, cellulose 38.02%, hemicellulose 25.01% DW.<br>Growth at 27.5 °C for 20 days | 115 U/g after 20 days | [114] |
| Tea residues | *T. versicolor* | Lignin 13.60%, cellulose 11.60%, hemicellulose 32.50% DW.<br>Growth at 26 °C for 8 days | 31.2 U/g after 8 days | [115] |
| Olive leaves and wheat straw | *Galactomyces geotrichum* | -<br>Growth at 30 °C for 26 days | 56 U/g after 14 days | [116] |

Different agro-wastes, specifically wheat- (WS) and bean-straw (BS) and reed grass (RG), were used as main ingredients for the production of LC by SSF with three strains of *Lentinula edodes*. In this study, lower lignin content and higher cellulose content gave the best results in terms of LC production and timing. WS was the best material, as it led to 579 U/g per dry weight after 25 days of cultivation [110]. Corn stalks were used for *T. versicolor* cultivation as raw materials and after a steam explosion. The pre-treatment was able to increase LC production by 2.1-fold (up to 2600 U/g after 15 days), likely as a result of polyphenol that was released from lignin (from 1.98 to 3.54 mg of gallic acid equivalent per g) and the increased accessibility of polysaccharides [111]. Similarly, rice straw was used raw and after a chemical pre-treatment with ammonia for LC production by *Funalia trogii*. Also in this case, the pretreatment boosted LC production up to 3.4 times due to the modification of the chemical composition of the rice straw [112]. When different pretreatments with ammonia were used, the harshest one, leading to the highest cellulose content, gave the highest LC yield [112]. The importance of the type of lignocellulose substrate in inducing LC was proven also in the work of Patel and Gupte [113]. They compared different raw materials for the production of LC by the fungus *Tricholoma giganteum*, namely wheat bran, wheat straw, rice bran, and rice straw. Wheat bran was the best material for fungal growth and LC production, which was increased by several fold (up to 89,800 U/g) under optimized conditions and by using copper sulphate and phenolic inducers.

*Coriolus versicolor* was cultivated in a mesh tray bioreactor using sorghum bagasse to produce different ligninolytic enzymes, including LC. The optimization of mesh sizes in the tray and the airflow rate led to a 1.9-fold increase in LC production [114], confirming the important effect of the reactor type and airflow system for the fungal growth and enzyme production.

Tea residues deriving from tea beverages, instant tea production, and polyphenol extraction can be used as a substrate for fungal growth and LC production, as described by Xu et al. [115]. Compared to the cereal derived wastes, the tea residues showed lower lignin and cellulose content in favor of higher hemicellulose. Nonetheless, the authors were able to produce up to 31.2 U/g of LC after only eight days. A different approach was used by Pourkhali et al., who isolated an LC-producing fungi from the olive oil mill wastewater to produce LC using the olive-leaf wastes [116]. In this way, they were able to use an already adapted fungal species and boosted LC production thanks to the addition of another waste, wheat straw. In this work, the reaction parameters were also optimized by the Taguchi method of an orthogonal array design of experiment (DOE) to obtain up to 56 U/g of LC after 14 days of fermentation.

These recent studies demonstrate that using sustainable substrate for LC production is feasible and allows the valorization of different kinds of wastes and reduce the costs of LC production and the environmental pollution that is derived from their disposal. Pretreatments, redox mediators, and parameter optimization are often needed to boost LC production or increase the rate of production.

Scaling up of LC production by SSF, enzyme immobilization, and purification technologies remain future activities to investigate for successful and simple industrial application.

### 3.2. Laccase Immobilization on Agro Industrial Wastes

Enzyme immobilization is defined as a process which binds a soluble enzyme to a carrier matrix to generate an insoluble species with improved catalysis and resistance to harsh environmental conditions [117]. Immobilization has been widely used in industry to enhance enzyme stability and its recovery in batch or continuous reactor operations [118,119]. The attachment to a matrix makes the enzymes insoluble and thus, easier to recover, and improves the general stability by making the enzyme more rigid, delaying its degradation or unfolding, and preventing the inhibition by the substrate [120].

Nowadays, increasing attention is given to carrier materials which are cost-effective, safe, biodegradable, and environmentally friendly. Together with the traditional biobased immobilization materials (agarose, starch, chitosan, alginate), agro-industrial wastes represent an important opportunity to improve circularity and sustainability in the food system [121]. Lignocellulosic waste, spent grains, eggshell membrane, biochars, and chicken feathers were successfully used to immobilize LCs, as well as other enzymes from fungal origin [122], as reported in Table 2.

Spent grains (SGs) are a waste that is generated after the mashing and lautering of beer. Due to the presence of many functional groups (carboxyl, hydroxyl, and amino) to which LC can be covalently linked or adsorbed, it can be used as a carrier for enzyme immobilization. da Silva et al. (2012) immobilized a commercial LC from *Aspergillus* spp. that is used in textile industry by covalent bonding on SGs to obtain higher storage, operational, and thermal stability [123]. Similarly, Girelli and Couto (2021) immobilized a commercial LC preparation from *T. versicolor* by adsorption, although this resultd in a significantly lower catalytic activity [124].

Eggshell membrane (ESM) is a by-product of egg transformation chain and represents an interesting immobilization support for LC immobilization. ESM is mainly composed of proteins (80%) as well as being a cross-linked, water insoluble, and a porous material. The main protein components are collagen, hyaluronic acid, and glucosamine, which are rich in amino, hydroxyl, carboxyl, and thiol groups that are needed for immobilization [125]. A commercial LC from *T. versicolor* was immobilized in ESM by means of both covalent bonding and adsorption [126]. In this preliminary study, different protocols were compared and the reusability was 30–40% after six cycles.

Immobilization of LC on coconut fibers (CFs) has been widely explored for LC in both food and non-food applications. Adsorption of a commercial LC from *T. versicolor* on CFs proved to extend the stability and reuse of LC over 10 cycles for fruit juice clarification [127]. Cristóvão et al. investigated two different immobilization techniques for a commercial LC from *Aspergillus* spp., namely adsorption and covalent bonding on CFs. Covalent bonding proved to be more efficient than adsorption to increase operational and storage stability, as well as activity towards textile dyes [128,129]. Nonetheless, adsorption of the dyes to the carrier was observed so reversibility of the reaction should be taken into account in further studies.

Biochar is the product that remains after the pyrolysis of a biomass, i.e., burning in the absence of or with limited air to temperatures of over 250 °C. Due to its chemical stability, high surface area, and porosity, it has been used as a carrier for enzyme immobilization, mostly by adsorption. With regard to LC immobilization, different studies were performed to remove environmental pollutants. Li et al. immobilized an LC from *Coprinus comatus* on two different wood biochars deriving from maple and spruce, obtaining the best results with maple biochar that had a higher surface area and pore volume. A higher stability, reusability (five to seven cycles) and enhanced enzymatic degradation towards chlorinated biphenyls were obtained [130]. Rice straw biochar was also used to immobilize an LC from *T. maxima* for an effective anthracene degradation [131]. The enzyme retained 66% of activity after immobilization and up to 60% activity was retained after six cycles of operational use.

Covalent bonding of an LC from *T. versicolor* to pine wood, pig manure, and almond shell biochars was also tested for diclofenac removal [132]. As reported in other studies, the most promising results were obtained for biochars with a higher surface area. In addition, functionalization with citric acid was shown to improve LC covalent immobilization up to 20%. Chicken feathers are a waste by-product of the poultry industry. Common disposal methods are harmful for the environment and include incineration or burial in landfills [133]. Suman et al. [134] covalently immobilized a LC from *T. maxima* with chicken feathers that were functionalized with amino 3-aminopropyltrimethoxysilane. They obtained a LC with higher thermal stability and no significant loss of enzyme activity after eight cycles. The complete degradation of veratril alcohol was obtained in 48 h.

**Table 2.** Laccase immobilization on sustainable carriers.

| Carrier Material | LC Origin | Immobilization Method | Results | Application | Reference |
|---|---|---|---|---|---|
| Brewer's spent grain | Commercial LC (DeniLite base) from *Aspergillus* spp. | Adsorption to acid/base treated spent grain | Recovered activity 99%, immobilization yield 95%, 75% activity retained after 10 cycles at pH 10 | - | [123] |
| Spent grains | Commercial LC (S igma-Aldrich) from *T. versicolor* | Imine binding with acid/base treated spent grain and adsorption to spent grain | Recovered activity 39%, immobilization yield 1.3%, 58% activity retained after 6 cycles | 58% removal of syringic acid after 6 cycles and complete removal in 4 h | [124] |
| Egg shell membrane | Commercial LC (S igma-Aldrich) from *T. versicolor* | Covalent bonding and adsorption | Recovered activity 57%, immobilization yield 41%, 40% activity retained after 6 cycles | 57% removal of syringic acid after 24 h | [126] |
| Green Coconut fibre (CF) | Commercial LC (DeniLite base) from *Aspergillus* spp. | Acid/oxidative pretreatment; Covalent attachment to functionalized CF | Immobilization yield 74%, 55% activity retained after 10 cycles | Up to 70% degradation of textile dyes. Operational activity is significantly reduced in following cycles. | [127] |
| CF | *T. versicolor* | CF was subjected to acid/base pretreatment associated with thermal decompression. Adsorption to functionalized CF; covalent bonding via glutaradehyde. | Recovered activity 59%, immobilization yield 98%, 16.5-fold increase in thermal stability, 80% activity retained after 10 cycles in operational conditions | 65% reduction of phenolic compounds from apple juice | [129] |
| Maple biochar | *Coprinus comatus* | Adsorption to maple biochar | Recovered activity 66.5%, immobilization yield 64%, 66% activity retained after 7 cycles. | 71.4% degradation of chlorinated biphenyls after 5 h of treatment (2.5 times higher than the free enzyme) | [130] |
| Rice straw biochar | *T. maxima* | Adsorption to acid-treated rice straw biochar | Recovered activity 66%, immobilization yield 100% at pH 3.1, 60% activity retained after 6 cycles | 98% degradation of anthracene after 24 h (comparable to the free enzyme) | [131] |
| Pine wood biochar | *T. versicolor* | Covalent immobilization via glutaraldehyde to citric acid pretreated biochars | Recovered activity 20.1%, 46% activity retained after 5 cycles | 98.9% removal of diclofenac after 5 h of treatment | [132] |
| Pig manure biochar | | | Recovered activity 40.2%, 40% activity retained after 5 cycles | 98.9% removal of diclofenac after 2 h of treatment | |
| Almond shell biochar | | | Recovered activity 31.8%, 43% activity retained after 5 cycles | 98.9% removal of diclofenac after 4 h of treatment | |
| Chicken feathers (Chf) | *T. maxima* | Covalent immobilization via glutaraldehyde to Chf functionalized with amino 3-aminopropyltrimethoxysilane | Recovered activity up to 93%, immobilization yield up to 74.24%, no significant loss of activity after 8 cycles | Complete oxidation of veratryl alcohol after 48 h | [134] |

Compared to the use of synthetic carriers, the use of agro-wastes reduces the environmental impact of enzyme immobilization. Nonetheless, in most cases the treatment with chemicals or functionalizing agents is needed [122].

These processes have been validated only at the laboratory level, thus further studies are needed to assess scale-up and efficacy in industrial operational conditions. An important issue that needs to be addressed in operational conditions is the possible enzyme leaking from the carrier material, which has to be evaluated also in accordance with the final use of the treated material (food, feed, raw materials for fermentation etc.). When using high performing enzymes, it is mandatory to use low dosages and reduce the risk of having high enzyme amounts being possibly released from the carrier. Nonetheless, the studies that were presented here showed promising results also in terms of enzyme leakage after multiple cycles.

A further step towards sustainability, scale-up, and economic feasibility has to be done to reduce the use of these chemicals that are needed to pretreat the material. Despite that, the impact of the disposal of such wastes, generally by burning, still remains more concerning.

### 3.3. Valorizarion of Agro-Wastes by Laccase Pretreatment

As already mentioned, biorefinery for energy production is one of the main pathways that LC pretreated agro-wastes may enter as raw materials. Indeed, the so-called second-generation bioethanol employs lignocellulosic biomass and wastes as a source of fermentable sugars [135]. Nonetheless, the pretreatment to remove or degrade lignin is mandatory. Effective delignification increases sugar accessibility and enhances the following saccharification step, while the removal of phenols allows the microorganisms to efficiently ferment and produce bioethanol or methane. Purified LC preparations were successfully used for the delignification of different vegetable biomasses (Table 3).

**Table 3.** Laccase application for the valorization of agro-wastes.

| LC Origin | Application | Results | Reference |
|---|---|---|---|
| *Trichoderma asperellum* | Pretreatment of sweet sorghum stover for biohydrogen production | Removal of up to 77% of lignin; 3.26-fold increase in biohydrogen production. | [136] |
| *Pleurotus ostreatus* | Pretreatment of apple pomace, potato peels, and coffee silverskin for energy production | Up to 83% saccharification yields and ~70% phenol reduction using 2.5% vanillin as redox mediator | [137] |
| *P. djamor* | Valorization of pineapple leaf waste for biofuel production | Removal of up to 78.57% of lignin, 2.6-fold increase in reducing sugars | [138] |
| *P. djamor* | Pretreatment of sugarcane tops | Decrease of up to 79.1% of lignin content; increase of 3.3 fold in fermentable sugars | [139] |
| *Pycnoporus cinnabarinus* | Valorization of sugarcane bagasse for energy production | Decrease of up to 27% of lignin content; increase of 39% of glucose release | [140] |
| | Valorization of straw for energy production | Decrease of up to 31% of lignin content; increase of 46% of glucose release | |
| *Trametes maxima* | Pretreatment of jute sticks for energy production | Decrease of up to 21.8% of lignin content; increase of 19.5% of glucose release using 5% HBT as redox mediator | [141] |
| *Lentinus squarrosulus* MR13 | Pretreatment of kans grass (Saccharum spontaneum) | Decrease of up to 87.8% of lignin content; increase of 9% of bioethanol production | [142] |
| | | Decrease of up to 81.2% of lignin content; production of up to 500 mg/g of fermentable sugars | [143] |
| *T. versicolor* | Pretreatment of coffee bean processing waste for composting | Increase in total plate count values | [144] |

A purified LC from *Trichoderma asperellum* was used to pretreat sweet sorgum stover [136]. Thanks to its high stability at high temperatures and low pH, the authors were able to remove of up to 77% of lignin and obtain a 3.26-fold increase in biohydrogen production, proving that this LC could represent an important tool to be used in biofuels conversion. Giacobbe et al. [137] used two different LC preparations from *P. ostreatus* to de-lignify three different agro-wastes, namely apple pomace, potato peels, and coffee silverskin. The treatments with and without the addition of vanillin as a redox mediator were effective in reducing the lignin content (73 and 83%, respectively), to remove phenols, and to obtain satisfactory saccharification yields. The authors highlight that using such enzymatic pretreatment allows the avoidance of filtration and washing steps, with important advantages in sugar retention for the subsequent saccharification step and wastewater production [136].

Pineapple leaves are an important agro-industrial waste which can be valorized for biofuel production after lignin removal. Banerjee et al. treated this waste with an LC preparation from *P. djamor* with optimized solid loading (20%, *w*/*v*), incubation time (six hours), temperature (40 °C), pH (7), and enzyme concentration (500 IU/mL) obtaining the removal of up to 78.57% of the lignin and a 2.6-fold increase in reducing sugars [138]. Similarly, Sherpa et al. optimized the sugarcane top pretreatment with an LC from *P. djamor*. In accordance with the study of Banerjee et al., they obtained optimized conditions for solid loading (21%, *w*/*v*), incubation time (six hours), temperature (40 °C), pH (7), and enzyme concentration (430 IU/mL) obtaining the removal of up to 79.1% of lignin and a 3.3-fold increase in reducing sugars [139].

The redox mediator HBT was used in combination with an LC from *P. cinnabarinus* for the pretreatment of sugarcane bagasse and straw, with the aim of removing lignin and reuse these wastes for energy production [140]. Thanks to the LC treatment, the lignin was reduced by 27% and 31% from sugarcane bagasse and straw, respectively, allowing the release of 39% and 46% of glucose for the saccharification step.

*T. maxima* LC was used in combination with HBT to pretreat the jute sticks, a widely abundant waste that is derived from jute fiber extraction [141]. By using the laccase-mediator treatment, the lignin was reduced up to 21.8% with a single reaction and the yield of fermentable sugars was increased by 19.5%.

Rajac and Banerjee studied the delignification of kans grass for bioethanol production in two different studies using different conditions for the enzymatic hydrolysis [142,143]. Promising results in terms of delignification, sugar release (up to 500 mg/g), and bioethanol production (increase of 9%) were obtained.

LC pre-treatment of vegetable biomass can be also helpful for the production of sustainable compost. Besides being hardly degraded by the microorganisms that are used for composting, lignin hinders humus formation. The addition of lignin-degrading microorganisms has been proven to boost the rate of composting, its maturity, stability, and final quality [145]. The use of whole microorganisms is widely spread, especially because the spent mushroom substrates can be used for composting, as fertilizers, or for remediation purposes [146]. So far, only the use of an LC preparation from *T. versilocolor* was reported by Nazilah et al., who treated the coffee bean processing waste to improve its composting quality. The reduction in lignin and phenols allowed it to have a higher and more stable microbial count and to improve the compost quality [146].

The use of enzymes in industry is a reality in many fields of application. In particular in biorefinery processes, cellulases have been mostly applied to recover the simple sugars from the polysaccharide fraction. Therefore, the implementation of the use of an LC for the pretreatment of the vegetable biomass is feasible. Nonetheless, to save costs, reduce the environmental impact, and maximize yields, it requires a careful analysis of technical solutions that enable the design of a single reactor to perform delignification, saccharification, and fermentation in a continuous or sequential mode [144]. Eventually, it is important to study the delignification kinetic and the enzyme stability/activity/adsorption to the lignocellulosic material in a bioreactor to obtain reproducible and satisfactory results also when material with different lignin/cellulose/hemicellulose ratio is used.

## 4. New Trends and Challenges for Laccase Application

### 4.1. Cold-Active Laccases

A novel trend in enzyme technology is the use of cold-active enzymes, namely extremozymes that are produced by psychrophiles which have a high catalytic activity at moderate-low temperature [147]. The use of such enzymes is of particular interest because they allow the treatment of agro-wastes and vegetable biomass in the site of production. This is because the process can be performed at room temperature, saving the energy that is required to heat the biomasses as well as without complex transportation procedures.

Moreover, such enzymes can be used to process heat-sensitive products in the food industry. LCs for phenolic compounds removal in chilled juices, wine, and ready to drink coffee and tea, or to improve the texture in surimi, yogurt and ice cream, and meat represent possible applications of cold-active LC in the food industry [148]. Eventually, the thermolability of cold-active enzymes will allow an easy control of the process since they can be selectively inactivated by mild heat [148].

The interest in cold-active and cold-adapted LCs has risen, in particular for the possibility of removing contaminants from effluents or wastes that are directly in the site of contamination or without the need for elaborate equipment. Table 4 summarizes the most recent studies on cold-active LCs.

**Table 4.** Cold-active and alkaline active laccases and their application.

| LC origin | Application | Results | Reference |
| --- | --- | --- | --- |
| *Botrytis* sp. FQ | Removal of antibiotics | 70% activity at temperatures between 0 °C and 30 °C and an optimal temperature of 15 °C on dimethoxyphenol and removal of up to 60% of triclosan at pH 6.8, 20 °C | [149] |
| *Kabatiella bupleuri* | Dye decolorization | Retention of 60% of the maximum activity at 10 °C and over 40% in ice; up to 48.6% degradation of crystal violet after 1-h reaction with ABTS | [150] |
| Lac-Q from *Pycnoporus* spp. | Removal of antibiotics | Retention of 51% of the maximum activity at in ice; degradation of 50 mg $L^{-1}$ of oxytetracycline at pH 6.0 and 0 °C after 5 min with ABTS | [151] |
| *Melanocarpus albomyces* | Oxidation of dimethoxyphenol | Optimum pH of 9 for the substrate 2,6-dimethoxyphenol | [152] |
| *Coprinopsis cinerea* | Dye decolorization | High activity towards 2,6-dimethoxyphenol at pH 8–8.5; decolorization of 87% of indigo dye at pH 7.0–7.5 | [153] |

Shi et al. [149] isolated a cold-active LC from *Botrytis* sp. FQ grown on tomato fruits that were stored at 4 °C. The enzyme showed 70% activity at temperatures between 0 °C and 30 °C and an optimal temperature of 15 °C on 2,6-dimethoxyphenol, a classic LC substrate. When tested in operational conditions, pH 6.8, 20 °C, the LC was able to remove up to 60% of triclosan. A cold-adapted LC from *Kabatiella bupleuri* was isolated and characterized by Wisniewska et al. [150]. The enzyme had an optimal temperature of 30 °C. Nonetheless it retained 60% of its maximum activity at 10 °C and over 40% in ice. Interestingly, the LC could be efficiently produced using a sustainable material (BSG) such as a carbon source in SSF and used for dye decolorization. Lac-Q from *Pycnoporus* spp. showed an optimum temperature of 70 °C but retained about 51% of the maximal activity when incubation in ice. Lac-Q was used in combination with ABTS to degrade oxytetracycline [151].

### 4.2. Alkaline Active Laccase

LCs are generally stable at alkaline pHs. Nonetheless at such pHs, their activity is barely detectable [152]. This pH profile of most fungal LCs was explained by an inhibitory effect of the OH- ions on the T2 and T3 centers, as well as by the redox potential difference between the reducing substrate and the T1 center at an alkaline pH [154]. Alkaline-active LCs are rare and can be found in alkaliphilic organisms or obtained by protein engineering. In the latter case, structure-based amino acid substitution near the T1 and TNC sites proved to be effective in shifting LC optimum pH activity [152].

Catalytic activity at a high pH is a desirable characteristic in certain industrial contexts such as bioethanol production, hair coloring, paper bio-bleaching, or pollutant removal from wastewaters. Lignin is insoluble at neutral-acidic pHs thus, an LC could be applied for its depolymerization only after solubilization in alkaline solution [155]. Wastewaters from textile industries generally possess a neutral to alkaline pH (7–10) because dye bleaching is traditionally performed using an alkali solution, such as sodium hypochlorite [156]. Therefore, the study of alkaline active LCs which could be potentially used in such industrial applications is an important topic of research that has been explored little (Table 4).

Protein engineering via error-prone PCR was used to obtain LC mutants which exhibited high activity towards 2,6-dimethoxyphenol at pH 8–8.5. When tested in operational conditions (pH 7.0–7.5) it efficiently decolorized 87% of indigo dye [153]. A knowledge-gaining directed evolution (KnowVolution) process was used to increase the alkaline tolerance of an LC from *Melanocarpus albomyces*. By means of two single mutations near the T1 site, the authors were able to obtain an LC with an optimum pH of 9 for the substrate 2,6-dimethoxyphenol [157]. Eventually, directed evolution was used to obtain a blood tolerant LC for LC-based bio-electronic devices application [158]. By introducing mutations near the T1 site, the authors were able to shift the pH activity profile of a thermostable LC to more alkaline values and to significantly enhance NaCl tolerance.

The research on these topics is still at an early age and further studies are needed to ensure the implementation of enzyme production at industrial level, enzyme improvement to avoid the use of synthetic mediators, and the toxicological assessment of the degradation products in the case of remediation applications.

The search for an LC with extended catalytic properties, stability, and activity in the harsh environmental conditions that are typical of industrial processes is a major topic of research. As shown by the studies reported here on cold- and alkaline-active LCs, extremozymes can be found in nature, but most likely can be improved or created in the laboratories thanks to the application of biotechnology.

Rational and random mutagenesis by error-prone PCR has represented the method of choice to improve LC catalytic activity in the past. Later, other techniques, such as site-directed mutagenesis, semi rational design, and directed evolution have been used and often combined in a stepwise mutagenesis approach [158]. Recently, DNA shuffling, directed evolution, site-directed, and site-saturation mutagenesis have been applied to obtain engineered, tailor-made LCs or chimeras with increased thermostability and activity under harsh environmental conditions (extreme temperatures, pHs, presence of surfactants, solvents, or enzyme inhibitors). Moreover, the use of computational methods to flank protein engineering is opening new opportunities to deliver tailored catalysts for specific applications [158].

## 5. Conclusions

Enzymatic catalysis is an essential pillar to ensure sustainability in industrial processing. Enzymes are safe, selective, specific, biodegradable, and require mild operational conditions. Nonetheless, their application in an industrial context is limited due to difficulties in scale-up, discouraging technology readiness, and process economics. Among the many industrial enzymes, LCs show a unique profile in terms of greenness, versatility, pollutants removal, and efficiency in lignin degradation for the exploitation and valorization of agro-wastes. Indeed, lignin removal is one of the main limitations in closing the loop of agricultural systems because it is the main obstacle for the reuse of wastes as raw materials for the production of valuable compounds or for fermentations. Nonetheless, in recent decades, research has made many steps forward to fully exploit this important enzyme, ensure the economic feasibility of its production, and to overcome the technical limitations of its application. In this review the most recent advances for LC production, immobilization, and use of ago-wastes were given. At the laboratory level these methods proved to be efficient and feasible. Nonetheless, a continuous crosstalk and collaboration with industries is needed to let the new era of LC-sustainable use in industry begin.

Still, after over 100 years from its discovery, LCs remain one of the most promising green enzymes for sustainability, reuse, valorization, and decontamination of vegetable biomasses.

**Author Contributions:** Conceptualization, M.L. and G.M.; writing—original draft preparation, M.L., O.G. and T.F.; writing—review and editing, M.L., O.G., T.F., G.M. and A.F.L.; funding A.F.L. All authors have read and agreed to the published version of the manuscript.

**Funding:** This work was financially supported by H2020-WIDESPREAD-2020-5 MycoTWIN "Enhancing Research and Innovation Capacity Of TÜBİTAK MAM Food Institute On Management Of Mycotoxigenic Fungi And Mycotoxins" Grant Agreement-952337.

**Conflicts of Interest:** The authors declare no conflict of interest.

## References

1. Chapman, J.; Ismail, A.E.; Dinu, C.Z. Industrial applications of enzymes: Recent advances, techniques, and outlooks. *Catalysts* **2018**, *8*, 238. [CrossRef]
2. European Commission. Communication from the Commission to the European Parliament, the Council, the European Economic and Social Committee and the Committee of the Regions Next Steps for a Sustainable European Future European Action for Sustainability. 2016. Available online: https://eur-lex.europa.eu/legal-content/EN/TXT/PDF/?uri=CELEX:52016DC0739&from=EN (accessed on 25 February 2021).
3. Mika, L.; Csefalvay, E.; Nemeth, A. Catalytic Conversion of Carbohydrates to Initial Platform Chemicals: Chemistry and Sustainability. *Chem. Rev.* **2018**, *118*, 505–613. [CrossRef] [PubMed]
4. Grand View Research. Available online: https://www.grandviewresearch.com/industry-analysis/enzymes-industry (accessed on 8 April 2021).
5. Global Market Insight. Available online: https://www.gminsights.com/industry-analysis/enzymes-market (accessed on 8 April 2021).
6. Fasim, A.; More, V.S.; More, S.S. Large-scale production of enzymes for biotechnology uses. *Curr. Opin. Biotechnol.* **2021**, *69*, 68–76. [CrossRef]
7. Contesini, F.J.; Frandsen, R.J.N.; Damasio, A. CAZymes in Biorefinery: From Genes to Application. *Front. Bioeng. Biotechnol.* **2021**, *9*, 86. [CrossRef]
8. Malhotra, M.; Suman, S.K. Laccase-mediated delignification and detoxification of lignocellulosic biomass: Removing obstacles in energy generation. *Environ. Sci. Pollut. Res.* **2021**, *28*, 58929–58944. [CrossRef] [PubMed]
9. FAO. The State of Food and Agriculture Moving Forward on Food Loss and Waste Reduction. 2019. Available online: http://www.fao.org/3/ca6030en/ca6030en.pdf (accessed on 2 April 2021).
10. Lange, L.; Connor, K.O.; Arason, S.; Bundgård-Jørgensen, U.; Canalis, A.; Carrez, D.; Gallagher, J.; Gøtke, N.; Huyghe, C.; Jarry, B.; et al. Developing a Sustainable and Circular Bio-Based Economy in EU: By Partnering Across Sectors, Upscaling and Using New Knowledge Faster, and For the Benefit of Climate, Environment & Biodiversity, and People & Business. *Front. Bioeng. Biotechnol.* **2021**, *8*, 1456. [CrossRef]
11. Cannatelli, M.D.; Ragauskas, A.J. Two Decades of Laccases: Advancing Sustainability in the Chemical Industry. *Chem. Rec.* **2017**, *17*, 122–140. [CrossRef]
12. Zucca, P.; Cocco, G.; Sollai, F.; Sanjust, E. Fungal laccases as tools for biodegradation of industrial dyes. *Biocatalysis* **2016**, *1*, 82–108. [CrossRef]
13. Moiseenko, K.V.; Savinova, O.S.; Vasina, D.V.; Kononikhin, A.S.; Tyazhelova, T.V.; Fedorova, T.V. Laccase Isoenzymes of Trametes hirsuta LE-BIN072: Degradation of industrial dyes and secretion under the different induction conditions. *Appl. Biochem. Biotechnol.* **2018**, *54*, 834–841. [CrossRef]
14. Zhang, Y.; Lin, D.F.; Hao, J.; Zhao, Z.H.; Zhang, Y.J. The crucial role of bacterial laccases in the bioremediation of petroleum hydrocarbons. *World J. Microbiol. Biotechnol.* **2020**, *36*, 1–10. [CrossRef]
15. Barrios-Estrada, C.; de Jesús Rostro-Alanis, M.; Muñoz-Gutiérrez, B.D.; Iqbal, H.M.; Kannan, S.; Parra-Saldívar, R. Emergent contaminants: Endocrine disruptors and their laccase-assisted degradation—A review. *Sci. Total. Environ.* **2018**, *612*, 1516–1531. [CrossRef]

16. Loi, M.; Fanelli, F.; Zucca, P.; Liuzzi, V.C.; Quintieri, L.; Cimmarusti, M.T.; Monaci, L.; Haidukowski, M.; Logrieco, A.F.; Sanjust, E.; et al. Aflatoxin $B_1$ and $M_1$ degradation by Lac2 from *Pleurotus pulmonarius* and redox mediators. *Toxins* **2016**, *8*, 245. [CrossRef]
17. Yoshida, H. Chemistry of lacquer (urushi). *J. Chem. Soc.* **1883**, *43*, 472–486. [CrossRef]
18. Bertrand, G. Simultaneous occurence of laccase and tyrosinase in the juice of some mushrooms. *C. R. Hebd. Séances Acad. Sci.* **1896**, *123*, 463–465.
19. Giardina, P.; Faraco, V.; Pezzella, C.; Piscitelli, A.; Vanhulle, S.; Sannia, G. Laccases: A never-ending story. *Cell. Mol. Life Sci.* **2010**, *67*, 369–385. [CrossRef]
20. Hoegger, P.J.; Kilaru, S.; James, T.Y.; Thacker, J.R.; K??es, U. Phylogenetic comparison and classification of laccase and related multicopper oxidase protein sequences. *FEBS J.* **2006**, *273*, 2308–2326. [CrossRef] [PubMed]
21. Munk, L.; Sitarz, A.K.; Kalyani, D.C.; Mikkelsen, J.D.; Meyer, A.S. Can laccases catalyze bond cleavage in lignin? *Biotechnol. Adv.* **2015**, *33*, 13–24. [CrossRef] [PubMed]
22. Alcalde, M. Laccases: Biological functions, molecular structure and industrial applications. In *Industrial Enzymes*; Springer: Dordrecht, The Netherlands, 2007; pp. 461–476. ISBN 978-1-4020-5376-4.
23. Sharma, P.; Goel, R.; Capalash, N. Bacterial laccases. *World J. Microbiol. Biotechnol.* **2007**, *23*, 823–832. [CrossRef]
24. Nordberg, H.; Cantor, M.; Dusheyko, S.; Hua, S.; Poliakov, A.; Shabalov, I.; Dubchak, I. The genome portal of the Department of Energy Joint Genome Institute: 2014 updates. *Nucleic Acids Res.* **2014**, *42*, 26–31. [CrossRef] [PubMed]
25. Piscitelli, A.; Giardina, P.; Lettera, V.; Pezzella, C.; Sannia, G.; Faraco, V. Induction and transcriptional regulation of laccases in fungi. *Curr. Genom.* **2011**, *12*, 104–112. [CrossRef] [PubMed]
26. Jones, S.M.; Solomon, E.I. Electron transfer and reaction mechanism of laccases. *Cell. Mol. Life Sci.* **2015**, *72*, 869–883. [CrossRef]
27. Xu, F. Oxidation of phenols, anilines, and benzenethiols by fungal laccases: Correlation between activity and redox potentials as well as halide inhibition. *Biochemistry* **1996**, *35*, 7608–7614. [CrossRef] [PubMed]
28. Glazunova, O.A.; Trushkin, N.A.; Moiseenko, K.V.; Filimonov, I.S.; Fedorova, T.V. Catalytic efficiency of basidiomycete laccases: Redox potential versus substrate-binding pocket structure. *Catalysts* **2018**, *8*, 152. [CrossRef]
29. Frasconi, M.; Favero, G.; Boer, H.; Koivula, A.; Mazzei, F. Kinetic and biochemical properties of high and low redox potential laccases from fungal and plant origin. *Biochim. Biophys. Acta* **2010**, *1804*, 899–908. [CrossRef] [PubMed]
30. Lahtinen, M.; Kruus, K.; Boer, H.; Kemell, M.; Andberg, M.; Viikari, L. The effect of lignin model compound structure on the rate of oxidation catalyzed by two different fungal laccases. *J. Mol. Catal.* **2009**, *57*, 204–210. [CrossRef]
31. Xu, F.; Berka, R.M.; Wahleithner, J.; Nelson, B. Site-directed mutations in fungal laccase: Effect on redox potential, activity and pH profile. *Biochem. J.* **1998**, *70*, 63–70. [CrossRef] [PubMed]
32. Rivera-Hoyos, C.M.; Morales-Alvarez, E.D.; Poutou-Pinales, R.A.; Pedroza-Rodriguez, A.M.; Rodriguez-Vazquez, R.; Delgado-Boada, J.M. Fungal laccases. *Fungal Biol. Rev.* **2013**, *27*, 67–82. [CrossRef]
33. Mate, D.M.; Alcalde, M. Laccase engineering: From rational design to directed evolution. *Biotechnol. Adv.* **2015**, *33*, 25–40. [CrossRef] [PubMed]
34. Glazunova, O.A.; Polyakov, K.M.; Moiseenko, K.V.; Kurzeev, S.A.; Fedorova, T.V. Structure-function study of two new middle-redox potential laccases from basidiomycetes *Antrodiella faginea* and *Steccherinum murashkinskyi*. *Int. J. Biol. Macromol.* **2018**, *118*, 406–418. [CrossRef]
35. Rodgers, C.J.; Blanford, C.F.; Giddens, S.R.; Skamnioti, P.; Armstrong, F.A.; Gurr, S.J. Designer laccases: A vogue for high-potential fungal enzymes? *Trends Biotechnol.* **2010**, *28*, 63–72. [CrossRef] [PubMed]
36. Shleev, S.; Christenson, A.; Serezhenkov, V.; Burbaev, D.; Yaropolov, A.; Gorton, L.; Ruzgas, T. Electrochemical redox transformations of T1 and T2 copper sites in native *Trametes hirsuta* laccase at gold electrode. *Biochem. J.* **2005**, *385*, 745–754. [CrossRef]
37. Reinhammar, B.R.M. Oxidation-reduction potentials of the electron acceptors in laccases and stellacyanin. *BBA—Bioenerg.* **1972**, *275*, 245–259. [CrossRef]
38. Farver, O.; Wherland, S.; Koroleva, O.; Loginov, D.S.; Pecht, I. Intramolecular electron transfer in laccases. *FEBS J.* **2011**, *278*, 3463–3471. [CrossRef]
39. Morozova, O.V.; Shumakovich, G.P.; Gorbacheva, M.A.; Shleev, S.V.; Yaropolov, A.I. "Blue" laccases. *Biochemistry* **2007**, *72*, 1136–1150. [CrossRef] [PubMed]
40. Baldrian, P. Fungal laccases—Occurrence and properties. *FEMS Microbiol. Rev.* **2006**, *30*, 215–242. [CrossRef] [PubMed]
41. Claus, H. Laccases: Structure, reactions, distribution. *Micron* **2004**, *35*, 93–96. [CrossRef] [PubMed]
42. Hildén, K.; Hakala, T.K.; Lundell, T. Thermotolerant and thermostable laccases. *Biotechnol. Lett.* **2009**, *31*, 1117–1128. [CrossRef]
43. Li, K.; Xu, F.; Eriksson, K.E.L. Comparison of fungal laccases and redox mediators in oxidation of a nonphenolic lignin model compound. *Appl. Environ. Microbiol.* **1999**, *65*, 2654–2660. [CrossRef]
44. Xu, Y.; Hong, Y.; Xiao, Y.; Fang, W. Preparation and application of polyclonal antibody against a recombinant laccase. *Cell. Mol. Immunol.* **2007**, *4*, 315–317. [PubMed]
45. Vite-Vallejo, O.; Palomares, L.A.; Dantán-González, E.; Ayala-Castro, H.G.; Martínez-Anaya, C.; Valderrama, B.; Folch-Mallol, J. The role of N-glycosylation on the enzymatic activity of a *Pycnoporus sanguineus* laccase. *Enzyme Microb. Technol.* **2009**, *45*, 233–239. [CrossRef]
46. Maestre-Reyna, M.; Liu, W.C.; Jeng, W.Y.; Lee, C.C.; Hsu, C.A.; Wen, T.N.; Wang, A.H.J.; Shyur, L.F. Structural and functional roles of glycosylation in fungal laccase from *Lentinus* sp. *PLoS ONE* **2015**, *10*, e0120601. [CrossRef]

47. Glazunova, O.A.; Moiseenko, K.V.; Kamenihina, I.A.; Isaykina, T.U.; Yaropolov, A.I.; Fedorova, T.V. Laccases with variable properties from different strains of *Steccherinum ochraceum*: Does glycosylation matter? *Int. J. Mol. Sci.* **2019**, *20*, 2008. [CrossRef]
48. Lisova, Z.A.; Lisov, A.V.; Leontievsky, A.A. Two laccase isoforms of the basidiomycete *Cerrena unicolor* VKMF-3196. Induction, isolation and properties. *J. Basic Microbiol.* **2010**, *50*, 72–82. [CrossRef]
49. Васина, Д.В.; Логинов, Д.С.; Королева, О.В. Сравнительныйанализ протеома базидиального гриба *Trametes hirsuta* при культивировании на средах различного состава. Биохимия **2013**, *78*, 627–636.
50. Glazunova, O.A.; Moiseenko, K.V.; Savinova, O.S.; Fedorova, T.V. Purification and Characterization of Two Novel Laccases from *Peniophora lycii*. *J. Fungi* **2020**, *6*, 340. [CrossRef] [PubMed]
51. Garzillo, A.M.; Colao, M.C.; Caruso, C. Laccase from the white-rot fungus *Trametes trogii*. *Appl. Microbiol. Biotechnol.* **1998**, *49*, 545–551. [CrossRef] [PubMed]
52. Reiss, R.; Ihssen, J.; Richter, M.; Eichhorn, E.; Schilling, B.; Thöny-Meyer, L. Laccase versus laccase-like multi-copper oxidase: A comparative study of similar enzymes with diverse substrate spectra. *PLoS ONE* **2013**, *8*, e65633. [CrossRef]
53. Astolti, P.; Brandi, P.; Galli, C.; Gentili, P.; Gerini, M.F.; Greci, L.; Lanzalunga, O. New mediators for the enzyme laccase: Mechanistic features and selectivity in the oxidation of non-phenolic substrates. *New J. Chem.* **2005**, *29*, 1308–1317. [CrossRef]
54. Lange, H.; Decina, S.; Crestini, C. Oxidative upgrade of lignin—Recent routes reviewed. *Eur. Polym. J.* **2013**, *49*, 1151–1173. [CrossRef]
55. D'Acunzo, F.; Galli, C.; Gentili, P.; Sergi, F. Mechanistic and steric issues in the oxidation of phenolic and non-phenolic compounds by laccase or laccase-mediator systems. the case of bifunctional substrates. *New J. Chem.* **2006**, *30*, 583–591. [CrossRef]
56. Munk, L.; Andersen, M.L.; Meyer, A.S. Influence of mediators on laccase catalyzed radical formation in lignin. *Enzyme Microb. Technol.* **2018**, *116*, 48–56. [CrossRef]
57. Kumar, S.V.S.; Phale, P.S.; Durani, S.; Wangikar, P.P. Combined sequence and structure analysis of the fungal laccase family. *Biotechnol. Bioeng.* **2003**, *83*, 386–394. [CrossRef]
58. Ducros, V.; Brzozowski, A.M.; Wilson, K.S.; Brown, S.H.; Ostergaard, P.; Shneider, P.; Yaver, D.S.; Pedersen, A.H.; Davies, G.J. Crystal structure of type-2 Cu depleted laccase from *Coprinus Cinereus* at 2,2 A resolution. *Nat. Struct. Biol.* **1998**, *5*, 310–316. [CrossRef]
59. Hakulinen, N.; Rouvinen, J. Three-dimensional structures of laccases. *Cell. Mol. Life Sci.* **2015**, *72*, 857–868. [CrossRef] [PubMed]
60. Glazunova, O.A.; Polyakov, K.M.; Fedorova, T.V.; Dorovatovskii, P.V.; Koroleva, O.V. Elucidation of the crystal structure of *Coriolopsis caperata* laccase: Restoration of the structure and activity of the native enzyme from the T2-depleted form by copper ions. *Acta Crystallogr. Sect. D Biol. Crystallogr.* **2015**, *71*, 854–861. [CrossRef]
61. Polyakov, K.M.; Fedorova, T.V.; Stepanova, E.V.; Cherkashin, E.A.; Kurzeev, S.A.; Strokopytov, B.V.; Lamzin, V.S.; Koroleva, O.V. Structure of native laccase from *Trametes hirsuta* at 1.8 A resolution. *Acta Crystallogr. D Biol. Crystallogr.* **2009**, *65*, 611–617. [CrossRef] [PubMed]
62. Durão, P.; Chen, Z.; Silva, C.S.; Soares, C.M.; Pereira, M.M.; Todorovic, S.; Hildebrandt, P.; Bento, I.; Lindley, P.F.; Martins, L.O. Proximal mutations at the type 1 copper site of CotA laccase: Spectroscopic, redox, kinetic and structural characterization of I494A and L386A mutants. *Biochem. J.* **2008**, *412*, 339–346. [CrossRef] [PubMed]
63. Osipov, E.; Polyakov, K.; Kittl, R.; Shleev, S.; Dorovatovsky, P.; Tikhonova, T.; Hann, S.; Ludwig, R.; Popov, V. Effect of the L499M mutation of the ascomycetous *Botrytis aclada* laccase on redox potential and catalytic properties. *Acta Crystallogr. Sect. D Biol. Crystallogr.* **2014**, *70*, 2913–2923. [CrossRef] [PubMed]
64. Xu, F.; Palmer, A.E.; Yaver, D.S.; Berka, R.M.; Gambetta, G.A.; Brown, S.H.; Solomon, E.I. Targeted mutations in a *Trametes villosa* laccase. *J. Biol. Chem.* **1999**, *274*, 12372–12375. [CrossRef]
65. Matera, I.; Gullotto, A.; Tilli, S.; Ferraroni, M.; Scozzafava, A.; Briganti, F. Crystal structure of the blue multicopper oxidase from the white-rot fungus *Trametes trogii* complexed with p-toluate. *Inorg. Chim. Acta* **2008**, *361*, 4129–4137. [CrossRef]
66. Macellaro, G.; Baratto, M.C.; Piscitelli, A.; Pezzella, C.; Fabrizi De Biani, F.; Palmese, A.; Piumi, F.; Record, E.; Basosi, R.; Sannia, G. Effective mutations in a high redox potential laccase from *Pleurotus ostreatus*. *Appl. Microbiol. Biotechnol.* **2014**, *98*, 4949–4961. [CrossRef]
67. Kallio, J.; Auer, S.; Jänis, J.; Andberg, M.; Kruus, K.; Rouvinen, J.; Koivula, A.; Hakulinen, N. Structure–function studies of a *Melanocarpus albomyces* laccase suggest a pathway for oxidation of phenolic compounds. *J. Mol. Biol.* **2009**, *392*, 895–909. [CrossRef] [PubMed]
68. Bertrand, T.; Jolivalt, C.; Briozzo, P.; Caminade, E.; Joly, N.; Madzak, C.; Mougin, C. Crystal structure of a four-copper laccase complexed with an arylamine: Insights into substrate recognition and correlation with kinetics. *Biochemistry* **2002**, *41*, 7325–7333. [CrossRef] [PubMed]
69. Xie, T.; Liu, Z.; Liu, Q.; Wang, G. Structural insight into the oxidation of sinapic acid by CotA laccase. *J. Struct. Biol.* **2015**, *190*, 155–161. [CrossRef] [PubMed]
70. Enguita, F.J.; Marçal, D.; Martins, L.O.; Grenha, R.; Henriques, A.O.; Lindley, P.F.; Carrondo, M.A. Substrate and dioxygen binding to the endospore coat laccase from *Bacillus subtilis*. *J. Biol. Chem.* **2004**, *279*, 23472–23476. [CrossRef] [PubMed]
71. Liu, Z.; Xie, T.; Zhong, Q.; Wang, G. Crystal structure of CotA laccase complexed with 2,2-azinobis-(3-ethylbenzothiazoline-6-sulfonate) at a novel binding site. *Acta Crystallogr. Sect. Struct. Biol. Commun.* **2016**, *72*, 328–335. [CrossRef] [PubMed]
72. Cole, J.L.; Clark, P.A.; Solomon, E.I. Spectroscopic and chemical studies of the laccase trinuclear copper active site: Geometric and electronic structure. *J. Am. Chem. Soc.* **1990**, *112*, 9534–9548. [CrossRef]

73. Komori, H.; Sugiyama, R.; Kataoka, K.; Miyazaki, K.; Higuchi, Y.; Sakurai, T. New insights into the catalytic active-site structure of multicopper oxidases. *Acta Crystallogr. Sect. D Biol. Crystallogr.* **2014**, *70*, 772–779. [CrossRef] [PubMed]
74. Polyakov, K.M.M.; Gavryushov, S.; Ivanova, S.; Fedorova, T.V.V.; Glazunova, O.A.A.; Popov, A.N.N.; Koroleva, O.V.V. Structural study of the X-ray-induced enzymatic reduction of molecular oxygen to water by *Steccherinum murashkinskyi* laccase: Insights into the reaction mechanism. *Acta Crystallogr. Sect. D Struct. Biol.* **2017**, *73*, 388–401. [CrossRef] [PubMed]
75. Polyakov, K.M.; Gavryushov, S.; Fedorova, T.V.; Glazunova, O.A.; Popov, A.N. The subatomic resolution study of laccase inhibition by chloride and fluoride anions using single-crystal serial crystallography: Insights into the enzymatic reaction mechanism. *Acta Crystallogr. Sect. D Struct. Biol.* **2019**, *75*, 804–816. [CrossRef] [PubMed]
76. Moreno, A.D.; Ibarra, D.; Eugenio, M.E.; Tomás-Pejó, E. Laccases as versatile enzymes: From industrial uses to novel applications. *J. Chem. Technol. Biotechnol.* **2020**, *95*, 481–494. [CrossRef]
77. Morozova, O.V.; Shumakovich, G.P.; Shleev, S.V.; Yaropolov, Y.I. Laccase-mediator systems and their applications: A review. *Appl. Biochem. Microbiol.* **2007**, *43*, 523–535. [CrossRef]
78. Cañas, A.I.; Camarero, S. Laccases and their natural mediators: Biotechnological tools for sustainable eco-friendly processes. *Biotechnol. Adv.* **2010**, *28*, 694–705. [CrossRef]
79. Yesilada, O.; Birhanli, E.; Geckil, H. Bioremediation and Decolorization of Textile Dyes by White Rot Fungi and Laccase Enzymes. In *Mycoremediation and Environmental Sustainability*, 1st ed.; Prasad, R., Ed.; Springer: Cham, Switzerland, 2018; Volume 2, pp. 121–153. [CrossRef]
80. Dhillon, G.S.; Kaur, S.; Brar, S.K.; Verma, M. Flocculation and haze removal from crude beer using in-house produced laccase from *Trametes versicolor* cultured on brewer's spent grain. *J. Agric. Food Chem.* **2012**, *60*, 7895–7904. [CrossRef] [PubMed]
81. Lettera, V.; Pezzella, C.; Cicatiello, P.; Piscitelli, A.; Giacobelli, V.G.; Galano, E.; Amoresano, A.; Sannia, G. fficient immobilization of a fungal laccase and its exploitation in fruit juice clarification. *Food Chem.* **2016**, *196*, 1272–1278. [CrossRef]
82. Loi, M.; Quintieri, L.; Fanelli, F.; Caputo, L.; Mulè, G. Application of a recombinant laccase-chlorogenic acid system in protein crosslink and antioxidant properties of the curd. *Food Res. Int.* **2018**, *106*, 763–770. [CrossRef]
83. Loi, M.; Quintieri, L.; De Angelis, E.; Monaci, L.; Logrieco, A.F.; Caputo, L.; Mule, G. Yield improvement of the Italian fresh Giuncata cheese by laccase–induced protein crosslink. *Int. Dairy J.* **2020**, *100*, 104555. [CrossRef]
84. Li, M.; Karboune, S.; Light, K.; Kermasha, S. Oxidative cross-linking of potato proteins by fungal laccases: Reaction kinetics and effects on the structural and functional properties. *Innov. Food Sci. Emerg. Technol.* **2021**, *71*, 102723. [CrossRef]
85. Zhu, Y.; Tao, H.; Janaswamy, S.; Zou, F.; Cui, B.; Guo, L. The functionality of laccase-or peroxidase-treated potato flour: Role of interactions between protein and protein/starch. *Food Chem.* **2021**, *341*, 128082. [CrossRef]
86. Manhivi, V.E.; Amonsou, E.O.; Kudanga, T. Laccase-mediated crosslinking of gluten-free amadumbe flour improves rheological properties. *Food Chem.* **2018**, *264*, 157–163. [CrossRef] [PubMed]
87. Loi, M.; Fanelli, F.; Cimmarusti, M.T.; Mirabelli, V.; Haidukowski, M.; Logrieco, A.F.; Caliandro, R.; Mule, G. *In vitro* single and combined mycotoxins degradation by Ery4 laccase from *Pleurotus eryngii* and redox mediators. *Food Control* **2018**, *90*, 401–406. [CrossRef]
88. Bilal, M.; Rasheed, T.; Nabeel, F.; Iqbal, H.M.; Zhao, Y. Hazardous contaminants in the environment and their laccase-assisted degradation–a review. *J. Env Manag.* **2019**, *234*, 253–264. [CrossRef] [PubMed]
89. Xu, P.; Du, H.; Peng, X.; Tang, Y.; Zhou, Y.; Chen, X.; Fei, J.; Meng, Y.; Yuan, L. Degradation of several polycyclic aromatic hydrocarbons by laccase in reverse micelle system. *Sci. Total Environ.* **2020**, *708*, 134970. [CrossRef]
90. Kudanga, T.; Nemadziva, B.; Le Roes-Hill, M. Laccase catalysis for the synthesis of bioactive compounds. *Appl. Microbiol. Biotechnol.* **2017**, *101*, 13–33. [CrossRef]
91. Bhusainahalli, V.M.; Spatafora, C.; Chalal, M.; Vervandier-Fasseur, D.; Meunier, P.; Latruffe, N.; Tringali, C. Resveratrol-Related Dehydrodimers: Laccase—Mediated Biomimetic Synthesis and Antiproliferative Activity. *Eur. J. Org. Chem.* **2012**, *27*, 5217–5224. [CrossRef]
92. Sun, K.; Cheng, X.; Yu, J.; Chen, L.; Wei, J.; Chen, W.; Wang, J.; Li, S.; Liu, Q.; Si, Y. Isolation of *Trametes hirsuta* La-7 with high laccase-productivity and its application in metabolism of 17β-estradiol. *Environ. Pollut.* **2020**, *263*, 114381. [CrossRef] [PubMed]
93. Mikolasch, A.; Manda, K.; Schlüter, R.; Lalk, M.; Witt, S.; Seefeldt, S.; Lindequist, U. Comparative analyses of laccase-catalyzed amination reactions for production of novel β-lactam antibiotics. *Biotechnol. Appl. Biochem.* **2012**, *59*, 295–306. [CrossRef] [PubMed]
94. Mikolasch, A.; Hammer, E.; Witt, S.; Lindequist, U. Laccase-catalyzed derivatization of 6-aminopenicillanic, 7-aminocephalosporanic and 7-aminodesacetoxycephalosporanic acid. *AMB Express* **2020**, *10*, 1–7. [CrossRef] [PubMed]
95. Lapinsonnière, L.; Picot, M.; Barrière, F. Enzymatic versus Microbial Bio-Catalyzed Electrodes in Bio-Electrochemical Systems. *ChemSusChem* **2012**, *5*, 995–1005. [CrossRef] [PubMed]
96. Rangelov, S.; Nicell, J.A. Modelling the transient kinetics of laccase-catalyzed oxidation of four aqueous phenolic substrates at low concentrations. *Biochem. Eng. J.* **2018**, *132*, 233–243. [CrossRef]
97. Slagman, S.; Zuilhof, H.; Franssen, M.C. Laccase-mediated grafting on biopolymers and synthetic polymers: A critical review. *ChemBioChem* **2018**, *19*, 288. [CrossRef] [PubMed]
98. Meza, J.C.; Auria, R.; Lomascolo, A.; Sigoillot, J.C.; Casalot, L. Role of ethanol on growth, laccase production and protease activity in *Pycnoporus cinnabarinus* ss3. *Enzym. Microb. Technol.* **2007**, *41*, 162–168. [CrossRef]

99. Rodríguez-Couto, S. Solid-State Fermentation for Laccases Production and Their Applications. In *Current Developments in Biotechnology and Bioengineering*, 1st ed.; Pandey, A., Larroche, C., Socco, C.R., Eds.; Elsevier: Amsterdam, The Netherlands, 2018; pp. 211–234. [CrossRef]
100. Rodrigues, E.M.; Karp, S.G.; Malucelli, L.C.; Helm, C.V.; Alvarez, T.M. Evaluation of laccase production by Ganoderma lucidum in submerged and solid-state fermentation using different inducers. *J. Basic Microb.* **2019**, *59*, 784–791. [CrossRef]
101. Wang, F.; Xu, L.; Zhao, L.; Ding, Z.; Ma, H.; Terry, N. Fungal Laccase Production from Lignocellulosic Agricultural Wastes by Solid-State Fermentation: A Review. *Microorganisms* **2019**, *7*, 665. [CrossRef] [PubMed]
102. Del Cerro, C.; Erickson, E.; Dong, T.; Wong, A.R.; Eder, E.K.; Purvine, S.O.; Mitchell, H.D.; Weitz, K.K.; Markillie, K.M.; Burnet, M.C.; et al. Intracellular pathways for lignin catabolism in white-rot fungi. *Proc. Natl. Acad. Sci. USA* **2021**, *118*. [CrossRef]
103. De Castro, R.J.S.; Sato, H.H. Enzyme production by solid state fermentation: General aspects and an analysis of the physicochemical characteristics of substrates for agro-industrial wastes valorization. *Waste Biomass Valoriz.* **2015**, *6*, 1085–1093. [CrossRef]
104. Isanapong, J.; Krailoekpaiboon, T.; Noiniyom, W.; Panchal, S. Utilization of Organic Wastes for Laccase Production by *Pleurotus ostreatus*. *Appl. Sci. Eng. Progr.* **2017**, *10*, 239–244. [CrossRef]
105. Sánchez, Ó.J.; Montoya, S. Assessment of polysaccharide and biomass production from three white-rot fungi by solid-state fermentation using wood and agro-industrial residues: A kinetic approach. *Forests* **2020**, *11*, 1055. [CrossRef]
106. Arora, S.; Rani, R.; Ghosh, S. Bioreactors in solid state fermentation technology: Design, applications and engineering aspects. *J. Biotech.* **2018**, *269*, 16–34. [CrossRef] [PubMed]
107. Tišma, M.; Jurić, A.; Bucić-Kojić, A.; Panjičko, M.; Planinić, M. Biovalorization of brewers' spent grain for the production of laccase and polyphenols. *J. Inst. Brew.* **2018**, *124*, 182–186. [CrossRef]
108. Dhillon, G.S.; Kaur, S.; Brar, S.K. *In-vitro* decolorization of recalcitrant dyes through an ecofriendly approach using laccase from *Trametes versicolor* grown on brewer's spent grain. *Int. Biodeter. Biodegr.* **2012**, *72*, 67–75. [CrossRef]
109. Philippoussis, A.; Diamantopoulou, P.; Papadopoulou, K.; Lakhtar, H.; Roussos, S.; Parissopoulos, G.; Papanikolaou, S. Biomass, laccase and endoglucanase production by *Lentinula edodes* during solid state fermentation of reed grass, bean stalks and wheat straw residues. *World J. Microb. Biotechnol.* **2011**, *27*, 285–297. [CrossRef]
110. Adekunle, A.E.; Zhang, C.; Guo, C.; Liu, C.Z. Laccase production from *Trametes versicolor* in solid-state fermentation of steam-exploded pretreated cornstalk. *Waste Biomass Valoriz.* **2017**, *8*, 153–159. [CrossRef]
111. Li, G.; Fu, Y.; Dang, W.; Hu, R.; Xue, H. The effects of aqueous ammonia-pretreated rice straw as solid substrate on laccase production by solid-state fermentation. *Bioprocess Biosyst. Eng.* **2019**, *42*, 567–574. [CrossRef] [PubMed]
112. Patel, H.; Gupte, A. Optimization of different culture conditions for enhanced laccase production and its purification from *Tricholoma giganteum* AGHP. *Bioresour. Bioprocess.* **2016**, *3*, 1–10. [CrossRef]
113. Mishra, V.; Jana, A.K. Sweet sorghum bagasse pretreatment by *Coriolus versicolor* in mesh tray bioreactor for selective delignification and improved saccharification. *Waste Biomass Valoriz.* **2019**, *10*, 2689–2702. [CrossRef]
114. Xu, L.; Sun, K.; Wang, F.; Zhao, L.; Hu, J.; Ma, H.; Ding, Z. Laccase production by *Trametes versicolor* in solid-state fermentation using tea residues as substrate and its application in dye decolorization. *J. Environ. Manag.* **2020**, *270*, 110904. [CrossRef]
115. Pourkhanali, K.; Khayati, G.; Mizani, F.; Raouf, F. Isolation, identification and optimization of enhanced production of laccase from *Galactomyces geotrichum* under solid-state fermentation. *Prep. Biochem. Biotechnol.* **2021**, *51*, 659–668. [CrossRef] [PubMed]
116. Bernal, C.; Rodriguez, K.; Martinez, R. Integrating enzyme immobilization and protein engineering: An alternative path for the development of novel and improved industrial biocatalysts. *Biotech. Adv.* **2018**, *36*, 1470–1480. [CrossRef] [PubMed]
117. Sigurdardóttir, S.B.; Lehmann, J.; Ovtar, S.; Grivel, J.C.; Negra, M.D.; Kaiser, A.; Pinelo, M. Enzyme immobilization on inorganic surfaces for membrane reactor applications: Mass transfer challenges, enzyme leakage and reuse of materials. *Adv. Synth. Catal.* **2018**, *360*, 2578–2607. [CrossRef]
118. Daronch, N.A.; Kelbert, M.; Pereira, C.S.; de Araújo, P.H.H.; de Oliveira, D. Elucidating the choice for a precise matrix for laccase immobilization: A review. *Chem. Eng. J.* **2020**, *397*, 125506. [CrossRef]
119. Huang, P.S.; Boyken, S.E.; Baker, D. The coming of age of de novo protein design. *Nature* **2016**, *537*, 320–327. [CrossRef] [PubMed]
120. Alvarado-Ramírez, L.; Rostro-Alanis, M.; Rodríguez-Rodríguez, J.; Castillo-Zacarías, C.; Sosa-Hernández, J.E.; Barceló, D.; Iqbal, H.M.N.; Parra-Saldívar, R. Exploring current tendencies in techniques and materials for immobilization of laccases–A review. *Int J. Biol. Macromol.* **2021**, *181*, 683–696. [CrossRef] [PubMed]
121. Girelli, A.M.; Astolfi, M.L.; Scuto, F.R. Agro-industrial wastes as potential carriers for enzyme immobilization: A review. *Chemosphere* **2020**, *244*, 125368. [CrossRef] [PubMed]
122. Da Silva, A.M.; Tavares, A.P.; Rocha, C.M.; Cristóvão, R.O.; Teixeira, J.A.; Macedo, E.A. Immobilization of commercial laccase on spent grain. *Process. Biochem.* **2012**, *47*, 1095–1101. [CrossRef]
123. Girelli, A.M.; Scuto, F.R. Spent grain as a sustainable and low-cost carrier for laccase immobilization. *Waste Manag.* **2021**, *128*, 114–121. [CrossRef]
124. Ahmed, T.A.; Suso, H.P.; Hincke, M.T. In-depth comparative analysis of the chicken eggshell membrane proteome. *J. Proteom.* **2017**, *155*, 49–62. [CrossRef] [PubMed]
125. Girelli, A.M.; Scuto, F.R. Eggshell membrane as feedstock in enzyme immobilization. *J. Biotechnol.* **2021**, *325*, 241–249. [CrossRef]
126. De Souza Bezerra, T.M.; Bassan, J.C.; de Oliveira Santos, V.T.; Ferraz, A.; Monti, R. Covalent immobilization of laccase in green coconut fiber and use in clarification of apple juice. *Process. Biochem.* **2015**, *50*, 417–423. [CrossRef]

127. Cristóvão, R.O.; Silvério, S.C.; Tavares, A.P.M.; Jose, M.L.; Boaventura, R.A.L.; Macedo, E.A.; Coelho, M.A.Z. Green coconut fiber: A novel carrier for the immobilization of commercial laccase by covalent attachment for textile dyes decolourization. *World J. Microbiol. Biotechnol.* **2012**, *28*, 2827–2838. [CrossRef] [PubMed]
128. Cristóvão, R.O.; Tavares, A.P.; Brígida, A.I.; Loureiro, J.M.; Boaventura, R.A.; Macedo, E.A.; Coelho, M.A.Z. Immobilization of commercial laccase onto green coconut fiber by adsorption and its application for reactive textile dyes degradation. *J. Mol. Catal. B Enzym.* **2011**, *72*, 6–12. [CrossRef]
129. Li, N.; Xia, Q.; Niu, M.; Ping, Q.; Xiao, H. Immobilizing laccase on different species wood biochar to remove the chlorinated biphenyl in wastewater. *Sci. Rep.* **2018**, *8*, 13947. [CrossRef]
130. Imam, A.; Suman, S.K.; Singh, R.; Vempatapu, B.P.; Ray, A.; Kanaujia, P.K. Application of laccase immobilized rice straw biochar for anthracene degradation. *Environ. Pollut.* **2021**, *268*, 115827. [CrossRef]
131. Lonappan, L.; Liu, Y.; Rouissi, T.; Brar, S.K.; Verma, M.; Surampalli, R.Y. Adsorptive immobilization of agro-industrially produced crude laccase on various micro-biochars and degradation of diclofenac. *Sci. Total Environ.* **2018**, *640*, 1251–1258. [CrossRef] [PubMed]
132. Tesfaye, T.; Sithole, B.; Ramjugernath, D. Valorisation of chicken feathers: Recycling and recovery routes. In Proceedings of the Sixteenth International Waste Management and Landfill Symposium S, Cagliari, Italy, 2–6 October 2017.
133. Suman, S.K.; Patnam, P.L.; Ghosh, S.; Jain, S.L. Chicken feather derived novel support material for immobilization of laccase and its application in oxidation of veratryl alcohol. *ACS Sustain. Chem. Eng.* **2018**, *7*, 3464–3474. [CrossRef]
134. Baig, K.S.; Wu, J.; Turcotte, G. Future prospects of delignification pretreatments for the lignocellulosic materials to produce second generation bioethanol. *Int. J. Energy Res.* **2019**, *43*, 1411–1427. [CrossRef]
135. Shanmugam, S.; Hari, A.; Ulaganathan, P.; Yang, F.; Krishnaswamy, S.; Wu, Y.R. Potential of biohydrogen generation using the delignified lignocellulosic biomass by a newly identified thermostable laccase from Trichoderma asperellum strain BPLMBT1. *Int. J. Hydrog.* **2018**, *43*, 3618–3628. [CrossRef]
136. Giacobbe, S.; Pezzella, C.; Lettera, V.; Sannia, G.; Piscitelli, A. Laccase pretreatment for agrofood wastes valorization. *Bioresour. Technol.* **2018**, *265*, 59–65. [CrossRef]
137. Banerjee, R.; Chintagunta, A.D.; Ray, S. Laccase mediated delignification of pineapple leaf waste: An ecofriendly sustainable attempt towards valorization. *BMC Chem.* **2019**, *13*, 58. [CrossRef]
138. Sherpa, K.C.; Ghangrekar, M.M.; Banerjee, R. A green and sustainable approach on statistical optimization of laccase mediated delignification of sugarcane tops for enhanced saccharification. *J. Environ. Manag.* **2018**, *217*, 700–709. [CrossRef]
139. Rencoret, J.; Pereira, A.; del Río, J.C.; Martínez, A.T.; Gutiérrez, A. Delignification and saccharification enhancement of sugarcane byproducts by a laccase-based pretreatment. *ACS Sustain. Chem. Eng.* **2017**, *5*, 7145–7154. [CrossRef]
140. Suman, S.K.; Malhotra, M.; Kurmi, A.K.; Narani, A.; Bhaskar, T.; Ghosh, S.; Iata Jain, S. Jute sticks biomass delignification through laccase-mediator system for enhanced saccharification and sustainable release of fermentable sugar. *Chemosphere* **2022**, *286*, 131687. [CrossRef] [PubMed]
141. Rajak, R.C.; Banerjee, R. An eco-friendly process integration for second generation bioethanol production from laccase delignified Kans grass. *Energy Conv. Manag.* **2018**, *157*, 364–371. [CrossRef]
142. Rajak, R.C.; Banerjee, R. Enzyme mediated biomass pretreatment and hydrolysis: A biotechnological venture towards bioethanol production. *RSC Adv.* **2016**, *6*, 61301–61311. [CrossRef]
143. Lu, H.; Lou, H.; Hu, J.; Liu, Z.; Chen, Q. Macrofungi: A review of cultivation strategies, bioactivity, and application of mushrooms. *Compr. Rev. Food Sci. Food Saf.* **2020**, *19*, 2333–2356. [CrossRef]
144. Olivieri, G.; Wijffels, R.H.; Marzocchella, A.; Russo, M.E. Bioreactor and bioprocess design issues in enzymatic hydrolysis of lignocellulosic biomass. *Catalysts* **2021**, *11*, 680. [CrossRef]
145. Branà, M.T.; Sergio, L.; Haidukowski, M.; Logrieco, A.F.; Altomare, C. Degradation of aflatoxin B1 by a sustainable enzymatic extract from spent mushroom substrate of *Pleurotus eryngii*. *Toxins* **2020**, *12*, 49. [CrossRef]
146. Nazilah, K.R.; Koentjoro, M.P.; Isdiantoni, E.I.; Muslihatin, W.; Prasetyo, E.N. Effect of laccase oxidation pre-treatment on coffee (Coffea arabica) bean processing waste for composting substrate. In *AIP Conference Proceedings*; AIP Publishing LLC: Melville, NY, USA, 2020; Volume 2215, p. 070008. [CrossRef]
147. Kuddus, M. Cold-active enzymes in food biotechnology: An updated mini review. *J. Appl. Biol. Biotechnol.* **2018**, *6*, 58–63. [CrossRef]
148. Hamid, B.; Mohiddin, F.A. Cold-Active Enzymes in Food Processing. In *Enzymes in Food Technology*; Kuddus, M., Ed.; Springer Nature: Singapore, 2018. [CrossRef]
149. Shi, Y.; Kong, D.; Liu, J.; Lu, J.; Yin, X.; Zhou, Q. Transformation of triclosan by a novel cold-adapted laccase from Botrytis sp. FQ. *Front. Environ. Sci Eng.* **2017**, *11*, 6. [CrossRef]
150. Wisniewska, K.M.; Twarda-Clapa, A.; Białkowska, A.M. Screening of Novel Laccase Producers—Isolation and Characterization of Cold-Adapted Laccase from *Kabatiella bupleuri* G3 Capable of Synthetic Dye Decolorization. *Biomolecules* **2021**, *11*, 828. [CrossRef] [PubMed]
151. Tian, Q.; Dou, X.; Huang, L.; Wang, L.; Meng, D.; Zhai, L.; Liao, X. Characterization of a robust cold-adapted and thermostable laccase from *Pycnoporus* sp. SYBC-L10 with a strong ability for the degradation of tetracycline and oxytetracycline by laccase-mediated oxidation. *J. Hazard. Mater.* **2020**, *382*, 121084. [CrossRef]

152. Novoa, C.; Dhoke, G.V.; Mate, D.M.; Martínez, R.; Haarmann, T.; Schreiter, M.; Eidner, J.; Schwerdtfeger, R.; Lorents, P.; Davari, M.D.; et al. KnowVolution of a fungal laccase toward alkaline pH. *ChemBioChem* **2019**, *20*, 1458–1466. [CrossRef]
153. Yin, Q.; Zhou, G.; Peng, C.; Zhang, Y.; Kües, U.; Liu, J.; Xiao, Y.; Fang, Z. The first fungal laccase with an alkaline pH optimum obtained by directed evolution and its application in indigo dye decolorization. *AMB Express* **2019**, *9*, 1–13. [CrossRef]
154. Xu, F. Effects of redox potential and hydroxide inhibition on the pH activity profile of fungal laccases. *J. Biolog. Chem.* **1997**, *272*, 924–928. [CrossRef]
155. Hämäläinen, V.; Grönroos, T.; Suonpää, A.; Heikkilä, M.W.; Romein, B.; Ihalainen, P.; Malandra, R.; Birikh, K.R. Enzymatic processes to unlock the lignin value. *Front. Bioeng. Biotechnol.* **2018**, *6*, 20. [CrossRef]
156. Du, W.; Zuo, D.; Gan, H.; Yi, C. Comparative study on the effects of laser bleaching and conventional bleaching on the physical properties of indigo kapok/cotton denim fabrics. *Appl. Sci.* **2019**, *9*, 4662. [CrossRef]
157. Mate, D.M.; Gonzalez-Perez, D.; Falk, M.; Kittl, R.; Pita, M.; De Lacey, A.L.; Ludwig, R.; Shleev, S.; Alcalde, M. Blood tolerant laccase by directed evolution. *Chem. Biol.* **2013**, *20*, 223–231. [CrossRef]
158. Stanzione, I.; Pezzella, C.; Giardina, P.; Sannia, G.; Piscitelli, A. Beyond natural laccases: Extension of their potential applications by protein engineering. *Appl. Microb. Biotechnol.* **2020**, *104*, 915–924. [CrossRef]

Journal of *Fungi*

MDPI

*Review*

# Biosynthesis Pathways, Transport Mechanisms and Biotechnological Applications of Fungal Siderophores

**Lorenzo Pecoraro [1,*,†], Xiao Wang [1,†], Dawood Shah [1,2], Xiaoxuan Song [1], Vishal Kumar [3], Abdul Shakoor [1,4], Keshawanand Tripathi [5], Pramod W. Ramteke [6] and Rupa Rani [1,7]**

1 School of Pharmaceutical Science and Technology, Tianjin University, 92 Weijin Road, Tianjin 300072, China; wang_xiao1996@163.com (X.W.); dawoodshah616@gmail.com (D.S.); harusong@163.com (X.S.); abdul_shakoor954@yahoo.com (A.S.); ruparani719@gmail.com (R.R.)
2 Institute of Biotechnology and Genetic Engineering, The University of Agriculture Peshawar, Peshawar 25000, Pakistan
3 Department of Food Science and Technology, Yeungnam University, Gyongsan 38541, Korea; vkaggarwal180@gmail.com
4 College of Geography and Environmental Science, Henan University, Kaifeng 475004, China
5 Center for Conservation and Utilization of Blue-Green Algae, ICAR-Indian Agricultural Research Institute, New Delhi 110012, India; tripathikn009@gmail.com
6 Faculty of Life Sciences, Mandsaur University, Mandsaur 458001, India; pwramteke@gmail.com
7 Department of Environmental Science and Engineering, Indian Institute of Technology (Indian School of Mines), Dhanbad 826004, India
* Correspondence: lorenzo.pecoraro@tju.edu.cn; Tel.: +86-185-2082-4550
† Co-first author.

**Citation:** Pecoraro, L.; Wang, X.; Shah, D.; Song, X.; Kumar, V.; Shakoor, A.; Tripathi, K.; Ramteke, P.W.; Rani, R. Biosynthesis Pathways, Transport Mechanisms and Biotechnological Applications of Fungal Siderophores. *J. Fungi* **2022**, *8*, 21. https://doi.org/10.3390/jof8010021

Academic Editor: Baojun Xu

Received: 8 December 2021
Accepted: 21 December 2021
Published: 28 December 2021

**Publisher's Note:** MDPI stays neutral with regard to jurisdictional claims in published maps and institutional affiliations.



**Abstract:** Iron (Fe) is the fourth most abundant element on earth and represents an essential nutrient for life. As a fundamental mineral element for cell growth and development, iron is available for uptake as ferric ions, which are usually oxidized into complex oxyhydroxide polymers, insoluble under aerobic conditions. In these conditions, the bioavailability of iron is dramatically reduced. As a result, microorganisms face problems of iron acquisition, especially under low concentrations of this element. However, some microbes have evolved mechanisms for obtaining ferric irons from the extracellular medium or environment by forming small molecules often regarded as siderophores. Siderophores are high affinity iron-binding molecules produced by a repertoire of proteins found in the cytoplasm of cyanobacteria, bacteria, fungi, and plants. Common groups of siderophores include hydroxamates, catecholates, carboxylates, and hydroximates. The hydroxamate siderophores are commonly synthesized by fungi. L-ornithine is a biosynthetic precursor of siderophores, which is synthesized from multimodular large enzyme complexes through non-ribosomal peptide synthetases (NRPSs), while siderophore-Fe chelators cell wall mannoproteins (FIT1, FIT2, and FIT3) help the retention of siderophores. *S. cerevisiae*, for example, can express these proteins in two genetically separate systems (reductive and nonreductive) in the plasma membrane. These proteins can convert Fe (III) into Fe (II) by a ferrous-specific metalloreductase enzyme complex and flavin reductases (FREs). However, regulation of the siderophore through Fur Box protein on the DNA promoter region and its activation or repression depend primarily on the Fe availability in the external medium. Siderophores are essential due to their wide range of applications in biotechnology, medicine, bioremediation of heavy metal polluted environments, biocontrol of plant pathogens, and plant growth enhancement.

**Keywords:** fungal cytoplasm proteins; iron uptake; iron-binding molecules; siderophore biosynthesis; hydroxamate; siderophore-Fe chelators mannoproteins; biotechnology; medicine; biocontrol; bioremediation

## 1. Introduction

Iron plays a vital role in the growth and development of living organisms, and it is one of the most abundant elements found on earth [1]. Mineral bioweathering is important in soil ecosystems because it increases the availability of iron for colonizing organisms, which is limited in oxygenic environments [2,3]. Indeed, Fe is converted into insoluble oxyhydroxide polymers under aerobic conditions at biological pH (oxidized form) [4]. Plants require ($10^{-5}$–$10^{-7}$ M) of Fe (II) for growth and development, whereas the solubility of Fe (III) in nature is $10^{-17}$ M at pH 7 [5]. During the evolution of life on earth, the majority of iron was in insoluble Fe (III) forms. Therefore, microorganisms including fungi, bacteria, and cyanobacteria adapted to low iron availability levels and overcame iron deficiency by synthesizing siderophores [6–8]. Siderophores are low molecular weight compounds (200–2000 Da) produced by different microbes, which chelate the iron from different habitats [9]. Scientific data have shown the ubiquitous presence of siderophores in plants and microbes. A significant number, i.e., 500 fungal and bacterial siderophores, are documented [10]. Furthermore, mammalian siderophores have also been reported [11]. Several laboratory methods, including spectrophotometric titration, electrophoretic mobility, mass spectrometry, acid hydrolysis, and biological activity tests are used to characterize siderophores [12].

Fe plays an essential role in vital functions, including photosynthesis, respiration, synthesis of DNA, RNA, proteins, and enzyme cofactors [13]. In human serum, transferrin is an iron transport protein that maintains Fe (III) concentration (10–24 M) and blocks the entry of pathogens [14]. Fungi and other microorganisms adopt various strategies for iron acquisition from the extracellular environment by (i) use of metal ion transporters [15], (ii) acquisition from heme and heme containing-proteins [16], (iii) acquisition from transferrin, lactoferrin, and ferritin [17,18], (iv) use of reductive systems of iron uptake, and (v) siderophore-mediated iron transport [8]. Siderophore-mediated Fe-scavenging is an essential process in soil ecosystems that improves the bioavailability of iron derived from mineral dissolution via bioweathering. Siderophores primarily scavenge iron through complex formation with other metals such as molybdenum and cobalt [19]. These compounds promote plant growth and play an important role in pathogen biocontrol [20] and bioremediation of metal-polluted environments [21].

Fungal siderophores are very diverse and show striking structures. Fungi usually produce hydroxamate and carboxylate siderophore types, which have been primarily studied in *Aspergillus* species. For instance, *A. fumigatus* and *A. nidulans* synthesize about 55 types of siderophores. Both species live as saprotrophs, contributing to maintaining carbon and nitrogen cycles in the environment [8]. Many fungi can produce more than one siderophore type, especially under low iron availability. *Aspergillus fumigatus* often produces a hydroxamate siderophore and triacetyl fusarine C (TAFC) for tapping extracellular iron [22]. This fungus can also secret a siderophore called ferricrocin for mobilization and distribution of hyphal iron, and intracellular iron storage. Besides, *A. fumigatus* synthesizes the conidial siderophore 'hydroxyferricrocin' for storing iron in its conidia to support the germination process and reduce oxidative stress [23]. *Aspergillus nidulans* has been reported to produce two main siderophores, ferricrocin (Frr) and ferrihordin. A twenty-four hour culture of *A. nidulans* produced an unacetylated form of TAFC, which is known as fusigen. In contrast, an older strain (48 h) yielded acetylated TAFC due to breakdown and uptake of fusigen [24]. Another fungus, *Wolfiporia cocos*, known as a brown-rot fungus, may produce different types of catecholate siderophores [25]. According to Haselwandter et al. [26], the two ubiquitous ectomycorrhizal basidiomycetes *Laccaria laccata* and *L. bicolor* can produce linear ester-containing Fsg siderophores, i.e., CPG, Frr and TAFC, apart from the hydroxamates.

Many researchers are now interested in knowing how some fungal strains have evolved to produce different types of siderophores, what are the suitable conditions for the production of siderophores, how siderophores, in general, can contribute to fungal survival, and what are the structural differences between fungal siderophores. Indeed, many studies have used structural and stereochemical analyses to describe the properties of fungal siderophores, while attempts have been made to understand the production, recognition, and transportation mechanism of siderophores in different fungal species. Some members of siderophore classes have been characterized by crystalline structures, the absolute metal center configuration, and conformation in solutions [8,27]. Understanding of chemical and structural properties and transport mechanisms of fungal siderophores has also inspired many researchers. Ferrichromes (FRC), for example, are the best studied siderophores and are now regarded as biomolecules. However, there is still a dearth of knowledge on how siderophores are designed for transport. Except in plants, nothing is known about the possible membrane-located transport system for siderophores in the fungal plasma membrane.

Further studies are still needed to describe in detail, or confirm, the protein nature of many fungal siderophores. However, some studies have already clarified the energy requirement for siderophore-mediated iron transport in fungi. There is evidence suggesting that the transportation of siderophores can be mechanically transferred across the cytoplasmic membrane. However, it is still not fully known yet to what extent this is possible.

## 2. Overview of Fungal Siderophores

Fungi are heterotrophic eukaryotic organisms [28]. Among them, saprobic fungal species are decomposers that break down and feed on decaying organic matter. Many fungi show two major responses to iron concentration in the environment: siderophore synthesis under iron stress, and a high-affinity ferric iron reductase [29]. On the basis of chemical interaction sites, siderophores have been classified into two major groups, i.e., Enterobactin and Hydroxamates. Enterobactin is regarded as a good iron chelator, which has the ability to interact with iron and catecholate hydroxyl groups. Hydroxamates are unique, due to the presence of N-hydroxylated amide bonds. The hydroxamate siderophores, such as FRC, are commonly produced by fungi. Siderophores that are synthesized by microbes and plants are classified based on coordinating groups and Fe binding system, including (i) phenolates, (ii) hydroxamates, (iii) polycarboxylates (Table 1). Common types of siderophores are hydroxamates, catecholates, and carboxylates [30]. Nitrogen, oxygen, and sulfur atoms can take part in iron coordination in the carboxylate group of siderophores [31,32]. Another group of siderophores is known as the mixed type. Mixed type siderophores do not belong to hydroxamate and the aromatic hydroxyl category. These hybrid types of siderophores are classified based on the position of the Fe (III) binding group. Mixed-type siderophores can bind salicylic acid and nitrogen [4]. Common examples of mixed type siderophores are (i) lysine derivatives, such as myobactin, (ii) ornithine derivatives, including pyoverdine, and (iii) histamine derivatives, such as anguibactin. Other examples are thiazoline, oxazoline, and pyoverdine. Pyoverdine, in particular, has been reported as a signaling molecule in the bacterium *Pseudomonas aeruginosa* [33] and as an inhibitor molecule against zinc-containing matrix metalloproteinases (MMPs), which often degrade in extracellular matrixes [34]. The main groups of siderophores are itemized in Figure 1. Many phytopathogenic fungi synthesize unique compounds to chelate iron, but also produce phytotoxins. Fungi mainly synthesize hydroxamate-type siderophores (derived from the nonproteinogenic ornithine amino acid) (Table 1).

**Table 1.** Types of siderophores and their characteristics.

| Siderophores | Types | Characteristics | References |
|---|---|---|---|
| Hydroxamates | Rhodotorulic acid | The diketopiperazine ring of N5-acetyl-N5-hydroxy-L-ornithine units linked head-to-head. Produced mainly by basidiomycetous yeasts such as *Rhodotorula* spp. | Haas [35]; Das et al. [36] |
| | Coprogens | The diketopiperazine ring (dimerum acid) of diketopiperazine ring (dimerum acid) units linked head-to-head. Produced generally by a number of plant pathogens, such as *H. capsulatum*, *B. dermatitidis*, *Fusarium dimerum* and *Curvularia lunata*. These are di or tri-hydroxamates derivatives of rhodotorulic acid with a linear structure composed of trans-fusarinine units. | Haas [35]; Das et al. [36] |
| | Ferrichromes | Cyclic hexapeptides consisting of tripeptide of *N*-acyl-*N*-hydroxyornithine and three amino acids, serine, glycine and alanine. Several different acyl groups have been found in this family such as acetyl, malonyl, transb-ethylglutaconyl, trans-anhydromevalonyl, and cis-anhydromevalonyl. Ferrichromes are produced by phytopathogenic fungi and by *Microsporum* sp., *Trichophyton* sp., and *Aspergillus* spp. Another function of ferrichromes is the intracellular storage of iron. | Haas [35]; Das et al. [36] |
| | Fusarinines | Linear or cyclic hydroxamates composed of *N*-hydroxyornithine, which is N-acylated by anhydromevalonic acid. Produced by *Fusarium* spp., *Paecilomyces* spp., and *Aspergillus* spp. | Das et al. [36] |
| Polycarboxylates | Rhizoferrin | A citric acid-containing polycarboxylate called rhizoferrin has been isolated from *Rhizopus microsporus* var. *rhizopodiformis*. The molecule is composed of two citric acid units linked to diaminobutane. Produced mainly by Mucoromycota, Mucorales (Mucoraceae, Thamnididiaceae, and Choanephoraceae) and Mortierellales (Mortierellaceae), and Entomophthoromycota, Entomophthorales. | Das et al. [35] |

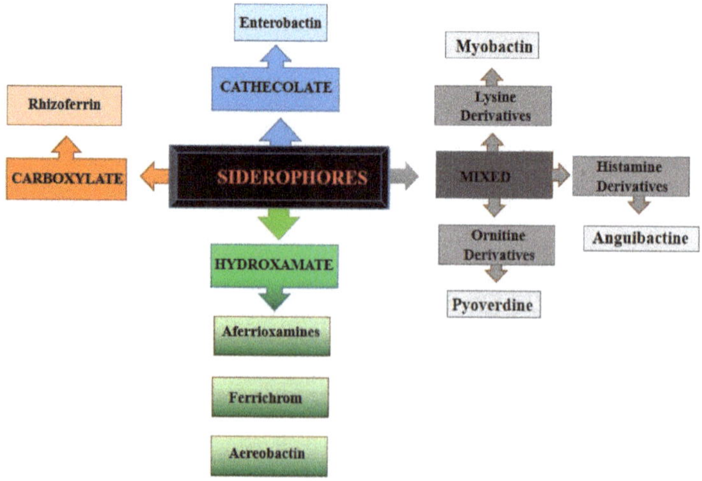


**Figure 1.** General groups of siderophores.

Hydroxamates are further categorized into three groups: (i) fusarinines (FSR), (ii) coprogens (CPG), (iii) FRC, with a few exceptions, such as the rhizoferrin (Figure 2). Rhizoferrin is a carboxylate-type siderophore produced by certain zygomycetes. Siderophore synthesis by fungi depends on the nutrient medium and culture conditions. Hydrophilic siderophores are derived from the common structural unit $N^{\delta}$-hydroxyornithine. In fungi, they consist of hydroxylated and alkylated ornithine amino acid, while in bacteria, they are acylated and hydroxylated alkylamines [37], represented by $N^6$-acyl-$N^6$-Hydroxylysine or $N^5$-acyl-$N^5$-Hydroxyornithine reported by Winkelmann [38]. All hydroxamate siderophores are characterized by peptide linkage [24], except fusarinine C (FsC), synthesized by *Aspergillus nidulans*, which shows ester bonds. Two $O_2$ molecules of these groups bind with Fe, known as bi-dentate ligand. Hydroxamate siderophores are capable of binding hexadentate octahedral complex with Fe (III) [39].

Fungi synthesize more than one type of siderophore belonging to a single structural family or different structural families. For instance, *Trichoderma pseudokoningii* and *T. longibrachiatum* synthesize all three structural families of siderophores [40]. The FSR (fusarinine) is a siderophore synthesized by young cultures of *Fusarium roseum*, whereas at the older stage of culture, FSR is replaced by malonichrome. Siderophore identity is a valuable trait in fungal taxonomy [41]. In fact, wide ranges of siderophores are found within a fungal genus. Mor et al. [42] found that siderophore production in *Trichophyton* and *Microsporum* species are similar. For instance, *T. rubrum* and *T. mentagrophytes* produced siderophores that are also synthesized by *Microsporum* taxa. The same siderophores, ferrichrome C and ferricrocin, are produced in *Microsporum* species, including *M. canis* and *M. gypseum*. In contrast, *T. mentagrophytes* and *T. Tonsurans* produce only ferrichrome. Other fungal siderophores are described in Table 2.

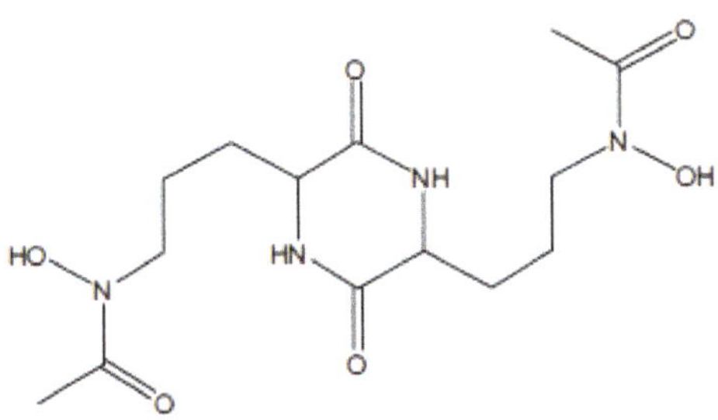

Rhodotorulic acid

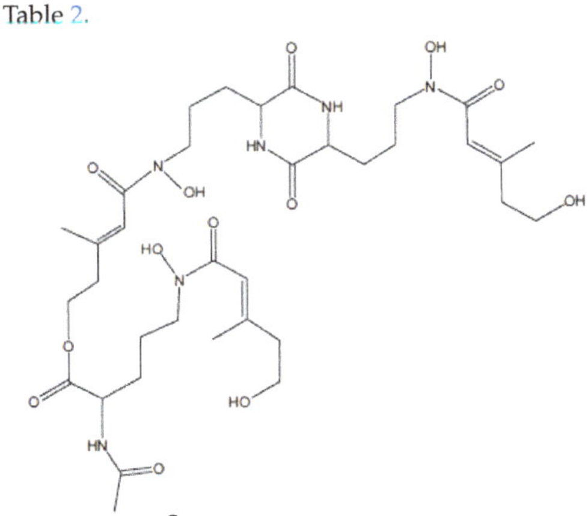

Coprogens

**Figure 2.** *Cont.*

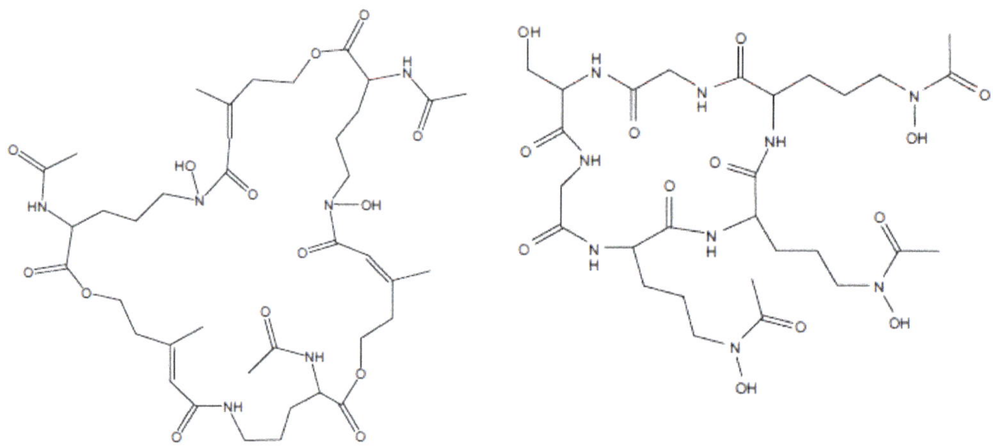

Triacetylfusarinine C

Ferricrocin

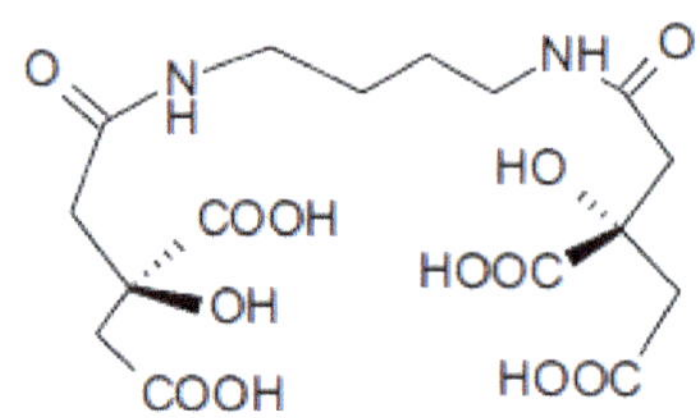

Rhizoferrin

**Figure 2.** Structure of fungal siderophores.

**Table 2.** Some examples of fungal siderophores.

| Fungal Source | Siderophores | References |
|---|---|---|
| *Aspergillus* sp., *Penicillium oxalicum*, *Aureobasidium pullulans*, *Phanerochaete chrysosporium* | Hydroxamates | Ghosh et al. [22] |
| *Alternaria longipes* | Trihydroxamate (Hydroxycoprogen I, Hydroxyneocoprogen I) | Jalal and Helm [43] |
| *A. longipes* | Trihydroxamate ($N^b$-dimethyl coprogen, $N^b$-dimethyl neocoprogen I and $N^b$-dimethyl isoneocoprogen | Jalal et al. [44] |
| *Candida* sp. | Ferrichrome, hydroxamates | Baakza et al. [37] |
| *Curvularia lunata* | Trihydroxamate (Neocoprogen II) | Hossain et al. [45] |
| *Epicoccum purpurascens* (Syn. *E. nigrum*) | Trihydroxamate (Isoneocoprogen I or Triornicin) | Frederick et al. [46] |
| *E. nigrum and C. lunata* | Trihydroxamate (Isotriornicin or Neocoprogen I) | Frederick et al. [46]; Chowdappa et al. [47] |
| *Fusarium dimerum* | Dihydroxamate (Dimerum acid) | Diekmann [48] |
| *F. roseum* | Cis-fusarinine | Emery [49] |
| *dimerum* | Trans-Fusarinine | Diekmann [48] |

**Table 2.** *Cont.*

| Fungal Source | Siderophores | References |
|---|---|---|
| *F. dimerum* | Trihydroxamate (Coprogen B) | Diekmann [48] |
| *Gliocladium virens* | Fusarinine A, Fusarinine B | Jalal et al. [50] |
| *Neurospora crassa* | Neurosporin | Eng-Wilmot et al. [51] |
| *Penicillium* sp. | Trihydroxamate (Coprogen) | Pidacks et al. [52] |
| *Penicillium* sp. | N, N′N′-triacetylfusarinine C | Moore and Emery [53] |
| *Paracoccidioides* sp. | Hydroxamates | Lesuisse et al. [54] |
| *Rhizopus microsporus* | Carboxylates (rhizoferrin) | Drechsel et al. [31] |
| *Rhodothamus chamaecistus* | Fusarinine C (FsC), Fusigen | Haselwandter et al. [55] |
| *Saccharomyces cerevisiae* | Catecholate, hydroxamate, ferrioxamine, ferricrocin | Senthilnithy [56] |
| *Trichoderma* sp. | Trihydroxamate (Pamitoylcorprogen) | Anke et al. [40] |
| *Trichoderma* sp. | Hydroxamates, carboxylates | Baila et al. [57] |
| *Ustilago inflorescentiae* | Dihydroxamate (Rhodotorulic acid) | Atkin and Neilands [58]; Muller et al. [59] |

## 3. Biosynthesis of Siderophores and Regulation

Biosynthesis of siderophores is divided into two major pathways: nonribosomal peptide synthetase (NRPS)-dependent, and NRPS-independent [60,61]. NRPSs are large multimodular enzyme complexes. They consist of adenylation domain (A), thiolation domain (T), condensation domain (C), and thioesterase domain (TE). Each module of the NRPS enzymes is responsible for adding amino acids (AAs) and forming peptide bonds. NRPS enzymes determine the sequence and number of AAs in the peptide chain [62]. NRPS recognizes and activates AAs by activating A-domain and acylating adenylate via an ATP-dependent reaction. The next steps are thiolation of (T) domain, in which the activated ester is covalently linked, followed by condensation of the (C) domain, and direct transfer of another acyl amino acid to form a peptide bond [63]. The thioesterase domain (TE) is present in the final unit. The last step consists of assembling or releasing chains from the NRPS by hydrolysis or cyclization. Cleavage of the acyl thioester, which binds to the T domain, is an NADH-dependent reaction [64]. Fungi commonly use four main mechanisms for iron uptake across the cytoplasmic membrane as follows: (i) A shuttle mechanism in which the siderophore bound iron can be taken into the cell and the iron is released by a reductase or by direct ligand exchange in which the recipient siderophore becomes the storage molecule. The gathering ligand is released to capture another iron molecule. This type of transfer has been used by coprogen and ferrichrome families. (ii) A direct transfer mechanism in which iron is taken up (iron is first reduced by the reductive pathway before taken up) while the ligand remains outside the cell. The iron transfer is not a membrane-reductive event but is a membrane-mediated exchange between the gathering siderophore and an internal chelating agent. The transfer mechanism may be by ligand exchange (nonenzymatic) to an internal pool of the chelating agent, which then serves as the storage compound. This type of transfer has been used by rhodotorulic acid. (iii) An esterase-reductase mechanism in which iron ligand is taken up and ester bonds of the iron ligand are split to excrete fusarinine moieties, followed by reduction and storage of ferric iron. This type of transfer has been used by ferric triacetylfusarinine C. (iv) A reductive mechanism with the transport of some ferrichromes, which do not to enter cells but give up ferric iron by reduction with transport of the ferrous iron [36].

Hydroxamate fungal siderophores have similar biosynthetic pathways in terms of their basic unit, characterized by hydroxyornithine [65,66]. Figure 3 represents a general schematic biosynthetic pathway [67]. The first step is the hydroxylation of L-ornithine. This is a precursor of siderophores, which is converted into *N*-hydroxy-L-ornithine in a reaction catalyzed by the enzymes L-ornithine and L-ornithine $N^5$-oxygenase [61,68]. The second step is the acylation of N-hydroxy L-ornithine to form *N*-acyl-*N*-hydroxy-L-ornithine in the presence of the enzyme transacetylase [67]. The acyl-CoA derivative is an acyl donor, and the reaction is catalyzed by acyl-CoA: N-hydroxy-L-ornithine N-acyl transferase. This step has been reported in *Ustilago sphaerogena* [69,70], while N-acetyltransferase activity in siderophore biosynthetic pathways has been reported in *Rhodotorula pilimanae* [71]. N-acetyltransferase has also been reported in other fungi such as *Fusarium cubense* [72], *Rhodotorula glutinis* [73], and *Aspergillus quadricinctus* [67]. The following steps are the condensation of several N-acyl-N-hydroxy-L-ornithines combined in two to three units and the formation of dipeptides and triesters such as FsC, rhodotorulic acid, and coprogen B [67]. Condensation of amino acid and formation of cyclic peptide FRC were reported in *Aspergillus quadricinctus* [74]. Peptide biosynthesis in siderophore production and respective genes (sid2) have been described in *Ustilago maydis* and *Trichoderma virens* (Psy1) [75,76]. Disruption of these genes makes *U. maydis* and *T. virens* unable to synthesize ferrichrome [76,77].

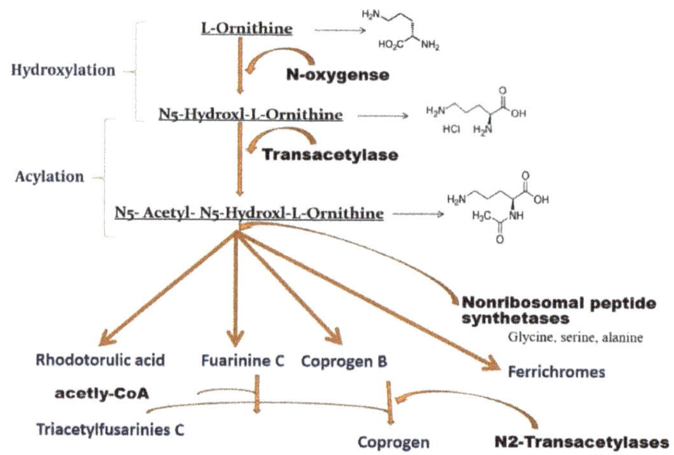


**Figure 3.** L-ornithine is the biosynthetic precursor of siderophores in fungi. Intermediate steps including hydroxylation, acylation and acetylation lead to the synthesis of Rhodotorulic acid, Fusarinine C and Coprogen (modified and adapted from Philpott [29]).

Transport of iron via siderophores is an investment of biosynthetic energy. The function of siderophores is mainly iron sequestration from an external medium. During excretion, siderophores are released, and a few of them are loaded with iron molecules for supporting the growth of the producing microorganism [78]. Therefore, the excretion of siderophores responds to the availability of iron in the external medium [79]. Fur is a ferric uptake regulatory protein produced in Gram-negative and positive bacteria [80]. A constitutive siderophore mutant in *Salmonella* was observed by Ernst et al. [80], and cloned [81]. The $Fur^+$ gene is repressed along with a number of genes involved in iron uptake [82]. The proposed repression model requires the Fur-protein's internal binding of ferrous iron and then binds to the Fur Box present on the target DNA. Inhibition of RNA polymerase is responsible for the search of the promoter region of the iron-regulatory region. Under limited iron conditions, transport systems and siderophore biosynthesis are activated, and Fur protein is separated from the Fur box on the DNA [83]. In fungi, transcriptional repressor (Fur) proteins are known as GATA factor proteins [84,85]. These proteins contain GATA-type zinc fingers that bind to the siderophore biosynthesis genes [86,87] (Figure 4). Table 3 represents negative fungal regulatory proteins in the biosynthesis and transport of siderophores.

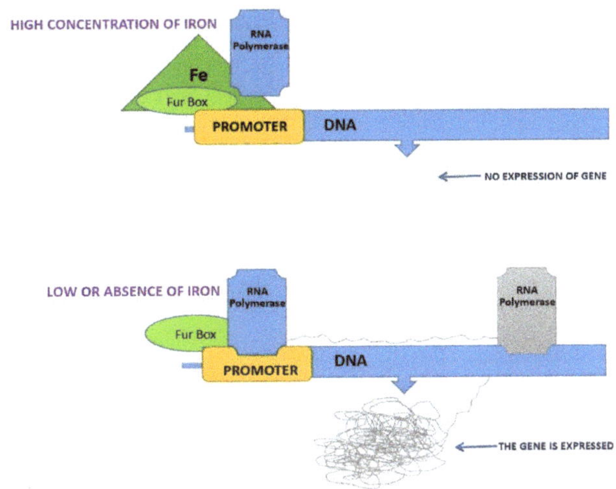


**Figure 4.** Regulatory model of biosynthesis of siderophore and Fur Box activation or repression: At the high iron concentration, formation of Fur box protein and iron complex in the promoter region of the DNA and RNA polymerases are unable to move forward due to the regression of the gene, while at low concentration of iron, Fur proteins release RNA polymerases leading to the expression of genes.

**Table 3.** Negative fungal regulatory proteins in the biosynthesis and transport of siderophores.

| S.No | Regulatory Protein Similar to GATA Factor | Organisms | References |
|---|---|---|---|
| 1. | URBS1 | *Ustilago maydis* | Voisard et al. [88]; An et al. [86] |
| 2. | SRE | *Neurospora crassa* | Zhou et al. [89] |
| 3. | SREP | *Penicillium chrysogenum* | Haas et al. [90] |
| 4. | SREA | *Aspergillus nidulans* | Haas et al. [91]; Oberegger et al. [24] |
| 5. | GAF2p | *Schizosaccharomyces pombe* | Hoe et al. [92]; Pelletier et al. [93] |

In iron-deficient conditions, *C. albicans* and *S. cerevisiae* use the Aft1 transcription factor to bind with the promoter region of siderophore biosynthetic genes and activate the expression of genes [94,95]. Therefore, external ion concentration is the regulatory factor for intracellular biosynthesis of siderophores and transport proteins in microorganisms. The regulation of siderophore production and transporter proteins are the most important ecological aspects of siderophores. Iron sensing is a crucial step for the survival of competing microorganisms in natural environments. Iron sensing assists the microbes in adapting to ever-changing iron metabolism in different habitats [96]. Therefore, those microorganisms highly sensitive to iron regulation are more resistant to environmental changes. Previous research has provided insight into the low concentrations of iron in environments. Usually, iron-deficient environments are colonized by aerobic microorganism through up-regulation of siderophore biosynthetic genes and transport proteins [97]. The important question here is: where is the low iron content in natural habitats? The marine region, especially in the open oceans, calcareous soils, and freshwater lakes contain iron concentration in surface water in the nano-molar 0,2 to 1 nM range [98] and inhibits growth of plankton, bacteria, and plants.

In the human body, free iron is absent, and if any iron is present, it is in protein-bound forms such as transferrin or lactoferrin and ferritin [99]. Pathogens can multiply in low iron environments by sequestering iron from host proteins using ferric-binding protein and transferrin-binding protein, as found in *Neisseria*. These transport proteins move from the periplasm to the cytosol [100]. *Pseudomonas aeruginosa* cell lysis and degradation of proteins by the proteases and subsequent iron scavenging by the excretion of siderophores is another method of iron acquisition. There are different routes by which pathogens can utilize iron from the host cells, but the most suitable system is siderophore mediated transport [101]. Another essential aspect of siderophore ecology is the energy saving. The siderophore biosynthesis requires energy, which is obtained from ATP and carbon sources. Siderophore production starts after the germination of conidiospores in fungi [102]. The conidiospores contain a certain amount of siderophores packed into the wall of the spore and secreted at the time of germination [103]. Siderophore genes are responsible for the sporulation in some fungal strains. Knock out of siderophore genes in fungi was shown to inhibit the sporulation process [104]. A notable example is the *Aspergillus fumigatus* conidial siderophore 'hydroxyferricrocin', which also aids in germination and oxidative stress tolerance [23].

## 4. Siderophore Mediated Iron Transport in Fungi

The fungal cell wall is made up of glucans, chitin, chitosan and glycosylated proteins, and shows a highly dynamic structure [105]. The outer layers of the fungal cell wall are composed of mannoproteins [106]. Mannoproteins affect cell permeability and are influenced by growth conditions. Indeed, mannoproteins allow the passage of nutrients across the cell wall to the periplasmic space and plasma membrane [54]. Regulation and uptake of iron molecules are essential in fungi for maintaining homeostatic processes. As a result, fungi commonly use four main mechanisms for iron uptake, including ferric iron ($Fe^{3+}$) uptake through the production of siderophores, iron assimilation through a redox reaction, heme uptake, and direct iron uptake [8,35]. Every fungal species exhibits an extracellular iron-uptake mechanism known as siderophore-iron transporter (SIT). This SIT, in principle, is constituted by a major protein family that facilitates iron uptake in fungi, acting through the help of the plasma membrane, with high solubility and energy, as a proton-coupled symporter, and releasing iron-chelated siderophore during cell growth. The triacetyl fusarine C (TAFC) and fusarinine C (FsC) have been found to enhance the iron release through partial hydrolysis by the esterase (Estb) enzyme [15].

The structural configuration and properties of siderophores have revealed their vast affinity for iron. All siderophores differ from each other; nonetheless, they share a common conserved structure with a similar functional unit and show an identical pattern of binding to other molecules, i.e., transferrin and lactoferrin. Siderophores typically consist of a peptide backbone that interacts with receptors present in the outer membrane of the cell surface [107]. The hydroxamate siderophores are more structurally complex and hydrophilic. However, denticity plays a more critical role in their affinity toward iron. Most siderophores exhibit a hexadentate structure that allows six coordination sites for ferric ions [10]. An example of hexadentate siderophore is dihydroxylbenzoylserine trimer. It is commonly produced from enterobactin and exhibits preorganized metal-binding via macro cyclization [108].

It has been revealed that the hexadentate siderophores usually have a higher affinity for Fe (III) than tetradentate siderophores. Each of their molecules contains three bidentate ligands fused to form a hexadentate complex. This characteristic also reduces entropic changes during the chelation of a single ferric ion, as compared to the bidentate siderophores, which have only two to three ligand molecules [8,10]. The denticity of siderophores also differs depending on their architecture, and varies from linear dimer to trimer, or cyclic trimer. The cyclization of siderophores, related to the TAFC and enterobactin groups, enhances stability and helps them to resist enzyme degradation [38].

Siderophore structure also varies based on the presence of different functional groups. According to Renshaw et al. [109], fungal siderophores are usually formed by the basic unit N$\delta$-acyl-N$\delta$-hydroxyornithine, i.e., L-isomer of N5-hydroxy-N5-acetylornithine. These siderophores are derivatives of L-ornithine, except for *Neurospora crassa* siderophore. *Neurospora crassa*, obtains a siderophore from neurosporin [51]. All siderophores produced by fungi belong to the hydroxamate group, except the polycarboxylate rhizoferrin. It was revealed by Huschka et al. [110] that the geometrical stability of any siderophore complex depends mainly on the kind and number of its N-acyl residues surrounding the iron coordination center. Well configured stable L-cis ferrichrome siderophores have been identified in *Aspergillus quandricinctus*, *Neurospora crassa*, and *Penicillium parvum*.

The mechanism of siderophore transport is specific or well defined in fungi, which may use multiple transport systems or produce more than one siderophore at a time for efficient recruitment and transportation of metal ions. For example, Howard [15] reported that *Agaricus bisporus* has various transport systems for FSR and FRC, while *N. crassa* has different recognition sites for CPG and the ferrichrome-type siderophore system. *Saccharomyces cerevisiae* retains significant amounts of Fe-chelating molecules, termed as siderophores, in its cell wall and periplasmic space. Under Fe-deficient condition, there is the expression of very high levels of three mannoproteins (FIT1p, FIT2p, and FIT3p) commonly regarded as facilitators of Fe-transport [111]. Siderophore-Fe chelators improve protein retention in the cell wall, whereas deletion of related genes controlling the mannoproteins fit1, fit2, and fit3 reduces ferrichrome and ferrioxamine uptake by 50%. Fe-transport of *S. cerevisiae* is expressed into two genetically separate systems known as reductive and nonreductive systems [29]. The two-step reductive system operates at the plasma membrane level, where the reduction of Fe III (oxidized) into Fe II (reduced) is done by a ferrous-specific complex also known as the high-affinity transporter [112,113].

### 4.1. Ferric Reductase Enzymes (FRE)

The ferric reductase enzymes (FREs) are metalloreductases encoded by FRE1 and FRE2 genes expressed on the plasma membrane, reducing or oxidizing iron and copper [114,115]. These reductases are integral membrane proteins (multiple) and possess binding sites for Fe, coenzymes (FAD, NADPH), and cytochrome (b-types) [116]. *Saccharomyces cerevisiae* exhibited reductases in different forms such as FRE1, 2, 3, 4, 5, and 6 under iron-deficient conditions, and expressed FRE1 and 7, under copper depleted conditions. Fre3p was expressed on the plasma membrane and showed siderophore reductase activity, while Fre6p was located on a vacuolar membrane and involved in reductive export of iron into the cytosol. Initially, ferric citrate reductases (Fre1 and Fre2p) of yeast are characterized by broad substrate specificity and Fe uptake from many ferric sources. Siderophores bind ferric iron with high affinity and convert ferric ion into a ferrous ion, allowing transport by a specific ferrous ion transporter (Figure 5).

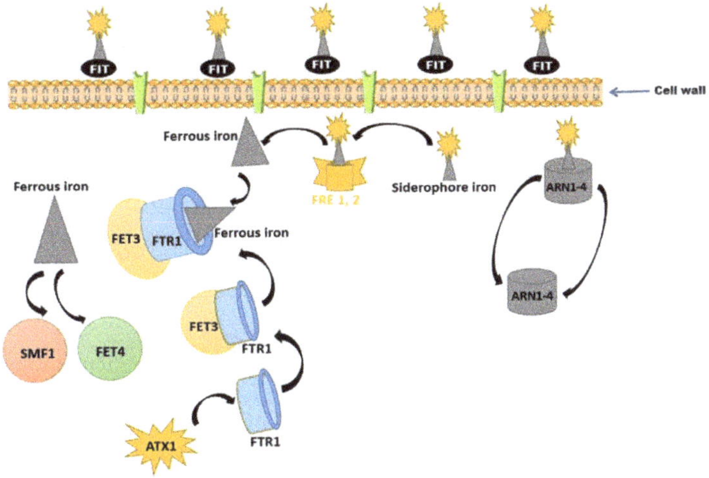


**Figure 5.** Mechanisms of ferric ion (Fe III) transport in the yeast cell, and subsequent transformation into ferrous ion (Fe II).

The plasma membrane of *S. cerevisiae* is bounded by a porous cell wall, providing the shape of the cells, protecting them from osmotic lysis, and preventing the entry of too large macromolecules. The siderophore-iron retention proteins (FIT mannoproteins) of the cell wall facilitate iron retention with siderophores and do not require siderophore uptake. In a few cases, siderophores cross the cell wall via nonspecific pores. FRE reductases (FRE1,2) reduce Fe (III) into Fe (II) before uptake, and reduced iron is taken up through high-affinity ferrous iron transporters (Fet3p and Ftr1p complex) or low-affinity iron transporters (Fet4 and Smf1). However, the Fet3p requires copper for optimum activity [29].

Fre1p and Fre2p catalyze the reduction of several iron-siderophore chelates, including ferrichrome and ferrioxamine B [117]. Fre3p encodes a plasma membrane reductase that catalyzes the reductive uptake of iron bound with hydroxamate siderophores, and Fre4p specifically catalyzes the reductive uptake of di-hydroxamate rhodotorulic acid-iron [118]. The standard reduction potentials of physiological reductants such as NADPH are lower than those of ferric siderophore complexes, which are kinetically unfavorable on the cell surface. However, by coupling reduction of ferric-siderophore with competitive ligand exchange with a ferrous iron-specific chelator at lower pH in hydrophobic environments, the reduction potential can be changed into the level of physiologic reductants [10,119]. Therefore, in vivo ferric siderophores are reduced into ferrous ions at reduced pH in a lipid rich environment. Similarly, transferrin, a high-affinity iron-binding glycoprotein, reduces ferric ion into ferrous ion and removes the reduced form of iron.

### 4.2. Multicopper Permease

Multicopper permease (Ftr1p) is a high-affinity transport complex responsible for transporting reduced forms of iron [120]. The apparent Km of the high affinity enzyme complex is 0.2μM [121] and allows transport at low iron concentration. The ferrous iron is oxidized by Fet3p and requires molecular oxygen [122,123]. The ferric iron is transported into the cytosol via the Ftr1p permease. The oxidase reaction is copper-dependent, and four copper ions are inserted into Fet3p during post-translation in the secretory pathway on the post-Golgi compartment. The copper chaperone Atx1p is responsible for binding and transporting copper to Ccc2p [124]. The copper transporter pumps copper into the post-Golgi vesicle lumen [125]. Both proteins are necessary to maintain adequate cellular copper levels and functioning secretory pathways [126]. These copper-binding proteins, Atx1p and Ccc2p, are synthesized during iron deprivation and not copper deficiency, indicating Fet3-dependent iron uptake. Assembled complexes (Fet3p and Ftr1p) of proteins are retained by quality control systems in the endoplasmic reticulum, if expressed in the absence of their protein partner [127]. The process of ubiquitin-mediated endocytosis rapidly degrades the Fet3p/Ftr1p complex in the presence of high levels of iron [127].

### 4.3. Siderophore-Iron Transporters

Most fungi synthesize and secrete siderophores and small organic compounds that specifically bind with iron molecules with high affinity [7]. Fungal siderophores bind iron with dissociation constants ($10^{-29}$ M), showing greater affinity than any iron-binding ligand in the biological systems. Hsiang [128] reported that all fungi express a nonreductive uptake system specific to siderophore iron chelates. Fungi express transporters with specificity for siderophores secreted by other species of fungi. When the siderophore is abundant, the reductive system of transport can catalyze the uptake of siderophore-bound iron. More than 50% of genes are transcriptionally activated under iron-deprived conditions and are involved in the uptake of iron chelators. The evolution of these uptake systems helps fungi and other microorganisms to compete during the limited availability of iron.

## 5. Biotechnological Applications of Siderophores

### 5.1. Treatment of Infectious Diseases and Anticancer Activity

Siderophores can be used for the treatment of thalassemia [129,130], a disease associated with inherent blood disorder due to abnormal hemoglobin formation. Many studies have shown that compounds such as desferrioxamine B (DFO) are used to reduce this disorder [131–134]. Rhodotorulic acid, a fungal siderophore, has been investigated as an alternative to DFO for iron and aluminum overload [135]. The mechanism of rhodotorulic acid is similar to the iron excretion of DFO, but increases zinc excretion, resulting in toxicity at the administration site [136]. DFO and other hydroxamate siderophores have been used to treat cancer, malaria actinide contamination, and other infectious diseases [137,138]. DFO was also reported for treatment of acute lymphoblastic leukemia by Estrov et al. [139]. Vergne et al. [140] reported that several other siderophores exhibit anticancerous and antitumor activity. For example, triornicin fungal siderophore, produced by *Epicoccum purpurascens* has an inhibitory effect on tumors in mice [46]. For treating infections, siderophore transport/uptake research can enhance movement of the drugs into the microbial cell conjugated with the siderophore-drug [140,141]. Actinides are radioactive elements known as potent carcinogens. Siderophores and their analogs may enhance the excretion and removal of actinides [142]. However, siderophore research in the medical field is still in progress. Most studies have been concentrated on siderophores of bacterial origin, mainly DFO, and on the siderophore analogs hydroxypyridinones.

### 5.2. Application of Siderophore as Drug Delivery Agents

Siderophore combined with antibiotic are used as a 'Trojan horse' for targeted drug delivery (Figure 6). Using siderophore receptors, this method enables antibiotic transfer across the membrane. *Escherichia coli* was treated with a solution of two arthrobactin-carbacephem conjugates that were created [143]. A siderophore cephalosporin conjugate was examined against several pathogenic bacteria such as *Pseudomonas aeruginosa* and *E. Coli* [144]. In addition, simultaneous treatment with conjugates containing hydroxamic and catechol resulted in bacterial growth inhibition. The sideromycins, which are connected to lorabid or ciprofloxacin, are also of interest [145]. Sideromycins-lorabid conjugate invades the periplasm, whereas sideromycins-ciprofloxacin attacks the cell wall. *Staphylococcus aureus* growth was effectively inhibited by sideromycins. Pyochelin-norfloxacin was another siderophore conjugate compound investigated [146]. Siderophore analogues were developed and connected to an antibiotic-norfloxacin. Among four conjugates, two were found effective against *P. aeruginosa*. A vanchrobactin-norfloxacin conjugate, which demonstrated antibacterial properties against *V. anguillarum* and its variants, is another example of siderophore analogs [147]. Some siderophore-antibiotic conjugates, however, have been shown to promote bacterial growth. For example, *Mycobacterium smegmatis* was not affected by spermexatol-carbacephalosporin conjugates [148].

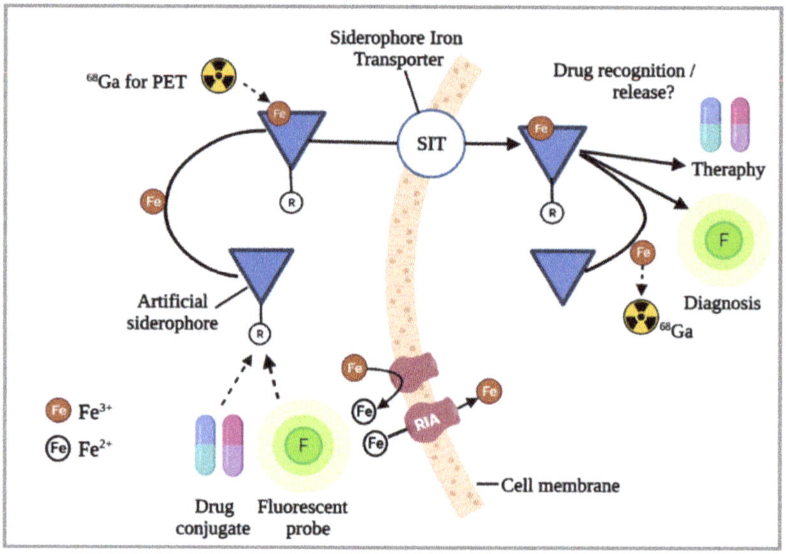


**Figure 6.** Siderophores for medical applications. The use of different siderophores (natural, artificial, and modified) linked to drugs or fluorescent probes, and $^{68}$Ga radiolabeled for molecular imaging and/or curative practices. Siderophores facilitated intracellular absorption potentially allows diagnosis through the use of fluorescent or radioactive signal transduction or therapy by introducing antifungal drugs, thus obeying a trojan horse strategy. The siderophores' metabolic path and translation target is shown by solid line arrows. Substitution of Fe with $^{68}$Ga (for PET) or drug molecules–or fluorescent probes–siderophores conjugate is indicated by dotted line arrows (modified and adapted from Petrik et al. [149].

### 5.3. Application of Siderophores in Vaccine Development

A vibriobactin analogue either connected to bovine serum albumin (BSA) or ovalbumin (OVA) was reported to stimulate the production of antibodies in a mouse model [150]. Based on a murine model of an uropathogenic *E. coli* (UPEC), conjugates of siderophores and antigens were recently exploited for the development of vaccines against urinary tract infections (UTIs) [151]. A decrease in bacterial concentration when mice were given siderophore-cBSA (aerobactin-cationized bovine serum albumin) conjugates, indicated protection via adaptive immunity [152]. Other studies have shown the importance of Fe transport receptors in pathogenic bacteria as vaccine components [153]. An assessment of *E. coli* (O157:H7) siderophore receptors, as well as porin proteins, was carried out on cattle. Two vaccine doses were found to reduce the bacterial prevalence in cattle [154]. As a potent component of vaccine, FhuD receptor of *S. aureus* was examined in another study. Because of the absence of conformational alteration, FhuD-ferrichrome was exploited as the model vaccine antigen [155].

### 5.4. Application of Siderophores in Diagnostics

#### 5.4.1. Radiolabeled Siderophores for Imaging Fungal Infections

Present diagnostic methods such as computed tomography (CT) have severe limitations regarding specificity and sensitivity. For instance, any radiological indication can be associated with fungal infection in CT; thus, novel and better diagnostic methods are required for invasive fungal infections [156]. Siderophore iron transporters (SITs) are potential targets of molecular imaging methods due to their higher upregulation during fungal infection, substrate specificity, and radiolabeled substrate accumulation in the target cell after their energy-driven uptake (Figure 6). In addition, lower molecular mass and high hydrophilicity of siderophores lead to better circulation in infected tissues, speedy clearance from nonspecific tissues and elimination through renal excretion. Another vital characteristic of SITs is the possibility of substituting Fe in the siderophores with an Fe-mimicking radionuclide. Gallium-68 is a diamagnetic isosteric substitution for $Fe^{3+}$ that has been used to describe siderophore complexes extensively [149].

Using only microgram of the siderophore, radiolabeling of a variety of desferri-siderophores with gallium-68 can be achieved, as shown by proof-of-concept experiments [157,158]. *Aspergillus fumigatus* absorption of the siderophore was elevated under iron-deficient conditions and could be inhibited when there was an abundant siderophore or sodium azide, showing that the uptake is selective and energy-dependent. Using PET (Positron emission tomography)/CT technology, pulmonary infection caused by *A. fumigatus* was imaged in a rat model, which revealed a significant buildup of [$^{68}$Ga] Ga-triacetyl fusarinine C (TAFC) in the fungal-infested areas. In sterile inflammations and tumour cells, a considerable accumulation of [$^{68}$Ga] Ga-TAFC was not observed [159]. It was also shown that the use of siderophores is specific to species to a certain level.

Some in vitro research has shown significant uptake of [$^{68}$Ga] Ga-TAFC by *Fusarium solani*, *Rhizopus oryzae*, and *A. fumigatus*, but negligible uptake by *A. flavus*, *Candida albicans*, *A. terreus*, or the bacteria *Klebsiella pneumoniae*, *P. aeruginosa*, and *S. aureus*. In contrast, *A. fumigatus* showed the highest uptake of [$^{68}$Ga] Ga-desferrioxamine-E (DFO-E), followed by *A. terreus*, *F. solani*, *R. oryzae*, *A. flavus*, and the bacterial species *S. aureus* [159]. Altogether, TAFC seems to be fungal-specific in comparison to DFO-E. Therefore, $^{68}$Ga-labeled siderophores, particularly [$^{68}$Ga] Ga-TAFC, have great potential for imaging invasive *A. fumigatus* infections in patients [149].

Chemical modification of siderophores has been designed for new promising applications. For example, a hybrid imaging compound may be created by binding fluorescent dyes, allowing PET/CT with gallium-68 and optical imaging [149]. To study TAFC identification in *A. fumigatus* by the MirB transporter, initial efforts were undertaken to chemically alter it with positive, negative or neutral charged functional groups [160]. Even without inhibiting fungal uptake, chemical alterations were possible, with encouraging results coming from the diacetylated version of fusarinine C (DAFC), in which functional groups were used to change the free amine. In order to explore the hybrid imaging concept, fluorescent dyes were combined to DAFC based on these results [161]. Using the optical signals, these fluorescent siderophores enable image-guided approaches, such as bronchoscopy or surgical probing. Furthermore, such compounds can be applied for fluorescent microscopy. Specifically, a FsC-Cy5.5 conjugate was used to image the skin infections resulting from *Trichophyton rubrum* [162].

#### 5.4.2. Diagnostics of Siderophore from Urine

The rapid clearance of [$^{68}$Ga] Ga-TAFC through the renal system limits the use of PET to detect infections in the kidney or bladder. These results, instead, sparked research into the use of TAFC as a urine biomarker to diagnose invasive *A. fumigatus* infections. Clinical trials [163] and clinical samples [164] both yielded promising findings. However, additional research is required to validate this strategy prior to its application in the clinic. Moreover, the mass spectrometry approach applied is not often practiced in diagnostic labs. Nonetheless, the presence of TAFC in aspergillosis patients demonstrates that the siderophore system is activated in human infections [149].

### 5.5. Bioremediation of Metal Polluted Environments

Metals have played a vital role in the development of human civilization. However, manufacturing, sludge application, nuclear power plants and mining have caused a serious increase of heavy metal pollution in the environment [165]. In particular, soil heavy metal pollution has become one of the environmental problems of global concern. Siderophores have a strong solubilizing effect on a variety of metals such as Cr, Cu, Ni, Pb, Zn and the actinides $Th^{4+}$, $U^{4+}$ and $Pu^{4+}$ [166]. For this reason, siderophores can contribute to the bioremediation of heavy metal contaminated soil. Although siderophores are mainly used to chelate $Fe^{3+}$, they can be used in the detoxification process of heavy metal pollutants by combining a variety of toxic metals. The ability of siderophores to chelate heavy metals mainly depends on the stability constant of the siderophore and the metal to form a complex [167]. The use of siderophores for remediation of heavy metal pollution has the advantages of low cost, high efficiency, and no environmentally hazardous collateral. In recent years, there has been an increasing interest in the application of siderophores in metal bioremediation. Phytoremediation is an emerging and practical technology in the field of bioremediation, but heavy metal stress can interfere with the absorption and utilization of iron by plants, cause iron deficiency, affect chlorophyll synthesis, turn young leaves yellow, and hinder plant growth. Siderophore-producing rhizosphere fungi are a group of plant rhizosphere growth-promoting microorganisms whose siderophore products can react with iron-containing minerals in the soil to generate soluble $Fe^{3+}$-siderophore chelates, which promote the dissolution of $Fe^{3+}$ in the soil [168]. These rhizosphere fungi can provide nutrients (especially iron) to plants under heavy metal stress, by inducing selective absorption by plants of the different metals available in the environment. In fact, siderophores combined with iron can effectively be absorbed by plant cells, whereas they cannot easily enter the cells when combined with other heavy metals. Therefore, siderophores can significantly alleviate the stress caused by metal toxicity and promote plant growth [169]. In addition, studies have shown that siderophore-producing rhizosphere fungi can maintain their growth-promoting effects on plants in heavy metal polluted soil because siderophores can alleviate the inhibitory effect of heavy metals on the synthesis of plant growth hormones, such as indoleacetic acid, by fungi [170]. Siderophores produced by rhizosphere fungi can chelate with $Fe^{3+}$ to inhibit the absorption of iron by plant pathogens, thus reducing the activity of pathogens, protecting plants from diseases and promoting plant growth [171]. In addition to their growth-promoting effect to improve the biomass of plants, fungal siderophores can also improve the activity of heavy metals in the rhizosphere environment and promote the absorption and accumulation of heavy metals in plants. The activity of heavy metals in the plant rhizosphere is an important factor that determines whether large amounts of heavy metals can be absorbed by plants [172]. The siderophores produced by rhizosphere fungi can combine with heavy metal ions in the soil to form a soluble heavy metal-siderophore chelate, thereby improving the activity of heavy metals in the rhizosphere environment, increasing the absorption and accumulation of heavy metals by plants, and improving the efficiency of phytoremediation [173]. Dahlheimer and colleagues used siderophores to react with oxides containing the heavy metal ions $Pt^{4+}$ and $Pd^{2+}$ and found that siderophores can form soluble chelates with $Pt^{4+}$ and $Pd^{2+}$, thus increasing the solubility of Pt and Pd [174]. Hong et al. reported the in vitro dissolution of copper and zinc via siderophores produced by the fungal species *Fusarium solani* [175]. Other studies have also confirmed that siderophores can promote the dissolution of many common heavy metals, and even metalloids, from their minerals. For instance, siderophores can promote the dissolution of manganese-containing minerals such as $Mn_3O_4$ [176]. At the same time, heavy metals can influence siderophore-producing microorganisms, and the total amount of siderophores increased in copper-contaminated sites [177]. Siderophores and heavy metals can also be stored in fungi after chelation. For example, in soil contaminated by metals, *Hypocrea lixii* can secrete siderophores to accumulate copper and zinc in the biomass [78].

The motor manufacturing industry, sewage sludge, vehicle emissions, and mining are common contributing sources for metal pollution [178,179]. Neptunium and plutonium are man-made actinides present in the environment as pollutants due to the testing of nuclear power stations and weapons production, posing a significant environmental hazard [180]. Siderophores are effective for the solubilization of actinides [181] and form stable tetravalent actinides. They also play a significant role in the mobilization of other metals, including zinc, copper, lead, and cadmium [182,183]. Siderophore-producing microorganisms are abundant in soil [184] and affect the bioavailability of metals and radionuclides present in the environment [183,185]. Siderophores could be used to develop metal recovery or remediation of waste sites, including radioactive waste, due to their complexing ability [167].

### 5.6. Plant Growth Enhancement and Biocontrol of Plant Pathogens

Microbial siderophores provide Fe-nutrition when the bioavailability of iron is limited to enhance plant growth [186], although the mechanism of siderophore mediated Fe-nutrition is still not fully known. However, two possible mechanisms by which plants could obtain Fe from microbial siderophores have been suggested: (i) high redox potential of microbial siderophores can be reduced by the donation of ferrous in the transport system, and (ii) microbial ferric ions are transported in plant root through the apoplast where the reduction of siderophore takes place [187], with consequent ferrous accumulation in the apoplast leading to a high concentration of Fe (II) in the root [188], and (iii) siderophores of microbial origin can chelate Fe from soils and perform ligand exchange with phytosiderophores [189]. These mechanisms depend on several parameters such as concentrations of phytosiderophores, microbial source, pH, and redox potential of root environment [186]. Schenk et al. [190] found that siderophores are a valid eco-friendly alternative to hazardous pesticides. Mycorrhizal fungi are used as biofertilizers for the enhancement of plant growth and development. Higher levels of Fe-sequestration occurred in plants associated with mycorrhizal fungi compared to nonmycorrhizal plants, suggesting that enhanced plant nutrition by mycorrhizal fungi depends on fungal siderophores [191,192]. Plant growth-promoting activities of fungi were previously investigated by Yadav et al. [193] who found that fungal species such as *Trichoderma harzianum*, *Penicillium citrinum*, and *Aspergillus niger* produced siderophores and increased root and shoot length of chickpeas. Siderophores play a significant role in biological control as competitors for Fe to reduce the Fe availability for the pathogens of plants [194]. Wilt diseases of potatoes caused by *Fusarium oxysporum* can be controlled by Pyoverdine siderophores produced by *Pseudomonas* sp. [195]. Apart from fungi, bacterial strains, mainly belonging to the genus *Pseudomonas*, have been extensively studied to improve plant growth by synthesizing siderophores or protecting the plant host from pathogens [196].

### 5.7. Enzyme-Inhibiting Activity

Siderophores are iron chelators capable of inhibiting the iron-dependent activity of enzymes by withdrawing iron. Several studies have shown that ribonucleotide reductase activity is reduced by synthetic siderophores [197] as a result of inhibition of biosynthesis of DNA. In proliferating neoplastic cells, iron delivering transferrin receptors were found to frequently occur on the cell surface and enhance iron requirement by the cell. Inhibition of iron supply by the siderophores reduced the growth of neoplastic cells [132,197]. Therefore, siderophores can be used as inhibitors of cell proliferation and help to design drugs with anticancer activity.

### 5.8. Computational Approaches for the Application of Siderophores

With the advancement in bioinformatics and computational approaches, it has become easier to explore gene clusters for siderophore biosynthetic pathways and their interactions with other proteins and peptides. Siderophores are complex structures of nonproteinogenic amino acids with huge structural variations. These structural variations make siderophores suitable for various applications. Computational and bioinformatics tools help to predict the affinity and properties of siderophores. Norine is a bioinformatics platform that is an easily available unique resource devoted to elucidating the structures of nonribosomal peptides. This tool helps to identify newly discovered siderophores, whether new nonribosomal peptides or variants of an existing family. Similarly, AntiSMASH (antibiotics & Secondary Metabolite Analysis Shell) is an internet-based bioinformatics tool which finds the region in plants, fungi, and bacteria responsible for the biosynthesis of secondary metabolites. AntiSMASH allocates a functional "siderophore" label for biological gene clusters which contain the lucA/lucC gene family specific to the siderophores biosynthetic pathways, which further helps to predict the siderophores activity [198]. These tools help to derive siderophore biosynthesis pathways and make them accessible for reference. The prediction of biosynthetic pathways for the production of siderophores facilitated the discovery of novel and exclusive siderophores such as thioquinolobactin [199]. Molecular dockings and dynamic simulations are additional techniques for analyzing the interaction of siderophores with other proteins. Samsonov et al. [200] successfully applied molecular docking and molecular dynamic simulations to analyze the potential binding interaction between nine bacterial siderophores and lipocaline, a member of human serum a1-acid glycoprotein. A comparative study of computational and experimental results indicated that serum a1-acid glycoprotein can effectively bind with the Fe-BisHaCam and petrobactin, which shows that serum a1-acid glycoprotein can be putatively involved in the nullification of bacterial infections by capturing iron-chelating compounds. Furthermore, Xie et al. [201] explored bioinformatics approaches such as mutasynthesis, genome mining, and activity screening to synthesize fluorinated amychelin fluoroamychelin I siderophores. The resulting fluorinated fluoroamychelin I was able to rescue *Caenorhabditis elegans* from *Pseudomonas aeruginosa*-mediated killing with greater efficiency than traditional antibiotics, including meropenem and ceftazidime. The study showed a successful implementation of bioinformatics approaches for the production of synthetic antibacterial compounds by modifying siderophores. In addition to these, Flux balance analysis (FBA) is a computational tool that is frequently used in metabolic engineering for the improvement of production yield. Siderophores are small-sized metal chelators that are usually secreted in very small quantities by their native microbial hosts. FBA can be applied to predict media composition to enhance production yield that can be further verified experimentally. Recently FBA has been applied to improve the heterologous expression of siderophores by *E. coli* K-12 MG1655 [202]. In conclusion, computational and bioinformatics approaches are helpful to identify novel siderophores, their biosynthetic pathways, and biological activities. Furthermore, these techniques are applicable to modify siderophore structures in order to improve their bioactivity and even their production yield.

## 6. Importance of Siderophores in Nature

### 6.1. Weathering of Soil Minerals

Soil microbes produce siderophores that can stimulate the dissolution of insoluble phases minerals [203,204]. Several mechanisms for siderophore-stimulated Fe dissolution have been proposed [205]. In general, the Fe (III) and siderophore complex forms on the mineral surface and is subsequently transported into the adjacent soil solution, where it is accessible for absorption by microbes and plants [203,206]. Because siderophores and Fe form more permanent complexes, their influence on soil mineral weathering may be greater than low molecular mass organic acids (LMMOAs). Siderophores and Fe (III) form 1:1 complexes with K constants ranging from of 1030 to 1052 [207], whereas, with Fe (III), oxalic acids have constants of $K = 10^8$ and citric acids have constants of $K = 10^{12}$ [208]. Nevertheless, when both siderophores and LMMOAs are present, the mineral dissolution rate is increased more than when the solitary siderophore is present [209]. Many studies have reported microbial siderophore involvement in dissolution of Fe minerals due to their relevance in weathering and soil formation. In a study, researchers revealed the effectiveness of hydroxamate siderophores formed by *Suillus granulatus* in dissolving goethite. Due to the continual synthesis of siderophores by *S. granulatus*, significant amounts of Fe ($10^9$ mol $m^2/h$) were mobilized [210]. Additionally, Sokolova et al. [211] found that fungal siderophores belonging to the family of ferrichrome, including ferricrocin and ferrichrome, contributed to altering the surface structure of biotite and promoting its dissolution in forest podzolic soil.

Compared to a synthetic siderophore, the presence of siderophore-producing actinobacteria, such as *Arthrobacter* and *Streptomyces*, significantly enhanced the Fe dissolution from hornblende [206]. Phytosiderophores have been shown to elevate Fe-containing mineral dissolution, thus contributing to weathering processes of minerals, for example goethite and ferrihydrate [212]. According to Reichard et al. [213], the highest goethite dissolution rate of 1.7 nmol $m^2/h$ at pH 6 was obtained in the presence of 2′-deoxymugineic acid (phytosiderophores). The transporter genes expression of Fe–phytosiderophore in barley has recently been shown to improve its capacity to remove Fe from soil minerals [214].

### *6.2. Oceanic Biogeochemical Cycle of Fe*

Trace metal biogeochemical cycling in the oceans has become a significant source of concern. Fe has gained the most attention out of all the trace metals found in marine waters since it is a critical micronutrient for marine life and affects phytoplankton production and community structure [215]. Marine bacteria are responsible for most organic Fe chelators found in seawater and hence play a central part in the Fe biogeochemical cycle in the ocean [216]. These bacteria compete for Fe with phytoplankton by creating several forms of siderophores, which significantly influence the abundance and solubility of Fe in the ocean [217]. In marine siderophores, either citrate or b-hydroxyaspartate contributes hydroxyl–carboxylate functional groups [198,218]. Siderophores contribute to the photochemical cycle of Fe in surface water by producing complexes of Fe (III)–siderophore that improve the availability of Fe for the phytoplankton [219,220]. Ferrioxamine G was discovered to be broadly dispersed in surface waters across the Atlantic Ocean, while ferrioxamine E was shown to have a more diverse distribution at different depths. These data imply that marine siderophores play a significant part in the biogeochemical cycle of Fe by elevating the abundance, as well as the availability, of Fe in the Atlantic Ocean [221,222].

## 7. Conclusions

Iron is an essential element for all living organisms. Fungi acquire iron from extracellular environments by secreting siderophores, which are low molecular weight, iron-binding molecules. Siderophores play a significant role in the iron homeostasis of fungi, which are similar to bacteria and plants for the mobilization of extracellular iron. Considerable progress has been made on siderophore uptake research and understanding iron assimilation mechanisms in *S. cerevisiae*. However, much more remains to be explored regarding biosynthetic pathways, iron assimilation, and regulation. In this regard, further studies based on genome sequence analysis of fungi and siderophore-mediated iron acquisition in a wide range of fungal species are still needed. Apart from fungal siderophore-type descriptions, several unexplored aspects need to be elucidated, including extracellular excretion mechanisms, details of the siderophore biosynthetic pathways, intracellular iron release, iron metabolism and storage of iron. Iron requirements of fungi open up new research areas for the development of novel antifungal treatments such as iron chelation therapy. Functional studies of siderophores may reveal novel non-ribosomal peptide synthetases that can open the way for the development of new compounds with pharmaceutical value. In fungal species, siderophore-mediated iron uptake is essential for survival as free-living organisms and for establishment of commensal and pathogenic relationships.

Studies conducted on low Fe bioavailability and siderophore activity in different environments may enhance our understanding of siderophore ecology and functions. Metagenomic analyses provide an excellent platform to clarify the structural diversity of siderophores among different fungal species. The knowledge of siderophore biosynthesis and utilization mechanisms in each fungal species could be essential for eradicating pathogenic fungi hiding and replicating in host macrophages.

With the advent of genome sequencing technologies and concurrent omics analysis, there has been a vast increase in our knowledge of siderophore biosynthesis over the past two decades. The first step was identifying gene clusters acting as sources of siderophores via bioinformatics. Prediction of substrate specificity combined with algorithms parsing metabolomic data to link the clusters to corresponding compounds may constitute the following steps. For each step, multiple new techniques have been developed in the last few years. In silico genome mining is an efficient high-throughput approach to uncover potential nonribosomal peptide synthetases (NRPS) genes. Analytical pipelines linking genomics with other omics data have been developed and can reveal much information on the synthesis of such natural products. Computational tools coupled with genome mining provide efficient methods to identify and characterize biosynthetic gene clusters BGCs [223]. Natural product research and siderophore research have been concentrated on bacterial species, and there is an obvious bias in data availability and algorithm development for fungal research. Therefore, it is essential to consider the differences and test the relevance of already developed tools on fungal data before blind usage. It would be ideal to generate, collect, and analyze fungal NRPS data, and particularly reorganize siderophore-producing data for fungal siderophore identification. The lack of such curated data is currently a shortcoming in developing and training prediction/classification models for fungal siderophores. Algorithms for the identification of siderophore-producing BGCs integrated with high-throughput proteomic and metabolomic product detection techniques can lead to the discovery and characterization of novel siderophores.

With the help of computational approaches, inhibitors can be designed against siderophore biosynthetic pathway enzymes and siderophore transporter proteins by applying the principles of structure-based drug design and/or fragment-based drug design. Similarly, the structure-based pharmacophores for these catalytic enzymes can be predicted and searched against the chemical databases for compounds suitable as siderophore biosynthetic enzyme inhibitors. In contrast, biosynthetic pathway enzyme competitive inhibitors can be identified using the shape and structural information of enzyme substrates. Applying shape-based and fingerprint-based similarity searches allows suitable competitive enzyme inhibitors to be predicted from chemical databases. In addition, a drug repurposing approach can be applied to find existing drugs with antifungal activity. This approach reduces drug discovery cost to a great extent, with a higher success rate than other approaches.

**Author Contributions:** L.P. conceived the idea and coordinated the contribution of all coauthors. L.P. and X.W. wrote large part of the manuscript. All authors (L.P., X.W., D.S., X.S., V.K., A.S., K.T., P.W.R., R.R.) significantly contributed to the work by writing part of the manuscript, preparing figures and tables, reviewing and editing. All authors have read and agreed to the published version of the manuscript.

**Funding:** This research received no external funding.

**Institutional Review Board Statement:** Not applicable.

**Informed Consent Statement:** Not applicable.

**Data Availability Statement:** Not applicable.

**Conflicts of Interest:** The authors declare no conflict of interest.

## Abbreviations

| | |
|---|---|
| TAFC | triacetyl fusarine C |
| FsC | fusarine C |
| FSR | fusarinine |
| CPG | coprogen |
| Frr | ferricrocin |
| FRC | ferrichromes |
| Fsg | fusigen |
| NRPS | nonribosomal peptide synthetases |
| SIT | siderophore-iron transporter |
| Estb | esterase |
| FIT | facilitator of iron transport |
| FREs | ferric reductase enzymes |
| Fet | ferrous iron transporter |
| DFO | desferrioxamine B |

## References


1. Winkelmann, G. Importance of siderophores in fungal growth, sporulation and spore germination. *Front. Mycol.* **1991**, *1991*, 49–65.
2. Ducklow, H.W.; Oliver, J.L.; Smith, W.O. The Role of Iron as a Limiting Nutrient for Marine Plankton Processes. 2018. Available online: pal.lternet.edu/docs/bibliography/Public/259lterc.pdf (accessed on 21 April 2021).
3. Mustoe, G.E. Biogenic weathering: Solubilization of iron from minerals by epilithic freshwater algae and cyanobacteria. *Microorganism* **2018**, *6*, 8. [CrossRef]
4. Paul, A.; Dubey, R. Characterization of protein involved in nitrogen fixation and estimation of Co-factor. *Appl. J. Curr. Res. Biosci. Plant Biol.* **2015**, *2*, 89–97.
5. Martinez, J.L.; Delgado-Iribarren, A.; Baquero, F. Mechanisms of iron acquisition and bacterial virulence. *FEMS Microb. Rev.* **1990**, *7*, 45–56. [CrossRef]
6. Wang, X.; Pecoraro, L. Analysis of soil fungal and bacterial communities in Tianchi Volcano crater, northeast China. *Life* **2021**, *11*, 280. [CrossRef]
7. Neilands, J. Siderophores: Structure and function of microbial iron transport compounds. *J. Biol. Chem.* **1995**, *270*, 26723–26726. [CrossRef]
8. Khan, A.; Singh, P.; Srivastava, A. Synthesis, nature and utility of universal iron chelator—Siderophore: A review. *Microbiol. Res.* **2018**, *212–213*, 103–111. [CrossRef]
9. Chincholkar, S.B.; Chaudhari, B.L.; Rane, M.R. Microbial Siderophore: A State of Art. In *Microbial Siderophores*; Soil Biology; Varma, A., Chincholkar, S.B., Eds.; Springer: Berlin/Heidelberg, Germany, 2007; Volume 12.
10. Boukhalfa, H.; Crumbliss, A. Chemical aspects of siderophore mediated iron transport. *Biometals* **2002**, *15*, 325–339. [CrossRef] [PubMed]
11. Devireddy, L.; Hart, D.; Goetz, D.; Green, M. A mammalian siderophore synthesized by an enzyme with a bacterial homolog involved in enterobactin production. *Cell* **2010**, *141*, 1006–1017. [CrossRef]
12. Pluháček, T.; Lemr, K.; Ghosh, D.; Milde, D.; Novák, J.; Havlíček, V. Characterization of microbial siderophores by mass spectrometry. *Mass Spectrom. Rev.* **2016**, *35*, 35–47. [CrossRef]
13. Aguado-Santacruz, G.A.A.; Moreno-Gomez, B.A.; Jimenez-Francisco, B.B.; Garcia-Moya, E.B.; Preciado-Ortiz, R.E. Impact of the microbial siderophores and phytosiderophores on the iron assimilation by plants: A synthesis. *Rev. Fitotec. Mex.* **2012**, *35*, 9–21.
14. Raymond, K.; Dertz, E.; Kim, S. Enterobactin: An archetype for microbial iron transport. *Proc. Natl. Acad. Sci. USA* **2003**, *100*, 3584–3588. [CrossRef] [PubMed]
15. Howard, D.H. Acquisition, transport, and storage of iron by pathogenic fungi. *Clin. Microbiol. Rev.* **1999**, *12*, 394–404. [CrossRef] [PubMed]
16. Lawen, A.; Lane, D.J.R. Mammalian iron homeostasis in health and disease: Uptake, storage, transport, and molecular mechanisms of action. *Antioxid. Redox Signal.* **2013**, *18*, 2473–2507. [CrossRef]
17. Mech, F.; Wilson, D.; Lehnert, T.; Hube, B.; Figge, M.T. Epithelial invasion outcompetes hypha development during *Candida albicans* infection as revealed by an image-based systems biology approach. *Cytom* **2014**, *85*, 126–139. [CrossRef] [PubMed]
18. Crawford, A.; Wilson, D. Essential metals at the host-pathogen interface: Nutritional immunity and micronutrient assimilation by human fungal pathogens. *FEMS Yeast Res.* **2015**, *15*, fov071. [CrossRef] [PubMed]
19. Bellenger, J.P.; Wichard, T.; Kraepiel, A.M.L. Vanadium requirements and uptake kinetics in the dinitrogen-fixing bacterium *Azotobacter vinelandii*. *Appl. Environ. Microbiol.* **2008**, *74*, 1478–1484. [CrossRef] [PubMed]
20. Verma, V.C.; Singh, S.K.; Prakash, S. Bio-control and plant growth promotion 926 potential of siderophore producing endophytic Streptomyces from *Azadirachta indica* A. *Juss J. Basic Microb.* **2011**, *51*, 550–556. [CrossRef] [PubMed]
21. Ishimaru, Y.; Takahashi, R.; Bashir, K.; Shimo, H.; Senoura, T.; Sugimoto, K.; Ono, K.; Yano, M.; Ishikawa, S.; Arao, T.; et al. Characterizing the role of 699 rice NRAMP5 in Manganese, Iron and Cadmium Transport. *Sci. Rep.* **2012**, *2*, 28. [CrossRef] [PubMed]
22. Ghosh, S.K.; Banerjee, S.; Sengupta, C. Bioassay, characterization and estimation of siderophores from some important antagonistic fungi. *J. Biopest.* **2017**, *10*, 105–112.
23. Schrettl, M.; Ibrahim-Granet, O.; Droin, S.; Huerre, M.; Latgé, J.; Haas, H. The crucial role of the *Aspergillus fumigatus* siderophore system in interaction with alveolar macrophages. *Microbes Infect.* **2010**, *12*, 1035–1041. [CrossRef]
24. Oberegger, H.; Schoeser, M.; Zadra, I.; Abt, B.; Haas, H. SREA is involved in regulation of siderophore biosynthesis: Utilization and uptake in *Aspergillus nidulans*. *Mol. Microbiol.* **2001**, *41*, 1077–1089. [CrossRef]
25. Arantes, V.; Milagres, A.M.F. Response of Wolfifiporia cocos to iron availability: Alterations in growth, expression of cellular proteins, Fe3+-reducing activity and Fe3+- chelators production. *J. Appl. Microbiol.* **2008**, *104*, 185–193.
26. Haselwandter, K.; Häninger, G.; Ganzera, M.; Haas, H.; Nicholson, G.; Winkelmann, G. Linear fusigen as the major hydroxamate siderophore of the ectomycorrhizal Basidiomycota *Laccaria laccata* and *Laccaria bicolor*? *Biometals* **2013**, *26*, 969–979. [CrossRef]
27. Osman, A.; Işıl, V.; Fatih, K. Microbial siderophores: Potential medicinal applications of the siderophores. *J. Biotechnol. Sci. Res.* **2019**, *6*, 32–40.
28. Hagen, J.B. Five Kingdoms, More or Less: Robert Whittaker and the Broad Classification of Organisms. *BioScience* **2012**, *62*, 67–74. [CrossRef]
29. Philpott, C.C. Iron uptake in fungi: A system for every source. *Biochim. Biophys. Acta* **2006**, *1763*, 636–645. [CrossRef] [PubMed]

30. Ali, S.S.; Vidhale, N.N. Bacterial siderophore and their application: A review. *Int. J. Curr. Microbiol. Appl. Sci.* **2013**, *2*, 303–312.
31. Drechsel, H.; Tschierske, M.; Thieken, A.; Jung, H.; Zähner, G.; Winkelmann, G. The carboxylate type siderophore rhizoferrin and its analogs produced by directed fermentation. *J. Ind. Microbiol.* **1995**, *14*, 105–112. [CrossRef]
32. Butler, A.; Theisen, R.M. Iron (III)–siderophore coordination chemistry: Reactivity of marine siderophores. *Coord. Chem. Rev.* **2010**, *254*, 288–296. [CrossRef]
33. Braud, A.; Geoffroy, V.; Hoegy, F.; Mislin, G.L.; Schalk, I.J. Presence of the siderophores pyoverdine and pyochelin in the extracellular medium reduces toxic metal accumulation in Pseudomonas aeruginosa and increases bacterial metal tolerance. *Environ. Microbiol. Rep.* **2010**, *2*, 419–425. [CrossRef] [PubMed]
34. Kessenbrock, K.; Plaks, V.; Werb, Z. Matrix metalloproteinases: Regulators of the tumor microenvironment. *Cell* **2010**, *141*, 52–67. [CrossRef] [PubMed]
35. Haas, H. Fungal siderophore metabolism with a focus on *Aspergillus fumigatus*. *Nat. Prod. Rep.* **2014**, *31*, 1266–1276. [CrossRef]
36. Das, A.; Prasad, R.; Srivastava, A.; Giang, P.H.; Bhatnagar, K.; Varma, A. Fungal siderophores: Structure, functions and regulation. In *Microbial Siderophores*; Springer: Berlin/Heidelberg, Germany, 2007; pp. 1–42.
37. Baakza, A.; Dave, B.P.; Dube, H.C. Chemical nature, ligand denticity and quantification of fungal siderophores. *NISCAIR* **2004**, *42*, 96–105.
38. Winkelmann, G. Microbial siderophore-mediated transport. *Biochem. Soc. Trans.* **2002**, *30*, 691–696. [CrossRef]
39. Saha, M.; Sarkar, S.; Sarkar, B.; Sharma, B.K.; Bhattacharjee, S.; Tribedi, P. Microbial siderophores and their potential applications: A review. *Environ. Sci. Pollut. Res. Int.* **2016**, *23*, 3984–3999. [CrossRef] [PubMed]
40. Anke, H.; Kinn, J.; Bergquist, K.E.; Sterner, O. Production of siderophores by strains of the genus Trichoderma—Isolation and characterization of the new lipophilic coprogen derivative, palmitoylcoprogen. *Biol. Met.* **1991**, *4*, 176–180. [CrossRef]
41. Charlang, G.; Ng, B.; Horowitz, N.H.; Horowitz, R.M. Cellular and extracellular siderophores of *Aspergillus nidulans* and *Penicillium chrysogenum*. *Mol. Cell Biol.* **1981**, *1*, 94–100.
42. Mor, H.; Kashman, Y.; Winkelmann, G.; Barash, I. Characterization of siderophores produced by different species of the dermatophytic fungi *Microsporum* and *Trichophyton*. *Biometals* **1992**, *5*, 213–216. [CrossRef]
43. Jalal, M.A.F.; Helm, D. Siderophores of highly phytopathogenic *Alternaria longipes*. *Biol. Met.* **1989**, *2*, 11–17. [CrossRef]
44. Jalal, M.A.F.; Love, S.K.; van der Helm, D. N alpha-dimethylcoprogens three novel trihydroxamate siderophores from pathogenic fungi. *Biol. Met.* **1988**, *1*, 4–8. [CrossRef]
45. Hossain, M.B.; Jalal, M.A.F.; Benson, B.A.; Barnes, C.L.; Van der Helm, D. Structure and conformation of two coprogentype siderophores: Neocoprogen I and neocoprogen II. *J. Am. Chem. Soc.* **1987**, *109*, 4948–4954. [CrossRef]
46. Frederick, C.B.; Bentley, M.D.; Shive, W. Structure of triornicin, a new siderophore. *Biochemistry* **1981**, *20*, 2436–2438. [CrossRef]
47. Chowdappa, S.; Jagannath, S.; Konappa, N.; Udayashankar, A.C.; Jogaiah, S. Detection and Characterization of Antibacterial Siderophores Secreted by Endophytic Fungi from *Cymbidium aloifolium*. *Biomolecules* **2020**, *10*, 1412. [CrossRef]
48. Diekmann, H. Metabolic products of microorganisms 81. Occurrence and structures of coprogen B and dimerum acid. *Arch. Mikrobiol.* **1970**, *73*, 65–76. [CrossRef]
49. Emery, T. Malonichrome, a new iron chelates from *Fusarium roseum*. *Biochim. Biophys. Acta* **1980**, *629*, 382–390. [CrossRef]
50. Jalal, M.A.F.; Love, S.K.; van der Helm, D. Siderophore Mediated Iron (III) Uptake in *Gliocladium virens*: 1. Properties of cis-Fusarinine, trans-Fusarmine, Dimerum Acid, and Their Ferric Complexes. *J. Inorg. Biochem.* **1986**, *28*, 417–430. [CrossRef]
51. Eng-Wilmot, D.L.; Rahman, A.; Mendenhall, J.V.; Grayson, S.L.; Dick, V.D.H. Molecular structure of ferric neurosporin, a minor siderophore-like compound containing Nδ-hydroxy-D-ornithine. *J. Am. Chem. Soc.* **1984**, *106*, 1285–1290. [CrossRef]
52. Pidacks, C.; Whitehill, A.R.; Pruess, L.M.; Hesseltine, C.W.; Hutchings, B.L.; Bohonos, N.; Williams, J.H. Coprogen, the isolation of a new growth factor required by *Pilobolus* species. *J. Am. Chem. Soc.* **1953**, *75*, 6064–6065. [CrossRef]
53. Moore, R.E.; Emery, T. N(alpha)-acetylfusarinines: Isolation, characterization, and properties. *Biochemistry* **1976**, *15*, 2719–2723. [CrossRef]
54. Lesuisse, E.; Raguzzi, F.; Crichton, R.R. Iron Uptake by the Yeast Saccharomyces cerevisiae: Involvement of a Reduction Step. *J. Gen. Microbiol.* **1987**, *133*, 3229–3236. [CrossRef]
55. Haselwandter, K.; Dobernigg, B.; Beck, W.; Jung, G.; Cansier, A.; Winkelmann, G. Isolation and identification of hydroxamate siderophores of ericoid mycorrhizal fungi. *Biometals* **1992**, *5*, 51–56. [CrossRef]
56. Senthilnithy, R.; De Costa, M.D.P.; Gunawardhana, H.D. The pKa values of ligands and stability constants of the complexes of Fe(III), Cu(II) and Ni(II) with some hydroxamic acids: A comparative study of three different potentiometric methods. *J. Natl. Sci. Found.* **2008**, *36*, 191–198.
57. Silva-Bailão, M.G.; Bailão, E.F.L.C.; Lechner, B.E.; Gauthier, G.M.; Lindner, H.; Bailão, A.M.; Haas, H.; de Almeida Soares, C.M. Hydroxamate production as a high affinity iron acquisition mechanism in *Paracoccidioides* spp. *PLoS ONE* **2014**, *9*, e105805. [CrossRef]
58. Atkin, C.L.; Neilands, J.B. Rhodotorulic acid, a diketopiperazine dihydroxamic acid with growth-factor activity. Isolation and characterization. *Biochemistry* **1968**, *7*, 3734–3739. [CrossRef]
59. Muller, G.; Barclay, S.J.; Raymond, K.N. The mechanism and specificity of iron transport in *Rhodotorula pilimanae* probed by synthetic analogs of Rhodotorulic acid. *J. Biol. Chem.* **1985**, *260*, 13916–13920. [CrossRef]
60. Challis, G.L. A widely distributed bacterial pathway for siderophore biosynthesis independent of nonribosomal peptide synthetases. *Chem. Biochem.* **2005**, *6*, 601–611. [CrossRef]

61. Sah, S.; Singh, R. Siderophore: Structural and functional characterization—A comprehensive review. *Agriculture* **2015**, *61*, 97–114. [CrossRef]
62. Crosa, J.H.; Walsh, C.T. Genetics and assembly line enzymology of siderophore biosynthesis in bacteria. *Microbiol. Mol. Biol. Rev.* **2002**, *66*, 223–249. [CrossRef]
63. Ravel, J.; Cornelis, P. Genomics of pyoverdine-mediated iron uptake in pseudomonads. *Trends Microbiol.* **2003**, *11*, 195–200. [CrossRef]
64. Nadia, K.; Challis, G.L. Complex enzymes in microbial natural product biosynthesis. *Methods Enzymol.* **2009**, *458*, 431–435.
65. Hider, R. Siderophore mediated absorption of iron. *Struct. Bond.* **1984**, *58*, 25–87.
66. Mei, B.; Leon, S.A. Molecular biology of iron transport. In *Metal Ions in Fungi*; Marcel Dekker: New York, NY, USA, 1994; pp. 117–148.
67. Plattner, H.J.; Diekmann, H. Enzymology of siderophore biosynthesis in fungi. In *Metal Ions in Fungi*; Winkelmann, G., Winge, D.R., Eds.; Decker: New York, NY, USA, 1994; pp. 99–116.
68. De Luca, N.G.; Wood, P.M. Iron uptake by fungi: Contrasted mechanisms with internal or external reduction. *Adv. Microb. Physiol.* **2001**, *43*, 39–74.
69. Ong, D.E.; Emery, T.F. Ferrichrome biosynthesis: Enzyme catalyzed formation of the hydroxamic acid group. *Arch. Biochem. Biophys.* **1972**, *148*, 77–83. [CrossRef]
70. Neilands, J.B.; Konopka, K.; Schwyn, B.; Coy, M.; Francis, R.T.; Paw, B.H.; Bagg, A. Comparative biochemistry of microbial iron assimilation. In *Iron Transport in Microbes, Plants and Animals*; Winkelmann, G., Winge, D.R., Eds.; VCH Weinheim: New York, NY, USA, 1987; pp. 3–34.
71. Schwyn, B.; Neilands, J.B. Universal chemical assay for the detection and 878 determinations of siderophores. *Anal. Biochem.* **1987**, *160*, 47–56. [CrossRef]
72. Anke, H.; Anke, T.; Diekmann, H. Biosynthesis of sideramines in fungi. Fusigen synthetase from extracts of *Fusarium cubense*. *FEBS Lett.* **1973**, *36*, 323–325. [CrossRef]
73. Calvente, V.; Benuzzi, D. Antagonistic action of siderophores from *Rhodotorula glutinis* upon the postharvest pathogen *Penicillium expansum*. *Int. Biodeterior. Biodegrad.* **1999**, *43*, 167–172. [CrossRef]
74. Hummel, W.; Diekmann, H. Preliminary characterization of ferrichrome synthetase from *Aspergillus quadricinctus*. *Biochim. Biophys. Acta* **1981**, *657*, 313–320. [CrossRef]
75. Yuan, W.M.; Gentil, G.D.; Budde, A.D.; Leong, S.A. Characterization of the *Ustilago maydis* sid2 gene, encoding a multidomain peptide synthetase in the ferrichrome biosynthetic gene cluster. *J. Bacteriol.* **2001**, *183*, 4040–4405. [CrossRef]
76. Wilhite, S.E.; Lumsden, R.D.; Straney, D.C. Peptide synthetase gene in *Trichoderma virens*. *Appl. Environ. Microbiol.* **2001**, *67*, 5055–5062. [CrossRef]
77. Leong, S.A.; Winkelmann, G. Molecular biology of iron transport in fungi. *Met. Ions Biol. Syst.* **1998**, *35*, 147–186.
78. Ahmad, E.; Holmstrom, S.J.M. Siderophores in environmental research: Roles and applications. *Microb. Biotechnol.* **2014**, *7*, 196–208. [CrossRef]
79. Sanchez, N.; Brown, E.A.; Olsen, Y.; Vadstein, O.L.; Humberto, E.; Gonzalez, I.; Van Ardelan, M. Effect of Siderophore on Iron Availability in a Diatom and a Dinoflagellate Species: Contrasting Response in Associated Bacteria. *Front. Mar. Biol.* **2018**, *5*, 1–18. [CrossRef]
80. Ernst, J.F.; Bennet, R.L.; Rothfield, L.I. Cunstitutive expression if the iron-enterochelin and ferrichrome uptake systems in a mutant strain of *Salmonella typhimurium*. *J. Bacteriol.* **1978**, *135*, 928–934. [CrossRef] [PubMed]
81. Hantke, K. Cloning of the repressor protein gene of iron regulated systems in *E. coli* K12. *Mol. Gen. Genet.* **1984**, *197*, 337–341. [CrossRef]
82. De Lorenzo, V.; Bindereif, A.; Paw, B.H.; Neilands, J.B. Aerobactin biosynthesis and transport genes of plasmid ColV-K30 in *Escherichia coli* K-12. *J. Bacteriol.* **1986**, *165*, 570–578. [CrossRef] [PubMed]
83. Tanui, C.K.; Shyntum, D.Y.; Priem, S.L.; Theron, J.; Moleleki, L.N. Influence of the ferric uptake regulator (Fur) protein on pathogenicity in *Pectobacterium Carotovorum* subsp. *Brasiliense*. *PLoS ONE* **2017**, *12*, e0177647. [CrossRef]
84. Scazzocchio, C. The fungal GATA factors. *Curr. Opin. Microbiol.* **2000**, *3*, 126–131. [CrossRef]
85. Kim, I.H.; Wen, Y.; Son, J.S.; Lee, K.H.; Kim, K.S. The Fur-Iron Complex Modulates Expression of the Quorum-Sensing Master Regulator, SmcR, To Control Expression of Virulence Factors in *Vibrio vulnificus*. *Infect. Innum.* **2013**, *81*, 2888–2898. [CrossRef] [PubMed]
86. An, Z.; Mei, B.; Yuan, W.M.; Leong, S.A. The distal GATA sequences of the sid1 promoter of Ustilago maydis mediate iron repression of siderophore production and interact directly with Urbs1, a GATA family transcription factor. *EMBO J.* **1997**, *16*, 1742–1750. [CrossRef]
87. Gauthier, G.M.; Sullivan, T.D.; Gallardo, S.S.; Brandhorst, T.T.; Wymelenberg, V.; Cuomo, C.A.; Suen, G.; Currie, C.R.; Klein, B.S. SREB, a GATA Transcription Factor That Directs Disparate Fates in *Blastomyces dermatitidis* Including Morphogenesis and Siderophore Biosynthesis. *PLoS Pathog.* **2010**, *6*, e1000846. [CrossRef] [PubMed]
88. Voisard, C.; Wang, J.; McEvoy, J.L.; Xu, P.; Leong, S.A. Urbsl, a gene regulating siderophore biosynthesis in *Ustilago maydis*, encodes a protein similar to the erythroid transcription factor GATA-1. *Mol. Cell. Biol.* **1993**, *13*, 7091–7100.
89. Zhou, L.W.; Haas, H.; Marzluf, G.A. Isolation and characterization of a new gene, sre, which encodes a GATAtype regulatory protein that controls iron transport in *Neurospora crassa*. *Mol. Gen. Genet.* **1998**, *259*, 532–540. [CrossRef]

90. Haas, H.; Angermayr, K.; Zadra, I.; Stoffler, G. Overexpression of nreB, a new GATA factor-encoding gene of *Penicillium chrysogenum*, leads to repression of the nitrate assimilatory gene cluster. *J. Biol. Chem.* **1997**, *272*, 22576–225882. [CrossRef] [PubMed]
91. Haas, H.; Zadra, I.; Stoffler, G.; Angermaryr, K. The *Aspergillus nidulans* GATA factor SREA is involved in regulation of siderophore biosynthesis and control of iron uptake. *J. Biol. Chem.* **1999**, *274*, 4613–4619. [CrossRef] [PubMed]
92. Hoe, K.L.; Won, M.S.; Yoo, O.J.; Yoo, H.S. Molecular cloning of GAF2, a *Schizosaccharomyces pombe* GATA factor, which has two zinc-finger sequences. *Biochem. Mol. Biol. Int.* **1996**, *39*, 127–135. [CrossRef] [PubMed]
93. Pelletier, B.; Trott, A.; Morano, K.A.; Labbe, S. Functional characterization of the iron-regulatory transcription factor Fep1 from *Schizosaccharomyces pombe*. *J. Biol. Chem.* **2005**, *280*, 25146–25161. [CrossRef] [PubMed]
94. Yamaguchi-Iwai, Y.; Dancis, A.; Klausner, R.D. AFT1: A mediator of iron regulated transcriptional control in *Saccharomyces cerevisiae*. *EMBO J.* **1995**, *14*, 1231–1239. [CrossRef]
95. Yamaguchi-Iwai, Y.; Stearman, R.; Dancis, A.; Klausner, R.D. Iron-regulated DNA binding by the AFT1 protein controls the iron regulon in yeast. *EMBO J.* **1996**, *15*, 3377–3384. [CrossRef] [PubMed]
96. McRose, D.L.; Baars, O.; Seyedsayamdost, M.R.; Morel, F. Quorum sensing and iron regulate a two-for-one siderophore gene cluster in *Vibrio harveyi*. *Proc. Natl. Acad. Sci. USA* **2018**, *115*, 7581–7586. [CrossRef] [PubMed]
97. Tshikantwa, T.S.; Ullah, M.W.; He, F.; Yanh, G. Current Trends and Potential Applications of Microbial Interactions for Human Welfare. *Front. Microbiol.* **2018**, *9*, 1156. [CrossRef]
98. Butler, A. Marine siderophores and microbial iron mobilization. *Biometals* **2005**, *18*, 369–374. [CrossRef] [PubMed]
99. Emerson, D. Biogeochemistry and microbiology of microaerobic Fe(II) oxidation. *Biochem. Soc. Trans.* **2012**, *40*, 1211–1216. [CrossRef] [PubMed]
100. Cornelissen, C.N.; Sparling, P.F. Neisseria. *Iron Transp. Bact.* **2004**, *2004*, 256–272.
101. Butt, A.T.; Thomas, M.S. Iron acquisition mechanisms and their role in the virulence of *Burkholderia* species. *Front. Cell Infect. Microbiol.* **2017**, *7*, 460. [CrossRef] [PubMed]
102. Li, Y.; Wang, Z.K.; Liu, X.; Song, Z.Y.; Li, R.; Shao, C.W.; Yin, Y.P. Siderophore Biosynthesis but Not Reductive Iron Assimilation Is Essential for the Dimorphic Fungus *Nomuraea rileyi* Conidiation, Dimorphism Transition, Resistance to Oxidative Stress, Pigmented Microsclerotium Formation, and Virulence. *Front. Microbiol.* **2016**, *7*, 931. [CrossRef] [PubMed]
103. Matzanke, B.F.; Bill, E.; Trautwein, A.X.; Winkelmann, G. Ferricrocin functions as the main intracellular iron-storage compound in mycelia of *Neurospora crassa*. *Biol. Met.* **1988**, *1*, 18–25. [CrossRef]
104. Eisendle, M.; Oberegger, H.; Zadra, I.; Haas, H. The siderophore system is essential for viability of Aspergillus nidulans: Functional analysis of two genes encoding 1-ornithine N5-monooxygenase (sidA) and a non-ribosomal peptide synthetase (sidC). *Mol. Microbiol.* **2003**, *49*, 359–375. [CrossRef]
105. Garcia-Rubio, R.; de Oliveira, H.C.; Rivera, J.; Trevijano-Contador, N. The Fungal Cell Wall: *Candida*, *Cryptococcus*, and *Aspergillus* Species. *Front. Microbiol.* **2020**, *10*, 2993. [CrossRef] [PubMed]
106. Orlean, P. *The Molecular and Cellular Biology of the Yeast Saccharomyces*; Cell Cycle and Cell, Biology; Pringle, J.R., Broach, j.R., Jones, E.W., Eds.; Cold Spring Harbor Laboratory Press: Cold Spring Harbor, NY, USA, 1997; pp. 229–362.
107. Marathe, R.; Phatake, Y.; Sonawane, A. Bioprospecting of *Pseudomonas aeruginosa* for their potential to produce siderophore: Process optimization and evaluation of its bioactivity. *Int. J. Bioassays* **2015**, *4*, 3667–3675.
108. Johnstone, T.; Nolan, E. Beyond iron: Non-classical biological functions of bacterial siderophores. *Dalton Trans.* **2015**, *44*, 6320–6339. [CrossRef] [PubMed]
109. Renshaw, J.; Robson, G.; Trinci, A.; Wiebe, M.; Livens, F.; Collison, D.; Taylor, R. Fungal siderophores: Structures, functions and applications. *Mycol. Res.* **2002**, *106*, 1123–1142. [CrossRef]
110. Huschka, H.G.; Jalal, M.A.F.; Helm, D.V.D.; Winkelmann, G. Molecular recognition of siderophores in fungi: Role of iron-surrounding N-acyl residues and the peptide backbone during membrane transport in *Neurospora crassa*. *J. Bacteriol.* **1986**, *167*, 1020–1024. [CrossRef] [PubMed]
111. Protchenko, O.; Ferea, T.; Rashford, J.; Tiedeman, J.; Brown, P.O.; Botstein, D.; Philpott, C.C. Three cell wall mannoproteins facilitate the uptake of iron in *Saccharomyces cerevisiae*. *J. Biol. Chem.* **2001**, *276*, 49244–49250. [CrossRef] [PubMed]
112. Van, H.O.; Ward, A.; Kaplan, J. Transition metal transport in yeast. *Ann. Rev. Microbiol.* **2002**, *56*, 237–261. [CrossRef] [PubMed]
113. Kosman, D.J. Molecular mechanisms of iron uptake in fungi. *Mol. Microbiol.* **2003**, *47*, 1185–1197. [CrossRef]
114. Hassett, R.; Kosman, D.J. Evidence for Cu(II) reduction as a component of copper uptake by *Saccharomyces cerevisiae*. *J. Biol. Chem.* **1995**, *270*, 128–134. [CrossRef] [PubMed]
115. Knight, S.A.B.; Lesuisse, E.; Stearman, R.; Klausner, R.D.; Dancis, A. Reductive iron uptake by *Candida albicans*: Role of copper, iron and the TUP1 regulator. *Microbiology-SGM* **2002**, *148*, 29–40. [CrossRef]
116. Finegold, A.A.; Shatwell, K.P.; Segal, A.W.; Klausner, R.D.; Dancis, A. Intramembrane bis-heme motif for transmembrane electron transport conserved in a yeast iron reductase and the human NADPH oxidase. *J. Biol. Chem.* **1996**, *271*, 31021–31024. [CrossRef]
117. Yun, C.W.; Ferea, T.; Rashford, J.; Ardon, O.; Brown, P.O.; Botstein, D.; Kaplan, J.; Philpott, C.C. Desferrioxamine-mediated Iron Uptake in *Saccharomyces cerevisiae*; evidence for two pathways of iron uptake. *J. Biol. Chem.* **2000**, *275*, 10709–10715. [CrossRef]
118. Yun, C.W.; Bauler, M.; Moore, R.E.; Klebba, P.E.; Philpott, C.C. The Role of the FRE Family of Plasma Membrane Reductases in the Uptake of Siderophore-Iron in *Saccharomyces cerevisiae*. *J. Biol. Chem.* **2001**, *276*, 10218–10223. [CrossRef]

119. Dhungana, S.; Crumbliss, A.L. Coordination chemistry and redox processes in siderophore-mediated iron transport. *Geomicrobiol. J.* **2005**, *22*, 87–98. [CrossRef]
120. Askwith, C.; Eide, D.; Van Ho, A.; Bernard, P.S.; Li, L.; Davis Kaplan, S.; Sipe, D.M.; Kaplan, J. The FET3gene of S. cerevisiae encodes a multicopper oxidase required for ferrous iron uptake. *Cell* **1994**, *76*, 403–410. [CrossRef]
121. Dancis, A.; Roman, D.G.; Anderson, G.J.; Hinnebusch, A.G.; Klausner, R.D. Ferric reductase of *Saccharomyces cerevisiae*: Molecular characterization, role in iron uptake, and transcriptional control by iron. *Proc. Natl. Acad. Sci. USA* **1992**, *89*, 3869–3873. [CrossRef] [PubMed]
122. De Silva, D.; Davis-Kaplan, S.; Fergestad, J.; Kaplan, J. Purification and characterization of Fet3 protein, a yeast homologue of ceruplasmin. *J. Biol. Chem.* **1997**, *272*, 14208–14213. [CrossRef] [PubMed]
123. Hassett, R.F.; Romeo, A.M.; Kosman, D.J. Regulation of high affinity iron uptake in the yeast *Saccharomyces cerevisiae*. *Role Dioxygen Fe J. Biol. Chem.* **1998**, *273*, 7628–7636. [CrossRef]
124. Lin, S.J.; Pufahl, R.A.; Dancis, A.; O'Halloran, T.V.; Culotta, V.C. A role for the *Saccharomyces cerevisiae* ATX1 gene in copper trafficking and iron transport. *J. Biol. Chem.* **1997**, *272*, 9215–9220. [CrossRef] [PubMed]
125. Yuan, D.S.; Stearman, R.; Dancis, A.; Dunn, T.; Beeler, T.; Klausner, R.D. The Menkes/Wilson disease gene homologue in yeast provides copper to a ceruloplasmin-like oxidase required for iron uptake. *Proc. Natl. Acad. Sci. USA* **1995**, *92*, 2632–2636. [CrossRef]
126. Dunaief, J.L.; Strober, B.E.; Guha, S.; Khavari, P.A.; Ålin, K.; Luban, J.; Begemann, M.; Crabtree, G.R.; Goff, S.P. The retinoblastoma protein and BRG1 form a complex and cooperate to induce cell cycle arrest. *Cell* **1994**, *79*, 119–130. [CrossRef]
127. Felice, M.R.; De Domenico, I.; Li, L.; Ward, D.M.; Bartok, B.; Musci, G.; Kaplan, J. Post-transcriptional regulation of the yeast high affinity iron transport system. *J. Biol. Chem.* **2005**, *280*, 22181–22190. [CrossRef] [PubMed]
128. Hsiang, T.; Baillie, D.L. Comparison of the Yeast Proteome to Other Fungal Genomes to Find Core Fungal Genes. *J. Mol. Evol.* **2005**, *60*, 475–483. [CrossRef]
129. Day, J.P.; Ackrill, P. The chemistry of desferrioxamine chelation for aluminium overload in renal dialysis patients. *Ther. Drug Monit.* **1993**, *15*, 598–601. [CrossRef] [PubMed]
130. Faa, G.; Crisponi, G. Iron chelating agents in clinical practice. *Coord. Chem. Rev.* **1999**, *184*, 291–310. [CrossRef]
131. Peterson, C.M.; Garziano, J.H.; Grady, R.W.; Jones, R.L.; Vlassara, H.V.; Canale, V.C.; Miller, D.R.; Cerami, A. Chelation studies with 2,3-dihydroxybenzoic acid in patients with thalassaemia major. *Br. J. Haematol.* **1976**, *33*, 477–485. [CrossRef] [PubMed]
132. Kontoghiorghes, G.J. Comparative efficacy and toxicity of desferrioxamine, deferriprone and other iron and aluminium che-lating drugs. *Toxicol. Lett.* **1995**, *80*, 1–18.
133. Dexter, D.T.; Ward, R.J.; Florence, A.; Jenners, P.; Crichton, R.R. Effects of desferrithiocin and its derivatives on peripheral iron and striatal dopamine and 5-hydroxytryptamine metabolism in ferrocene-loaded rat. *Biochem. Pharmacol.* **1999**, *58*, 151–155. [CrossRef]
134. Martell, S.J.D.; Beattie, A.I.; Walters, C.J.; Nayar, T.; Briese, R. Simulating fisheries management strategies in the Strait of Georgia ecosystem using Ecopath and Ecosim. *Fish. Cent. Res. Rep.* **2002**, *10*, 16–23.
135. Bergeron, R.J.; Wollenweber, M.; Wiegand, J. An investigation of desferrithiocin metabolism. *J. Med. Chem.* **1994**, *37*, 2889–2895. [CrossRef]
136. Grady, R.W.; Peterson, C.M.; Jones, R.L.; Graziano, J.H.; Bhargava, K.K.; Berdoukas, V.A.; Kokkini, G.; Loukopoulos, D.; Cerami, A. Rhodotorulic acid - investigation of its potential as an iron-chelating drug. *J. Pharmacol. Exp. Ther.* **1979**, *209*, 342–348.
137. Stradling, G.N. Decorporation of actinides: A review of recent research. *J. Alloys Compd.* **1998**, *271/273*, 11–12. [CrossRef]
138. Weinberg, E.D. The role of iron in protozoan and fungal infectious diseases. *J. Eukaryot. Microbiol.* **1999**, *46*, 231–238. [CrossRef]
139. Estrov, Z.; Tawa, A.; Wang, X.H.; Pube, I.D.; Cohen, A.; Gelfand, E.W.; Freedman, M.H. In vitro and in vivo effects of deferoxamine in neonatal acute leukemia. *Blood* **1987**, *69*, 757–761. [CrossRef] [PubMed]
140. Vergne, A.F.; Walz, A.J.; Miller, M.J. Iron chelators from mycobacteria and potential therapeutic applications. *Nat. Prod. Rep.* **2000**, *17*, 99–116. [CrossRef] [PubMed]
141. Lu, Y.; Miller, M.J. Syntheses and studies of multi warhead siderophore-5-fluorouridine conjugates. *Bioorg. Med. Chem.* **1999**, *7*, 3025–3038. [CrossRef]
142. Durbin, P.W.; Kullgren, B.; Xu, J.; Raymon, K.N. New agents for in vivo chelation of uranium (VI): Efficacy and toxicity in mice of multidentate catecholate and hydroxypyridinonate ligands. *Health Phys.* **1997**, *72*, 865–879. [CrossRef]
143. Ghosh, A.; Miller, M.J. Synthesis of novel citrate-based siderophores and siderophore-beta-lactam conjugates. Iron transport-mediated drug delivery systems. *J. Org. Chem.* **1993**, *58*, 7652–7659. [CrossRef]
144. Minnick, A.A.; McKee, J.A.; Dolence, E.K.; Miller, M.J. Iron transport-mediated antibacterial activity and development of resistance to hydroxamate and catechol siderophore-carbacephalosporin conjugates. *Antimicrob. Agents Chemother.* **1992**, *36*, 840–850. [CrossRef]
145. Wencewicz, T.A.; Long, T.E.; Möllmann, U.; Miller, M.J. Trihydroxamate siderophore–fluoroquinolone conjugates are selective sideromycin antibiotics that target *Staphylococcus aureus*. *Bioconjugate Chem.* **2013**, *24*, 473–486. [CrossRef] [PubMed]
146. Rivault, F.; Liébert, C.; Burger, A.; Hoegy, F.; Abdallah, M.A.; Schalk, I.J.; Mislin, G.L.A. Synthesis of pyochelin–norfloxacin conjugates. *Bioorg. Med. Chem. Lett.* **2007**, *17*, 640–644. [CrossRef]
147. Souto, A.; Montaos, M.A.; Balado, M.; Osorio, C.R.; Rodríguez, J.; Lemos, M.L.; Jiménez, C. Synthesis and antibacterial activity of conjugates between norfloxacin and analogues of the siderophore vanchrobactin. *Bioorg. Med. Chem.* **2013**, *21*, 295–302. [CrossRef]

148. Möllmann, U.; Ghosh, A.; Dolence, E.K.; Dolence, J.A.; Reissbrodt, R. Selective growth promotion and growth inhibition of Gram-negative and Gram-positive bacteria by synthetic siderophore-β-lactam conjugates. *Biometals* **1998**, *11*, 1–12. [CrossRef] [PubMed]
149. Petrik, M.; Pfister, J.; Misslinger, M.; Decristoforo, C.; Haas, H. Siderophore-Based Molecular Imaging of Fungal and Bacterial Infections—Current Status and Future Perspectives. *J. Fungi* **2020**, *6*, 73. [CrossRef]
150. Bergeron, R.J.; Bharti, N.; Singh, S.; Mcmanis, J.S.; Wiegand, J.; Green, L.G. Vibriobactin antibodies: A vaccine strategy. *J. Med. Chem.* **2009**, *52*, 3801–3813. [CrossRef]
151. Mike, L.; Smith, S.; Sumner, C.; Mobley, H. LB-S & T-35 siderophore vaccine conjugates protect against uropathogenic *Escherichia coli* in a murine model of urinary tract infection. *J. Urol.* **2016**, *195*, e351.
152. Mike, L.A.; Smith, S.N.; Sumner, C.A.; Eaton, K.A.; Mobley, H.L.T. Siderophore vaccine conjugates protect against uropathogenic *Escherichia coli* urinary tract infection. *Proc. Natl. Acad. Sci. USA* **2016**, *113*, 13468–13473. [CrossRef]
153. Brumbaugh, A.R.; Smith, S.N.; Mobley, H.L. Immunization with the yersiniabactin receptor, FyuA, protects against pyelonephritis in a murine model of urinary tract infection. *Infect. Immun.* **2013**, *81*, 3309–3316. [CrossRef] [PubMed]
154. Fox, J.T.; Thomson, D.U.; Drouillard, J.S.; Thornton, A.B.; Nagaraja, T.G. Efficacy of *Escherichia coli* O157: H7 siderophore receptor/porin proteins–based vaccine in feedlot cattle naturally shedding *E. coli* O157. *Foodborne Pathog. Dis.* **2009**, *6*, 893–899. [CrossRef] [PubMed]
155. Mariotti, P.; Malito, E.; Biancucci, M.; Surdo, P.L.; Bottomley, M.J. Structural and functional characterization of the *Staphylococcus aureus* virulence factor and vaccine candidate FhuD2. *Biochem. J.* **2013**, *449*, 683–693. [CrossRef] [PubMed]
156. Sanguinetti, M.; Posteraro, B.; Beigelman-Aubry, C.; Lamoth, F.; Dunet, V.; Slavin, M.; Richardson, M.D. Diagnosis and treatment of invasive fungal infections: Looking ahead. *J. Antimicrob. Chemother.* **2019**, *74*, ii27–ii37. [CrossRef]
157. Petrik, M.; Haas, H.; Dobrozemsky, G.; Lass-Florl, C.; Helbok, A.; Blatzer, M.; Dietrich, H.; Decristoforo, C. $^{68}$Ga-siderophores for PET imaging of invasive pulmonary aspergillosis: Proof of principle. *J. Nucl. Med.* **2010**, *51*, 639–645. [CrossRef]
158. Petrik, M.; Haas, H.; Schrettl, M.; Helbok, A.; Blatzer, M.; Decristoforo, C. In vitro and in vivo evaluation of selected $^{68}$Ga-siderophores for infection imaging. *Nucl. Med. Biol.* **2012**, *39*, 361–369. [CrossRef] [PubMed]
159. Petrik, M.; Haas, H.; Laverman, P.; Schrettl, M.; Franssen, G.M.; Blatzer, M.; Decristoforo, C. $^{68}$Ga-triacetylfusarinine C and $^{68}$Ga-ferrioxamine e for aspergillus infection imaging: Uptake specificity in various microorganisms. *Mol. Imaging Biol.* **2014**, *16*, 102–108. [CrossRef]
160. Kaeopookum, P.; Summer, D.; Pfister, J.; Orasch, T.; Lechner, B.E.; Petrik, M.; Novy, Z.; Matuszczak, B.; Rangger, C.; Haas, H.; et al. Modifying the siderophore triacetylfusarinine C for molecular imaging of fungal infection. *Mol. Imaging Biol.* **2019**, *21*, 1097–1106. [CrossRef]
161. Pfister, J.; Summer, D.; Petrik, M.; Khoylou, M.; Lichius, A.; Kaeopookum, P.; Kochinke, L.; Orasch, T.; Haas, H.; Decristoforo, C. Hybrid imaging of aspergillus fumigatus pulmonary infection with fluorescent, $^{68}$Ga-labelled siderophores. *Biomolecules* **2020**, *10*, 168. [CrossRef]
162. Lu, Y.; Wang, H.Y.; Wang, Z.Z.; Cong, Y.; Zhang, P.; Liu, G.L.; Liu, C.G.; Chi, Z.M.; Chi, Z. Metabolic rewiring improves the production of the fungal active targeting molecule Fusarinine C. *ACS Synth. Biol.* **2019**, *8*, 1755–1765. [CrossRef]
163. Skriba, A.; Pluhacek, T.; Palyzova, A.; Novy, Z.; Lemr, K.; Hajduch, M.; Petrik, M.; Havlicek, V. Early and non-invasive diagnosis of aspergillosis revealed by infection kinetics monitored in a rat model. *Front. Microbiol.* **2018**, *9*, 2356. [CrossRef]
164. Hoenigl, M.; Orasch, T.; Faserl, K.; Prattes, J.; Loeffler, J.; Springer, J.; Gsaller, F.; Reischies, F.; Duettmann, W.; Raggam, R.B.; et al. Triacetylfusarinine C: A urine biomarker for diagnosis of invasive aspergillosis. *J. Infect.* **2019**, *78*, 150–157. [CrossRef]
165. Wasi, S.; Tabrez, S.; Ahmad, M. Toxicological effects of major environmental pollutants: An overview. *Environ. Monit. Assess.* **2013**, *185*, 2585–2593. [CrossRef]
166. Schalk, I.J.; Hannauer, M.; Braud, A. New roles for bacterial siderophores in metal transport and tolerance. *Environ. Microbiol.* **2011**, *13*, 2844–2854. [CrossRef]
167. Hernlem, B.J.; Vane, L.M.; Sayles, G.D. The application of siderophores for metal recovery and waste remediation: Examination of correlations for prediction of metal affinities. *Water Res.* **1999**, *33*, 951–960. [CrossRef]
168. Barton, L.E.; Quicksall, A.N.; Maurice, P.A. Siderophore-mediated dissolution of hematite (α-$Fe_2O_3$): Effects of nanoparticle size. *Geomicrobiol. J.* **2012**, *29*, 314–322. [CrossRef]
169. Sinha, S.; Mukherjee, S.K. Cadmium–induced siderophore production by a high cd-resistant bacterial strain relieved cd toxicity in plants through root colonization. *Curr. Microbiol.* **2008**, *56*, 55–60. [CrossRef]
170. Dimkpa, C.O.; Merten, D.; Svato, A.; Buchel, G.; Kothe, E. Metal-induced oxidative stress impacting plant growth in contaminated soil is alleviated by microbial siderophores. *Soil Biol. Biochem.* **2009**, *41*, 154–162. [CrossRef]
171. Compant, D.; Nowak, C.; Barka, E.A. Use of plant growth-promoting bacteria for biocontrol of plant diseases: Principles, mechanisms of action, and future prospects. *Appl. Environ. Microbiol.* **2005**, *71*, 4951–4959. [CrossRef]
172. Kayser, A.; Wenger, K.; Keller, A.; Attinger, W.; Felix, H.R.; Gupta, S.K.; Schulin, R. Enhancement of phytoextraction of Zn, Cd, and Cu from calcareous soil: The use of NTA and sulfur amendments. *Environ. Sci. Technol.* **2000**, *34*, 1778–1783. [CrossRef]
173. Braud, A.; Jézéquel, K.; Bazot, S.; LEb Eau, T. Enhanced phytoextraction of an agricultural Cr and Pb contaminated soil by bioaugmentation with siderophore-producing bacteria. *Chemosphere* **2008**, *74*, 280–286. [CrossRef]
174. Dahlheimer, S.R.; Neal, C.R.; Fein, J.B. Potential mobilization of platinum-group elements by siderophores in surface environments. *Environ. Sci. Technol.* **2007**, *41*, 870–875. [CrossRef]

175. Hong, J.W.; Park, J.Y.; Gadd, G.M. Pyrene degradation and copper and zinc uptake by *Fusarium solani* and *Hypocrea lixii* isolated from petrol station soil. *J. Appl. Microbiol.* **2010**, *108*, 2030–2040. [PubMed]
176. Duckworth, O.W.; Sposito, G. Siderophore−Manganese(III) Interactions II. Manganite Dissolution Promoted by Desferrioxamine B. *Environ. Sci. Technol.* **2005**, *39*, 6045–6051. [CrossRef]
177. Hesse, E.; O'Brien, S.; Tromas, N.; O'Brien, S.; Tromas, N.; Bayer, F.; Luján, A.M.; van Veen, E.M.; Hodgson, D.J.; Buckling, A.; et al. Ecological selection of siderophore-producing microbial taxa in response to heavy metal contamination. *Ecol. Lett.* **2018**, *21*, 117–127. [CrossRef]
178. Ledin, M.; Pedersen, K. The environmental impact of mine wastes-roles of microorganisms and their significance in treatment of mine wastes. *Earth Sci. Rev.* **1996**, *41*, 67–108. [CrossRef]
179. Gray, S.N. Fungi as potential bioremediation agents in soil contaminated with heavy or radioactive metals. *Biochem. Soc. Trans.* **1998**, *26*, 666–670. [CrossRef]
180. Gopalan, A.S.; Huber, V.J.; Zincircioglu, O.; Smith, P.H. Novel tetrahydroxamate chelators for actinide complexation: Synthesis and binding studies. *J. Chem. Soc. Chem. Commun.* **1992**, *17*, 1266–1268. [CrossRef]
181. Brainard, J.R.; Strietelmeier, B.A.; Smith, P.H.; Langsten-Unkefer, P.F.; Barr, M.E.; Ryan, R.R. Actinide binding and solubilization by microbial siderophores. *Radiochim. Acta* **1992**, *58/59*, 357–363. [CrossRef]
182. Neubauer, U.; Nowak, B.; Furrer, G.; Schulin, R. Heavy metal sorption on clay minerals affected by the siderophore desferrioxamine B. *Environ. Sci. Technol.* **2000**, *34*, 2749–2755. [CrossRef]
183. Kraemer, D.; Tepe, N.; Pourret, O.; Bau, M. Negative cerium anomalies in manganese (hydr)oxide precipitates due to cerium oxidation in the presence of dissolved siderophores. *Geochim. Cosmochim. Acta* **2016**, *196*, 197–208. [CrossRef]
184. Crowley, D.E.; Reid, C.P.; Szaniszlo, P.J. Utilization of microbial siderophores in iron acquisition by oat. *Plant Physiol.* **1988**, *87*, 680–685. [CrossRef]
185. Ruggiero, C.E.; Neu, M.P.; Vanderberg, L.A. Plutonium and uranium interactions with siderophores of aerobic soil microbes. *J. Inorg. Biochem.* **1999**, *74*, 282.
186. Crowley, D.A. Microbial siderophores in the plant rhizosphere. In *Iron Nutrition in Plants and Rhizospheric Microorganisms*; Barton, L.L., Abadía, J., Eds.; Springer: Berlin/Heidelberg, Germany, 2006; Volume 614, pp. 169–189.
187. Mengel, K. Iron availability in plant tissues—Iron chlorosis on calcareous soils. In *Iron Nutrition in Soils and Plants*; Abadia, J., Ed.; Kluwer: Dordrecht, The Netherlands, 1995; pp. 389–397.
188. Kosegarten, H.U.; Hoffmann, B.; Mengel, K. Apoplastic pH and Fe3+ reduction in intact sunflower leaves. *Plant Physiol.* **1999**, *121*, 1069–1079. [CrossRef]
189. Masalha, J.; Kosegarten, H.; Elmacio, M.K. The central role of microbial activity for iron acquisition in maize and sunflower. *Biol. Fertil. Soils* **2000**, *30*, 433–439. [CrossRef]
190. Schenk, P.M.; Carvalhais, L.C.; Kazan, K. Unraveling plant–microbe interactions: Can multi-species transcriptomics help? *Trends Biotechnol.* **2012**, *30*, 177–184. [CrossRef]
191. Caris, C.; Hordt, W.; Hawkins, H.J.; Römheld, V.; George, E. Studies of iron transport by arbuscular mycorrhizal hyphae from soil to peanut and sorghum plants. *Mycorrhiza* **1998**, *8*, 35–39. [CrossRef]
192. Sagar, A.; Rathore, P.; Ramteke, P.W.; Ramakrishna, W.; Reddy, M.S.; Pecoraro, L. Plant Growth Promoting Rhizobacteria, Arbuscular Mycorrhizal Fungi and Their Synergistic Interactions to Counteract the Negative Effects of Saline Soil on Agriculture: Key Macromolecules and Mechanisms. *Microorganisms* **2021**, *9*, 1491. [CrossRef]
193. Yadav, S.; Kaushik, R.; Saxena, A.K.; Arora, D.K. Diversity and phylogeny of plant growth-promoting bacilli from moderately acidic soil. *J. Basic Microbiol.* **2011**, *51*, 98–106. [CrossRef]
194. Shah, D.; Khan, M.S.; Aziz, S.; Ali, H.; Pecoraro, L. Molecular and Biochemical Characterization, Antimicrobial Activity, Stress Tolerance, and Plant Growth-Promoting Effect of Endophytic Bacteria Isolated from Wheat Varieties. *Microorganisms* **2022**, *10*, 21. [CrossRef]
195. Schippers, B.; Lugtenberg, B.; Weisbeek, P.J. Plant growth control by fluorescent pseudomonads. In *Innovative Approaches to Plant Disease Control*; Chet, J., Ed.; Wiley: New York, NY, USA, 1987; pp. 19–36.
196. Gamalero, E.; Glick, B.R. Mechanisms used by plant growth-promoting bacteria. In *Bacteria in Agrobiology: Plant Nutrient Management*; Maheshwari, M.K., Ed.; Springer: Berlin/Heidelberg, Germany, 2011; pp. 17–46.
197. Bergeron, R.J.; Braylan, R.; Goldey, S.; Ingeno, M. Effects of the Vibrio cholerae siderophore vibriobactin on the growth characteristics of L1210 cells. *Biochem. Biophys. Res. Commun.* **1986**, *136*, 273–280. [CrossRef]
198. Hider, R.C.; Kong, X. Chemistry and biology of siderophores. *Nat. Prod. Rep.* **2010**, *27*, 637–657. [CrossRef] [PubMed]
199. Sazinas, P.; Hansen, M.L.; Aune, M.I.; Fischer, M.H.; Jelsbak, L. A rare thioquinolobactin siderophore present in a bioactive *Pseudomonas* sp. DTU12.1. *Genome Biol. Evol.* **2019**, *11*, 3529–3533. [CrossRef]
200. Samsonov, S.A.; Zsila, F.; Maszota-Zieleniak, M. Acute phase α1-acid glycoprotein as a siderophore-capturing component of the human plasma: A molecular modeling study. *J. Mol. Graph. Model.* **2021**, *105*, 107861. [CrossRef]
201. Xie, F.; Dai, S.; Zhao, Y.; Huang, P.; Yu, S.; Ren, B.; Wang, Q.; Ji, Z.; Alterovitz, G.; Zhang, Q.; et al. Generation of fluorinated amychelin siderophores against *Pseudomonas aeruginosa* infections by a combination of genome mining and mutasynthesis. *Cell Chem. Biol.* **2020**, *27*, 1532–1543. [CrossRef]
202. Swayambhu, G.; Moscatello, N.; Atilla-Gokcumen, G.E.; Pfeifer, B.A. Flux balance analysis for media optimization and genetic targets to improve heterologous siderophore production. *Iscience* **2020**, *23*, 101016. [CrossRef] [PubMed]

203. Kraemer, S.M. Iron oxide dissolution and solubility in the presence of siderophores. *Aquat. Sci.* **2004**, *66*, 3–18. [CrossRef]
204. Shirvani, M.; Nourbakhsh, F. Desferrioxamine-B adsorption to and iron dissolution from palygorskite and sepiolite. *Appl. Clay Sci.* **2010**, *48*, 393–397. [CrossRef]
205. Holmén, B.A.; Casey, W.H. Hydroxamate ligands, surface chemistry, and the mechanism of ligand-promoted dissolution of goethite [α-FeOOH (s)]. *Geochim. Cosmochim. Acta* **1996**, *60*, 4403–4416. [CrossRef]
206. Kalinowski, B.; Liermann, L.J.; Brantley, S.L.; Barnes, A.; Pantano, C.G. X-ray photoelectron evidence for bacteria-enhanced dissolution of hornblende. *Geochim. Cosmochim. Acta* **2000**, *64*, 1331–1343. [CrossRef]
207. Matzanke, B.F. Structures, coordination chemistry and functions of microbial iron chelates. In *CRC Handbook of Microbial Iron Chelates*; CRC Press: Boca Raton, FL, USA, 2017; pp. 15–64.
208. Perrin, D. *Stability Constants: Part B*; IUPAC: Pergamon, Turkey, 1979.
209. Reichard, P.; Kretzschmar, R.; Kraemer, S.M. Dissolution mechanisms of goethite in the presence of siderophores and organic acids. *Geochim. Cosmochim. Acta* **2007**, *71*, 5635–5650. [CrossRef]
210. Watteau, F.; Berthelin, J. Microbial dissolution of iron and aluminium from soil minerals: Efficiency and specificity of hydroxamate siderophores compared to aliphatic acids. *Eur. J. Soil Biol.* **1994**, *30*, 1–9.
211. Sokolova, T.; Tolpeshta, I.; Topunova, I. Biotite weathering in podzolic soil under conditions of a model field experiment. *Eurasian Soil Sci.* **2010**, *43*, 1150–1158. [CrossRef]
212. Hiradate, S.; Inoue, K. Dissolution of iron from iron (hydr) oxides by mugineic acid. *Soil Sci. Plant. Nutr.* **1998**, *44*, 305–313. [CrossRef]
213. Reichard, P.; Kraemer, S.M.; Frazier, S.W.; Kretzschmar, R. Goethite dissolution in the presence of phytosiderophores: Rates, mechanisms, and the synergistic effect of oxalate. *Plant Soil* **2005**, *276*, 115–132. [CrossRef]
214. Gómez-Galera, S.; Sudhakar, D.; Pelacho, A.M.; Capell, T.; Christou, P. Constitutive expression of a barley Fe phytosiderophore transporter increases alkaline soil tolerance and results in iron partitioning between vegetative and storage tissues under stress. *Plant Physiol. Biochem.* **2012**, *53*, 46–53. [CrossRef]
215. Gledhill, M.; Buck, K.N. The organic complexation of iron in the marine environment: A review. *Front. Microbiol.* **2012**, *3*, 69. [CrossRef]
216. Boyd, P.W.; Jickells, T.; Law, C.S.; Blain, S.; Boyle, E.A.; Buesseler, K.O.; Coale, K.H.; Cullen, J.J.; de Baar, H.J.W.; Follows, M.; et al. Mesoscale iron enrichment experiments 1993-2005: Synthesis and future directions. *Science* **2007**, *315*, 612–617. [CrossRef]
217. Cordero, O.X.; Ventouras, L.A.; DeLong, E.F.; Polz, M.F. Public good dynamics drive evolution of iron acquisition strategies in natural bacterioplankton populations. *Proc. Natl. Acad. Sci. USA* **2012**, *109*, 20059–20064. [CrossRef]
218. Martinez, J.S.; Butler, A. Marine amphiphilic siderophores: Marinobactin structure, uptake, and microbial partitioning. *J. Inorg. Biochem.* **2007**, *101*, 1692–1698. [CrossRef]
219. Barbeau, K.; Rue, E.L.; Bruland, K.W.; Butler, A. Photochemical cycling of iron in the surface ocean mediated by microbial iron (III)-binding ligands. *Nature* **2001**, *413*, 409–413. [CrossRef]
220. Hunter, K.A.; Boyd, P.W. Iron-binding ligands and their role in the ocean biogeochemistry of iron. *Environ. Chem.* **2007**, *4*, 221–232. [CrossRef]
221. Mawji, E.; Gledhill, M.; Milton, J.A.; Tarran, G.A.; Ussher, S.; Thompson, A.; Wolff, G.A.; Worsfold, P.J.; Achterberg, E.P. Hydroxamate siderophores: Occurrence and importance in the Atlantic Ocean. *Environ. Sci. Technol.* **2008**, *42*, 8675–8680. [CrossRef]
222. Amin, S.A.; Amin, S.A.; Green, D.H.; Waheeb, D.A.; Gärdes, A.; Carrano, C.J. Iron transport in the genus Marinobacter. *Biometals* **2012**, *25*, 135–147. [CrossRef]
223. Asemoloye, M.D.; Marchisio, M.A.; Gupta, V.K.; Pecoraro, L. Genome-based engineering of ligninolytic enzymes in fungi. *Microb. Cell Factories* **2021**, *20*, 20. [CrossRef]

Journal of *Fungi*

MDPI

*Article*

# A Simple and Low-Cost Strategy to Improve Conidial Yield and Stress Resistance of *Trichoderma guizhouense* through Optimizing Illumination Conditions

**Yifan Li [1,†], Xiya Meng [1,†], Degang Guo [1], Jia Gao [2], Qiwei Huang [1], Jian Zhang [1], Reinhard Fischer [2], Qirong Shen [1] and Zhenzhong Yu [1,*]**

1 Laboratory of Bio-Interactions and Crop Health, Jiangsu Provincial Key Lab for Organic Solid Waste Utilization, National Engineering Research Center for Organic-Based Fertilizers, Jiangsu Collaborative Innovation Center for Solid Organic Waste Resource Utilization, Nanjing Agricultural University, Nanjing 210095, China; liyifan@njau.edu.cn (Y.L.); 2020803234@stu.njau.edu.cn (X.M.); 2020103137@njau.edu.cn (D.G.); qwhuang@njau.edu.cn (Q.H.); zhangjian@njau.edu.cn (J.Z.); qirongshen@njau.edu.cn (Q.S.)

2 Department of Microbiology, Institute for Applied Biosciences, Karlsruhe Institute of Technology (KIT)—South Campus, 76131 Karlsruhe, Germany; jia.gao@kit.edu (J.G.); reinhard.fischer@kit.edu (R.F.)

* Correspondence: yuzhenzhong@njau.edu.cn

† These authors contributed equally to this work.

**Abstract:** Light is perceived by photoreceptors in fungi and further integrated into the stress-activated MAPK HOG pathway, and thereby potentially activates the expression of genes for stress responses. This indicates that the precise control of light conditions can likely improve the conidial yield and stress resistance to guarantee the low cost and long shelf life of *Trichoderma*-based biocontrol agents and biofertilizers. In this study, effects of wavelengths and intensities of light on conidial yield and stress tolerance to osmotic, oxidative and pH stresses in *Trichoderma guizhouense* were investigated. We found that 2 μmol photons/($m^2 \times s$) of blue light increased the conidial yield more than 1000 folds as compared to dark condition and simultaneously enhanced conidial stress resistance. The enhanced conidial stress resistance is probably due to the upregulated stress-related genes in blue light, which is under the control of the blue light receptor BLR1 and the MAP kinase HOG1.

**Keywords:** biofertilizer; light intensity; light wavelength; conidial yield; conidial stress resistance

**Citation:** Li, Y.; Meng, X.; Guo, D.; Gao, J.; Huang, Q.; Zhang, J.; Fischer, R.; Shen, Q.; Yu, Z. A Simple and Low-Cost Strategy to Improve Conidial Yield and Stress Resistance of *Trichoderma guizhouense* through Optimizing Illumination Conditions. *J. Fungi* **2022**, *8*, 50. https://doi.org/10.3390/jof8010050

Academic Editor: Baojun Xu

Received: 18 November 2021
Accepted: 29 December 2021
Published: 5 January 2022



## 1. Introduction

Fungi adapt continuously to the ever-changing environment to survive in nature. Light, as one of the most common environmental cues, regulates various morphological and physiological processes of fungi [1]. In *Aspergillus nidulans*, red light stimulates conidiation (asexual development) and represses the formation of cleistothecia (sexual development) [2–4], whereas in *Neurospora crassa* and *Trichoderma atroviride* blue light promotes conidiation [5,6]. However, in the plant pathogen *Alternaria alternata*, blue and green light repress conidiation [7]. Sporangiophores of the zygomycete fungus *Phycomyces blakesleeanus* exhibit phototropism in response to blue light [8]. White light also accelerates biofilm formation and melanin accumulation in *A. niger* [9]. Hence, light responses vary from fungus to fungus.

Fungi utilize different photoreceptors to perceive a wide range of light wavelengths [1,6,10,11]. The most representative fungal photoreceptors studied at the molecular level are the light, oxygen, and voltage (LOV) domain-containing blue-light receptor white collar 1 (WC-1) and vivid (VVD) in *N. crassa*, and the red-light receptor phytochrome FphA in *A. nidulans*. The blue- and red-light signaling pathways have been well studied in *N. crassa* and *A. nidulans*, respectively [1,5,12]. In addition, fungi can also sense blue light with cryptochrome and green light with the retinal-binding protein opsin [13–17]. On the other hand, although responses of many fungi to monochromatic light, which is perceived by specific photoreceptors, have been studied, the effects of the light intensities of specific wavelength on fungal development have received comparatively little attention.

*Trichoderma* species are widely used in agriculture as biocontrol agents and biofertilizers because of their outstanding performances in antagonizing plant pathogenic fungi and promoting plant growth [18,19]. *T. guizhouense*, initially isolated from mature compost, defeats plant pathogenic fungi through reactive oxygen species (ROS) and secondary metabolites [20,21]. It can also secrete the swollenin protein to benefit plant growth [22]. These properties make *T. guizhouense*-based biocontrol agents and biofertilizers increasingly popular. In the production of these agents and biofertilizers, high conidial yield and stress resistance are the guarantees of cost reduction and shelf-life extension.

In the entomopathogenic fungus *Metarhizium robertsii*, light modulates the expression of stress-related genes differentially and therefore affects conidial tolerance [23]. We also proved previously that in *A. alternata* white light regulates the expression of catalase-encoding genes [7]. Recently, we provided evidence that the blue light receptor BLR1, the white collar 1 orthologue, dominates conidiation in *T. guizhouense* and the blue light receptor ENV1, the orthologue of VVD, modulates photoadaptation [24]. BLR1 employs the stress-activated MAPK HOG pathway, which is also responsible for the sensing of osmotic, oxidative, and heat-shock stresses in *A. nidulans*, to activate a large proportion of blue light-regulated genes [24–29]. However, how other monochromatic light affects conidiation in *T. guizhouense* remains unknown.

Here, we investigated the effects of light wavelengths and intensities on the conidial yield and resistance to several abiotic stresses in *T. guizhouense*. The mechanism for the enhancement of stress resistance of conidia produced in light was also elucidated.

## 2. Materials and Methods

### 2.1. Strains and Culture Conditions

Wild type of *Trichoderma guizhouense* NJAU 4742 and three mutants, Δ*blr1*, Δ*env1* and Δ*hog1*, which were constructed previously [24], were used in this study. The blue light receptors BLR1 and ENV1 are absent in the Δ*blr1*- and the Δ*env1*-mutant strains, respectively. The Δ*hog1*-mutant strain lacks the gene encoding MAP kinase HOG1. All strains were cultivated on potato dextrose agar medium (PDA, BD Difco, Germany) at 28 °C. pH, osmotic and oxidative stresses imposed or illumination conditions during the cultivation were indicated additionally in the following experiments.

### 2.2. Light System

The light system used in this study was developed previously [30]. The LED panel hanged in the metal box is 39 cm long and 28 cm wide and emits blue (450 nm), green (525 nm), red (700 nm), far-red (760 nm) and white light. LEDs soldered on the panel were purchased from Ushio Inc. (Tokyo, Japan). The quality of the LEDs was measured by JAZ-COMBO S/N:JAZA0503 with a QP400-1-VIS-NIR and CC-3-UV-S spectrometer unit (Ocean Optics). The wavelengths, intensities and specific timings of light were automatically controlled by the self-made software coded for Photon P1 WIFI module. An ocean optics JAZA0503 spectrometer was used to calibrate the software. The photon flux of each LED type was measured at a distance of 20 cm and the maximum photon flux was limited to 17 μmol photons/($m^2 \times s$). Six light boxes were placed in an incubator in the dark room to make sure that the conditions except for illumination in each box are identical.

### 2.3. Quantification of Conidia

Conidial yield was quantified as previously described by Li et al. [24]. The wild type strain was revived on PDA medium for two days at 28 °C in the dark and afterwards fresh mycelia were inoculated on PDA medium (Ø 6 cm petri dish) using a hole puncher. The inoculated plates were incubated in darkness or different light conditions at 28 °C for four days in light boxes. Conidia of each plate were harvested with 20 mL distilled water and the suspension was filtered with Miracloth (Millipore, Merck KGaA, Darmstadt, Germany) to remove the mycelia. The filtrate was diluted properly, and then conidial concentration was determined using a hemocytometer. Three biological replicates of each culture condition were analyzed, and the assay was repeated three times.

### *2.4. Assessment of Conidial Stress Resistance*

Assessment of conidial resistance to abiotic stresses was performed as previously described by Dias et al. with minor modifications [23,31]. Conidia produced in different wavelengths or light intensities were harvested as above and spore suspensions were diluted to a final concentration of $1 \times 10^3$ conidia/mL. 100 µL of a spore suspension was spread on PDA medium (Ø 9 cm petri dish) supplemented with different concentrations of NaCl (0, 0.2, 0.4, 0.6, 0.8 and 1.0 M), KCl (0, 0.2, 0.4, 0.6, 0.8 and 1.0 M) and sorbitol (0, 0.4, 0.8, 1.2 and 1.6 M). Oxidative stress was imposed with different concentrations of hydrogen peroxide (0, 1, 2 and 3 mM). To analyze the conidial resistance to different pH, the pH values of the PDA medium were adjusted to 6.0, 7.0, 8.0, 9.0 and 10.0 before autoclaving. The germination rate was then calculated after 30 h of incubation at 28 °C in the dark. Each culture condition has three biological replicates and the experiment was performed in triplicate.

### *2.5. RNA Isolation and Quantitative Real-Time PCR (qPCR)*

The wild type, the Δ*blr1*-, the Δ*env1*-, and the Δ*hog1*-mutant strains were incubated on the PDA medium (Ø 6 cm petri dish) covered with cellophane at 28 °C in the dark for 24 h. The mycelium was harvested in dim-green light immediately or after 45 min illumination of blue light and frozen immediately in liquid nitrogen. Mycelia were disrupted in a homogenizer (Precellys® Evolution Super Homogenizer, Bertin, France) with glass beads at 5500 rpm for 2 min. Total RNA was extracted using the SteadyPure® Plant RNA Extraction Kit (Accurate Biotechnology (Hunan) Co., Ltd., Shenzhen, China). ~1 µg of total RNA was used for cDNA synthesis with HiScript® II Q RT SuperMix for qPCR (+gDNA wiper) (Vazyme Biotechnology Co., Ltd., Nanjing, China). The cDNA synthesis included 2 min of the removal of gDNA at 42 °C, followed by 15 min at 50 °C for the reverse transcription reaction. The cDNA samples were diluted to a final concentration of 100 ng/µL in DEPC water for qPCR.

qPCR was performed with the ChamQ SYBR qPCR Master Mix (Vazyme Biotechnology Co., Ltd., Nanjing, China). Each reaction of 20 µL contained 10 µL of 2 × ChamQ SYBR qPCR Master Mix, 0.2 µM of primers and 100 ng of cDNA. The program started with 3 min of the inactivation of the reverse transcription reaction at 95 °C, followed by 40 cycles of PCR reaction (10 s at 95 °C and then 30 s at 60 °C). After each PCR, melting curve analyses were carried out to assess the dissociation characteristics of double-stranded DNA. The translation elongation factor 1 alpha (*tef1*) gene (OPB38715) was used to normalize the expression level of each gene. Primers used in this study are listed in Supplementary Table S1. Each expression level is the average of three biological replicates.

### *2.6. Transcriptome Analysis*

Transcriptome data of *T. guizhouense* wild type and the mutants Δ*blr1*, Δ*env1* and Δ*hog1* in response to blue light are available online in the NCBI database Sequencing Read Archive (SRA) under the accession number PRJNA743899. The genes encoding proteins putatively involved in stress resistance were screened from the differentially expressed genes (DEGs) identified in a previous study [24]. The reliability of the transcriptome data was validated by qPCR. The heatmap was generated in the R environment.

### 2.7. Statistical Analysis

Data are generally presented as means ± S.D. (standard deviation) with the biological replicates as indicated in each figure legend. The statistical analysis was carried out using One-way ANOVA analysis in the IBM SPSS Statistics 25 program.

## 3. Results

### 3.1. Specific Wavelengths of Light Activate Conidiation of T. guizhouense

Fungi potentially respond to UV, blue, green, red, far-red, and white light, depending on their specific photoreceptors or if the light condition is stressful for them [1]. Hence, we first analyzed the effects of blue, green, red, far-red, and white light on the development of *T. guizhouense* using a self-designed light system. We cultured *T. guizhouense* wild type on PDA medium at 28 °C in blue (450 nm), green (525 nm), red (700 nm), far-red (760 nm) and white light and also in darkness. The light intensity was set to 1.7 µmol photons/($m^2$ × s). After four days of cultivation, the colonies grown in white and blue light were covered with green conidiophores, whereas in green light and darkness the colonies stayed white and only few conidiophores were observed (Figure 1A). Red and far-red light also promoted the formation of conidiophores to some extent.

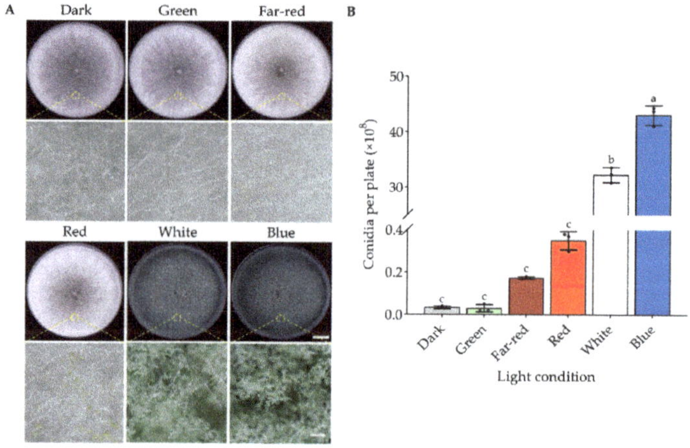


**Figure 1.** Effects of different wavelengths of light on *T. guizhouense* conidiation. (**A**) Phenotype of the wild type strain cultured under different wavelengths of light. The wild type strain was grown on PDA medium at 28 °C for four days. Scale bar, 1 cm. Colony surface was zoomed in under a stereoscopic micro-scope and the scale bar is 300 µm. (**B**) Quantification of conidia produced under different wavelengths of light after four days of cultivation. The mean values for the three biological replicates are displayed. The error bar represents the standard deviation (SD) of three biological replicates. One-way ANOVA was used for the statistical analysis ($p \leq 0.05$). Means significantly different from each other do not share the same letter.

The conidial yield for each condition was further quantified (Figure 1B). In darkness and green light, the strain produced equal amounts of conidia (0.3 × $10^7$ conidia/plate) suggesting that *T. guizhouense* does not respond to green light as far as the conidiation is concerned. The conidial yield (4.3 × $10^9$ conidia/plate) in blue light was the highest and 1433 folds of that in darkness. Compared with white light (3.2 × $10^9$ conidia/plate), blue light still increased the conidial yield by 34.4%. Conidial yields in red light (3.5 × $10^7$ conidia/plate) and far-red light (1.7 × $10^7$ conidia/plate) were also far lower than that in blue light.

*3.2. Low Intensity Blue Light Can Promote Conidial Yield Effectively*

Next, we analyzed the effects of blue-light intensities on *T. guizhouense* conidiation. The intensities of light inside the lightboxes were set to 0, 1, 2, 3, 4, 5, 7, 11 and 15 μmol photons/($m^2 \times s$) separately. Wild type was cultured in light boxes for four days and then the conidial yield was quantified. We found that although the conidial yield ($4.1 \times 10^9$ conidia/plate) in 2 μmol photons/($m^2 \times s$) blue light was the highest, the differences between tested intensities were not significant (Figure 2A). Notably, the vegetative growth of the strain was inhibited when the intensity was equal to or more than 4 μmol photons/($m^2 \times s$) (Figure 2B). The diameter of the colony under the light intensity of 4 μmol photons/($m^2 \times s$) was inhibited by 4% in comparison to that in darkness. The diameters of colonies further decreased along with the increase of the light intensities. It is also notable that strong blue light (15 μmol photons/($m^2 \times s$)) activated the conidiation process earlier than the low-intensity blue light (2 μmol photons/($m^2 \times s$)) (Figure 2C). After 2 days of cultivation, the colony grown in strong blue light was smaller but much greener than the ones in darkness and in low-intensity blue light.

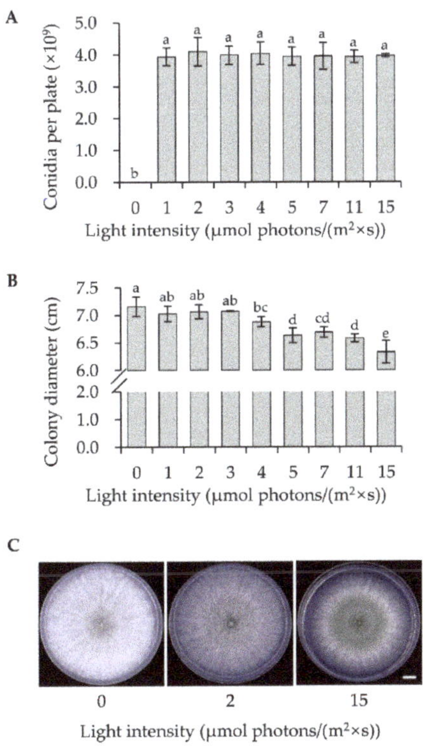


**Figure 2.** Effects of the intensities of blue light on vegetative growth and conidiation. (**A**) Quantification of the conidia produced in different intensities of blue light. Wild type was grown on the PDA medium at 28 °C for four days. The error bar represents the SD of three biological replicates. (**B**) Colony diameter of wild type in different intensities of blue light. The error bar represents the SD of three biological replicates. One-way ANOVA was used for the statistical analysis ($p \leq 0.05$). Means significantly different from each other do not share the same letter. (**C**) Phenotype of wild type grown in low and high intensities of blue light. The wild type strain was grown on the PDA medium at 28 °C for two days. Scale bar, 1 cm.

### 3.3. Blue and White Light Significantly Enhance the Conidial Resistance to Oxidative and pH Stresses

To analyze the conidial stress resistance, conidia produced in darkness or blue, red, far-red, or white light were harvested, and fresh conidia were directly used for the following experiments. To estimate the oxidative stress resistance, conidia were spread on PDA medium with increasing concentrations (up to 3 mM) of hydrogen peroxide. The germination rate of the conidia on each plate was calculated after 30 h of incubation in the dark. At all tested concentrations of hydrogen peroxide, conidia produced under blue or white light were more tolerant to oxidative stress than those produced under other conditions (Figure 3A). After the concentration of $H_2O_2$ was increased to 3 mM, the germination rate of conidia formed under blue light (60.0% germination rate) was about 1.8 folds of that of conidia generated in the dark (34.2%), 1.4 folds of that produced under red light (41.4%), and 2.9 folds of that under far-red light (20.6%).

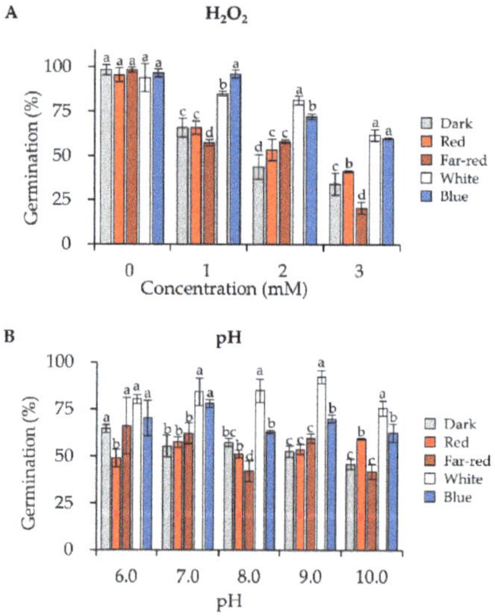


**Figure 3.** Tolerance to oxidative stress and high pH of conidia formed under different wavelengths of light. Germination rates of conidia under oxidative (**A**) or pH (**B**) stresses. For all experiments, conidia were spread on the PDA medium with increasing concentrations of $H_2O_2$ or pH as indicated, and the germination rates were calculated after 30 h of incubation in the dark. Error bars indicate SD of three biological replicates. One-way ANOVA was used for the statistical analysis ($p \leq 0.05$). Means significantly different from each other do not share the same letter.

We also assessed the resistance of conidia to different pH, ranging from 6.0 to 10.0 and conidia formed in white and blue light still exhibited increased tolerance to different pH (Figure 3B). When the pH was above 7.0, conidia produced under white light were more tolerant than conidia produced under blue light.

### 3.4. Blue and White Light Enhance the Conidial Stress Resistance to Several Osmotic Stresses

To determine the osmotic stress resistance, we first analyzed the germination rates of the conidia produced under different light when incubated on PDA medium supplemented with different concentrations of NaCl. In comparison to darkness, blue and white light significantly enhanced the conidial tolerance towards NaCl (Figure 4A). When the concentration of NaCl increased to 1 M, the germination rate of conidia formed under blue light (58.65%) was 2.5 folds of that in the dark (23.0%). Conidia formed in red light exhibited similar stress resistance to that in the dark, whereas conidia formed in far-red light were more tolerant than that in red light. We also exposed the conidia to different concentrations of KCl and sorbitol. Likewise, conidia formed in blue and white light were always the most tolerant to these stresses (Figure 4B,C).

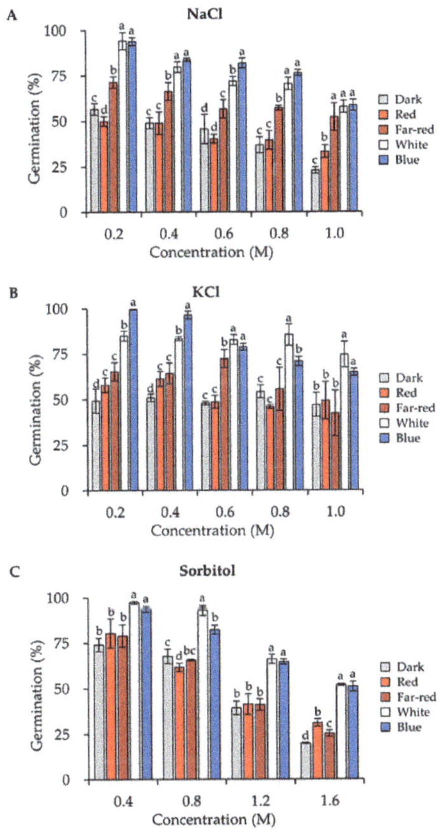


**Figure 4.** Osmotic stress tolerance of conidia formed under different wavelengths of light. Conidia were spread on the PDA medium with increasing concentrations of NaCl (**A**), KCl (**B**) or sorbitol (**C**) as indicated, and the germination rates were calculated after 30 h of incubation in the dark. Error bars indicate SD of three biological replicates. One-way ANOVA was used for the statistical analysis ($p \leq 0.05$). Means significantly different from each other do not share the same letter.

### *3.5. Blue Light Upregulates the Expression of the Genes Encoding HOG Pathway Components*

MAPK HOG pathway is crucial for blue light signaling in *T. guizhouense* [24]. By reanalyzing the transcriptome published previously [24], we found that transcript abundance of *ssk2* (OPB39222), *pbs2* (OPB41576) and *hog1* (OPB38173) encoding components of HOG pathway were all upregulated after growth under blue light (Figure 5A). Moreover, the transcript abundance of *ssk1* (OPB38780) encoding an orthologue of *A. nidulans* SskA, the response regulator of the two-component system, was also significantly upregulated. This means in blue light the signaling pathway is probably more sensitive to environmental stresses.

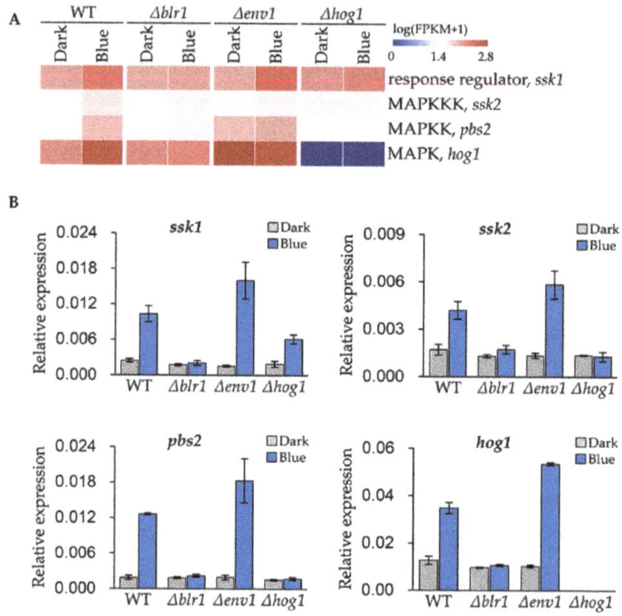


**Figure 5.** Analysis of transcript abundances of the HOG pathway genes in different strains. (**A**) Transcriptional abundances of the HOG pathway genes derived from the transcriptome data. (**B**) Expression levels of the HOG pathway genes in wild type, the Δ*blr1*-, the Δ*env1*- and the Δ*hog1*-mutant strains under blue light. The expression level of each gene was normalized to the *tef1* gene. Error bars indicate SD of three biological replicates.

To validate the reliability of the transcriptome data and further analyze the regulatory pattern of the photoreceptors on these genes, we measured the expression levels of these genes in wild type, the Δ*blr1*-, the Δ*env1*- and the Δ*hog1*-mutant strains by real-time quantitative PCR (qPCR). In wild type, the expression levels of *ssk1*, *ssk2*, *pbs2* and *pbs2* in blue light were 4.3, 2.5, 6.7 and 2.8 folds of that in the dark respectively (Figure 5B). And the fold changes of *ssk1*, *ssk2*, *pbs2* and *pbs2* in the Δ*env1*-mutant strain were 10.6, 4.4, 9.6 and 5.2, respectively, which were higher than that in wild type. However, these genes could not be induced anymore in the Δ*blr1*-mutant strain. In the Δ*hog1*-mutant strain, the expression levels of *ssk2*, *pbs2* and *hog1* did not change in blue light. Although *ssk1* was upregulated 3.2 folds after blue light exposure, the expression level was lower than that in wild type.

### *3.6. Blue Light Receptors Regulate the Stress Responsive Genes of T. guizhouense*

Superoxide dismutase (SOD) and catalase (CAT) are crucial for fungi to scavenge reactive oxygen species (ROS) [32]. In our previous study, 1615 blue light responsive genes in *T. guizhouense* wild type strain were identified through genome-wide gene expression analysis [24]. By reanalyzing the transcriptome data, we screened three CAT1-encoding genes, *cat1-1* (OPB39159), *cat1-2* (OPB42210) and *cat1-3* (OPB40299) and one SOD2-encoding gene, *sod2* (OPB46299) in wild type, which were upregulated significantly in blue light (Figure 6A). This result was confirmed again with qPCR in wild type, the Δ*blr1*-, the Δ*env1*-, and the Δ*hog1*-mutant strains. Surprisingly, in wild type the expression level of *cat1-1* in response to blue light was increased by 690 folds compared to darkness and in the Δ*env1*-mutant strain the expression level was even higher. Likewise, in blue light the expression levels of other CAT- and SOD-encoding genes were upregulated significantly in wild type and the Δ*env1*-mutant strain, but the increase was not observed in the Δ*blr1*- and the Δ*hog1*-mutant strains. Therefore, the upregulation of the expression of CAT- and SOD-encoding genes are controlled by blue light through BLR1 and HOG1.

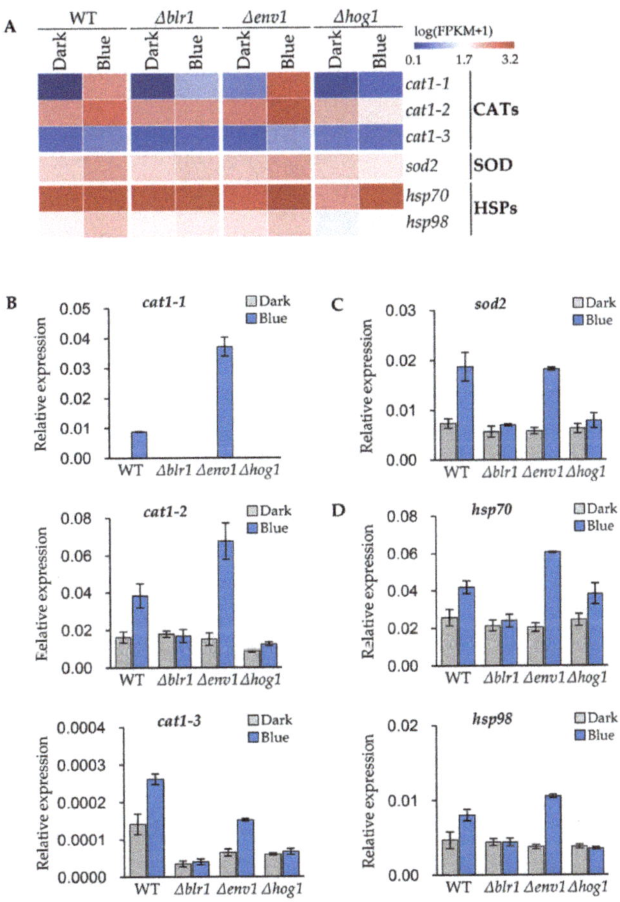


**Figure 6.** Blue-light-regulated stress response genes identified in *T. guizhouense*. (**A**) Transcriptional abundance of the catalase (CAT)-, SOD- and HSP-encoding genes. (**B–D**) Expression levels of the catalase- (**B**), SOD- (**C**) and HSP- (**D**) encoding genes in wild type, and the Δ*blr1*-, the Δ*env1*- and the Δ*hog1*-mutant strains. All strains were cultured on PDA medium at 28 °C for 24 h and then kept in the dark or exposed to blue light for 45 min. The expression level of the gene was normalized to the *tef1* gene. Error bars indicate SD of three biological replicates.

Heat shock proteins (HSPs), which function in refolding misfolded proteins, can protect the cell from damages caused by a variety of environmental stresses [33]. We identified two upregulated HSP-encoding genes *hsp70* (OPB39845) and *hsp98* (OPB41524) (Figure 6A). After blue light exposure, the expression levels of *hsp70* and *hsp98* increased 63% and 72% respectively in wild type, while in the Δ*env1*-mutant strain, the expression levels of *hsp70* and *hsp98* in blue light were 3.0 and 2.8 folds of that in the dark (Figure 6D). Like the CAT- and SOD-encoding genes, the deletion of BLR1 did not cause the expression upregulation of these genes. In the Δ*hog1*-mutant strain, the expression of *hsp70* only increased 57% in blue light, while *hsp98* did not change in response to blue light.

## 4. Discussion

The regulatory role of light on fungal development provided a possibility for us to enhance the conidial yield and stress resistance of *T. guizhouense*. Here, we show that blue light, superior to red, far-red and green light, can efficiently promote conidiation of

*T. guizhouense*. The conidiation-related genes in *T. guizhouense*, indeed, were drastically induced by blue light [24]. The fact that conidial yield in white light was lower than in blue light also suggests some interference of different wavelengths with the blue-light response [12]. Furthermore, red and far-red light slightly promoted the conidiation, which implies that phytochrome is probably functional in *T. guizhouense*. The biological function of *Trichoderma* phytochrome is worthy of further study.

Notably, strong blue light activated the early conidiation, but inhibited the growth of vegetative hyphae of *T. guizhouense*. In *A. nidulans*, strong blue light induces the generation of Reactive Oxygen Species (ROS) in hyphae [30]. More than a moderate light signal, the high intensity blue light ($\geq$4 μmol photons/($m^2 \times s$)) is stressful for vegetative hyphae of *T. guizhouense*, which forces the shift of development from vegetative growth to asexual reproduction. However, low intensities (<4 μmol photons/($m^2 \times s$)) of blue light not only increase the conidial yield of *T. guizhouense*, but these intensities do not retard the vegetative growth. During the production of conidia in the factory, in order to lower the cost and meanwhile reutilize agricultural wastes, the media for solid fermentation are commonly the discarded crop straws or other plant-derived wastes, which normally contain cellulose as main carbon source [34]. Therefore, it is important not to reduce vegetative growth to get the maximal biomass as a prerequisite for efficient cellulose degradation. Low-intensity blue light also consumes less power. Hence, low-intensity blue light stimulates conidiation, does not affect vegetative growth and is more economic. Low intensity blue light is more practical for large-scale solid fermentation.

Fungi utilize two-component system and the MAPK HOG pathway for sensing and responding to environmental signals [24–29]. The former comprises the hybrid histidine kinases (HHKs), which are potential sensors for different environmental signals, a phosphotransferase and several response regulators (RRs); the latter consists of a MAP kinase kinase kinase (MAPKKK), a MAP kinase kinase (MAPKK) and a MAP kinase (MAPK). The corresponding orthologues of the HOG pathway components in *T. guizhouense* are the MAPKKK SSK2, the MAPKK PBS2, and the MAPK HOG1. The results here reveal that in blue light, the genes encoding the components of HOG pathway were all upregulated, suggesting that the amounts of these components in conidia are probably increased, which then strengthens the HOG pathway. Hence, the conidia can also respond to other environmental signals more efficiently. The conidia formed in blue and white light were more resistant to osmotic, oxidative and pH stresses than those formed in other light conditions. Similar effects of light on conidial resistance were also observed in *M. robertsii* [23]. Our results demonstrate that light can significantly upregulate the expression of the stress-related genes, which depends on the light receptor BLR1 and the MAPK HOG pathway. Fungal hydrophobins (HFBs), the small, cysteine-rich, secreted proteins, are involved in conidial stress resistance [35]. Our recent study proved that in *T. guizhouense*, the production of hydrophobin HFB10 can be induced by blue light through BLR1 and HOG pathway. Moreover, the upregulation of cryptochrome and photolyase family proteins, severing as repair enzymes for UV–induced DNA lesions, are also under the control of BLR1 [24,36].

The blue light receptor BLR1 is essential for conidiation of *T. guizhouense* in blue light and controls 80% of light-regulated genes [24]. Although HOG1 controls more than 60% of light-regulated genes, the genes related to conidiation such as *brl1*, *aba1* and *wet1* are independent of it. It seems that during the process of conidiation, the role of HOG pathway is to specially modulate conidial fitness according to the intracellular and extracellular signals by equipping the conidia with stress-related proteins. Likewise, in *A. nidulans*, the viability of conidia was decreased when the MAP kinase SakA/HogA was absent [27]. HOG pathway is a hub of various environmental signals, but how it distinguishes these signals such as osmotic and oxidative stresses, light and heat shock, remains enigmatic. However, it is reasonable that a set of stress-related genes can be awoken not only by one environmental signal but others plugging into HOG pathway. That is why conidia produced in blue light are more resistant to different abiotic stresses. Recently, Rangel et al. showed that in the entomopathogenic fungi *Lecanicillium aphanocladii* and *Simplicillium lanosoniveum*, the effect of the nutritional stress on conidial resistance is stronger than that of illumination [31]. Conidia formed under nutritional stress are more resistant to several abiotic stresses than those formed in light. Whether the signal of nutritional stress is integrated into HOG pathway or there is an alternative to trigger the expression of stress-related genes remains to be elucidated in the future. The stronger effect of nutritional stress on conidial resistance also arises the question if one of the signals is overwhelming when light and other stresses coexist. Therefore, a larger regulatory network for gene expression depicting the crosstalk between light and stress signaling needs to be deciphered.

## 5. Conclusions

The strategy to improve the conidial yield and stress resistance of *T. guizhouense* through optimizing light conditions is feasible. In comparison to darkness, the conidial yield was increased by more than 1400 folds under 450 nm blue light and the conidial resistance was significantly enhanced. Low-intensity blue light (<4 μmol photons/($m^2 \times s$)) improved the conidial yield and stress resistance simultaneously, but strong blue light (≥4 photons/($m^2 \times s$)) delayed the vegetative growth of *T. guizhouense*. The enhanced conidial stress resistance is probably attributed to the upregulation of stress-related genes, which is controlled by the light receptor BLR1 and the MAPK HOG pathway.

**Supplementary Materials:** The following supporting information can be downloaded at: https://www.mdpi.com/article/10.3390/jof8010050/s1. Table S1: Oligonucleotides used in this study.

**Author Contributions:** Conceptualization, Y.L., Q.H., Q.S. and Z.Y.; Investigation, Y.L., X.M. and D.G.; Methodology, Y.L. and X.M.; Supervision, Q.H., J.Z., Q.S. and Z.Y.; Writing—original draft, Y.L.; Writing—review & editing, J.G., R.F. and Z.Y. All authors have read and agreed to the published version of the manuscript.

**Funding:** This work was supported by the National Natural Science Foundation of China (NSFC) (Grant Nos. 32070101 and 32072674), the Natural Science Foundation of Jiangsu Province (Grant No. BK20200542), the Jiangsu Agriculture Science and Technology Innovation fund (JASTIF) (Grant Nos. CX(21)2018 and CX(18)3059) and the Fundamental Research Funds for the Central Universities (Grant No. KYRC2021004).

**Data Availability Statement:** Not applicable.

**Conflicts of Interest:** The authors declare no conflict of interest.

## References

1. Yu, Z.Z.; Fischer, R. Light sensing and responses in fungi. *Nat. Rev. Microbiol.* **2019**, *17*, 25–36. [CrossRef] [PubMed]
2. Blumenstein, A.; Vienken, K.; Tasler, R.; Purschwitz, J.; Veith, D.; Frankenberg-Dinkel, N.; Fischer, R. The *Aspergillus nidulans* phytochrome FphA represses sexual development in red light. *Curr. Biol.* **2005**, *15*, 1833–1838. [CrossRef]
3. Mooney, J.L.; Yager, L.N. Light is required for conidiation in *Aspergillus nidulans*. *Genes Dev.* **1990**, *4*, 1473–1482. [CrossRef]
4. Purschwitz, J.; Mueller, S.; Kastner, C.; Schoser, M.; Haas, H.; Espeso, E.A.; Atoui, A.; Calvo, A.M.; Fischer, R. Functional and physical interaction of blue- and red-light sensors in *Aspergillus nidulans*. *Curr. Biol.* **2008**, *18*, 255–259. [CrossRef] [PubMed]
5. Dasgupta, A.; Fuller, K.K.; Dunlap, J.C.; Loros, J.J. Seeing the world differently: Variability in the photosensory mechanisms of two model fungi. *Environ. Microbiol.* **2016**, *18*, 5–20. [CrossRef]
6. Schmoll, M.; Esquivel-Naranjo, E.U.; Herrera-Estrella, A. *Trichoderma* in the light of day—Physiology and development. *Fungal Genet. Biol.* **2010**, *47*, 909–916. [CrossRef] [PubMed]
7. Igbalajobi, O.; Yu, Z.Z.; Fischer, R. Red- and blue-light sensing in the plant pathogen *Alternaria alternata* depends on phytochrome and the white-collar protein LreA. *Mbio* **2019**, *10*, e00371-19. [CrossRef] [PubMed]
8. Sanz, C.; Rodriguez-Romero, J.; Idnurm, A.; Christie, J.M.; Heitman, J.; Corrochano, L.M.; Eslava, A.P. *Phycomyces* MADB interacts with MADA to form the primary photoreceptor complex for fungal phototropism. *Proc. Natl. Acad. Sci. USA* **2009**, *106*, 7095–7100. [CrossRef] [PubMed]
9. Sun, W.J.; Yu, Y.; Chen, J.; Yu, B.; Chen, T.P.; Ying, H.J.; Zhou, S.M.; Ouyang, P.K.; Liu, D.; Chen, Y. Light signaling regulates *Aspergillus niger* biofilm formation by affecting melanin and extracellular polysaccharide biosynthesis. *Mbio* **2021**, *12*, e03434-20. [CrossRef]
10. Fuller, K.K.; Loros, J.J.; Dunlap, J.C. Fungal photobiology: Visible light as a signal for stress, space and time. *Curr. Genet.* **2015**, *61*, 275–288. [CrossRef] [PubMed]

11. Schumacher, J.; Gorbushina, A.A. Light sensing in plant- and rock-associated black fungi. *Fungal Biol.* **2020**, *124*, 407–417. [CrossRef] [PubMed]
12. Corrochano, L.M. Light in the fungal world: From photoreception to gene transcription and beyond. *Annu. Rev. Genet.* **2019**, *53*, 149–170. [CrossRef] [PubMed]
13. Bayram, O.; Biesemann, C.; Krappmann, S.; Galland, P.; Braus, G.H. More than a repair enzyme: *Aspergillus nidulans* photolyase-like CryA is a regulator of sexual development. *Mol. Biol. Cell* **2008**, *19*, 3254–3262. [CrossRef]
14. Bieszke, J.A.; Braun, E.L.; Bean, L.E.; Kang, S.C.; Natvig, D.O.; Borkovich, K.A. The *nop-1* gene of *Neurospora crassa* encodes a seven transmembrane helix retinal-binding protein homologous to archaeal rhodopsins. *Proc. Natl. Acad. Sci. USA* **1999**, *96*, 8034–8039. [CrossRef] [PubMed]
15. Cohrs, K.C.; Schumacher, J. The two cryptochrome/photolyase family proteins fulfill distinct roles in DNA photorepair and regulation of conidiation in the gray mold fungus *Botrytis cinerea*. *Appl. Environ. Microbiol.* **2017**, *83*, e00812-17. [CrossRef] [PubMed]
16. Garcia-Martinez, J.; Brunk, M.; Avalos, J.; Terpitz, U. The CarO rhodopsin of the fungus *Fusarium fujikuroi* is a light-driven proton pump that retards spore germination. *Sci. Rep.* **2015**, *5*, 7798. [CrossRef] [PubMed]
17. Wang, Z.; Wang, J.R.; Li, N.; Li, J.G.; Trail, F.; Dunlap, J.C.; Townsend, J.P. Light sensing by opsins and fungal ecology: Nop-1 modulates entry into sexual reproduction in response to environmental cues. *Mol. Ecol.* **2018**, *27*, 216–232. [CrossRef] [PubMed]
18. Ahmed, H.F.A.; Seleiman, M.F.; Al-Saif, A.M.; Alshiekheid, M.A.; Battaglia, M.L.; Taha, R.S. Biological control of celery powdery mildew disease caused by *Erysiphe heraclei* DC in vitro and in vivo conditions. *Plants* **2021**, *10*, 2342. [CrossRef] [PubMed]
19. Druzhinina, I.S.; Seidl-Seiboth, V.; Herrera-Estrella, A.; Horwitz, B.A.; Kenerley, C.M.; Monte, E.; Mukherjee, P.K.; Zeilinger, S.; Grigoriev, I.V.; Kubicek, C.P. *Trichoderma*: The genomics of opportunistic success. *Nat. Rev. Microbiol.* **2011**, *9*, 749–759. [CrossRef]
20. Pang, G.; Sun, T.T.; Yu, Z.Z.; Yuan, T.; Liu, W.; Zhu, H.; Gao, Q.; Yang, D.Q.; Kubicek, C.P.; Zhang, J.; et al. Azaphilones biosynthesis complements the defence mechanism of *Trichoderma guizhouense* against oxidative stress. *Environ. Microbiol.* **2020**, *22*, 4808–4824. [CrossRef] [PubMed]
21. Zhang, J.; Miao, Y.Z.; Rahimi, M.J.; Zhu, H.; Steindorff, A.; Schiessler, S.; Cai, F.; Pang, G.; Chenthamara, K.; Xu, Y.; et al. Guttation capsules containing hydrogen peroxide: An evolutionarily conserved nadph oxidase gains a role in wars between related fungi. *Environ. Microbiol.* **2019**, *21*, 2644–2658. [CrossRef] [PubMed]
22. Meng, X.H.; Miao, Y.Z.; Liu, Q.M.; Ma, L.; Guo, K.; Liu, D.Y.; Ran, W.; Shen, Q.R. *Tg*SWO from *Trichoderma guizhouense* NJAU4742 promotes growth in cucumber plants by modifying the root morphology and the cell wall architecture. *Microb. Cell Fact.* **2019**, *18*, 148. [CrossRef] [PubMed]
23. Dias, L.P.; Pedrini, N.; Braga, G.U.L.; Ferreira, P.C.; Pupin, B.; Araujo, C.A.S.; Corrochano, L.M.; Rangel, D.E.N. Outcome of blue, green, red, and white light on *Metarhizium robertsii* during mycelial growth on conidial stress tolerance and gene expression. *Fungal Biol.* **2020**, *124*, 263–272. [CrossRef]
24. Li, Y.F.; Sun, T.T.; Guo, D.G.; Gao, J.; Zhang, J.A.; Cai, F.; Fischer, R.; Shen, Q.R.; Yu, Z.Z. Comprehensive analysis of the regulatory network of blue-light-regulated conidiation and hydrophobin production in *Trichoderma guizhouense*. *Environ. Microbiol.* **2021**, *23*, 6241–6256. [CrossRef]
25. Esquivel-Naranjo, E.U.; Garcia-Esquivel, M.; Medina-Castellanos, E.; Correa-Perez, V.A.; Parra-Arriaga, J.L.; Landeros-Jaime, F.; Cervantes-Chavez, J.A.; Herrera-Estrella, A. A *Trichoderma atroviride* stress-activated MAPK pathway integrates stress and light signals. *Mol. Microbiol.* **2016**, *100*, 860–876. [CrossRef] [PubMed]
26. Furukawa, K.; Hoshi, Y.; Maeda, T.; Nakajima, T.; Abe, K. *Aspergillus nidulans* HOG pathway is activated only by two-component signalling pathway in response to osmotic stress. *Mol. Microbiol.* **2005**, *56*, 1246–1261. [CrossRef]
27. Lara-Rojas, F.; Sanchez, O.; Kawasaki, L.; Aguirre, J. *Aspergillus nidulans* transcription factor AtfA interacts with the MAPK SakA to regulate general stress responses, development and spore functions. *Mol. Microbiol.* **2011**, *80*, 436–454. [CrossRef]
28. Yu, Z.Z.; Armant, O.; Fischer, R. Fungi use the SakA (HogA) pathway for phytochrome-dependent light signalling. *Nat. Microbiol.* **2016**, *1*, 16019. [CrossRef]
29. Yu, Z.Z.; Ali, A.; Igbalajobi, O.A.; Streng, C.; Leister, K.; Krauss, N.; Lamparter, T.; Fischer, R. Two hybrid histidine kinases, TcsB and the phytochrome FphA, are involved in temperature sensing *Aspergillus nidulans*. *Mol. Microbiol.* **2019**, *112*, 1814–1830. [CrossRef]
30. Yu, Z.Z.; Streng, C.; Seibeld, R.F.; Igbalajobi, O.A.; Leister, K.; Ingelfinger, J.; Fischer, R. Genome-wide analyses of light-regulated genes in *Aspergillus nidulans* reveal a complex interplay between different photoreceptors and novel photoreceptor functions. *PLoS Genet.* **2021**, *17*, e1009845. [CrossRef]
31. Dias, L.P.; Souza, R.K.F.; Pupin, B.; Rangel, D.E.N. Conidiation under illumination enhances conidial tolerance of insect-pathogenic fungi to environmental stresses. *Fungal Biol.* **2021**, *125*, 891–904. [CrossRef] [PubMed]
32. Heller, J.; Tudzynski, P. Reactive oxygen species in phytopathogenic fungi: Signaling, development, and disease. *Annu. Rev. Phytopathol.* **2011**, *49*, 369–390. [CrossRef]
33. Rangel, D.E.N. Stress induced cross-protection against environmental challenges on prokaryotic and eukaryotic microbes. *World J. Microbiol. Biotechnol.* **2011**, *27*, 1281–1296. [CrossRef]
34. Quiroz, R.D.; Roussos, S.; Hernandez, D.; Rodriguez, R.; Castillo, F.; Aguilar, C.N. Challenges and opportunities of the bio-pesticides production by solid-state fermentation: Filamentous fungi as a model. *Crit. Rev. Biotechnol.* **2015**, *35*, 326–333. [CrossRef]

35. Cai, F.; Gao, R.W.; Zhao, Z.; Ding, M.Y.; Jiang, S.Q.; Yagtu, C.; Zhu, H.; Zhang, J.; Ebner, T.; Mayrhofer-Reinhartshuber, M.; et al. Evolutionary compromises in fungal fitness: Hydrophobins can hinder the adverse dispersal of conidiospores and challenge their survival. *ISME J.* **2020**, *14*, 2610–2624. [CrossRef] [PubMed]
36. Tagua, V.G.; Pausch, M.; Eckel, M.; Gutierrez, G.; Miralles-Duran, A.; Sanz, C.; Eslava, A.P.; Pokorny, R.; Corrochano, L.M.; Batschauer, A. Fungal cryptochrome with DNA repair activity reveals an early stage in cryptochrome evolution. *Proc. Natl. Acad. Sci. USA* **2015**, *112*, 15130–15135. [CrossRef] [PubMed]

Journal of *Fungi*

MDPI

*Review*

# Fungal Proteases as Emerging Biocatalysts to Meet the Current Challenges and Recent Developments in Biomedical Therapies: An Updated Review

**Muhammad Naeem [1], Saba Manzoor [2], Mashhud-Ul-Hasan Abid [3], Muhammad Burhan Khan Tareen [4], Mirza Asad [5], Sajida Mushtaq [6], Nazia Ehsan [7], Dua Amna [8], Baojun Xu [9,*] and Abu Hazafa [5,*]**

1 College of Life Science, Hebei Normal University, Shijiazhuang 050025, China; naeemsaleem413@gmail.com
2 Department of Zoology, University of Sialkot, Sialkot 51310, Pakistan; saba.manzoor@uskt.edu.pk
3 Department of Biochemistry, Bahauddin Zakariya University, Multan 60800, Pakistan; mashhoodabid@gmail.com
4 College of Food Science & Technology, Huazhong Agricultural University, Wuhan 430070, China; burhan.tareen786@yahoo.com
5 Department of Biochemistry, University of Agriculture Faisalabad, Faisalabad 38040, Pakistan; asadmirza074@gmail.com
6 Department of Zoology, Government College Women University, Sialkot 51040, Pakistan; sajida.mushtaq@gcwus.edu.pk
7 Department of Zoology, Wildlife and Fisheries, University of Agriculture Faisalabad, Faisalabad 38040, Pakistan; naziaeuaf@gmail.com
8 Institute of Food Science & Nutrition, Bahauddin Zakariya University, Multan 60800, Pakistan; fatimadua11233@gmail.com
9 Food Science and Technology Program, Beijing Normal University-Hong Kong Baptist University (BNU-HKBU) United International College, Zhuhai 519087, China
* Correspondence: baojunxu@uic.edu.cn (B.X.); abuhuzaifavirk@gmail.com (A.H.)

**Citation:** Naeem, M.; Manzoor, S.; Abid, M.-U.-H.; Tareen, M.B.K.; Asad, M.; Mushtaq, S.; Ehsan, N.; Amna, D.; Xu, B.; Hazafa, A. Fungal Proteases as Emerging Biocatalysts to Meet the Current Challenges and Recent Developments in Biomedical Therapies: An Updated Review. *J. Fungi* **2022**, *8*, 109. https://doi.org/10.3390/jof8020109

Academic Editors: David S. Perlin and Craig Faulds

Received: 29 November 2021
Accepted: 5 January 2022
Published: 24 January 2022



**Abstract:** With the increasing world population, demand for industrialization has also increased to fulfill humans' living standards. Fungi are considered a source of essential constituents to produce the biocatalytic enzymes, including amylases, proteases, lipases, and cellulases that contain broad-spectrum industrial and emerging applications. The present review discussed the origin, nature, mechanism of action, emerging aspects of genetic engineering for designing novel proteases, genome editing of fungal strains through CRISPR technology, present challenges and future recommendations of fungal proteases. The emerging evidence revealed that fungal proteases show a protective role to many environmental exposures and discovered that an imbalance of protease inhibitors and proteases in the epithelial barriers leads to the protection of chronic eosinophilic airway inflammation. Moreover, mitoproteases recently were found to execute intense proteolytic processes that are crucial for mitochondrial integrity and homeostasis function, including mitochondrial biogenesis, protein synthesis, and apoptosis. The emerging evidence revealed that CRISPR/Cas9 technology had been successfully developed in various filamentous fungi and higher fungi for editing of specific genes. In addition to medical importance, fungal proteases are extensively used in different industries such as foods to prepare butter, fruits, juices, and cheese, and to increase their shelf life. It is concluded that hydrolysis of proteins in industries is one of the most significant applications of fungal enzymes that led to massive usage of proteomics.

**Keywords:** fungal enzymes; proteases; catalytical properties; recent developments; genetic engineering; mitoproteases

## 1. Introduction

Excessive use of chemicals in different industries has increased tremendously in the past few years, including phenols, pesticides, polyaromatic hydrocarbons (PAHs), polychlorinated compounds, polychlorinated biphenyls, and arsenic. Exposure to these toxic chemicals has become a threat to the environment and public health concerns [1]. There is a need to replace these chemical compounds with fungal enzymes that are considered alternatives to the toxic chemicals and act as ecofriendly indicators to meet industrial demands [1]. The global market of fungal enzymes is drastically increasing in different sectors due to the high production rate, smooth downstream processing, and low costs [2,3].

Fungi are considered GRAS (Generally Regarded as Safe) organisms as compared to other microorganisms because they fulfill the criteria of industrial demands such as efficient growth on culture media in short duration and continuous supply of desired products [4,5]. Fungi also secreted a large variety of proteases, lipases, amylases, and amylases that play an important role in physiological processes such as germination, as defensins against other pathogens or for nutritional requirements for development [6,7]. Secretions of fungal enzymes occur from the cells present at the top of hyphae. These secreted enzymes can be used for industrial preparations of valuable products [8].

Fungal proteases have been widely studied due to their wide diversity [9]. Proteases have been isolated from different fungi such as *Schizophyllum commune, Pleurotus ostreatus, Phanerochaete chrysosporium, Thermomyces lanuginosus, Sporotrichum thermophiles, Myceliophthora thermophile, Thermomyces ibadanensis, Candida mogii, Saccharomyces pombe, Aspergillus flavus*, and *Neurospora crassa* [10–12]. Fungal proteases can be isolated through the fermentation process exhibiting high catalytic and specificity for the substrate [1,13].

Fungal proteases have diverse importance in corporate sectors such as the pharmaceutical, detergent, leather, waste, and food industries [14]. In the food industry, they are used to make beer, wine, and vinegar. Acidic proteases are used to improve wheat gluten's structural and functional properties [15]. In medical sectors, fungal proteases are used as therapeutic agents for the treatment of a variety of diseases such as cancer, HIV, inflammatory diseases, diabetes, and hepatic cancer [16,17]. In the textile and laundry industries, they are used to prepare enzyme-based detergents to remove the tough stains from clothes due to developing excellent washing performance compared to other microbial enzymes [13,18]. They are also involved in degrading lignocellulosic biomass, and products can be utilized as biofuels for the production of energy at the commercial level [8,19].

Fungal proteases hold a pivotal position in cellular signaling and physiological processes occurring in the human body [20]. They possess antiviral, anti-inflammatory, antioxidant, and antitumor activities due to the presence of bioactive compounds [21]. Through recombinant DNA technology, different genes from fungal proteases have been cloned and sequenced in order to enhance their production under optimum conditions [22,23]. For instance, a selected gene that encodes target enzyme shows its expression in the fungal cell factory, such as *T. reesei, A. nidulans and A. oryzae, A. oryzae* FG76, *Pseudomonas aeruginosa* CTM50182 and *P. chrysogenum* (Pg222) [24–28]. The accumulative data also revealed that CRISPR/Cas9 technology had been successfully developed in different fungi, including *Trichoderma reesei, Aspergillus stains, A. fumigatus*, and *Ustilago maydis* for genome editing [29–31].

Many fungal strains have been isolated, but their diversity remains unexplored and a barrier for cost production prior to use for different industrial operations. Due to vast biodiversity, many fungal strains have been reported that secreted proteases, but their nature, secretary pathways, molecular structures in 3D, mechanism of action, and roles in cellular and physiological processes remain unclear. The impact of COVID-19 on the global market affecting fungal proteases, genetic expression, and genome editing of many strains through CRISPR technology needs to be explored before using them in different sectors.

However, this review highlights the origin, nature, mechanism of action, emerging aspects of genetic engineering for designing novel proteases, genetic expression and genome editing of fungal strains through CRISPR technology, and roles in cellular and physiological processes. This review also presents the impact of COVID-19 on global market trends affecting fungal proteases.

## 2. Fungal Enzymes

Different enzymes are isolated from different fungus strains and are used as detergents for softening and washing clothes to maintain their quality, color, and other cloth properties. These different fungal enzymes are proteases, amylases, cellulases, and lipases used as detergents to increase their efficiency [32–35]. The characteristics of different fungal enzymes are presented in Table 1. These enzyme-based detergents also have advantages over synthetic chemicals because they are biodegradable, environment-friendly, and are used in a minimal amount to remove strains [36].

**Table 1.** The characteristics of different fungal enzymes.

| Enzyme Class | Type of Strain | Production Method | Mol. Weight (kDa) | Opt. pH | Opt. Temperature (°C) | Substrate | Inhibitors | Enzyme Activity (%) | Reference |
|---|---|---|---|---|---|---|---|---|---|
| Fungal amylase | *Thermomyces lanuginosus* | SSF | 33 | 5 | 48 | Wheat straw, guayule bagasse | SDS | 88 | [37,38] |
| | *Aspergillus fumigatus* | SmF | 28 | 4.5 | 30 | Pomegranate peel, wheat bran | Ebselen | 90 | [39,40] |
| | *Cryptococcus flavus* | SSF | 70 | 5 | 45 | Starch, amylose | Mercury | 80–84 | [41] |
| | *Aspergillus niger* | SSF | 115 | 5 | 75 | Cowpeas, chickpeas | | 88 | [42,43] |
| | *Mucor* sp. | SSF | 38 | 4 | 30 | Kidney beans, lupine | EDTA | 84–86 | [44,45] |
| | *Aspergillus oryzae* | SSF | 102 | 8 | 30 | Groundnut oil, sesame oil | Copper | 85 | [46,47] |
| | *Aspergillus kawachii* | SSF | 110 | 3 | 32 | Pearl millet | Mercury | 88–90 | [48,49] |
| | *Penicillium flerreus* | SSF | 25 | 6 | 28 | Cowpeas, chickpeas | Lead | 78 | [43,50] |
| Fungal cellulase | *Trichoderma viride* | SSF | 55 | 7 | 52 | CMC | Mercury | 86–88 | [51,52] |
| | *Peniophora* sp. | SmF | 30 | 4 | 58 | SKT | EDTA | 80 | [53] |
| | *Aspergillusniger IMMIS1* | SSF | 70 | 3.5 | 32 | RW, bread | Mercury | 90 | [54,55] |
| | *T. harzianum* | SSF | 40 | 5 | 70 | Sugarcane bagasse | EDTA | 85 | [13,56] |
| | *Aspergillus niger VTCC-F021* | SSF | 29 | 4 | 52 | Sugar cane, CMC | Zinc | 95 | [57,58] |
| | *Aspergillus terreus* | SmF | 52 | 3.5 | 13 | Cowpeas, chickpeas | Mercury | 78 | [59] |
| Fungal lipase | *Aspergiillus niger* | SSF | 30 | 3 | 40–48 | Coir waste, RH | Zinc | 88 | [42,60] |
| | *A. terreus* | SmF | 35 | 3.5 | 48 | MOC | α-Glucosidase | 85 | [59,61] |
| | *A. versicolor* | SmF | 93 | 7 | 60 | EMO | Lipstatin | 90 | [62,63] |
| | *A. tamarii* | SSF | 5 | 4.5 | 55 | GOC, agrowastes | Ebelactone | 92 | [64,65] |
| | *A. japonicus* | SmF | 9 | 4.6 | 25 | SFO, casein | Caulerpenyne | 88 | [66,67] |
| | *Mucor* sp. | SmF | 7 | 4.2 | 28 | Kidney beans, lupine | Percyquinin | 66 | [45,68] |
| Fungal protease | *Scopulariopsis* sp. | SSF | 38 | 8 | 56 | Glucose, peptone | NBS | 88 | [69,70] |
| | *Aspergillus niger* | SSF | 47 | 7 | 45 | Cowpeas, chickpeas | EDTA | 88 | [42,71] |
| | *Aspergillus fumigatus* | SSF | 40 | 8 | 31 | PL, casein | DTT | 90 | [72,73] |
| | *Rhizopus oryza* | SmF | 55 | 5 | 32 | WBW | DTT | 75 | [74,75] |
| | *Mucpr* Sp. | SSF | 35 | 7 | 28 | Kidney beans, lupine | EDTA | 88 | [45,76] |
| | *G. putredinis* | SmF | 48 | 7–8 | 29 | Soya bean meal | IAA | 85 | [13] |
| | *T. harzianum* | SSF | 45 | 7 | 19 | Glutamine | PMSF | 80–82 | [13] |

Note: SDS = Sodium Dodecyl Sulfate, EDTA = Ethylenediaminetetraacetic acid, DTT = Dithiothreitol, NBS = *N*-bromosuccinimide, PMSF = Phenylmethylsulphonylfluoride, IAA = Indole Acetic Acid, SSF = Solid-state fermentation, SmF = submerged fermentation, RH = Rice Husk, MOC = Mustard Oil Cake, GOC = Gingili Oil Cake, EMO = Edible Oil Mill, SFO = Sunflower Oil, PL = Pig Hung, CMC = Carboxy Methyl Cellulose, SCT = Silk Cotton Tree, RW = Rose Wood, and WBW = White Bread Waste.

### *Coproduction of Fungal Enzymes*

Fungi secrete different enzymes and peptides that play a significant role in cellular processes, growth, and sporulation. Fungi are classified as psychrophilic, mesophilic, and thermophilic on the basis of their particular habitat as these fungi produce a large variety of enzymes [77,78]. Some species of thermophilic fungi can tolerate high temperature and secrete a variety of proteases. Due to their longer shelf life, these proteases can be used in different industries, including food and pharmaceuticals [79]. Some fungi found in deep sea are also a major source of psychrophilic protease [80]. Different species of mesophilic fungi also contribute to the large-scale production of proteases [22]. *Thermomyces lanuginosus* shows optimum growth at 40–50 °C, and produces proteases, lipases, and amylases used to treat wastewater and the pharmaceutical industries [81,82]. *Neurospora crassa* shows optimum growth at 20–30 °C and produces alkaline, serine proteases, and cellulases used as a model organism to analyze genetic recombination [83].

Similarly, *Myceliophthora thermophile* shows optimum growth at 45–50 °C, produces alkaline protease and cellulases used for textile industries and bioremediation [84]. *Thermomyces ibadanensis* shows optimum growth at 55 °C, produces serine proteases and lipases used for wastewater treatment [85]. The detailed information about the coproduction of fungal enzymes from different fungal strains with their applications is presented in Table 2.

**Table 2.** The coproduction of fungal enzymes from different strains of fungus based on temperature.

| Fungal Strain | Nature | Genus | Opt. Temperature (°C) | Fungal Amylase | Fungal Cellulase | Fungal Lipase | Fungal Protease | Application | Reference |
|---|---|---|---|---|---|---|---|---|---|
| *Thermomyces lanuginosus* | *Thermophilic fungus* | *Thermomyces* | 40–50 | √ | × | √ | √ | Wastewater and pharmaceuticals waste treatment | [81,82] |
| *Sporotrichum thermophile* | *Thermophilic fungus* | *Sporotrichum* | 45–50 | × | √ | × | √ | Biomass degradation | [86,87] |
| *Myceliophthora thermophila* | *Thermophilic fungus* | *Myceliophthora* | 45–50 | × | √ | × | √ | Textile industries and bioremediation | [84] |
| *Thermomyces ibadanensis* | *Thermophilic fungus* | *Thermomyces* | 46–55 | × | × | √ | √ | Wastewater treatment | [85] |
| *Neurospora crassa* | *Mesophilic fungus* | *Neurospora* | 20–30 | × | √ | × | √ | As a model organism in an analysis of genetic recombination | [83] |
| *Aspergillus niger* | *Mesophilic fungus* | *Aspergillus* | 20–30 | √ | × | × | √ | Food industries | [88] |
| *Aspergillus flavus* | *Mesophilic fungus* | *Aspergillus* | 25–30 | √ | √ | × | √ | Textile, detergent, and paper industries | [88] |
| *Candida mogii* | *Psychrophilic fungus* | *Candida* | 5–10 | √ | × | √ | √ | Food industries | [89] |

Note: The tick (√) represents 'produce', and cross (×) represents 'not produce'.

## 3. Fungal Proteases

Proteases are obtained from plants, animals, and microorganisms. Plants and animals' proteases are more complex as compared to microorganisms. Fungal proteases are used as detergents for the removal of stains by hydrolyzing the peptide bond among the protein molecules. Different strains of fungus produce the proteases, including *Aspergillus niger* [90]. Proteases have advantages over other fungus enzymes as detergents. Their demand increased due to low compatible detergent due to washing performance as compared to other enzymes. They catalyzed the removal of protein strains by breaking the peptide bonds in all pH ranges, such as acidic, neutral, and alkaline [91]. Their use in industrial and households increased due to the high production rate, enzyme recovery from the respective media, environmental safety, the whiteness of clothes, and maintenance of fibers.

Among these enzymes, protease is mainly used in detergent formulations due to its immense use in different industries. Proteases are referred to as proteolytic enzymes, which are unique in nature due to their presence in all living organisms and their significant applications in cell growth and differentiation. The proteases may be divided into three main groups based on the pH range alkaline protease, acidic protease, and neutral protease. Acidic protease has an optimum pH between 2.0 to 5.0. The ideal pH condition of neutral protease is 7.0, and alkaline protease has an optimum pH of more than 7.0. The acidic protease mainly has a fungus origin. Alkaline proteases are most important due to their immense applications in the food and detergent industries [92].

Protease enzymes are classified into two types based on their mechanism of action: extracellular protease and intracellular protease. The extracellular enzyme catalyses the hydrolysis of large protein molecules into smaller molecules that are absorbed into the cells. The intracellular protease plays a significant role in the regulation of metabolism. The extracellular protease enzyme is mostly used at a commercial scale to degrade protein molecules in many industrial processes [93]. The structures of different proteases are presented in Figure 1.

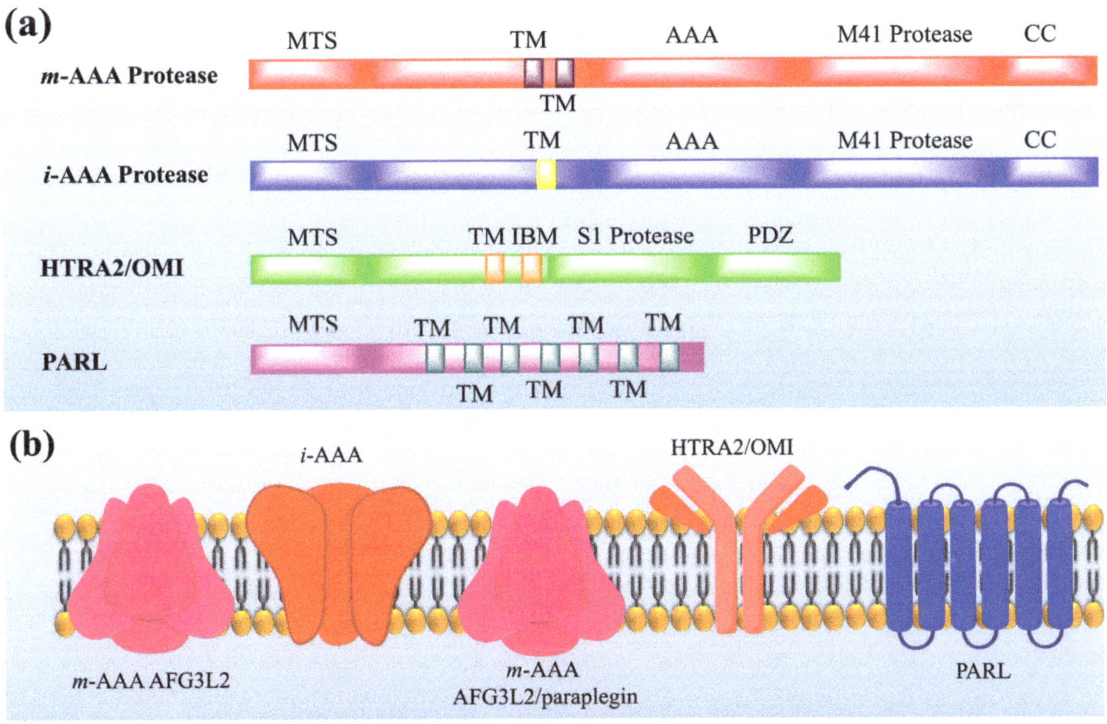


**Figure 1.** The representation of (**a**) structure and (**b**) topology of proteases. TM: Transmembrane domain, IBM: Inhibitor of apoptosis (IAP)-binding motif, MTS: Mitochondrial targeting sequencing, IMS: Intermembrane space, CC: Coiled-coil, AAA: Triple-A domain, M41: Protease metal-binding proteolytic domain, S1 protease: Trypsin-like protease domain. This figure is reproduced from Martinelli and Rugarli [94] after permission from Elsevier (License No. 5197711293142).

Alkaline proteases are used as detergents in the laundry industry to remove the dirty stains that are portentous in nature, in many food industries such as cheese making, baking meat industries, soaking processes, and many others in waste management from many food industries and also have some application in household activities [95]. Alkaline proteases act as an active ingredient in laundry detergent; they are considered a significant application of this enzyme. In the past, when detergent protease was produced at an industrial scale, it caused some allergic problems in workers from the dust of enzymes [96].

Among all groups or classes of proteases, the serine proteases are most effective in detergent action due to their stability and comparability in the existence of ingredients and other bleaching agents. It is expected that in 2020, the global market of industrial reach will climb to 7.5 billion and the growth rate will be around 8.2%. It is estimated that the growth rate will be maximum in detergents' enzymes. One type of such enzyme, protease, remained a well-known enzyme of 2015; its growth rate was 27.5%. It is expected to enhance its growth rate because of its flexibility in different sectors of pharmaceutical foods and detergents [97]. The information about novel protease enzymes isolated from different sources is presented in Table 3.

**Table 3.** The representation of novel fungal protease enzymes isolated from different sources.

| Enzyme Isolated | Enzyme Class | Active Site Residue (s) | Isolated Source | Reference |
|---|---|---|---|---|
| Clostripain, Streptopain | Cysteine proteases | Cysteine and histidine residues | *C. histolyticum*, *S. griseus* | [98,99] |
| Pepsins, proteases, rennet like proteases | Aspartic endoproteases | Two aspartate residues | *A. niger*, *M. miehei* | [100,101] |
| Chymotrypsins, subtilisins | Serine proteases | Serine residues | *B. sphaericus* | [102,103] |
| Collagenases, elastase | Metalloendoproteases | Metal ions | *C. histolyticum*, *P. aeruginosa* | [104,105] |
| Eqolisin protease | Glutamic proteases | Glutamate residues | *S. lignicola*, *A. niger* | [106,107] |
| Pepsins (A1), retropepsin (A2) | Acidic proteases | - | *A. niger*, *A. saitoi.* | [108,109] |
| Subtilisin, carlsberg | Alkaline proteases | - | *A. salinivibrio*, *C. aureus* | [109] |
| Neutrase, thermolysin | Neutral proteases | - | *Bacillus* sp. | [109] |

### 3.1. Origin of Fungal Proteases

The demand for fungal proteases has increased in recent decades. Different species of fungi secrete proteases, but the origin of proteases from basidiomycetes remains unclear [110]. *Aspergillus* species are considered an excellent source of proteases. Some other fungal species such as Penicillium and Rhizopus also produce proteases [111,112]. Proteases are also produced from the basidiomycetes such as *Schizophyllum commune*, *Armillariella mellea*, *Pleurotus ostreatus*, and *Phanerochaete chrysosporium* [113–115].

Mycelial secretion in saprophytic basidiomycetes led to discovery of proteases such as subtilases from *Serpula lacrymans*, *Pleurotus ostreatus*, and *Irpex lacteus* [116,117]. *Pleurotus* sp., such as *P. ostreatus* and *P. chrysosporium*, also produced the proteases involved in ligninolytic mechanism through fragmented degradation of laccase enzyme during fugal growth [118]. Hemolytic proteases secreted by *Pleurotus* sp. possess different activities in cellular processes. Secretion of hemolysin by *P. nebrodensis* shows apoptosis and antiproliferative activities and is involved in targeting cancerous cells [119]. These proteases tightly bind to the receptor proteins of HIV and inhibit them [120]. Some proteases isolated from *P. ostreatus* possess activities against different carcinomas [121].

### 3.2. Classification of Fungal Proteases

Fungal proteases are classified into different categories based on amino acids. A few are presented in the following sections.

#### 3.2.1. Protease Classes Based on Amino Acids

Proteases are categorized into the following classes based on amino acids' residue in their active site.

Serine Protease

Serine proteases are the most important class of proteases that contain the amino acid serine in their active site (see Figure 2). Serine residues in the active site make a catalytic triad with aspartate and histidine. This catalytic triad is conserved among all serine proteases. These residues are essential for their catalytic activity to cleave the substrate protein [122]. Serine proteases can be further divided based on substrate specificities such as elastase-like proteases, which possess the smaller S1 cleft than the chymotrypsin-like protease. They are hydrophobic in nature and possess the specificity for valine and glycine. These elastase-like proteases act on elastin by breaking them into smaller fragments, thus playing an essential role in forming connective tissues [3]. Fungal species such as *Aspergillus* sp., *Aspergillus oryzae, Aspergillus fischeri, Penicillium citrinum, Penicillium corylophilum. Penicillium waksmanii,* and *Neurospora Conidiobol* produce serine proteases [123,124].

Threonine Protease

Threonine proteases hold the threonine residues in their active site for their catalytic activity to cleave the substrate protein. These proteases show substrate specificity for bulky amino acids [109,125]. *Saccharomyces cerevisiae* are production sources of threonine peptidase with appreciable capacity for the production of threonine and potential for industrial application [123,126].

Cysteine Protease

Cysteine proteases hold the two amino residues in their active sites, such as cysteine and histidine (see Figure 2). Their catalytic activity can be maintained in the presence of reducing agents [109]. The substrate specificity of the cysteine proteases can be determined through the interaction between the side-chain amino acids of the particular substrate that can be accommodated into the S2 cleft, which is hydrophobic in nature and shows the specificity for the leucine and tyrosine. One of the best examples is the Cathepsin K that shows the specificity for kinins by cleaving the peptide bonds in the collagen tissue [98,127]. Fungi species such as *Aspergillus oryzae* produce cysteine proteases. Only a little information about the secretion of fungal cysteine proteases is known [128].

Aspartate Protease

Aspartate proteases hold the two aspartate residues in their active site. These aspartate residues are essential for their catalytic activity to cleave the substrate protein (see Figure 2). Aspartate proteases show specificity for aromatics such as phenylalanine, tyrosine, and tryptophan on both sides of the peptide bond [101]. Rennin-like aspartate proteases cause the cleavage of casein into the smaller peptides. Pepsins are isolated from *Aspergillus*, and rennet-like enzymes are isolated from *M. miehei* [100]. Fungi species such as Aspergillus, Penicillium, Rhizopus, and Neurospora produce aspartic proteases [100,129]. *E. parasitica* and *R. miehei* of fungi are used as production sources of aspartic peptidase with appreciable capacity for the production of acidic peptidase and potential for industrial application [129,130].

Glutamic Acid Protease

Glutamic acid protease holds the glutamate residues in their active site. These proteases also show substrate specificity for bulky amino acids. For instance, eqolisin can work at pH 2.0 when the casein is used as a substrate. Glutamic proteases were identified in the fungi *Scytalidium lignicola* and *Aspergillus niger* [106,107]. *Scytalidium lignicolum, Aspergillus niger, Cryphonectria parasitica, Talaromyces emersonii* and *Sclerotina sclerotiorum* secreted glutamic proteases [131–133].

Metalloprotease

Metalloproteases are the diverse classes of proteases containing metal ions in their active sites (see Figure 2). These metalloproteases are highly specific in their action [104]. Neutral proteases show the specificity for hydrophobic amino acids. The well-known metalloproteases such as collagenases and elastase are isolated from *C. histolyticum* and *P. aeruginosa* [105]. Fungi species such as Aspergillus, Penicillium, *Fusarium oxysporum*, and *A. fumigatus* produce metalloproteases [100,126,134].

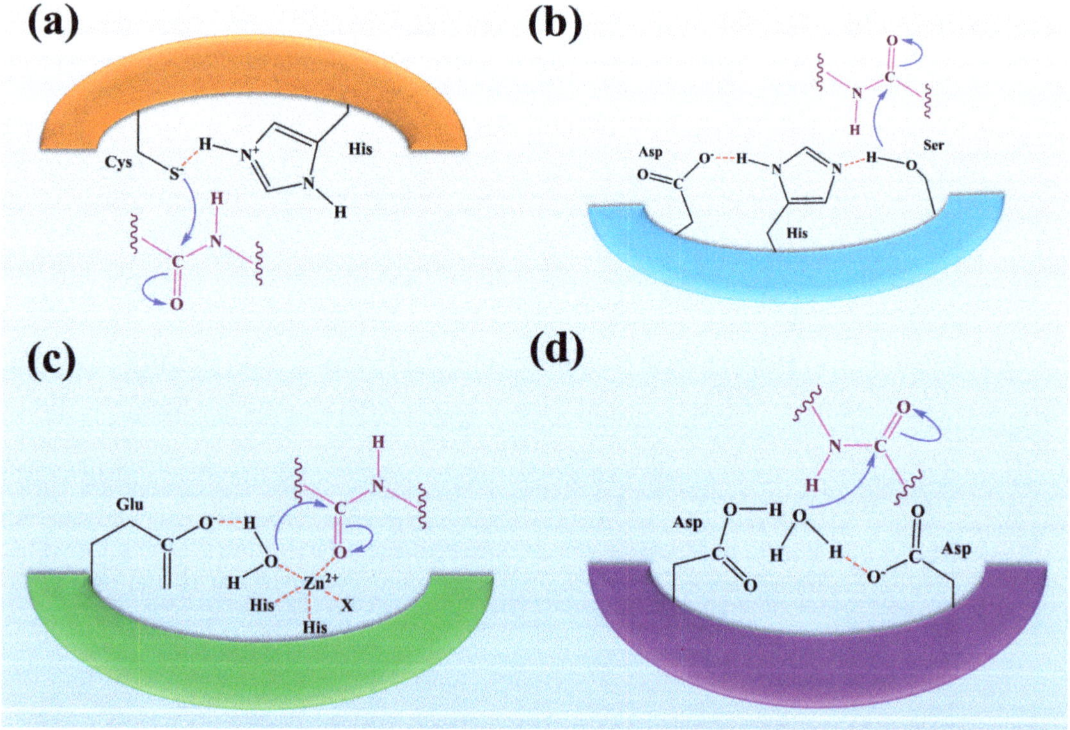


**Figure 2.** The representation of fungal protease mechanisms of (**a**) serine proteases (**b**) aspartyl proteases, (**c**) metalloproteases, and (**d**) cysteine proteases. It was reported that the eponymous residue is commonly formed as a pair with a proton withdrawing group in the active sites of cysteine and serine proteases to promote a nucleophilic attack on the peptide bond. In contrast, metalloproteases and aspartyl proteases activate water molecules as nucleophiles. Overall, it was observed that the process of peptide bond scission is the same for all classes of proteases. This figure is reproduced from Erez et al. [135] after permission from Springer Nature (License No. 5223461087907).

The mechanism of action of fungal proteases involves the formation of an intermediate acyl-enzyme that covalently links the enzyme to the N-terminal of the substrate. In the second step, the water molecule completes the hydrolysis by attacking the intermediate. It results in the release of the second-half product with the regeneration of free enzyme (see Figure 3) [136].

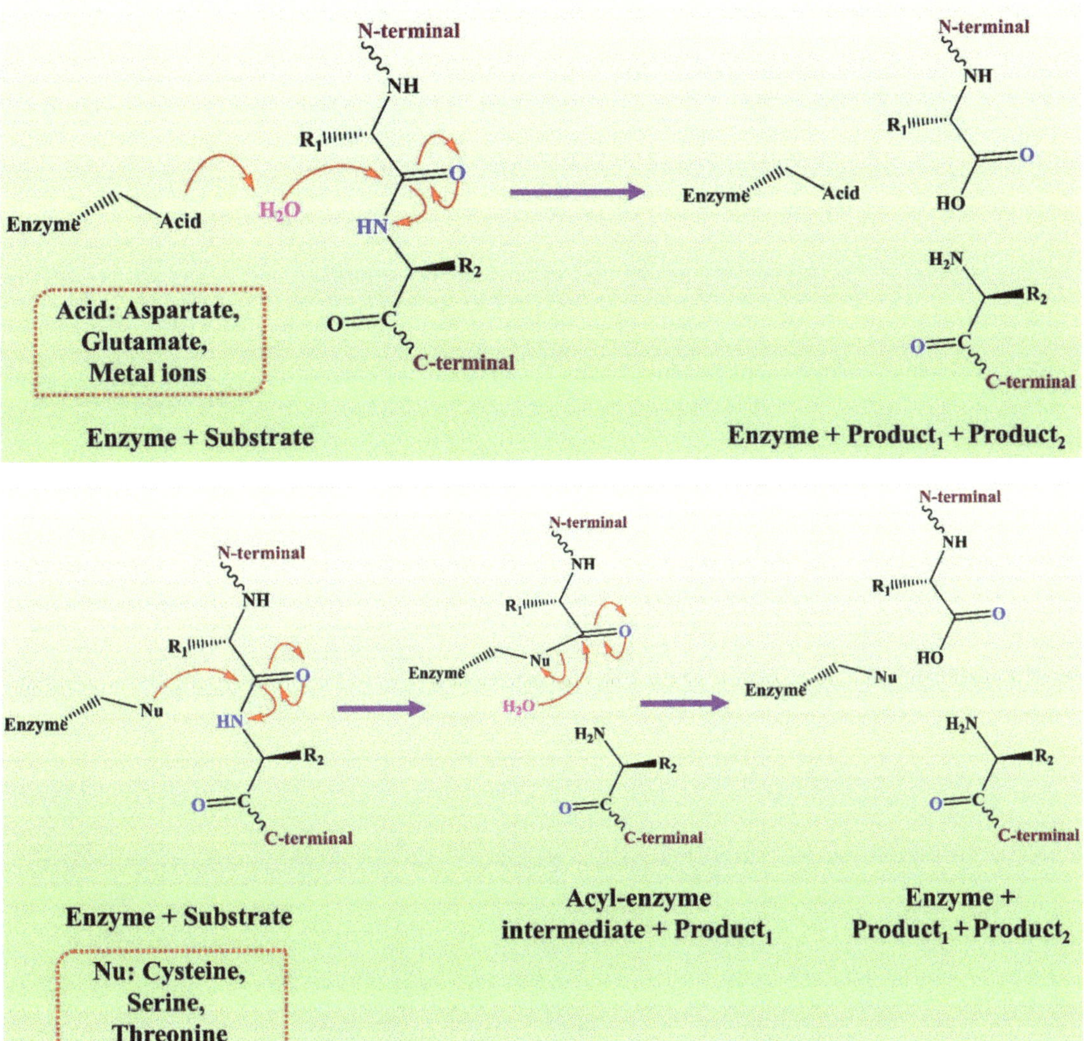


**Figure 3.** The representation of a comparison of the two hydrolytic mechanisms used for proteolysis. This figure is reproduced from Shafee [136] (Attribution NonCommercial 2.0 UK: England & Wales, CC BY-NY 2.0 UK).

## 4. Market Value of Fungal Proteases

Fungal proteases are one of the largest groups of industrial enzymes, and their global market is drastically increasing annually. Of the 60% of enzymes marketed worldwide, fugal proteases account for 20% [137,138]. The global market of fungal proteases is gaining more attention compared to other enzymes due to their high demand, catalytic properties, and low cost [139,140]. The global fungal enzymes market is highly consolidated owing to the presence of several key players operating in the global fungal enzymes market, namely Novozymes A/S, DSM, Chr. Hansen, DuPont, and BASF. The share market is divided into pharmaceuticals, bakery, beverages, sweet, and animal feed based on application [141]. Companies worldwide adopt evolutionary acquisition strategies by expanding the business network of proteases by considering the geographical demands, source type (especially microbial), and role in different industries [140].

*Global Proteases Market Segmentation*

The global protease market is segmented based on applications, forms, and types. The market is divided into renin, trypsin, pepsin, and others based on type. The renin segment leads the share market, and alkaline protease is the second segment that accounts for the largest market share due to usage in the food and dairy industry [142]. Based on form, the protease market has been divided into liquid and powder. The powder segments contribute to the largest market share due to the usage of many products that improve the half-life of food products [90]. The share market is divided into bakery and beverages, sweet, animal feed, and others based on application. The bakery segment contributes to the largest share of the market. It reflects the use of proteases in different industries for syntheses of food-based products [143].

## 5. Molecular Approaches, Cloning and Expression of Fungal Proteases

Recombinant DNA technology is used for fungal proteases to understand their gene expression at cellular and molecular levels [144]. Protein engineering of fungal proteases can be evaluated through directed mutagenesis for introducing the changes in the sequence of amino acid 3D structure [22]. Based on a particular sequence at the N-terminal, novel proteases were isolated from *P. ostreatus*. A new primer was designed for cloning and amplification of DNA sequence that showed regions with homology to the proteases isolated from *Neurospora crassa* [111]. *XPR2* gene encoding for an alkaline protease from *Yarrowia lipolytica* was cloned into *P. pastoris*, and its genomic sequence showed two glycosylation sites [145]. A serine protease gene (*Spr1*) encoding from an alkaline protease from *Monacrosporium megalosporum* showed homology to serine proteases PII and Azo1 isolated from the cuticle of nematode-trapping fungus and *Arthrobotrys oligospora* [146]. Gene cloning of *A. nidulans* gene (*prtA*) encoding the alkaline protease can be carried out in the presence of probing of the *A. nidulans* library through fragmented amplification of the *Aspergillus oryzae Alp*-encoding gene. The amino acid sequence in the *prtA* gene showed homology to the protease isolated for *A. oryzae* and *A. flavus* [147,148].

Recently, a novel neutral protease was produced from *Aspergillus oryzae Y1*. This protease was purified about 10-folds, and its recovery yield was about 45% [149]. This protease was purified as a potential candidate for industrial applications, especially in food industries [149,150]. The genetic recombination of the *A. oryzae* and *Aspergillus* led to the production of *A. oryzae* FG76, which secreted acid protease with 85% activity [28]. Novel acid protease was produced from *A. oryzae* FG76 from a culturing medium with 17-activity and showed high stability as compared to commercial enzymes [28]. Another novel protease was purified from basidiomycete fungus CTM10057 that acted as a potential candidate for industrial applications, especially in the detergent and laundry industries [150,151]. *Pseudomonas aeruginosa* CTM50182 secreted the AMPP protease with 70% activity that showed high stability as compared to commercial enzymes [28,152]. *Cryptococcus* sp. S-2 that produced the novel aspartic protease was purified in synthetic substrates similar to the purified pig pepsin A [153].

## 6. Role of CRISPR Technology in Fungal Enzymes

CRISPR/Cas9 technology has been successfully developed in fungi for genome editing such as *Trichoderma reesei*, *Aspergillus stains*, *A. fumigatus*, and *Ustilago maydis* (see Table 4) [29–31]. For example, genome editing of multiple genes was successfully achieved in *T. reesei* through CRISPR/Cas9 technology by following the co-transformation via in vitro synthesis of sgRNAs and donor DNA with the 200 bp homologous arms. This promising approach applies to filamentous fungi that degrade the wood lignocelluloses [154]. The CRISPR/Cas9 system also successfully developed in *A. fumigatus* following in vivo synthesis of sgRNA through the introduction of *A. fumigatus* U6 snRNA promoter, leading to 95–100% precise integration via the 35-bp homologous arms. This versatile genome editing approach is helpful for the investigation of mutated genes and clinical studies of *Aspergillus* species [155]. Genome editing through CRISPR/Cas9 technology is under validation for thermophilic fungi [156]. *Myceliophthora thermophile* is one of the thermophilic fungi involved in the degrading of biomass; therefore, it can be used for the production of thermostable enzymes as biofuels in different industries. Consequently, it is essential to reveal the genome editing of Myceliophthora through metabolic engineering important in reconstructing the lignocelluloses [157].

**Table 4.** The applications and efficiencies of different fungal enzymes isolated from fungal strains recently modified through CRISPR technology.

| Fungal Strain | Isolated Proteases | CRISPR System | Selective Marker | Promoter for sgRNA | Promoter for Cas9 | Delivery Method | Editing Method | Application | Efficiency (%) | Reference |
|---|---|---|---|---|---|---|---|---|---|---|
| *A. oryzae* | Aspartic acid protease | Cas-9 | *pyrG* | U6 | *amyB* | PMT | NHEJ | Genetic engineering, food industries | 10–30 | [158,159] |
| *M. thermophila* | Alkaline protease | Cas-9 | *bar* | U6 | *tef1* | AMT | HDR | pharmaceuticals biomass/waste treatment | 20–95 | [86,156] |
| *T. lanuginosus* | Serine alkaline protease | Cas-9 | *hph* | U6 | *act1* | AMT | NHEJ | Wastewater and pharmaceuti-cals waste treatment | 44–90 | [82,160] |
| *C. militaris* | Serine alkaline protease | Cas-9 | 5-FOA/blpR | U6 | *tef1* | AMT | NHEJ | Pharmacecurials | 82–89 | [161,162] |
| *F. graminearum* | Acid proteases | Cas-9 | *Fludioxonil* | U6 | *gpdA* | AMT | HDR | Food indsuries, pests conrol | 2–12 | [163,164] |
| *U. maydis* | Metalloproteases | Cas-9 | *Ip* | U6 | *otef* | PMT | NHEJ | Biofuels, pharmacuetcals | 50–90 | [165,166] |
| *N. crassa* | Serine proteases | Cas-9 | *bar* | SNR52 | *TrpC* | AMT | HDR | Genetic recombination | 60–80 | [83,167] |
| *G. lucidum* | Metalloproteases | Cas-9 | *ura3* | T7 | *gpdA* | PMT | NHEJ | Medicine | 28–33 | [168,169] |

Note: NHEJ = Non-homologous end joining, HDR = High-fidelity homology-directed repair, Cas = CRISPR-associated genes, gRNA = Guide RNA, sgRNA = Single guide RNA, AMT = Agrobacterium mediated transformation, and PMT = Protoplast transformation.

However, the off-target effects in fungi have been studied through CRISPR/Cas9 technology. Mutations were observed in *U. maydis* and *A. fumigatus* through whole-genome sequencing after CRISPR/Cas9 genome editing [170,171]. There is a need to elucidate the mechanism behind mutations/off-target events through CRISPR/Cas genome editing in fungi [172]. Some strategies have been developed in order to reduce off-target effects caused by CRISPR/Cas-9 genome editing in fungi. Based on genomic sequence, algorithms can be used for the selection of target sequences for predicting the off-target and on-targets sites [173]. In fungi, joining non-homologous regions can be carried out via a null mutant strain to reduce off-target effects [174]. High homologous recombination was observed through deletion of genes that particularly involved non-homologous end-joining. Therefore, mutants have been used in the form of hosts to process CRISPR genomic editing [175].

## 7. Recent Developments

In this modern era, biocatalysts are considered important progress in environmental protection. The production of an industrially important enzyme by using a cheap carbon source involves the continuous isolation of new strains [176]. Protease activity depends on many factors, including the pH of the medium, mechanical handling, and temperature. Genetic engineering plays an essential role in producing novel enzymes with unique properties and alteration of protease properties that can improve the industrial process [139]. Proteases can overcome the traditional methods of biomass conversion into biofuels. Enzymes are the best alternatives for methyl ter-butyl ether (MTBE) because MTBE is dangerous to human health. Proteases have the potential to use unrefined feedstock without the separation of fatty acids that are mainly present in unrefined feedstock [90]. The complete genome of many industrial important *Aspergillus* species that produce various enzymes has been sequenced [177]. Despite limitations at transcriptional and translation sites, with the help of enzyme engineering, the *Aspergillus* species has the potential to produce a high level of heterologous proteins [15].

## 8. Role of Fungal Proteases in Living Organisms

Mitoproteases process the proteins imported into mitochondria and help degrade the damaged and misfolded proteins to improve the quality of proteins [178]. However, with the help of recent technologies, a mutation in genes that encodes mitoproteases and results in different diseases can be found, providing additional information regarding the biological role of these proteases in other processes [179].

### 8.1. Mitoproteases Function in Protein Processing and Activation

Mitoproteases mainly participate in the proteins trafficking into mitochondria from cytosol with the help of chaperones and help in the activation and folding of proteins imported into mitochondria [180]. Proteases are synthesized in the cytosol and move the processed proteins into mitochondria, which helps in the correct entry of proteins into mitochondria by removing mitochondrial import signals (see Figure 4) [178,181].

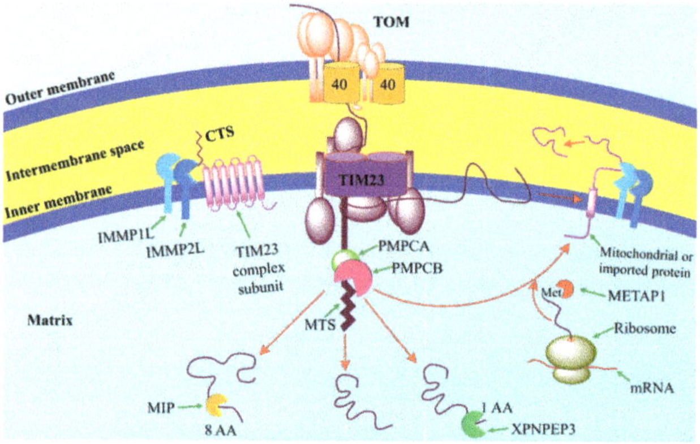


**Figure 4.** The mechanism of action of mitoprotease peptidases. Mitoproteases are the diverse group of enzymes that transport the biologically active proteins from cytosol to the inner mitochondrion to carry out the correct signals necessary for cellular processes [182]. Mitochondrial processing peptidase (MPP) is the part of the mitochondrial matrix that comprises two subunits such as PMPCA and PMPCB (protease mitochondrial processing peptidase subunit a and b) [183]. Mitochondrial intermediate peptidase (MIP) promotes the cleavage of octapeptide and X-Pro aminopeptidase 3 (XPNPEP3) that removed the amino acids from the amino-terminal of the MPP. Some proteins also pass through further photolytic cleavage via IMMP1L and IMMP2L. IMMPs attack the carboxy-terminal sequence (CTS) to promote cellular assembly [184]. Met aminopeptidase 1D (METAP1D) attacks on amino terminals of the initial Met of some of the polypeptides in order to get functionally active proteins [180]. This figure is reproduced from Quiros et al. [181] after permission from Springer Nature (License No. 5197691283486).

### 8.2. Cysteine Proteases in Atherosclerosis

Analysis of human tissues has determined that overexpression of cathepsin activity can lead to various inflammatory conditions, including rheumatoid arthritis and atherosclerosis [185]. During atherosclerosis, cathepsins' overexpression can be found in almost every cell that contains plaque tissues, mainly in macrophages and endothelial cells (see Figure 5) [186].

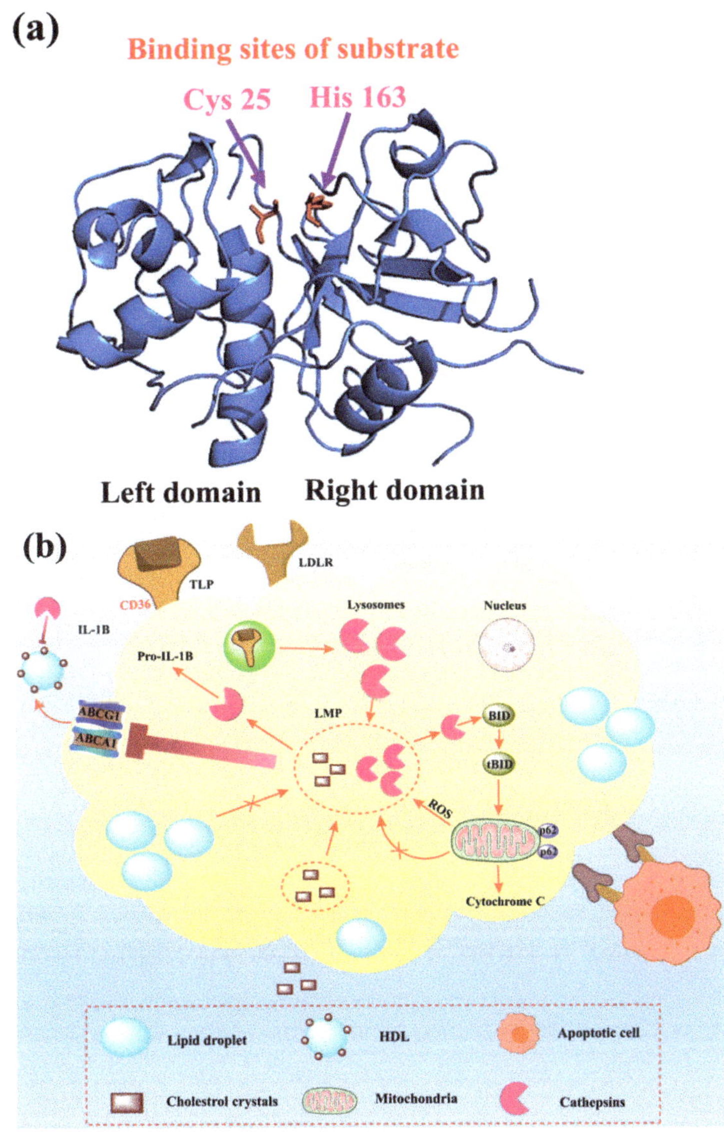


**Figure 5.** The representative structure of (**a**) fold of mature cathepsin L and (**b**) role of cathepsins in regulating inflammation in atherosclerosis. Cathepsins are involved in the regulation of inflammation in atherosclerosis by clearing the apoptotic cell [187]. Cathepsin function in atherosclerosis is catalyzed by suppressing the oily foam formation that leads to atherosclerosis in arteries due to excess cholesterol that increases the chances of reactive oxygen species and hence strong activity by clearance of the apoptotic cell. Cathepsins move into the cytosol by promoting apoptosis by degrading excess high-density lipoproteins (HDL) by processing the IL-1$\beta$ [188,189]. Lysosomes bind to Low-density lipoprotein receptors (LDLR), causing them to degrade fatty acids and lipids through acid hydrolases. Impairment in cathepsin's catalytic functioning leads to the deposition of excess cholesterol that causes stress and free radicals due to reactive oxygen species [185]. tBID: Truncated BID, LMP: Lysosomal membrane permeabilization, TLR: Toll-like receptor, ABCA1: ATP-binding cassette transporter, ABCG1: ATP-binding cassette sub-family G member 1, BID: BH3-interacting domain death. This figure is reproduced from Weiss-Sadan et al. [185] (Attribution 4.0 International, CC BY 4.0).

### 8.3. Role of Fungal Proteases in the Pathogenesis of Chronic Rhinosinusitis with Nasal Polyps

*Staphylococcus aureus*, fungi, cockroaches, and some airborne mites have serine proteases [190]. Interaction of proteases derived through allergens occurs with epithelial cells through three pathways: direct effects of protease on junctional proteins, epithelial activation through toll-like receptor 4, and reaction with protease-activated receptors present on the cell surface (see Figure 6) [191].

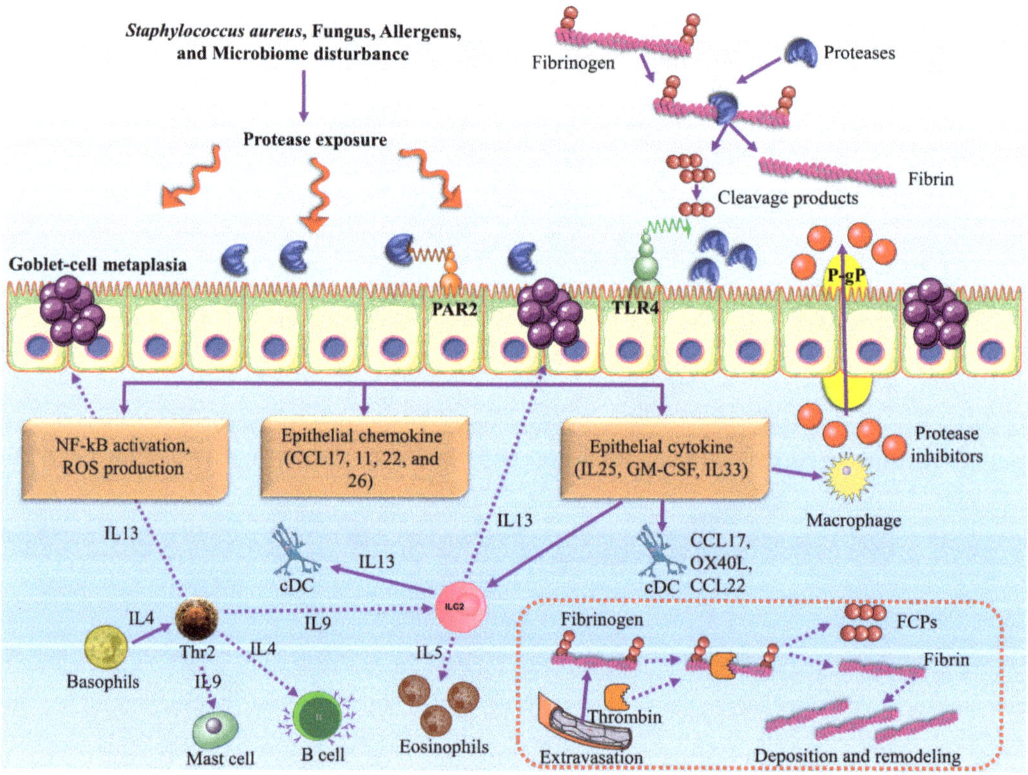


**Figure 6.** The mechanism of action of proteases to inhibit the allergen and other microbe attacks. Proteases are released in response to allergens due to a microbial attack that causes severe damage to epithelial cells by disrupting the junctional proteins [192]. Allergen proteases bind to protease-activated receptor 2 (PAR2) by cleaving the fibrinogen into the fibrin cleavage products (FCPs) that activate the toll-like receptor 4 (TLR4) and ILC2s. These ILC2s lead to activation of NF-kB and excess production of reactive oxygen species (ROS). It activates epithelial cells to release the pro-Th2 cell chemokines and cytokines that activated the instruct immature dendritic cells (iDCs). P-glycoproteins (P-gp) helps in removing the protease inhibitors in epithelial cells that cause the suppression of allergen proteases [190,193]. This figure is reproduced from Wu et al. [190] (Attribution 4.0 International, CC BY 4.0).

### 8.4. Fungal Protease as a New Therapeutic Strategy for Colorectal Cancer

Colorectal carcinoma is the most common cancer and has a high prevalence rate in developed countries due to issues related to diet. In the past decades, despite several drugs being used to treat colorectal cancer, the disease under the metastatic stages remains lethal. New enzymes therapy is urgently needed to treat the defective proteins during cell cycle stages. Proteases have been used for the suppression of tumour formation at the cellular level. A ubiquitin-proteasome system is a protease-based therapy for colorectal cancer [194].

The ubiquitin-proteasome system is the multiplex protease enzyme system essential for the survival of normal cells that causes the degradation of misfolded proteins. The ubiquitin-proteasome system also plays a critical role in controlling the turnover of the aberrant proteins or short-lived regulatory proteins involved in essential cellular processes, cell death, and cell signalling pathways, apoptosis, metastasis, and cell proliferation [195]. Protein substrates are mainly linked to the chain of the protein ubiquitin in the presence of ubiquitin-activating enzymes or E1, ubiquitin-conjugating enzymes or E2, and ubiquitin ligases or E3 to form the ubiquitin linkage. Polymerized ubiquitin chain provides the signal to transport the target proteins to the proteasome, proteolysis of the aberrant or target proteins broken down [196].

### 8.5. Protease as a New Therapeutic Strategy for Coeliac Disease

Celiac disease is one of the autoimmune disorders of the digestive system that affects the small intestine due to improper digestion of gluten-based toxic peptides such as gliadin. Recently, different protease-based therapies with a healthy diet plan and a gluten-free diet have effectively treated celiac disease [197]. Microbial peptidases also have been used to detoxify gluten by digesting them at the glutamine and proline sites. This approach cannot be practical due to the inactivation of most enzymes in the stomach wall through acidic pH [198].

Protease therapy is the best approach for protecting patients who have celiac disease from toxic or unwanted peptides. A recent study investigated the therapeutic use of the protease inhibitor elafin for the treatment of celiac disease [199]. Patients who have celiac disease expressed a lower amount of protease inhibitor elafin. Researchers performed experiments on mice models and found that delivery of elafin could reduce the inflammation in the small intestine. Elafin also protected the stomach wall by inhibiting the transformation of the gliadin peptides into the less toxic form [200].

### 8.6. Protease as a New Therapeutic Strategy for Neurological Disorders

Different disorders caused by prions are currently untreatable and universally fatal. Prions are the misfolded proteins that cause different fatal diseases in animals such as Gerstmann–Straussler–Scheinker syndrome and Creutzfeldt–Jakob disease [201]. Gerstmann–Straussler–Scheinker syndrome is a neurodegenerative disorder characterized by mutations in the PRNP gene due to prions that affect brain tissues. Creutzfeldt–Jakob disease is also a neurodegenerative disorder characterized by high prions in the nervous tissues that ultimately cause irreversible damage to the nerves. Prions disrupt brain functioning by changes in the neural tissues [201,202].

Enzyme-based therapy using the proteases is helpful for degrading the misfolded proteins that cause alterations in brain tissues. Recently, the *Bacillus* strain was successfully isolated and purified and was helpful in degrading the scrapie $PrP^{Sc}$. *B. pumilus* KS12 produces the keratinase that plays a vital role in degrading the Sup35NM. Genetically engineered protease (MSK 103) is also effective against the $PrP^{Sc}$. However, novel proteases should be discovered that can be used for targeting the prion diseases in animals [203,204].

## 9. Industrial Applications

Fungul protease has many applications in different sectors such as medicines and food industries, and it acts as a magic tool in biological research. It has diverse applications at industrial scales. Due to specificity in its mechanism of action, the protease is primarily involved in the formulations of medicines [205].

### 9.1. Food Industry

Proteases are used in food industries to manufacture wine, bread, cheese, and butter. Proteases in the dairy industry mainly act on the peptide bonds in cheese. Acid protease is used for the production of ethanol, which can be used as a source of nutrients for yeast. Acid protease acts on peptide bonds among amino acids, thus increasing the formation of ethanol from protease [90]. It resulted in the formation of casein, a protein found in milk. Some fungal strains such as Mucor michei are used to produce acidic proteases that can be used in cheese formation, thus replacing renin. Alkaline proteases are used in soya sauce, and enzymatic reactions that lead to the production of a high-quality protein called hydrolysates have different applications in nutrients and dairy-based products (see Figure 7) [15]. Fungal proteases that are used in the fruit juice and beverage industry mainly act on the peptide bonds among the proteins by degrading the complex compounds. It formed a turbidity complex from protein that ultimately improved gelatin hydrolysis, whey protein hydrolysis, gelatin hydrolysis, and meat mineralization [109]. Fungal-based proteases are utilized in food industries due to high specificity and catalytic activities as compared to the other fungal-based enzymes [176].

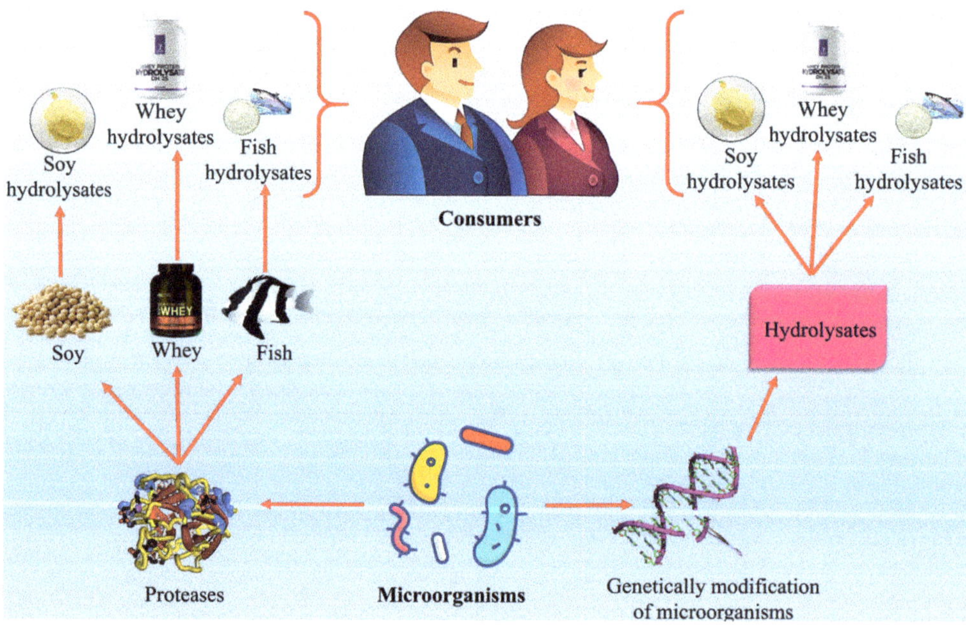


**Figure 7.** The application of fungal proteases in the food industry.

### 9.2. Waste Management and Bioremediation

Poultry industry waste and animal feathers could cause the deaths of livestock and other animals. Alkaline serine proteases are used to degrade keratin protein in the poultry industry [206]. Keratin is a structural protein with a long chain of amino acids linked through hydrogen and hydrophobic interaction [207]. Keratin released from the poultry industry releases waste and hazardous chemicals that can be controlled; alkaline proteases mainly act on the hydrogen bonds and hydrophobic linkages among the keratin [208]. It helps control the waste and hence plays a vital role in controlling environmental pollution. Different strains of fungi are used for degrading the keratin, namely *Pseudomonas* sp., *Aspergillus oryzae*, *Fusarium oxysporum*, *Trichophyton* sp., *Aspergillus terreus*, and *Scopulariopsis* sp., [209].

Fungal enzymes also play an essential role in bioremediation to control environmental pollution. Different strains of fungi such as *Hanerochaete chysosporium*, and *Pleurotus* sp. are used for bioremediation to maintain the ecological pollutant, thus helping in reducing the pollution [210]. Other applications of alkaline protease are found in degrading toxic proteins and chemicals released through the household. This alkaline protease acts on the toxic proteins and helps them to degrade into smaller fragments that can be released into the environment [211]. Thus, alkaline protease helps clean the environment by controlling the pollution from different sources of land and poultry through fungal-specific strains.

### 9.3. Medical and Pharmaceutical Industry

In medicine, some fugal proteases are isolated from *Paecilomyces marquandii* and *Doratomyces microsporus* as source of potent keratinases that might be utilized in the elimination of keratin in acne or psoriasis, and degradation of keratinized skin, depilation, preparation of vaccine for dermatophytosis therapy, and in the increase of fungal drug delivery [212,213]. These keratinases can remove the scar and regenerate the epithelia, accelerate healing processes, and might act also in the medicine of trauma. A new semi-alkaline protease with high collagenolytic activity was produced by *Aspergillus niger LCF9*. The enzyme hydrolyzed various collagen types and liberated low molecular weight peptides of potential therapeutic use [212].

Thus, the demand for fungal proteases increases in the pharmaceutical and medical fields. Fungal aspartic proteases from *A. niger* are used for the preparation of digestive powders [22]. L-glutaminase from *Aspergillus fumigatus* and *Penicillium allii* is used for the treatment of leukaemia by inhibiting the growth of cancerous cells [214]. Alkaline protease is also used as a digestible in the pharmaceutical industry to treat cystic fibrosis [215,216]. These proteases control the abnormal concentrations of bile salts and pancreatic selections [109]. Alkaline protease is used to treat the cancerous cells as these enzymes particularly target the fibrin and promote their degradation, thus they are used as an anticancer agent in thrombolytic therapy [90]. Matrix metalloproteinase catalyzes the breakdown of the peptide bond in the clot formation molecule by adding a water molecule. These proteins showed the expression against tumor cells [217].

### 9.4. Proteases in Silver Recovery

X-ray and photographic films are generally composed of 2% silver embedded in the gelatin layer. This silver can be recovered through different approaches. Conventionally used methods retrieved the silver recovery through the burning of the X-ray films that cause environmental issues. It also increased the risk of respiratory infections in nearby areas due to the explosion of carbon monoxide. Therefore, conventionally used methods for silver recovery are not effective, and an urgently advanced approach is needed to reduce the environmental and safety issues [218]. Alkaline proteases have been effectively used for silver recovery with no environmental hazards. Proteases from the *Bacillus* species can catalyze the hydrolysis of the gelatin layer with maximum retrieval of the silver by ensuring no damage from polyester base film that can be recycled [219].

### 9.5. Proteases in Silk-Degumming

Silk fibers are covered by sericin, the fibrous protein that surrounds them in the form of rough texture. Different conventional methods have been used for the removal of sericin in order to organize the structure of silk fibers [220]. These methods are expensive due to the high-cost machinery required to remove the sericin. Therefore, conventionally used methods for silk-degumming are not reliable. Proteases have been used for degumming silk in order to remove the sericin with the maintenance of fiber structure. Alkaline protease obtained from the *Bacillus* sp., RGR-14 is used for degumming silk prior to the conventional methods [221].

## 10. Novel Protease Inhibitors

Different inhibitors have been isolated and purified in searching for novel proteasse inhibitors—the novel protease inhibitor, PPF-BBI, that is separated from the skin of *Pelophylax fukienensis*. PPF-BBI as a novel inhibitor showed antimicrobial activity against *E. coli* and *C. albicans* [222]. Another study investigated the role of the novel serine protease inhibitor GP205 in the NS3/4A protein. The pharmacological analysis of the GP205 inhibitor showed the biological activities in targeting the HCV virus, and ultimately, this novel inhibitor could be a possible treatment for the Hepatitis C virus [223]. Different compounds quercetin 3-β-d-glucoside, helichrysetin, and herbacetin have been screened by flavonoid library. The biochemical analysis of these compounds showed biological activities in suppressing the MERS-COV 3Cl protease. These chemical compounds could be used as a possible treatment in targeting coronaviruses [224].

There are several protease inhibitors that showed broad spectrum inhibitory specificity and antifungal activity by tight binding [225] (see Table 5). These inhibitors are survivin (cysteine inhibitor) [226,227], diosgenin (metalloprotease inhibitor) [228,229], serpin (serine inhibitor) [230], saccharopepsin (aspartic acid inhibitor) (IA3) [231], streptomyces (metallopeptidase inhibitor; SMI) [232], RFLP-1 (Rhamnus frangula inhibitor) proteases [233]. These protease inhibitors have gained special importance in biomedicine.

**Table 5.** Fungal protease inhibitors for biomedical applications.

| Type of Fungal Protease Inhibitors | Proteases Inhibited | Family | Activity | Inhibitory Mechanism | Application | Reference |
|---|---|---|---|---|---|---|
| Survivin (Cysteine inhibitor) | Caspases- specific cysteine proteases | C14 | Antifungal | Tight binding | Role as a mitotic regulator of cell division and as an inhibitor of caspase activation in the process of apoptosis. | [226,227] |
| Diosgenin (Metalloprotease inhibitor) | Metalloproteases | M15 | Antifungal | Tight binding | These inhibit the secreted metallopeptidase relevant in brain invasion by cryptococcal cells, causing meningoencephalitis. | [228,229] |
| Serpin (Serine inhibitor) | Serine Proteases | C1 and C14 | Antifungal | Trapping traps the serine protease in a covalent complex | Fungal serpins as a therapeutic benefit toward several inflammation-related complications. | [230] |
| Saccharo (Pepsin, aspartic acid inhibitor) (IA3) | Aspartic acid proteases | C1 | Antifungal | Tight binding | Role as an inhibitor in the process of apoptosis and cancer. | [231] |
| Streptomyces (Metallopeptidase inhibitor) | Metalloproteases | C14 | Antifungal | Tight binding | Role as an inhibitor in tumor invasion and metastasis (the most validated target for cancer). | [228,232] |
| RflP-1 (Rhamnus frangula inhibitor proteases) | Serine protease | C14 | Antifungal | Trapping | Role in free radical scavenging activities. | [233] |

### 10.1. *Protease Inhibitors in Clinical Trials*

#### 10.1.1. HIV

Protease inhibitors have been used as therapeutic agents to target the human immunodeficiency virus that usually attacks immune cells. FDA has approved many protease inhibitors for HIV. These inhibitors included ritonavir, amprenavir, and indinavir. These inhibitors bind to HIV and inhibit viral replication [234,235]. Some protease inhibitors are currently under clinical trials; RO033-4649 (Roche) entered the phase I stage and TMC-114

(Tibotec) is under clinical phase III. These therapeutic agents have reduced the mortality rate and are helpful for treatment [236–238].

#### 10.1.2. HCV

Hepatitis C virus has become the most common cause of liver cirrhosis. Many protease inhibitors for HCV are currently in the clinical trials stage. These inhibitors include SCH 6 (Shering), phase I, and VX-950 (Vertex), which has entered the clinical trial phase II [239].

#### 10.1.3. Cancer

Protease inhibitors have been designed for the treatment of cancer. They are currently used as therapeutic agents to control the different mutations. Many protease inhibitors are currently in clinical trials. These inhibitors are COL-3 (Collagenex), which has entered the phase II stage and AG3340 (Agouron), which is in phase III [240,241].

### *10.2. Mechanisms of Action for Protease Inhibitors*

Inhibitors are those molecules that inhibit the activity of the proteases by blocking them. It can be explained via two different mechanisms (see Figure 8). In the case of a reversible reaction called tight binding interaction between protease and inhibitor, the inhibitor binds to the protease's active site. It results in forming a protease-inhibitor complex that can be broken down into the modified and unmodified form of the inhibitor. In the case of irreversible reaction that is also called trapping interaction between protease and inhibitor, the inhibitor binds to the protease's active site. This results in the cleavage of the peptide bond of the inhibitor and causes the conformational change of the inhibitor [242,243].

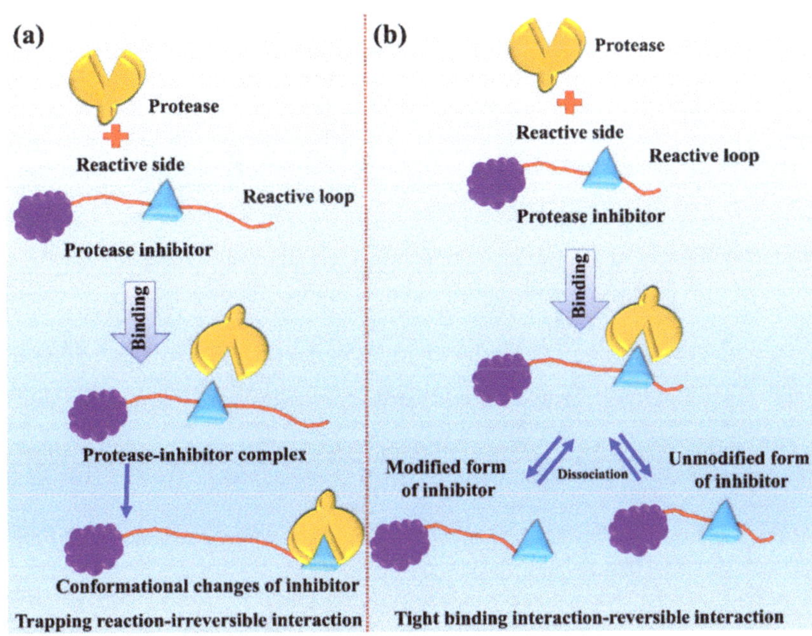


**Figure 8.** Mechanism of action of fungal protease inhibitors. (**a**) Trapping reaction-irreversible interaction and (**b**) Tight binding interaction-reversible interaction. This figure is reproduced from Rudzinska et al. [216] (Attribution-NonCommercial 3.0 Unported, CC BY-NY 3.0).

### *10.3. Discovery of Protease Inhibitors against COVID-19*

Coronavirus Main protease (Mpro) inhibitors receive great attention in targeting COVID-19 because of their role in processing the replicase during the post-translational processes. Their active site comprises two catalytic regions, C145 and H41 [244]. PLpro has been involved in bioprocessing the viral polypeptides into functional proteins. Their active site is composed of three catalytic regions including C111, H272, D286 [244]. These two proteases play a vital role in transcription by processing the two polypeptide proteins, pp1a and pp1ab, of the *Coronaviridae* genome. Therefore, these proteases can be used for antiviral drugs and discovery. These inhibitors possess poor absorption, excretion problems, and toxicological effects. Hence, the need to discover protease inhibitors remains a challenge [245].

In the current scenario, approaches for fragment-based drug design can help design protease inhibitor candidates that follow the screening and identification of those compounds with low molecular weight, advanced computational methods for the identification and sequencing of good fragments from the peptidomimetic compounds, and optimization of the [246] potential coronavirus protease inhibitor candidate against the COVID-19 [247,248].

The potential roles of Mpro and PLpro inhibitors for the treatment of SARS-CoV-2 are under clinical trials. For designing possible protease inhibitors for COVID-19, only those inhibitors that can inhibit the early viral replication by binding with the original substrate can reach the local target sites and provide high bioavailability. These inhibitors must be tested for antiviral activity through cellular and molecular pathways [249].

### 10.4. NP-Delivery Systems for Discovery of Protease Inhibitors

Most of the protease inhibitors have been tested for clinical trials for the treatment of cancers. These inhibitors included the prinomastat, tanomasta, and neovastat. These inhibitors have developed poor efficacy for clinical outcomes. Conventional chemotherapies for cancer treatment are not reliable due to poor diagnosis of cancerous cells at an early stage of cancer [216,250]. However, present challenges include designing protease inhibitors with a formulation of different types of nanoparticles for the treatment of cancers. Tumor cells also show chemoresistance to the conventionally developed chemotherapeutic agents. Protease inhibitors are promising candidates combined with NP-based delivery systems that can overcome chemoresistance [216].

Through advances in nanotechnology, the protease inhibitors carfilzomib and bortezomib have been designed with NP-delivery systems such as gold, PEGlycated, silica, and liposomes. These combinations of protease inhibitors demonstrate a high efficacy rate, increased circulation to the target cells, biocompatibility, and decreased systemic toxicity for breast, colon, and lung cancer [251,252]. This nanotechnology-based approach could help reduce the side effects of drugs, protect the normal tissues, and improve patients' quality. The detailed information about protease inhibitors with mechanism is presented in Table 6.

**Table 6.** Representation of protease inhibitors under clinical trials with the mechanism of action and therapeutic uses.

| Name of Inhibitor | Targeted Enzyme | Target Disease | Clinical Trials Stage | Reference |
|---|---|---|---|---|
| RO033-4649 | Therapeutic agents have reduced the rate of mortality and are helpful for treatment | AIDS | Under the clinical phase I | [238] |
| VX-950 | Become the most common cause of liver cirrhosis | HCV | Under the clinical trial phase II | [239] |
| COL-3 | Therapeutic agents to control the different mutations in colon cancer | Colon cancer | Entered the phase II stage | [240] |
| AG3340 | Therapeutic agents to control the different mutations in lung cancer | Lung cancer | Entered the phase II stage | [241] |

**Table 6.** *Cont.*

| Name of Inhibitor | Targeted Enzyme | Target Disease | Clinical Trials Stage | Reference |
|---|---|---|---|---|
| TMC-114 | Therapeutic agents have reduced the rate of mortality and are helpful for the treatment | AIDS | Under clinical phase III | [237] |
| Indinavir | These inhibitors bind to the HIV and inhibit the viral replication | HIV | FDA has approved this inhibitor | [234,235] |
| Ritonavir | These inhibitors bind to the HIV and inhibit the viral replication | HIV | FDA has approved this inhibitor | [234,235] |
| GP205 | GP205 inhibitor showed the biological activities in targeting the HCV virus, and ultimately, this novel inhibitor could be for possible treatment for Hepatitis C virus | HCV | GP205 inhibitor showed the biological activities in targeting the HCV virus, and ultimately, this novel inhibitor could be for possible treatment for Hepatitis C virus | [223] |
| Quercetin 3-β-d-glucoside and helichrysetin | The biochemical analysis of these compounds showed biological activities in the suppression of the MERS-COV 3Cl protease | Coronaviruses | These chemical compounds could be used as a possible treatment in targeting the coronaviruses | [224] |
| Mpro inhibitors | These inhibitors play a role in processing the replicase during the post-translational process's viruses of the *Coronaviridae* family | Coronaviruses | These proteases can be used for antiviral drug and discovery | [244–246] |
| PLpro inhibitors | These inhibitors play a vital role in transcription by the processing of the two polyproteins, pp1a and pp1ab | Coronaviruses | These proteases also can be used for the discovery of novel protease inhibitors | [244–246] |
| NP-delivery systems-based carfilzomib and bortezomib | This nanotechnology-based approach could help reduce the side effects of drugs | These protease inhibitors can be designed with combinations with gold, PEGlycated, silica, liposomes and demonstrate the high efficacy rate | Future discovery of novel inhibitors based on modified NP-systems protecting the normal tissues and improving the quality of patients | [252] |

## 11. Current Challenges and Future Perspective

Fungal proteases can be used as a target for therapeutic interventions because they are involved in many pathologic processes. Fungal proteases are challenging to target due to small information about substrate specificity and natural inhibitors [20]. The development of advanced techniques such as activity-based probes allows monitoring protease activity in living cells and helps understanding of protease function [253]. Fungi produce a wide variety of proteolytic enzymes used in biotechnology, food, detergents, and soil bioremediation.

The complete multiple genome sequencing has enabled us to identify the different classes of proteases, but there are still many gaps in understanding the role of many proteases [254]. Some significant challenges are addressed by the protease group in the development and usage of technologies at GNF in understanding protease biology to develop protease-based targeted therapeutics [255]. The GNF protease platform enables the use of comprehensive technologies that help identify native and optimal substrates [256]. Due to industrial development, environmental pollution has become a major concern, and chemically thermostable protease has taken over the conventional protease [257]. Toxic chemicals, including sodium sulphide, salt, lime, and solvents, are released from these industries and can pollute the environment [210,258]. Chemical industries are the first target to eliminate environmental pollution, and pre-tanning procedures produce a higher amount of pollution than post-tanning procedures [259].

The industrial and therapeutic use of proteases has grown in the past two decades. The expansion in protease markets has occurred due to unique protein engineering techniques [260]. For clinical applications, the success of apoptotic caspase activation with small, engineered protease can control the activity of the human protease. Due to proteolytic activities, tumor-imaging and drug-targeting a specific site can be enhanced in disease tissues with the help of proteases. Due to advancements in screening techniques, crystallography, and synthetic biology, we can hope that in the future, proteases will gain dramatic success [257,261].

## 12. Conclusions

Enzymes from fungi are used in a wide range of industries, including the pharmaceutical and agricultural sectors. Using enzymes in many industrial applications is a more cost-effective and ecologically friendly alternative. It is common practice in the biotechnology industry to employ fungi-derived proteolytic enzymes in the food and leather industries, as well as in ecological bioremediation processes, and to manufacture medicinal peptides. All living creatures rely on fungi and their proteases for a wide range of physiological, metabolic, and regulatory functions. Since fungal proteases are implicated in a wide range of diseases, they might be useful therapeutic targets. For fungal proteases, the lack of knowledge on substrate specificity and endogenous and natural inhibitors makes it challenging to build selective inhibitors. Transcriptional control of fungal extracellular protease expression may be a good candidate for genetic modification. CRISPR/Cas9 technology is anticipated to play a bigger role in genetic modification studies for filamentous fungus in the future as the technology continues to advance. Tools for functional genomics in filamentous fungal species will make it easier to implement fast advancements in CRISPR gene-editing technology.

**Author Contributions:** Conceptualization, A.H. and M.N.; software, A.H., M.B.K.T., D.A. and S.M. (Saba Manzoor); investigation, A.H.; resources, B.X.; data curation, M.N. and A.H.; writing—original draft preparation, A.H., S.M. (Saba Manzoor), M.A. and M.N.; writing—review and editing, M.-U.-H.A., N.E., A.H. and S.M. (Sajida Mushtaq); visualization, A.H., M.N., S.M. (Saba Manzoor) and B.X.; supervision, B.X. All authors have read and agreed to the published version of the manuscript.

**Funding:** This research received no external funding.

**Institutional Review Board Statement:** Not applicable.

**Informed Consent Statement:** Not applicable.

**Data Availability Statement:** Not applicable.

**Acknowledgments:** We extend our gratitude to the A. Q. Research Group, Pakistan for reviewing and editing the manuscript.

**Conflicts of Interest:** The authors declare no conflict of interest.

## References

1. Monod, M.; Capoccia, S.; Léchenne, B.; Zaugg, C.; Holdom, M.; Jousson, O. Secreted proteases from pathogenic fungi. *Int. J. Med. Microbiol.* **2002**, *292*, 405–419. [CrossRef] [PubMed]
2. Singh, R.; Mittal, A.; Kumar, M.; Mehta, P.K. Microbial proteases in commercial applications. *J. Pharm. Chem. Biol. Sci.* **2016**, *4*, 365–374.
3. Rao, M.B.; Tanksale, A.M.; Ghatge, M.S.; Deshpande, V.V. Molecular and biotechnological aspects of microbial proteases. *Microbiol. Mol. Biol. Rev.* **1998**, *62*, 597–635. [CrossRef] [PubMed]
4. Singh, N.; Gaur, S. GRAS Fungi: A New Horizon in Safer Food Product. In *Fungi in Sustainable Food Production*; Springer: Cham, Switzerland, 2021; pp. 27–37.
5. Novelli, P.K.; Barros, M.M.; Fleuri, L.F. Novel inexpensive fungi proteases: Production by solid state fermentation and characterization. *Food Chem.* **2016**, *198*, 119–124. [CrossRef]

6. Souza, P.M.d.; Bittencourt, M.L.d.A.; Caprara, C.C.; Freitas, M.d.; Almeida, R.P.C.d.; Silveira, D.; Fonseca, Y.M.; Ferreira, E.X.; Pessoa, A.; Magalhães, P.O. A biotechnology perspective of fungal proteases. *Braz. J. Microbiol.* **2015**, *46*, 337–346. [CrossRef]
7. Kumar, A.; Gautam, A.; Dutt, D. Screening of fungal resources for the production of cellulases and xylanases. *Biotechnol. J. Int.* **2015**, *9*, 1–13. [CrossRef]
8. Lange, L. Fungal enzymes and yeasts for conversion of plant biomass to bioenergy and high-value products. *Microbiol. Spectr.* **2017**, *5*. [CrossRef]
9. Damare, S.; Raghukumar, C.; Muraleedharan, U.D.; Raghukumar, S. Deep-sea fungi as a source of alkaline and cold-tolerant proteases. *Enzym. Microb. Technol.* **2006**, *39*, 172–181. [CrossRef]
10. Kumar, A. Aspergillus nidulans: A Potential Resource of the Production of the Native and Heterologous Enzymes for Industrial Applications. *Int. J. Microbiol.* **2020**, *2020*, 8894215. [CrossRef]
11. Gnanadoss, J.J.; Devi, S.K. Optimization of nutritional and culture conditions for improved protease production by Aspergillus nidulans and Aspergillus flavus. *J. Microbiol. Biotechnol. Food Sci.* **2021**, *2021*, 518–523. [CrossRef]
12. Hoffmeister, D.; Keller, N.P. Natural products of filamentous fungi: Enzymes, genes, and their regulation. *Nat. Prod. Rep.* **2007**, *24*, 393–416. [CrossRef]
13. Savitha, S.; Sadhasivam, S.; Swaminathan, K.; Lin, F.H. Fungal protease: Production, purification and compatibility with laundry detergents and their wash performance. *J. Taiwan Inst. Chem. Eng.* **2011**, *42*, 298–304. [CrossRef]
14. Liu, E.; Li, M.; Abdella, A.; Wilkins, M.R. Development of a cost-effective medium for submerged production of fungal aryl alcohol oxidase using a genetically modified Aspergillus nidulans strain. *Bioresour. Technol.* **2020**, *305*, 123038. [CrossRef]
15. Sharma, M.; Gat, Y.; Arya, S.; Kumar, V.; Panghal, A.; Kumar, A. A review on microbial alkaline protease: An essential tool for various industrial approaches. *Ind. Biotechnol.* **2019**, *15*, 69–78. [CrossRef]
16. Sabotič, J.; Kos, J. Microbial and fungal protease inhibitors—current and potential applications. *Appl. Microbiol. Biotechnol.* **2012**, *93*, 1351–1375. [CrossRef]
17. Agbowuro, A.A.; Huston, W.M.; Gamble, A.B.; Tyndall, J.D. Proteases and protease inhibitors in infectious diseases. *Med. Res. Rev.* **2018**, *38*, 1295–1331. [CrossRef]
18. Bezerra, V.H.S.; Cardoso, S.L.; Fonseca-Bazzo, Y.; Silveira, D.; Magalhães, P.O.; Souza, P.M. Protease Produced by Endophytic Fungi: A Systematic Review. *Molecules* **2021**, *26*, 7062. [CrossRef]
19. Gupta, V.K.; Kubicek, C.P.; Berrin, J.-G.; Wilson, D.W.; Couturier, M.; Berlin, A.; Edivaldo Filho, X.; Ezeji, T. Fungal enzymes for bio-products from sustainable and waste biomass. *Trends Biochem. Sci.* **2016**, *41*, 633–645. [CrossRef]
20. Yike, I. Fungal proteases and their pathophysiological effects. *Mycopathologia* **2011**, *171*, 299–323. [CrossRef] [PubMed]
21. Manganyi, M.C.; Ateba, C.N. Untapped potentials of endophytic fungi: A review of novel bioactive compounds with biological applications. *Microorganisms* **2020**, *8*, 1934. [CrossRef]
22. Srilakshmi, J.; Madhavi, J.; Lavanya, S.; Ammani, K. Commercial potential of fungal protease: Past, present and future prospects. *J. Pharm. Chem. Biol. Sci.* **2015**, *2*, 218–234.
23. Adrio, J.L.; Demain, A.L. Fungal biotechnology. *Int. Microbiol.* **2003**, *6*, 191–199. [CrossRef] [PubMed]
24. Guillemette, T.; van Peij, N.N.; Goosen, T.; Lanthaler, K.; Robson, G.D.; van den Hondel, C.A.; Stam, H.; Archer, D.B. Genomic analysis of the secretion stress response in the enzyme-producing cell factory Aspergillus niger. *BMC Genom.* **2007**, *8*, 158. [CrossRef] [PubMed]
25. Nevalainen, K.H.; Te'o, V.S.; Bergquist, P.L. Heterologous protein expression in filamentous fungi. *Trends Biotechnol.* **2005**, *23*, 468–474. [CrossRef]
26. Paloheimo, M.; Haarmann, T.; Mäkinen, S.; Vehmaanperä, J. Production of industrial enzymes in Trichoderma reesei. In *Gene Expression Systems in Fungi: Advancements and Applications*; Springer: Berlin/Heidelberg, Germany, 2016; pp. 23–57.
27. Benito, M.J.; Connerton, I.F.; Córdoba, J.J. Genetic characterization and expression of the novel fungal protease, EPg222 active in dry-cured meat products. *Appl. Microbiol. Biotechnol.* **2006**, *73*, 356–365. [CrossRef]
28. Li, C.; Xu, D.; Zhao, M.; Sun, L.; Wang, Y. Production optimization, purification, and characterization of a novel acid protease from a fusant by Aspergillus oryzae and Aspergillus niger. *Eur. Food Res. Technol.* **2014**, *238*, 905–917. [CrossRef]
29. Salazar-Cerezo, S.; Kun, R.S.; de Vries, R.P.; Garrigues, S. CRISPR/Cas9 technology enables the development of the filamentous ascomycete fungus Penicillium subrubescens as a new industrial enzyme producer. *Enzym. Microb. Technol.* **2020**, *133*, 109463. [CrossRef]
30. Satish, L.; Shamili, S.; Muthubharathi, B.C.; Ceasar, S.A.; Kushmaro, A.; Singh, V.; Sitrit, Y. CRISPR-Cas9 System for Fungi Genome Engineering Toward Industrial Applications. In *Genome Engineering via CRISPR-Cas9 System*; Elsevier: Amsterdam, The Netherlands, 2020; pp. 69–81.
31. Liu, R.; Chen, L.; Jiang, Y.; Zhou, Z.; Zou, G. Efficient genome editing in filamentous fungus Trichoderma reesei using the CRISPR/Cas9 system. *Cell Discov.* **2015**, *1*, 15007. [CrossRef]
32. Farooq, M.A.; Ali, S.; Hassan, A.; Tahir, H.M.; Mumtaz, S.; Mumtaz, S. Biosynthesis and industrial applications of α-amylase: A review. *Arch. Microbiol.* **2021**, *203*, 1281–1292. [CrossRef]
33. Singh, A.; Bajar, S.; Devi, A.; Pant, D. An overview on the recent developments in fungal cellulase production and their industrial applications. *Bioresour. Technol. Rep.* **2021**, *14*, 100652. [CrossRef]
34. Mehta, A.; Guleria, S.; Sharma, R.; Gupta, R. The Lipases and Their Applications with Emphasis on Food Industry. In *Microbial Biotechnology in Food and Health*; Elsevier: Amsterdam, The Netherlands, 2021; pp. 143–164.

35. Anitha, T.; Palanivelu, P. Purification and characterization of an extracellular keratinolytic protease from a new isolate of Aspergillus parasiticus. *Protein Expr. Purif.* **2013**, *88*, 214–220. [CrossRef]
36. Zdarta, J.; Jędrzak, A.; Klapiszewski, Ł.; Jesionowski, T. Immobilization of cellulase on a functional inorganic–organic hybrid support: Stability and kinetic study. *Catalysts* **2017**, *7*, 374. [CrossRef]
37. Khan, F.I.; Bisetty, K.; Singh, S.; Permaul, K.; Hassan, M.I. Chitinase from Thermomyces lanuginosus SSBP and its biotechnological applications. *Extremophiles* **2015**, *19*, 1055–1066. [CrossRef]
38. Sikandar, S.; Ujor, V.C.; Ezeji, T.C.; Rossington, J.L.; Michel, F.C., Jr.; McMahan, C.M.; Ali, N.; Cornish, K. Thermomyces lanuginosus STm: A source of thermostable hydrolytic enzymes for novel application in extraction of high-quality natural rubber from Taraxacum kok-saghyz (Rubber dandelion). *Ind. Crops Prod.* **2017**, *103*, 161–168. [CrossRef]
39. Dagenais, T.R.; Keller, N.P. Pathogenesis of Aspergillus fumigatus in invasive aspergillosis. *Clin. Microbiol. Rev.* **2009**, *22*, 447–465. [CrossRef]
40. Singh, S.; Singh, S.; Bali, V.; Sharma, L.; Mangla, J. Production of fungal amylases using cheap, readily available agriresidues, for potential application in textile industry. *BioMed Res. Int.* **2014**, *2014*, 215748. [CrossRef]
41. Wanderley, K.J.; Torres, F.A.; Moraes, L.M.; Ulhoa, C.J. Biochemical characterization of α-amylase from the yeast Cryptococcus flavus. *FEMS Microbiol. Lett.* **2004**, *231*, 165–169. [CrossRef]
42. Stojanović, J.; Jakovljević, V.; Matović, I.; Gajović, O.; Mijušković, Z.; Nedeljković, T. Influence of detergent on metabolic activity of fungi Aspergillus niger. *Nat. Sci.* **2011**, *3*, 466–470. [CrossRef]
43. Saleem, A.; Ebrahim, M.K. Production of amylase by fungi isolated from legume seeds collected in Almadinah Almunawwarah, Saudi Arabia. *J. Taibah Univ. Sci.* **2014**, *8*, 90–97. [CrossRef]
44. Souza, P.M.d.; Magalhães, P.d.O. Application of microbial α-amylase in industry-A review. *Braz. J. Microbiol.* **2010**, *41*, 850–861. [CrossRef]
45. Alves, M.H.; Campos-Takaki, G.M.; Porto, A.L.F.; Milanez, A.I. Screening of Mucor spp. for the production of amylase, lipase, polygalacturonase and protease. *Braz. J. Microbiol.* **2002**, *33*, 325–330. [CrossRef]
46. Rahardjo, Y.S.; Weber, F.J.; Le Comte, E.P.; Tramper, J.; Rinzema, A. Contribution of aerial hyphae of Aspergillus oryzae to respiration in a model solid-state fermentation system. *Biotechnol. Bioeng.* **2002**, *78*, 539–544. [CrossRef] [PubMed]
47. Balakrishnan, M.; Jeevarathinam, G.; Kumar, S.K.S.; Muniraj, I.; Uthandi, S. Optimization and scale-up of α-amylase production by Aspergillus oryzae using solid-state fermentation of edible oil cakes. *BMC Biotechnol.* **2021**, *21*, 33. [CrossRef] [PubMed]
48. Saranraj, P.; Stella, D. Fungal amylase—a review. *Int. J. Microbiol. Res.* **2013**, *4*, 203–211.
49. Sethi, B.K.; Jana, A.; Nanda, P.K.; DasMohapatra, P.K.; Sahoo, S.L.; Patra, J.K. Production of α-amylase by Aspergillus terreus NCFT 4269.10 using pearl millet and its structural characterization. *Front. Plant Sci.* **2016**, *7*, 639. [CrossRef]
50. Khajuria, R.; Singh, S. Fungal Amylases for the Detergent Industry. In *Microbes in Agriculture and Environmental Development*; CRC Press: Boca Raton, FL, USA, 2020; pp. 153–164.
51. Niyonzima, F.N.; More, S. Purification and properties of detergent-compatible extracellular alkaline protease from Scopulariopsis spp. *Prep. Biochem. Biotechnol.* **2014**, *44*, 738–759. [CrossRef]
52. Zahra, T.; Irfan, M.; Nadeem, M.; Ghazanfar, M.; Ahmad, Q.; Ali, S.; Siddique, F.; Yasmeen, Z.; Franco, M. Cellulase Production by Trichoderma viride in Submerged Fermentation using Response Surface Methodology. *Punjab Univ. J. Zool.* **2020**, *35*, 223–228. [CrossRef]
53. Niyonzima, F.N. Detergent-compatible fungal cellulases. *Folia Microbiol.* **2020**, *66*, 25–40. [CrossRef]
54. Imran, M.; Anwar, Z.; Zafar, M.; Ali, A.; Arif, M. Production and characterization of commercial cellulase produced through Aspergillus niger IMMIS1 after screening fungal species. *Pak. J. Bot.* **2018**, *50*, 1563–1570.
55. Imran, M.; Anwar, Z.; Irshad, M.; Javid, A.; Hussain, A.; Ali, S. Optimization of cellulase production from a novel strain of Aspergillus tubingensis IMMIS2 through response surface methodology. *Biocatal. Agric. Biotechnol.* **2017**, *12*, 191–198. [CrossRef]
56. da Silva Delabona, P.; Lima, D.J.; Robl, D.; Rabelo, S.C.; Farinas, C.S.; da Cruz Pradella, J.G. Enhanced cellulase production by Trichoderma harzianum by cultivation on glycerol followed by induction on cellulosic substrates. *J. Ind. Microbiol. Biotechnol.* **2016**, *43*, 617–626. [CrossRef]
57. Pham, T.H.; Quyen, D.T.; Nghiem, N.M. Purification and properties of an endoglucanase from Aspergillus niger VTCC-F021. *Turk. J. Biol.* **2012**, *36*, 694–701.
58. Pham, T.H.; Quyen, D.T.; Nghiem, N.M. Optimization of endoglucanase production by Aspergillus niger VTCC-F021. *Aust. J. Basic Appl. Sci.* **2010**, *6*, 4151–5157.
59. Niyonzima, F.N.; More, S.S. Purification and characterization of detergent-compatible protease from Aspergillus terreus gr. *3 Biotech* **2015**, *5*, 61–70. [CrossRef]
60. Hernández, M.S.; Rodríguez, M.R.; Guerra, N.P.; Rosés, R.P. Amylase production by Aspergillus niger in submerged cultivation on two wastes from food industries. *J. Food Eng.* **2006**, *73*, 93–100. [CrossRef]
61. Sethi, B.K.; Rout, J.R.; Das, R.; Nanda, P.K.; Sahoo, S.L. Lipase production by Aspergillus terreus using mustard seed oil cake as a carbon source. *Ann. Microbiol.* **2013**, *63*, 241–252. [CrossRef]
62. Choudhary, V. Production, isolation and characterization of alkaline protease from Aspergillus versicolor PF/F/107. *J. Acad. Indus. Res.* **2012**, *1*, 272–277.
63. Gopinath, S.C.; Hilda, A.; Anbu, P. Extracellular enzymatic activity profiles in fungi isolated from oil-rich environments. *Mycoscience* **2005**, *46*, 119–126. [CrossRef]

64. Da Silva, O.S.; de Almeida, E.M.; de Melo, A.H.F.; Porto, T.S. Purification and characterization of a novel extracellular serine-protease with collagenolytic activity from Aspergillus tamarii URM4634. *Int. J. Biol. Macromol.* **2018**, *117*, 1081–1088. [CrossRef]
65. Dayanandan, A.; Rani, S.; Shanmugavel, M.; Gnanamani, A.; Rajakumar, G.S. Enhanced production of Aspergillus tamarii lipase for recovery of fat from tannery fleshings. *Braz. J. Microbiol.* **2013**, *44*, 1089–1095. [CrossRef]
66. Pasin, T.; Benassi, V.; Moreira, E.; Jorge, J.; Polizeli, M. Prospecting filamentous fungi for amylase production: Standardization of Aspergillus japonicus culture conditions. *Biotechnol. J. Int.* **2014**, 482–498. [CrossRef]
67. Souza, L.T.A.; Oliveira, J.S.; dos Santos, V.L.; Regis, W.C.; Santoro, M.M.; Resende, R.R. Lipolytic potential of Aspergillus japonicus LAB01: Production, partial purification, and characterisation of an extracellular lipase. *BioMed Res. Int.* **2014**, *2014*, 108913. [CrossRef]
68. Karanam, S.K.; Medicherla, N.R. Enhanced lipase production by mutation induced Aspergillus japonicus. *Afr. J. Biotechnol.* **2008**, *7*. [CrossRef]
69. Niyonzima, F.N.; Veena, S.; More, S.S. Industrial Production and Optimization of Microbial Enzymes. In *Microbial Enzymes: Roles and Applications in Industries*; Springer: Singapore, 2020; pp. 115–135.
70. Tamminen, A.; Kramer, A.; Labes, A.; Wiebe, M.G. Production of scopularide A in submerged culture with Scopulariopsis brevicaulis. *Microb. Cell Factories* **2014**, *13*, 89. [CrossRef]
71. Macchione, M.M.; Merheb, C.W.; Gomes, E.; Da Silva, R. Protease production by different thermophilic fungi. *Appl. Biochem. Biotechnol.* **2008**, *146*, 223–230. [CrossRef] [PubMed]
72. Larcher, G.; Bouchara, J.-P.; Annaix, V.; Symoens, F.; Chabasse, D.; Tronchin, G. Purification and characterization of a fibrinogenolytic serine proteinase from Aspergillus fumigatus culture filtrate. *FEBS Lett.* **1992**, *308*, 65–69. [CrossRef]
73. Farnell, E.; Rousseau, K.; Thornton, D.J.; Bowyer, P.; Herrick, S.E. Expression and secretion of Aspergillus fumigatus proteases are regulated in response to different protein substrates. *Fungal Biol.* **2012**, *116*, 1003–1012. [CrossRef] [PubMed]
74. Kumar, S.; Sharma, N.S.; Saharan, M.R.; Singh, R. Extracellular acid protease from Rhizopus oryzae: Purification and characterization. *Process Biochem.* **2005**, *40*, 1701–1705. [CrossRef]
75. Benabda, O.; M'hir, S.; Kasmi, M.; Mnif, W.; Hamdi, M. Optimization of protease and amylase production by Rhizopus oryzae cultivated on bread waste using solid-state fermentation. *J. Chem.* **2019**, *2019*, 3738181. [CrossRef]
76. Abraham, L.D.; Breuil, C. Isolation and characterization of a subtilisin-like serine proteinase secreted by the sap-staining fungus Ophiostoma piceae. *Enzym. Microb. Technol.* **1996**, *18*, 133–140. [CrossRef]
77. de Oliveira, T.B.; Gomes, E.; Rodrigues, A. Thermophilic fungi in the new age of fungal taxonomy. *Extremophiles* **2015**, *19*, 31–37. [CrossRef]
78. Zucconi, L.; Pagano, S.; Fenice, M.; Selbmann, L.; Tosi, S.; Onofri, S. Growth temperature preferences of fungal strains from Victoria Land, Antarctica. *Polar Biol.* **1996**, *16*, 53–61. [CrossRef]
79. Lasa, I.; Berenguer, J. Thermophilic enzymes and their biotechnological potential. *Microbiologia* **1993**, *9*, 77–89.
80. Mikhailova, A.G.; Khairullin, R.F.; Demidyuk, I.V.; Kostrov, S.V.; Grinberg, N.V.; Burova, T.V.; Grinberg, V.Y.; Rumsh, L.D. Cloning, sequencing, expression, and characterization of thermostability of oligopeptidase B from Serratia proteamaculans, a novel psychrophilic protease. *Protein Expr. Purif.* **2014**, *93*, 63–76. [CrossRef]
81. Wancura, J.H.; Rosset, D.V.; Tres, M.V.; Oliveira, J.V.; Mazutti, M.A.; Jahn, S.L. Production of biodiesel catalyzed by lipase from Thermomyces lanuginosus in its soluble form. *Can. J. Chem. Eng.* **2018**, *96*, 2361–2368. [CrossRef]
82. Pathak, A.P.; Rathod, M.G. *A Review on Alkaline Protease Producers and Their Biotechnological Perspectives*; NISCAIR-CSIR: New Delhi, India, 2018; pp. 1113–1119.
83. Huynh, H.H.; Arioka, M. Functional expression and characterization of a glucuronoyl esterase from the fungus Neurospora crassa: Identification of novel consensus sequences containing the catalytic triad. *J. Gen. Appl. Microbiol.* **2016**, *62*, 217–224. [CrossRef]
84. Singh, B. Myceliophthora thermophila syn. Sporotrichum thermophile: A thermophilic mould of biotechnological potential. *Crit. Rev. Biotechnol.* **2016**, *36*, 59–69. [CrossRef]
85. De Oliveira, T.B.; Rodrigues, A. *Ecology of Thermophilic Fungi. Fungi in Extreme Environments: Ecological Role and Biotechnological Significance*; Springer International Publishing: Berlin/Heidelberg, Germany, 2019; pp. 39–57.
86. Singh, B.; Satyanarayana, T. Phytase production by a thermophilic mould Sporotrichum thermophile in solid state fermentation and its potential applications. *Bioresour. Technol.* **2008**, *99*, 2824–2830. [CrossRef]
87. Dilokpimol, A.; Mäkelä, M.R.; Cerullo, G.; Zhou, M.; Varriale, S.; Gidijala, L.; Brás, J.L.; Jütten, P.; Piechot, A.; Verhaert, R. Fungal glucuronoyl esterases: Genome mining based enzyme discovery and biochemical characterization. *New Biotechnol.* **2018**, *40*, 282–287. [CrossRef]
88. Salwan, R.; Sharma, V. Proteases from Extremophilic Fungi: A Tool for White Biotechnology. In *Recent Advancement in White Biotechnology Through Fungi*; Springer: Berlin/Heidelberg, Germany, 2019; pp. 299–317.
89. Kieliszek, M.; Kot, A.M.; Bzducha-Wróbel, A.; BŁażejak, S.; Gientka, I.; Kurcz, A. Biotechnological use of Candida yeasts in the food industry: A review. *Fungal Biol. Rev.* **2017**, *31*, 185–198. [CrossRef]
90. Naveed, M.; Nadeem, F.; Mehmood, T.; Bilal, M.; Anwar, Z.; Amjad, F. Protease—a versatile and ecofriendly biocatalyst with multi-industrial applications: An updated review. *Catal. Lett.* **2021**, *151*, 307–323. [CrossRef]

91. El-Khonezy, M.I.; Elgammal, E.W.; Ahmed, E.F.; Abd-Elaziz, A.M. Detergent stable thiol-dependant alkaline protease produced from the endophytic fungus Aspergillus ochraceus BT21: Purification and kinetics. *Biocatal. Agric. Biotechnol.* **2021**, *35*, 102046. [CrossRef]
92. Niyonzima, F.N.; More, S.S. Screening and optimization of cultural parameters for an alkaline protease production by Aspergillus terreus gr. under submerged fermentation. *Int. J. Pharm. Bio. Sci.* **2013**, *4*, 1016–1028.
93. Yadav, V.K.; Singh, V.; Mishra, V. Alkaline Protease: A Tool to Manage Solid Waste and Its Utility in Detergent Industry. In *Microbial Genomics in Sustainable Agroecosystems*; Springer: Berlin/Heidelberg, Germany, 2019; pp. 231–254.
94. Martinelli, P.; Rugarli, E.I. Emerging roles of mitochondrial proteases in neurodegeneration. *Biochim. Et Biophys. Acta (BBA)-Bioenerg.* **2010**, *1797*, 1–10. [CrossRef]
95. Hariharan, A.; Rajadurai, U.M.; Palanivel, I. Isolation, Purification and Mass Production of Protease from Bacillus subtilis. 2019. Available online: https://ssrn.com/abstract=3370124 (accessed on 15 July 2021).
96. Devi, M.K.; Banu, A.R.; Gnanaprabhal, G.; Pradeep, B.; Palaniswamy, M. Purification, characterization of alkaline protease enzyme from native isolate Aspergillus niger and its compatibility with commercial detergents. *Indian J. Sci. Technol.* **2008**, *1*, 1–6. [CrossRef]
97. Shahid, T.; Muhammad, S.; Ahmed, K. Enzyme Proteases Used in Laundry Detergents Engineering a Review. *Sci. Int.* **2016**, *28*, 2711–2717.
98. Corvo, I.; Ferraro, F.; Merlino, A.; Zuberbühler, K.; O'Donoghue, A.J.; Pastro, L.; Pi-Denis, N.; Basika, T.; Roche, L.; McKerrow, J.H. Substrate specificity of cysteine proteases beyond the S2 Pocket: Mutagenesis and molecular dynamics investigation of Fasciola hepatica Cathepsins L. *Front. Mol. Biosci.* **2018**, *5*, 40. [CrossRef]
99. Ou, J.-F.; Zhu, M.-J. An overview of current and novel approaches for microbial neutral protease improvement. *Int. J. Mod. Biol. Med* **2012**, *2*, 1–31.
100. Theron, L.W.; Divol, B. Microbial aspartic proteases: Current and potential applications in industry. *Appl. Microbiol. Biotechnol.* **2014**, *98*, 8853–8868. [CrossRef]
101. Dunn, B.M. Introduction to the aspartic proteinase family. *Aspartic Acid Proteases Ther. Targets* **2010**, *45*, 1–21.
102. Jiang, L.; Zhang, X.; Zhou, Y.; Chen, Y.; Luo, Z.; Li, J.; Yuan, C.; Huang, M. Halogen bonding for the design of inhibitors by targeting the S1 pocket of serine proteases. *RSC Adv.* **2018**, *8*, 28189–28197. [CrossRef]
103. Jiang, L.; Yuan, C.; Huang, M. A general strategy to inhibit serine protease by targeting its autolysis loop. *FASEB J.* **2021**, *35*, e21259. [CrossRef]
104. Rawlings, N.D.; Barrett, A.J. [13] Evolutionary families of metallopeptidases. *Methods Enzymol.* **1995**, *248*, 183–228.
105. Cheng, M.; Takenaka, S.; Aoki, S.; Murakami, S.; Aoki, K. Purification and characterization of an eggshell membrane decomposing protease from Pseudomonas aeruginosa strain ME-4. *J. Biosci. Bioeng.* **2009**, *107*, 373–378. [CrossRef]
106. Sims, A.H.; Dunn-Coleman, N.S.; Robson, G.D.; Oliver, S.G. Glutamic protease distribution is limited to filamentous fungi. *FEMS Microbiol. Lett.* **2004**, *239*, 95–101. [CrossRef]
107. Kuan, C.S.; Ng, K.P.; Yew, S.M.; Meleh, H.U.; Seow, H.F.; How, K.N.; Yeo, S.K.; Jee, J.M.; Tan, Y.-C.; Yee, W.-Y. Comparative Genomic and Phenotypic Analyses of Pathogenic Fungi Neoscytalidium Dimidiatum and Bipolaris Papendorfii, Isolated From Human Skin Scraping. *Res. Sq.* **2020**, 1–32. [CrossRef]
108. Mamo, J.; Assefa, F. The role of microbial aspartic protease enzyme in food and beverage industries. *J. Food Qual.* **2018**, *2018*, 7957269. [CrossRef]
109. Razzaq, A.; Shamsi, S.; Ali, A.; Ali, Q.; Sajjad, M.; Malik, A.; Ashraf, M. Microbial proteases applications. *Front. Bioeng. Biotechnol.* **2019**, *7*, 110. [CrossRef]
110. Erjavec, J.; Kos, J.; Ravnikar, M.; Dreo, T.; Sabotič, J. Proteins of higher fungi–from forest to application. *Trends Biotechnol.* **2012**, *30*, 259–273. [CrossRef]
111. Shaba, A.; Baba, J. Screening of Pleurotus ostreatus and Gleophylum sepiarium strains for extracellular protease enzyme production. *Bayero J. Pure Appl. Sci.* **2012**, *5*, 187–190. [CrossRef]
112. Papagianni, M. Fungal morphology and metabolite production in submerged mycelial processes. *Biotechnol. Adv.* **2004**, *22*, 189–259. [CrossRef] [PubMed]
113. Wang, H.; Ng, T. Pleureryn, a novel protease from fresh fruiting bodies of the edible mushroom Pleurotus eryngii. *Biochem. Biophys. Res. Commun.* **2001**, *289*, 750–755. [CrossRef]
114. Musatti, A.; Ficara, E.; Mapelli, C.; Sambusiti, C.; Rollini, M. Use of solid digestate for lignocellulolytic enzymes production through submerged fungal fermentation. *J. Environ. Manag.* **2017**, *199*, 1–6. [CrossRef] [PubMed]
115. Burdsall, H.H., Jr.; Volk, T.J. Armillaria solidipes, an older name for the fungus called Armillaria ostoyae. *North Am. Fungi* **2008**, *3*, 261–267. [CrossRef]
116. Faraco, V.; Palmieri, G.; Festa, G.; Monti, M.; Sannia, G.; Giardina, P. A new subfamily of fungal subtilases: Structural and functional analysis of a Pleurotus ostreatus member. *Microbiology* **2005**, *151*, 457–466. [CrossRef] [PubMed]
117. Nurika, I.; Suhartini, S.; Barker, G.C. Biotransformation of tropical lignocellulosic feedstock using the brown rot fungus Serpula lacrymans. *Waste Biomass Valorization* **2020**, *11*, 2689–2700. [CrossRef]
118. Cha, W.-S.; Park, S.-S.; Kim, S.-J.; Choi, D. Biochemical and enzymatic properties of a fibrinolytic enzyme from Pleurotus eryngii cultivated under solid-state conditions using corn cob. *Bioresour. Technol.* **2010**, *101*, 6475–6481. [CrossRef] [PubMed]

119. Ng, T.B.; Wong, J.H.; Cheung, R.C.F.; Tse, T.F.; Tam, T.; Chan, H. Mushrooms: Proteins, polysaccharidepeptide complexes and polysaccharides with antiproliferative and anticancer activities. *Int. J. Cancer Res. Prev.* **2014**, *7*, 287.
120. Lv, H.; Kong, Y.; Yao, Q.; Zhang, B.; Leng, F.-w.; Bian, H.-j.; Balzarini, J.; Van Damme, E.; Bao, J.-k. Nebrodeolysin, a novel hemolytic protein from mushroom Pleurotus nebrodensis with apoptosis-inducing and anti-HIV-1 effects. *Phytomedicine* **2009**, *16*, 198–205. [CrossRef]
121. Berne, S.; Križaj, I.; Pohleven, F.; Turk, T.; Maček, P.; Sepčić, K. Pleurotus and Agrocybe hemolysins, new proteins hypothetically involved in fungal fruiting. *Biochim. Et Biophys. Acta (BBA) Gen. Subj.* **2002**, *1570*, 153–159. [CrossRef]
122. Sumantha, A.; Larroche, C.; Pandey, A. Microbiology and industrial biotechnology of food-grade proteases: A perspective. *Food Technol. Biotechnol.* **2006**, *44*, 211.
123. Sandhya, C.; Nampoothiri, K.M.; Pandey, A. Microbial Proteases. In *Microbial Enzymes and Biotransformations*; Springer: Berlin/Heidelberg, Germany, 2005; pp. 165–179.
124. dos Santos Aguilar, J.G.; Sato, H.H. Microbial proteases: Production and application in obtaining protein hydrolysates. *Food Res. Int.* **2018**, *103*, 253–262. [CrossRef]
125. Baird, T.T., Jr.; Wright, W.D.; Craik, C.S. Conversion of trypsin to a functional threonine protease. *Protein Sci.* **2006**, *15*, 1229–1238. [CrossRef]
126. Jashni, M.K.; Dols, I.H.; Iida, Y.; Boeren, S.; Beenen, H.G.; Mehrabi, R.; Collemare, J.; de Wit, P.J. Synergistic action of a metalloprotease and a serine protease from Fusarium oxysporum f. sp. lycopersici cleaves chitin-binding tomato chitinases, reduces their antifungal activity, and enhances fungal virulence. *Mol. Plant-Microbe Interact.* **2015**, *28*, 996–1008. [CrossRef]
127. Barrett, A.J. [1] Classification of peptidases. *Methods Enzymol.* **1994**, *244*, 1–15. [CrossRef]
128. Brocklehurst, K.; Philpott, M.P. Cysteine proteases: Mode of action and role in epidermal differentiation. *Cell Tissue Res.* **2013**, *351*, 237–244. [CrossRef]
129. Yegin, S.; Fernandez-Lahore, M.; Salgado, A.J.G.; Guvenc, U.; Goksungur, Y.; Tari, C. Aspartic proteinases from Mucor spp. in cheese manufacturing. *Appl. Microbiol. Biotechnol.* **2011**, *89*, 949–960. [CrossRef]
130. Da Silva, R.R.; Souto, T.B.; de Oliveira, T.B.; de Oliveira, L.C.G.; Karcher, D.; Juliano, M.A.; Juliano, L.; de Oliveira, A.H.; Rodrigues, A.; Rosa, J.C. Evaluation of the catalytic specificity, biochemical properties, and milk clotting abilities of an aspartic peptidase from Rhizomucor miehei. *J. Ind. Microbiol. Biotechnol.* **2016**, *43*, 1059–1069. [CrossRef]
131. Dietrich, F.S.; Voegeli, S.; Brachat, S.; Lerch, A.; Gates, K.; Steiner, S.; Mohr, C.; Pöhlmann, R.; Luedi, P.; Choi, S. The Ashbya gossypii genome as a tool for mapping the ancient Saccharomyces cerevisiae genome. *Science* **2004**, *304*, 304–307. [CrossRef]
132. Feldmann, H. *Génolevures—A Novel Approach to 'Evolutionary Genomics'*; Elsevier: Amsterdam, The Netherlands, 2000; Volume 487, pp. 1–2.
133. Kellis, M.; Birren, B.W.; Lander, E.S. Proof and evolutionary analysis of ancient genome duplication in the yeast Saccharomyces cerevisiae. *Nature* **2004**, *428*, 617–624. [CrossRef]
134. Neto, Y.A.A.H.; de Souza Motta, C.M.; Cabral, H. Optimization of metalloprotease production by Eupenicillium javanicum in both solid state and submerged bioprocesses. *Afr. J. Biochem. Res.* **2013**, *7*, 146–157.
135. Erez, E.; Fass, D.; Bibi, E. How intramembrane proteases bury hydrolytic reactions in the membrane. *Nature* **2009**, *459*, 371–378. [CrossRef]
136. Shafee, T. *Evolvability of a Viral Protease: Experimental Evolution of Catalysis, Robustness and Specificity*; University of Cambridge: Cambridge, UK, 2014.
137. Madhavan, A.; Arun, K.; Binod, P.; Sirohi, R.; Tarafdar, A.; Reshmy, R.; Awasthi, M.K.; Sindhu, R. Design of novel enzyme biocatalysts for industrial bioprocess: Harnessing the power of protein engineering, high throughput screening and synthetic biology. *Bioresour. Technol.* **2021**, *325*, 124617. [CrossRef]
138. Veloorvalappil, N.J.; Robinson, B.S.; Selvanesan, P.; Sasidharan, S.; Kizhakkepawothail, N.U.; Sreedharan, S.; Prakasan, P.; Moolakkariyil, S.J.; Sailas, B. Versatility of microbial proteases. *Adv. Enzym. Res.* **2013**, *2013*, 36957.
139. Shankar, R.; Upadhyay, P.K.; Kumar, M. Protease Enzymes: Highlights on Potential of Proteases as Therapeutics Agents. *Int. J. Pept. Res. Ther.* **2021**, *27*, 1281–1296. [CrossRef]
140. Chapman, J.; Ismail, A.E.; Dinu, C.Z. Industrial applications of enzymes: Recent advances, techniques, and outlooks. *Catalysts* **2018**, *8*, 238. [CrossRef]
141. Timson, D.J. Four challenges for better biocatalysts. *Fermentation* **2019**, *5*, 39. [CrossRef]
142. Banerjee, G.; Ray, A.K. Impact of microbial proteases on biotechnological industries. *Biotechnol. Genet. Eng. Rev.* **2017**, *33*, 119–143. [CrossRef] [PubMed]
143. Ademosun, M.T.; Omoba, O.S.; Olagunju, A.I. Antioxidant properties, glycemic indices, and carbohydrate hydrolyzing enzymes activities of formulated ginger-based fruit drinks. *J. Food Biochem.* **2021**, *45*, e13324. [CrossRef] [PubMed]
144. Kumar, D.; Bhalla, T.C. Microbial proteases in peptide synthesis: Approaches and applications. *Appl. Microbiol. Biotechnol.* **2005**, *68*, 726–736. [CrossRef] [PubMed]
145. Ni, X.; Yue, L.; Chi, Z.; Li, J.; Wang, X.; Madzak, C. Alkaline protease gene cloning from the marine yeast Aureobasidium pullulans HN2-3 and the protease surface display on Yarrowia lipolytica for bioactive peptide production. *Mar. Biotechnol.* **2009**, *11*, 81–89. [CrossRef]

146. Kanda, S.; Aimi, T.; Kano, S.; Ishihara, S.; Kitamoto, Y.; Morinaga, T. Ambient pH signaling regulates expression of the serine protease gene (spr1) in pine wilt nematode-trapping fungus, Monacrosporium megalosporum. *Microbiol. Res.* **2008**, *163*, 63–72. [CrossRef]
147. Tzean, Y.; Chou, T.-H.; Hsiao, C.-C.; Shu, P.-Y.; Walton, J.D.; Tzean, S.-S. Cloning and characterization of cuticle-degrading serine protease from nematode-trapping fungus Arthrobotrys musiformis. *Mycoscience* **2016**, *57*, 136–143. [CrossRef]
148. Yang, J.; Zhang, K.-Q. Biological Control of Plant-Parasitic Nematodes by Nematophagous Fungi. In *Nematode-Trapping Fungi*; Springer: Berlin/Heidelberg, Germany, 2014; pp. 231–262.
149. Ao, X.-L.; Yu, X.; Wu, D.-T.; Li, C.; Zhang, T.; Liu, S.-L.; Chen, S.-J.; He, L.; Zhou, K.; Zou, L.-K. Purification and characterization of neutral protease from Aspergillus oryzae Y1 isolated from naturally fermented broad beans. *AMB Express* **2018**, *8*, 1–10. [CrossRef]
150. Benmrad, M.O.; Mechri, S.; Jaouadi, N.Z.; Elhoul, M.B.; Rekik, H.; Sayadi, S.; Bejar, S.; Kechaou, N.; Jaouadi, B. Purification and biochemical characterization of a novel thermostable protease from the oyster mushroom Pleurotus sajor-caju strain CTM10057 with industrial interest. *BMC Biotechnol.* **2019**, *19*, 96. [CrossRef]
151. Illuri, R.; Kumar, M.; Eyini, M.; Veeramanikandan, V.; Almaary, K.S.; Elbadawi, Y.B.; Biraqdar, M.; Balaji, P. Production, partial purification and characterization of ligninolytic enzymes from selected basidiomycetes mushroom fungi. *Saudi J. Biol. Sci.* **2021**, *28*, 7207–7218. [CrossRef]
152. Jaouadi, B.; Jaouadi, N.Z.; Rekik, H.; Naili, B.; Beji, A.; Dhouib, A.; Bejar, S. Biochemical and molecular characterization of Pseudomonas aeruginosa CTM50182 organic solvent-stable elastase. *Int. J. Biol. Macromol.* **2013**, *60*, 165–177. [CrossRef]
153. Mandujano-González, V.; Villa-Tanaca, L.; Anducho-Reyes, M.A.; Mercado-Flores, Y. Secreted fungal aspartic proteases: A review. *Rev. Iberoam. De Micol.* **2016**, *33*, 76–82. [CrossRef]
154. Rantasalo, A.; Vitikainen, M.; Paasikallio, T.; Jäntti, J.; Landowski, C.P.; Mojzita, D. Novel genetic tools that enable highly pure protein production in Trichoderma reesei. *Sci. Rep.* **2019**, *9*, 5032. [CrossRef]
155. Jiang, C.; Lv, G.; Tu, Y.; Cheng, X.; Duan, Y.; Zeng, B.; He, B. Applications of CRISPR/Cas9 in the Synthesis of Secondary Metabolites in Filamentous Fungi. *Front. Microbiol.* **2021**, *12*, 164. [CrossRef]
156. Liu, Q.; Gao, R.; Li, J.; Lin, L.; Zhao, J.; Sun, W.; Tian, C. Development of a genome-editing CRISPR/Cas9 system in thermophilic fungal Myceliophthora species and its application to hyper-cellulase production strain engineering. *Biotechnol. Biofuels* **2017**, *10*, 1. [CrossRef]
157. Adalsteinsson, B.T.; Kristjansdottir, T.; Merre, W.; Helleux, A.; Dusaucy, J.; Tourigny, M.; Fridjonsson, O.; Hreggvidsson, G.O. Efficient genome editing of an extreme thermophile, Thermus thermophilus, using a thermostable Cas9 variant. *Sci. Rep.* **2021**, *11*, 9586. [CrossRef]
158. Katayama, T.; Tanaka, Y.; Okabe, T.; Nakamura, H.; Fujii, W.; Kitamoto, K.; Maruyama, J.-i. Development of a genome editing technique using the CRISPR/Cas9 system in the industrial filamentous fungus Aspergillus oryzae. *Biotechnol. Lett.* **2016**, *38*, 637–642. [CrossRef]
159. Jin, F.-J.; Hu, S.; Wang, B.-T.; Jin, L. Advances in genetic engineering technology and its application in the industrial fungus Aspergillus oryzae. *Front. Microbiol.* **2021**, *12*, 353. [CrossRef]
160. Zou, Z.; Liu, F.; Selin, C.; Fernando, W. Generation and characterization of a virulent Leptosphaeria maculans isolate carrying a mutated AvrLm7 gene using the CRISPR/Cas9 system. *Front. Microbiol.* **2020**, *11*, 1969. [CrossRef]
161. Chen, B.-X.; Wei, T.; Ye, Z.-W.; Yun, F.; Kang, L.-Z.; Tang, H.-B.; Guo, L.-Q.; Lin, J.-F. Efficient CRISPR-Cas9 gene disruption system in edible-medicinal mushroom Cordyceps militaris. *Front. Microbiol.* **2018**, *9*, 1157. [CrossRef]
162. Das, S.K.; Masuda, M.; Sakurai, A.; Sakakibara, M. Medicinal uses of the mushroom Cordyceps militaris: Current state and prospects. *Fitoterapia* **2010**, *81*, 961–968. [CrossRef]
163. Gardiner, D.M.; Kazan, K. Selection is required for efficient Cas9-mediated genome editing in Fusarium graminearum. *Fungal Biol.* **2018**, *122*, 131–137. [CrossRef]
164. Joshi, R. A review of Fusarium oxysporum on its plant interaction and industrial use. *J. Med. Plants Stud.* **2018**, *6*, 112–115. [CrossRef]
165. Schuster, M.; Schweizer, G.; Kahmann, R. Comparative analyses of secreted proteins in plant pathogenic smut fungi and related basidiomycetes. *Fungal Genet. Biol.* **2018**, *112*, 21–30. [CrossRef]
166. Olicón-Hernández, D.R.; Araiza-Villanueva, M.G.; Pardo, J.P.; Aranda, E.; Guerra-Sánchez, G. New insights of Ustilago maydis as yeast model for genetic and biotechnological research: A review. *Curr. Microbiol.* **2019**, *76*, 917–926. [CrossRef] [PubMed]
167. Matsu-Ura, T.; Baek, M.; Kwon, J.; Hong, C. Efficient gene editing in Neurospora crassa with CRISPR technology. *Fungal Biol. Biotechnol.* **2015**, *2*, 4. [CrossRef] [PubMed]
168. Qin, H.; Xiao, H.; Zou, G.; Zhou, Z.; Zhong, J.-J. CRISPR-Cas9 assisted gene disruption in the higher fungus Ganoderma species. *Process Biochem.* **2017**, *56*, 57–61. [CrossRef]
169. Yang, H.; Wu, T.; Zhang, K. Effects of Extracts of Chinese Medicines on Ganoderma lucidum in Submerged Culture. *Acta Microbiol. Sin.* **2003**, *43*, 519–522.
170. Schuster, M.; Schweizer, G.; Reissmann, S.; Kahmann, R. Genome editing in Ustilago maydis using the CRISPR–Cas system. *Fungal Genet. Biol.* **2016**, *89*, 3–9. [CrossRef]

171. Al Abdallah, Q.; Souza, A.C.O.; Martin-Vicente, A.; Ge, W.; Fortwendel, J.R. Whole-genome sequencing reveals highly specific gene targeting by in vitro assembled Cas9-ribonucleoprotein complexes in Aspergillus fumigatus. *Fungal Biol. Biotechnol.* **2018**, *5*, 11. [CrossRef]
172. Zhang, Q.; Xing, H.-L.; Wang, Z.-P.; Zhang, H.-Y.; Yang, F.; Wang, X.-C.; Chen, Q.-J. Potential high-frequency off-target mutagenesis induced by CRISPR/Cas9 in Arabidopsis and its prevention. *Plant Mol. Biol.* **2018**, *96*, 445–456. [CrossRef]
173. Doench, J.G.; Fusi, N.; Sullender, M.; Hegde, M.; Vaimberg, E.W.; Donovan, K.F.; Smith, I.; Tothova, Z.; Wilen, C.; Orchard, R. Optimized sgRNA design to maximize activity and minimize off-target effects of CRISPR-Cas9. *Nat. Biotechnol.* **2016**, *34*, 184–191. [CrossRef]
174. Song, L.; Ouedraogo, J.-P.; Kolbusz, M.; Nguyen, T.T.M.; Tsang, A. Efficient genome editing using tRNA promoter-driven CRISPR/Cas9 gRNA in Aspergillus niger. *PLoS ONE* **2018**, *13*, e0202868. [CrossRef]
175. Huang, L.; Dong, H.; Zheng, J.; Wang, B.; Pan, L. Highly efficient single base editing in Aspergillus niger with CRISPR/Cas9 cytidine deaminase fusion. *Microbiol. Res.* **2019**, *223*, 44–50. [CrossRef]
176. Sharma, K.M.; Kumar, R.; Panwar, S.; Kumar, A. Microbial alkaline proteases: Optimization of production parameters and their properties. *J. Genet. Eng. Biotechnol.* **2017**, *15*, 115–126. [CrossRef]
177. Ward, O.P. Production of recombinant proteins by filamentous fungi. *Biotechnol. Adv.* **2012**, *30*, 1119–1139. [CrossRef]
178. Ravanelli, S.; den Brave, F.; Hoppe, T. Mitochondrial quality control governed by ubiquitin. *Front. Cell Dev. Biol.* **2020**, *8*, 270. [CrossRef]
179. Jadiya, P.; Tomar, D. Mitochondrial protein quality control mechanisms. *Genes* **2020**, *11*, 563. [CrossRef]
180. Deshwal, S.; Fiedler, K.U.; Langer, T. Mitochondrial proteases: Multifaceted regulators of mitochondrial plasticity. *Annu. Rev. Biochem.* **2020**, *89*, 501–528. [CrossRef]
181. Quiros, P.M.; Langer, T.; Lopez-Otin, C. New roles for mitochondrial proteases in health, ageing and disease. *Nat. Rev. Mol. Cell Biol.* **2015**, *16*, 345–359. [CrossRef]
182. Hofsetz, E.; Huesgen, P.F.; Trifunovic, A. Identification of Putative Mitochondrial Protease Substrates. In *Mitochondrial Gene Expression*; Springer: Berlin/Heidelberg, Germany, 2021; pp. 313–329.
183. Voos, W. Chaperone–protease networks in mitochondrial protein homeostasis. *Biochim. Et Biophys. Acta (BBA) Mol. Cell Res.* **2013**, *1833*, 388–399. [CrossRef]
184. Teixeira, P.F.; Glaser, E. Processing peptidases in mitochondria and chloroplasts. *Biochim. Et Biophys. Acta (BBA) Mol. Cell Res.* **2013**, *1833*, 360–370. [CrossRef]
185. Weiss-Sadan, T.; Gotsman, I.; Blum, G. Cysteine proteases in atherosclerosis. *FEBS J.* **2017**, *284*, 1455–1472. [CrossRef]
186. Liu, J.; Sukhova, G.K.; Yang, J.-T.; Sun, J.; Ma, L.; Ren, A.; Xu, W.-H.; Fu, H.; Dolganov, G.M.; Hu, C. Cathepsin L expression and regulation in human abdominal aortic aneurysm, atherosclerosis, and vascular cells. *Atherosclerosis* **2006**, *184*, 302–311. [CrossRef]
187. Jaffer, F.A.; Vinegoni, C.; John, M.C.; Aikawa, E.; Gold, H.K.; Finn, A.V.; Ntziachristos, V.; Libby, P.; Weissleder, R. Real-time catheter molecular sensing of inflammation in proteolytically active atherosclerosis. *Circulation* **2008**, *118*, 1802–1809. [CrossRef] [PubMed]
188. Weitoft, T.; Larsson, A.; Manivel, V.A.; Lysholm, J.; Knight, A.; Rönnelid, J. Cathepsin S and cathepsin L in serum and synovial fluid in rheumatoid arthritis with and without autoantibodies. *Rheumatology* **2015**, *54*, 1923–1928. [CrossRef] [PubMed]
189. Appelqvist, H.; Wäster, P.; Kågedal, K.; Öllinger, K. The lysosome: From waste bag to potential therapeutic target. *J. Mol. Cell Biol.* **2013**, *5*, 214–226. [CrossRef] [PubMed]
190. Wu, D.; Wei, Y.; Bleier, B.S. Emerging role of proteases in the pathogenesis of chronic rhinosinusitis with nasal polyps. *Front. Cell. Infect. Microbiol.* **2018**, *7*, 538. [CrossRef]
191. Stentzel, S.; Teufelberger, A.; Nordengrün, M.; Kolata, J.; Schmidt, F.; Van Crombruggen, K.; Michalik, S.; Kumpfmüller, J.; Tischer, S.; Schweder, T. Staphylococcal serine protease–like proteins are pacemakers of allergic airway reactions to Staphylococcus aureus. *J. Allergy Clin. Immunol.* **2017**, *139*, 492–500.e8. [CrossRef]
192. Kale, S.L.; Agrawal, K.; Gaur, S.N.; Arora, N. Cockroach protease allergen induces allergic airway inflammation via epithelial cell activation. *Sci. Rep.* **2017**, *7*, 42341. [CrossRef] [PubMed]
193. Teufelberger, A.R.; Nordengrün, M.; Braun, H.; Maes, T.; De Grove, K.; Holtappels, G.; O'Brien, C.; Provoost, S.; Hammad, H.; Gonçalves, A. The IL-33/ST2 axis is crucial in type 2 airway responses induced by Staphylococcus aureus–derived serine protease–like protein D. *J. Allergy Clin. Immunol.* **2018**, *141*, 549–559.e7. [CrossRef]
194. López-Otín, C.; Matrisian, L.M. Emerging roles of proteases in tumour suppression. *Nat. Rev. Cancer* **2007**, *7*, 800–808. [CrossRef]
195. Kwon, Y.T.; Ciechanover, A. The ubiquitin code in the ubiquitin-proteasome system and autophagy. *Trends Biochem. Sci.* **2017**, *42*, 873–886. [CrossRef]
196. Liu, J.; Shaik, S.; Dai, X.; Wu, Q.; Zhou, X.; Wang, Z.; Wei, W. Targeting the ubiquitin pathway for cancer treatment. *Biochim. Et Biophys. Acta (BBA)-Rev. Cancer* **2015**, *1855*, 50–60. [CrossRef]
197. Green, P.H.; Lebwohl, B.; Greywoode, R. Celiac disease. *J. Allergy Clin. Immunol.* **2015**, *135*, 1099–1106. [CrossRef]
198. Makharia, G.K.D. Current and emerging therapy for celiac disease. *Front. Med.* **2014**, *1*, 6. [CrossRef]
199. Caruso, J.A.; Akli, S.; Pageon, L.; Hunt, K.K.; Keyomarsi, K. The serine protease inhibitor elafin maintains normal growth control by opposing the mitogenic effects of neutrophil elastase. *Oncogene* **2015**, *34*, 3556–3567. [CrossRef]

200. Galipeau, H.J.; Wiepjes, M.; Motta, J.-P.; Schulz, J.D.; Jury, J.; Natividad, J.M.; Pinto-Sanchez, I.; Sinclair, D.; Rousset, P.; Martin-Rosique, R. Novel role of the serine protease inhibitor elafin in gluten-related disorders. *Am. J. Gastroenterol.* **2014**, *109*, 748. [CrossRef]
201. Ghetti, B.; Tagliavini, F.; Kovacs, G.G.; Piccardo, P. 37 Gerstmann–Str ä ussler–Scheinker Disease. In *Neurodegeneration: The Molecular Pathology of Dementia and Movement Disorders*; Wiley: Hoboken, NJ, USA, 2011; p. 364.
202. Iwasaki, Y. Creutzfeldt-Jakob disease. *Neuropathology* **2017**, *37*, 174–188. [CrossRef]
203. Yoshioka, M.; Miwa, T.; Horii, H.; Takata, M.; Yokoyama, T.; Nishizawa, K.; Watanabe, M.; Shinagawa, M.; Murayama, Y. Characterization of a proteolytic enzyme derived from a Bacillus strain that effectively degrades prion protein. *J. Appl. Microbiol.* **2007**, *102*, 509–515. [CrossRef]
204. Rajput, R.; Gupta, R. Thermostable keratinase from Bacillus pumilus KS12: Production, chitin crosslinking and degradation of Sup35NM aggregates. *Bioresour. Technol.* **2013**, *133*, 118–126. [CrossRef]
205. Chauhan, B.; Gupta, R. Application of statistical experimental design for optimization of alkaline protease production from Bacillus sp. RGR-14. *Process Biochem.* **2004**, *39*, 2115–2122. [CrossRef]
206. Reddy, C.C.; Khilji, I.A.; Gupta, A.; Bhuyar, P.; Mahmood, S.; AL-Japairai, K.A.S.; Chua, G.K. Valorization of keratin waste biomass and its potential applications. *J. Water Process Eng.* **2021**, *40*, 101707. [CrossRef]
207. Matkawala, F.; Nighojkar, S.; Kumar, A.; Nighojkar, A. Microbial alkaline serine proteases: Production, properties and applications. *World J. Microbiol. Biotechnol.* **2021**, *37*, 63. [CrossRef]
208. Chilakamarry, C.R.; Mahmood, S.; Saffe, S.N.B.M.; Arifin, M.A.B.; Gupta, A.; Sikkandar, M.Y.; Begum, S.S.; Narasaiah, B. Extraction and application of keratin from natural resources: A review. *3 Biotech* **2021**, *11*, 220. [CrossRef] [PubMed]
209. Qiu, J.; Wilkens, C.; Barrett, K.; Meyer, A.S. Microbial enzymes catalyzing keratin degradation: Classification, structure, function. *Biotechnol. Adv.* **2020**, *44*, 107607. [CrossRef] [PubMed]
210. Bhandari, S.; Poudel, D.K.; Marahatha, R.; Dawadi, S.; Khadayat, K.; Phuyal, S.; Shrestha, S.; Gaire, S.; Basnet, K.; Khadka, U. Microbial Enzymes Used in Bioremediation. *J. Chem.* **2021**, *2021*, 8849512. [CrossRef]
211. Saranya, P.; Selvi, P.; Sekaran, G. Integrated thermophilic enzyme-immobilized reactor and high-rate biological reactors for treatment of palm oil-containing wastewater without sludge production. *Bioprocess Biosyst. Eng.* **2019**, *42*, 1053–1064. [CrossRef]
212. Gradisar, H.; Friedrich, J.; Krizaj, I.; Jerala, R. Similarities and specificities of fungal keratinolytic proteases: Comparison of keratinases of Paecilomyces marquandii and Doratomyces microsporus to some known proteases. *Appl. Environ. Microbiol.* **2005**, *71*, 3420–3426.
213. Espersen, R.; Huang, Y.; Falco, F.C.; Hägglund, P.; Gernaey, K.V.; Lange, L.; Svensson, B. Exceptionally rich keratinolytic enzyme profile found in the rare actinomycetes Amycolatopsis keratiniphila D2T. *Appl. Microbiol. Biotechnol.* **2021**, *105*, 8129–8138. [CrossRef]
214. Barzkar, N.; Sohail, M.; Jahromi, S.T.; Nahavandi, R.; Khodadadi, M. Marine microbial L-glutaminase: From pharmaceutical to food industry. *Appl. Microbiol. Biotechnol.* **2021**, *105*, 4453–4466. [CrossRef]
215. Bond, J.S. Proteases: History, discovery, and roles in health and disease. *J. Biol. Chem.* **2019**, *294*, 1643–1651. [CrossRef]
216. Rudzińska, M.; Daglioglu, C.; Savvateeva, L.V.; Kaci, F.N.; Antoine, R.; Zamyatnin, A.A., Jr. Current Status and Perspectives of Protease Inhibitors and Their Combination with Nanosized Drug Delivery Systems for Targeted Cancer Therapy. *Drug Des. Dev. Ther.* **2021**, *15*, 9. [CrossRef]
217. Awad, M.F.; El-Shenawy, F.S.; El-Gendy, M.M.A.A.; El-Bondkly, E.A.M. Purification, characterization, and anticancer and antioxidant activities of l-glutaminase from Aspergillus versicolor Faesay4. *Int. Microbiol.* **2021**, *24*, 169–181. [CrossRef]
218. Qamar, S.A.; Asgher, M.; Bilal, M. Immobilization of alkaline protease from Bacillus brevis using Ca-alginate entrapment strategy for improved catalytic stability, silver recovery, and dehairing potentialities. *Catal. Lett.* **2020**, *150*, 3572–3583. [CrossRef]
219. Bholay, A.; More, S.; Patil, V.; Niranjan, P. Bacterial extracellular alkaline proteases and its industrial applications. *Int. Res. J. Biol. Sci.* **2012**, *1*, 1–5.
220. Abad, P.; Gouzy, J.; Aury, J.-M.; Castagnone-Sereno, P.; Danchin, E.G.; Deleury, E.; Perfus-Barbeoch, L.; Anthouard, V.; Artiguenave, F.; Blok, V.C. Genome sequence of the metazoan plant-parasitic nematode Meloidogyne incognita. *Nat. Biotechnol.* **2008**, *26*, 909–915. [CrossRef]
221. Nakpathom, M.; Somboon, B.; Narumol, N. Papain enzymatic degumming of Thai Bombyx mori silk fibers. *J. Microsc. Soc. Thail.* **2009**, *23*, 142–146.
222. Miao, Y.; Chen, G.; Xi, X.; Ma, C.; Wang, L.; Burrows, J.F.; Duan, J.; Zhou, M.; Chen, T. Discovery and rational design of a novel bowman-birk related protease inhibitor. *Biomolecules* **2019**, *9*, 280. [CrossRef]
223. Zhai, P.-b.; Qing, J.; Li, B.; Zhang, L.-q.; Ma, L.; Chen, L. GP205, a new hepatitis C virus NS3/4A protease inhibitor, displays higher metabolic stability in vitro and drug exposure in vivo. *Acta Pharmacol. Sin.* **2018**, *39*, 1746–1752. [CrossRef]
224. Goris, T.; Pérez-Valero, Á.; Martínez, I.; Yi, D.; Fernández-Calleja, L.; San Leon, D.; Bornscheuer, U.T.; Magadán-Corpas, P.; Lombo, F.; Nogales, J. Repositioning microbial biotechnology against COVID-19: The case of microbial production of flavonoids. *Microb. Biotechnol.* **2021**, *14*, 94–110. [CrossRef]
225. Cotabarren, J.; Lufrano, D.; Parisi, M.G.; Obregón, W.D. Biotechnological, biomedical, and agronomical applications of plant protease inhibitors with high stability: A systematic review. *Plant Sci.* **2020**, *292*, 110398. [CrossRef]
226. Rawlings, N.D. Peptidase inhibitors in the MEROPS database. *Biochimie* **2010**, *92*, 1463–1483. [CrossRef]

227. Garg, H.; Suri, P.; Gupta, J.C.; Talwar, G.; Dubey, S. Survivin: A unique target for tumor therapy. *Cancer Cell Int.* **2016**, *16*, 49. [CrossRef]
228. Rawlings, N.D.; Barrett, A.J.; Finn, R. Twenty years of the MEROPS database of proteolytic enzymes, their substrates and inhibitors. *Nucleic Acids Res.* **2016**, *44*, D343–D350. [CrossRef]
229. Gutierrez-Gongora, D.; Geddes-McAlister, J. From Naturally-Sourced Protease Inhibitors to New Treatments for Fungal Infections. *J. Fungi* **2021**, *7*, 1016. [CrossRef]
230. Kantyka, T.; Rawlings, N.D.; Potempa, J. Prokaryote-derived protein inhibitors of peptidases: A sketchy occurrence and mostly unknown function. *Biochimie* **2010**, *92*, 1644–1656. [CrossRef]
231. Greenbaum, L.M.; Sutherland, J.H. Host cathepsin D response to tumor in the normal and pepstatin-treated mouse. *Cancer Res.* **1983**, *43*, 2584–2587.
232. Kalchev, K.; Rabadjiev, Y.; Ganchev, D.; Tsenova, M.; Iliev, I.; Ivanova, I. Study of proteases and protease inhibitors from Streptomyces strains. *Bulg. J. Agric. Sci.* **2013**, *19*, 65–67.
233. Manojlovic, N.T.; Solujic, S.; Sukdolak, S.; Milosev, M. Antifungal activity of Rubia tinctorum, Rhamnus frangula and Caloplaca cerina. *Fitoterapia* **2005**, *76*, 244–246. [CrossRef]
234. Menéndez-Arias, L.; Tözsér, J. HIV-1 protease inhibitors: Effects on HIV-2 replication and resistance. *Trends Pharmacol. Sci.* **2008**, *29*, 42–49. [CrossRef]
235. Chandwani, A.; Shuter, J. Lopinavir/ritonavir in the treatment of HIV-1 infection: A review. *Ther. Clin. Risk Manag.* **2008**, *4*, 1023.
236. Fear, G.; Komarnytsky, S.; Raskin, I. Protease inhibitors and their peptidomimetic derivatives as potential drugs. *Pharmacol. Ther.* **2007**, *113*, 354–368. [CrossRef]
237. Purohit, R.; Rajendran, V.; Sethumadhavan, R. Studies on adaptability of binding residues flap region of TMC-114 resistance HIV-1 protease mutants. *J. Biomol. Struct. Dyn.* **2011**, *29*, 137–152. [CrossRef]
238. van Maarseveen, N.M.; Andersson, D.; Lepšík, M.; Fun, A.; Schipper, P.J.; de Jong, D.; Boucher, C.A.; Nijhuis, M. Modulation of HIV-1 Gag NC/p1 cleavage efficiency affects protease inhibitor resistance and viral replicative capacity. *Retrovirology* **2012**, *9*, 29. [CrossRef] [PubMed]
239. Liu-Young, G.; Kozal, M.J. Hepatitis C protease and polymerase inhibitors in development. *AIDS Patient Care STDs* **2008**, *22*, 449–457. [CrossRef] [PubMed]
240. Onoda, T.; Ono, T.; Dhar, D.K.; Yamanoi, A.; Nagasue, N. Tetracycline analogues (doxycycline and COL-3) induce caspase-dependent and-independent apoptosis in human colon cancer cells. *Int. J. Cancer* **2006**, *118*, 1309–1315. [CrossRef] [PubMed]
241. Sridhar, S.S.; Shepherd, F.A. Targeting angiogenesis: A review of angiogenesis inhibitors in the treatment of lung cancer. *Lung Cancer* **2003**, *42*, 81–91. [CrossRef]
242. Rawlings, N.D.; Tolle, D.P.; Barrett, A.J. Evolutionary families of peptidase inhibitors. *Biochem. J.* **2004**, *378*, 705–716. [CrossRef] [PubMed]
243. Clemente, M.; Corigliano, M.G.; Pariani, S.A.; Sánchez-López, E.F.; Sander, V.A.; Ramos-Duarte, V.A. Plant serine protease inhibitors: Biotechnology application in agriculture and molecular farming. *Int. J. Mol. Sci.* **2019**, *20*, 1345. [CrossRef]
244. Peele, K.A.; Durthi, C.P.; Srihansa, T.; Krupanidhi, S.; Ayyagari, V.S.; Babu, D.J.; Indira, M.; Reddy, A.R.; Venkateswarulu, T. Molecular docking and dynamic simulations for antiviral compounds against SARS-CoV-2: A computational study. *Inform. Med. Unlocked* **2020**, *19*, 100345. [CrossRef]
245. Macchiagodena, M.; Pagliai, M.; Procacci, P. Identification of potential binders of the main protease 3CLpro of the COVID-19 via structure-based ligand design and molecular modeling. *Chem. Phys. Lett.* **2020**, *750*, 137489. [CrossRef]
246. Rut, W.; Groborz, K.; Zhang, L.; Sun, X.; Zmudzinski, M.; Pawlik, B.; Wang, X.; Jochmans, D.; Neyts, J.; Młynarski, W. SARS-CoV-2 M pro inhibitors and activity-based probes for patient-sample imaging. *Nat. Chem. Biol.* **2021**, *17*, 222–228. [CrossRef]
247. Drag, M.; Salvesen, G.S. Emerging principles in protease-based drug discovery. *Nat. Rev. Drug Discov.* **2010**, *9*, 690–701. [CrossRef]
248. Amin, S.A.; Banerjee, S.; Ghosh, K.; Gayen, S.; Jha, T. Protease targeted COVID-19 drug discovery and its challenges: Insight into viral main protease (Mpro) and papain-like protease (PLpro) inhibitors. *Bioorganic Med. Chem.* **2020**, *29*, 115860. [CrossRef]
249. Lee, T.-W.; Cherney, M.M.; Huitema, C.; Liu, J.; James, K.E.; Powers, J.C.; Eltis, L.D.; James, M.N. Crystal structures of the main peptidase from the SARS coronavirus inhibited by a substrate-like aza-peptide epoxide. *J. Mol. Biol.* **2005**, *353*, 1137–1151. [CrossRef]
250. Shen, S.; Wu, Y.; Liu, Y.; Wu, D. High drug-loading nanomedicines: Progress, current status, and prospects. *Int. J. Nanomed.* **2017**, *12*, 4085. [CrossRef]
251. Shen, S.; Du, X.-J.; Liu, J.; Sun, R.; Zhu, Y.-H.; Wang, J. Delivery of bortezomib with nanoparticles for basal-like triple-negative breast cancer therapy. *J. Control. Release* **2015**, *208*, 14–24. [CrossRef]
252. Park, J.E.; Park, J.; Jun, Y.; Oh, Y.; Ryoo, G.; Jeong, Y.-S.; Gadalla, H.H.; Min, J.S.; Jo, J.H.; Song, M.G. Expanding therapeutic utility of carfilzomib for breast cancer therapy by novel albumin-coated nanocrystal formulation. *J. Control. Release* **2019**, *302*, 148–159. [CrossRef]
253. Gotou, T.; Shinoda, T.; Mizuno, S.; Yamamoto, N. Purification and identification of proteolytic enzymes from Aspergillus oryzae capable of producing the antihypertensive peptide Ile-Pro-Pro. *J. Biosci. Bioeng.* **2009**, *107*, 615–619. [CrossRef]
254. Song, R.; Qiao, W.; He, J.; Huang, J.; Luo, Y.; Yang, T. Proteases and Their Modulators in Cancer Therapy: Challenges and Opportunities. *J. Med. Chem.* **2021**, *64*, 2851–2877. [CrossRef]

255. Cavaco, M.; Andreu, D.; Castanho, M.A. The challenge of peptide proteolytic stability studies: Scarce data, difficult readability, and the need for harmonization. *Angew. Chem. Int. Ed.* **2021**, *60*, 1686–1688. [CrossRef]
256. Verma, S.; Goand, U.K.; Husain, A.; Katekar, R.A.; Garg, R.; Gayen, J.R. Challenges of peptide and protein drug delivery by oral route: Current strategies to improve the bioavailability. *Drug Dev. Res.* **2021**, *82*, 927–944. [CrossRef]
257. Sharma, A.; Gupta, G.; Ahmad, T.; Mansoor, S.; Kaur, B. Enzyme engineering: Current trends and future perspectives. *Food Rev. Int.* **2021**, *37*, 121–154. [CrossRef]
258. Nyika, J.M. The Use of Microorganism-Derived Enzymes for Bioremediation of Soil Pollutants. In *Recent Advancements in Bioremediation of Metal Contaminants*; IGI Global: Hershey/Pennsylvania, PA, USA, 2021; pp. 54–71.
259. Saravanan, A.; Kumar, P.S.; Vo, D.-V.N.; Jeevanantham, S.; Karishma, S.; Yaashikaa, P. A review on catalytic-enzyme degradation of toxic environmental pollutants: Microbial enzymes. *J. Hazard. Mater.* **2021**, *419*, 126451. [CrossRef]
260. Li, Q.; Yi, L.; Marek, P.; Iverson, B.L. Commercial proteases: Present and future. *FEBS Lett.* **2013**, *587*, 1155–1163. [CrossRef]
261. Chen, X.; Shen, H.; Shao, Y.; Ma, Q.; Niu, Y.; Shang, Z. A narrative review of proteolytic targeting chimeras (PROTACs): Future perspective for prostate cancer therapy. *Transl. Androl. Urol.* **2021**, *10*, 954. [CrossRef]

Journal of *Fungi*

MDPI

*Review*

# Recent Advances in *Alternaria* Phytotoxins: A Review of Their Occurrence, Structure, Bioactivity, and Biosynthesis

**He Wang [1], Yanjing Guo [1], Zhi Luo [1], Liwen Gao [1], Rui Li [2], Yaxin Zhang [1], Hazem M. Kalaji [3,4], Sheng Qiang [1] and Shiguo Chen [1,*]**

1 Weed Research Laboratory, College of Life Science, Nanjing Agricultural University, Nanjing 210095, China; 2018216001@njau.edu.cn (H.W.); 2021216003@stu.njau.edu.cn (Y.G.); 2020116003@stu.njau.edu.cn (Z.L.); gaoliwen0905@163.com (L.G.); 2019116003@njau.edu.cn (Y.Z.); wrl@njau.edu.cn (S.Q.)

2 Agricultural and Animal Husbandry Ecology and Resource Protection Center, Ordos Agriculture and Animal Husbandry Bureau, Ordos 017010, China; lirui01@163.com

3 Department of Plant Physiology, Institute of Biology, Warsaw University of Life Sciences SGGW, 159 Nowoursynowska 159, 02-776 Warsaw, Poland; hazem@kalaji.pl

4 Institute of Technology and Life Sciences—National Research Institute, Falenty, Al. Hrabska 3, 05-090 Raszyn, Poland

* Correspondence: chenshg@njau.edu.cn; Tel.: +86-25-84395117

**Abstract:** *Alternaria* is a ubiquitous fungal genus in many ecosystems, consisting of species and strains that can be saprophytic, endophytic, or pathogenic to plants or animals, including humans. *Alternaria* species can produce a variety of secondary metabolites (SMs), especially low molecular weight toxins. Based on the characteristics of host plant susceptibility or resistance to the toxin, *Alternaria* phytotoxins are classified into host-selective toxins (HSTs) and non-host-selective toxins (NHSTs). These *Alternaria* toxins exhibit a variety of biological activities such as phytotoxic, cytotoxic, and antimicrobial properties. Generally, HSTs are toxic to host plants and can cause severe economic losses. Some NHSTs such as alternariol, altenariol methyl-ether, and altertoxins also show high cytotoxic and mutagenic activities in the exposed human or other vertebrate species. Thus, *Alternaria* toxins are meaningful for drug and pesticide development. For example, AAL-toxin, maculosin, tentoxin, and tenuazonic acid have potential to be developed as bioherbicides due to their excellent herbicidal activity. Like altersolanol A, bostrycin, and brefeldin A, they exhibit anticancer activity, and ATX V shows high activity to inhibit the HIV-1 virus. This review focuses on the classification, chemical structure, occurrence, bioactivity, and biosynthesis of the major *Alternaria* phytotoxins, including 30 HSTs and 50 NHSTs discovered to date.

**Keywords:** *Alternaria* toxins; HSTs; NHSTs; biological activities; biosynthesis

**Citation:** Wang, H.; Guo, Y.; Luo, Z.; Gao, L.; Li, R.; Zhang, Y.; Kalaji, H.M.; Qiang, S.; Chen, S. Recent Advances in *Alternaria* Phytotoxins: A Review of Their Occurrence, Structure, Bioactivity, and Biosynthesis. *J. Fungi* **2022**, *8*, 168. https://doi.org/10.3390/jof8020168

Academic Editor: Baojun Xu

Received: 31 December 2021
Accepted: 7 February 2022
Published: 9 February 2022



## 1. Introduction

The fungal genus *Alternaria* is a widespread and successful group growing in diverse environments worldwide, ranging from saprophytes to pathogens and even endophytes. The genus *Alternaria* was identified in the year 1816 [1]. Currently, about 300 species have been described based on phylogenetic and morphological studies, which have been further divided into 26 sections [2–4]. As an outstanding group of fungal pathogens, *Alternaria* species can either cause diseases in a wide range of economically important crops [1], resulting in significant economic losses, or affect human and animal health, such as through upper respiratory tract infections and asthma [4,5].

To date, over 70 toxins with different chemical structures and behaviors are known to be produced by *Alternaria* species [6]. These toxins often exhibit a variety of bioactivities, such as phytotoxic, cytotoxic, and antimicrobial properties, etc. Generally, *Alternaria* phytotoxins are divided into host-selective toxins (HSTs) and non-host-selective toxins (NHSTs) based on the susceptibility or resistance of the host. HSTs are toxic only to host plants. In contrast, NHSTs can affect many plants, regardless of whether they are a host or non-host of the pathogen producing them [7]. Most HSTs have been considered as pathogenicity factors required for fungi to invade tissues and cause disease. On the other hand, NHSTs may contribute to the development of symptoms and the proliferation of plant pathogens [8,9].

Here, we review the toxins produced by *Alternaria* spp. and summarize the classification, occurrence, mode of action, biological activity, biosynthesis, and development value of each toxin. The phytotoxins presented in the paper will be termed "toxins", and those toxic to animals will be termed "mycotoxins".

## 2. Host-Selective Toxins

In this section, we reviewed 30 HSTs of *Alternaria* and summarised the related pathotypes, diseases caused, chemical properties, targets in plant organelles, and biosynthetic pathways of these toxins (Table 1). Based on their chemical structures, the HSTs of *Alternaria* can be classified into seven classes: (1) epoxy-decatrienoic acid (AK-toxins, AF-toxins, and ACT-toxins); (2) sphinganine analogue (AAL-toxins); (3) pyranones (ACR-toxins); (4) cyclic peptide (AM-toxins, destruxin B, and HC-toxin); (5) tetrapeptide (AS-I toxin); (6) diketopiperazine (maculosin); and (7) ribosomal peptide (ABR-toxin). In fact, the classes (4), (5), and (6) also fall into the larger family of non-ribosomal peptides.

**Table 1.** Host-selective toxins produced by *Alternaria* species.

| Toxins | *Alternaria* Species | Host Range | References |
|---|---|---|---|
| AK-toxins (AK-toxin I, II) | *A. alternata* f. sp. *kikuchana* (Japanese pear pathotype) | Japanese pear | [9–11] |
| AF-toxins (AF-toxin I, II, III) | *A. alternata* f. sp. *Fragariae* (Strawberry pathotype) | Strawberry | [12] |
| ACT-toxins (ACT-toxin I, II) | *A. alternata* f. sp. *citri tangerine* (Tangerine pathotype) | Tangerine | [13–15] |
| AAL-toxins ($TA_1$, $TA_2$, $TB_1$, $TB_2$, $TC_1$, $TC_2$, $TD_1$, $TD_2$, $TE_1$, $TE_2$) | *A. alternata* f. sp. *lycopersici* (Tomato pathotype) | Tomato | [16,17] |
| ACR-toxins (ACR-toxin I, II, III, IV, IV′) | *A. alternata* f. sp. *citri jambhiri* (Rough lemon pathotype) | Rough lemon | [18,19] |
| AM-toxins (AM-toxin I, II, III) | *A. alternata* f. sp. mali (Apple pathotype) | Apple | [20,21] |
| Destruxin B | *A. brassicae* | *Brassica* spp. | [22,23] |
| HC-toxin | *C. carbonum* and *A. jesenskae* | Maize | [24–26] |
| Maculosin | *A. alternata* (Spotted knapweed pathotype) | knapweed | [27,28] |
| AS-I toxin | *A. alternata* (Sunflower Pathotype) | Sunflower | [29] |
| ABR-toxin | *A. brassicae* | *Brassica* spp. | [23] |

### 2.1. AK-Toxins, AF-Toxins, and ACT-Toxins

AK-toxins produced by the Japanese pear pathotype of *A. alternata* f. sp. *Kikuchana* were first described in Japanese pear black spot disease [9–11]. The same researchers identified the chemical structure, absolute configuration, and biological activity of these toxins [10]. AK-toxins are the esters of 9,10-epoxy-8-hydroxy-9-methyl-decatrienoic acid (EDA), which are the derivative of phenylalanine and hydroxyldecartienoic acid. AK-toxins consist of two types, AK-toxins I and II. Both are also mixtures of three geometric isomers, namely type-a (*2E*, *4E*, *6Z*), type-b (*2E*, *4Z*, *6E*), and type-c (*2E*, *4E*, *6E*). For each compound, the main geometry is type-b (Figure 1a) [30]. Both toxins showed toxicity only in susceptible pear cultivars, and AK-toxin I was more abundant and showed higher biological activity [31,32]. In Nijisseike, a susceptible Japanese pear cultivar, the concentration that caused venous necrosis was 5 nM of AK-toxin I or 100 nM of AK-toxin II. However, at

0.1 mM of AK-toxins I and II, there was no effect on the leaves of a resistant cultivar such as Chojuro [31].

*Alternaria* black spot disease of strawberry was first reported in 1977 and the causal pathogen was identified as *A. alternata* strawberry pathotype (*A. alternata* f. sp. *fragariae*) [12]. The pathogen produces three key molecules, AF-toxins I, II, and III (Figure 1b). AF-toxins I and II were isolated in 1979 and AF-toxin III was isolated in 1984. The chemical structures of these three toxins were first determined in 1986 [33]. The three AF-toxins have the same EDA structures, which are very similar to the AK-toxins. The conformation of the EDA parts of the AF-toxin is type-a (*2E*, *4E*, *6Z*). Of these three toxins, AF-toxin I is toxic to strawberries and pears, AF-toxin II shows toxicity to pears only, and AF-toxin III shows high toxicity to strawberries but low toxicity to pears [34].

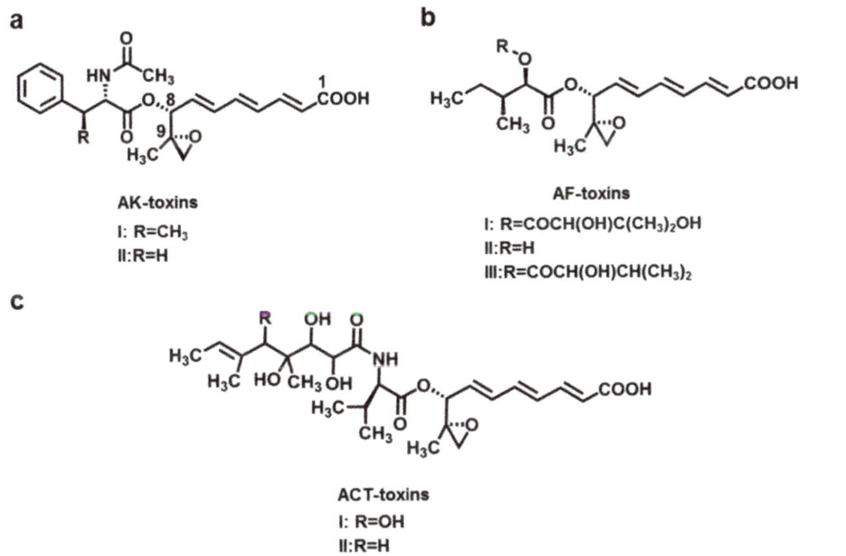


**Figure 1.** Chemical structures of AK-toxins (**a**), AF-toxins (**b**), and ACT-toxins (**c**).

*Alternaria* brown spot disease of the emperor mandarin was first reported in Australia in 1903, and the pathogen was identified as the mandarin pathotype of *A. alternata* f. sp. *citri tangerine* in 1966 [13]. This pathotype is highly toxic to mandarins, tangerines, grapefruit, and hybrids of grapefruit and tangerine, as well as mandarin and sweet orange [14]. The crucial pathogenicity depends on the action of ACT-toxins (Figure 1c). It can cause brown to black spots on young leaves, twigs, and fruits of tangerines. ACT-toxins can also be transmitted through the veins and cause more severe lesions [15]. ACT-toxins at a concentration of $2 \times 10^{-8}$ M can cause necrotic lesions on citrus leaves with rapid electrolyte loss from the host cells. The ACT-toxins consist of three components, EDA, valine, and polyketide. ACT-toxins have two types that differ only in the R group. The conformation of the EDA component of ACT-toxins is the type B form (*2E*, *4Z*, *6E*). ACT-toxins are more abundant and toxic to citrus [35].

The target of action of AK-, AF-, and ACT-toxins is the plasma membrane of susceptible cells [34–37]. They cause a sudden, and markedly increased, $K^+$ loss from the plasma membrane after a few minutes of toxin treatment, resulting in membrane invagination, vesiculation, fragmentation, and depolarization, which causes a decrease in the membrane potential gradient [37,38]. Within 1–3 h after toxin treatment, Golgi vesicles fuse with the damaged plasma membrane [37]. No damage was observed in intracellular organelles, except for the plasma membrane of host cells. Of these three toxins, AK-toxins and AF-toxins irreversibly depolarized the plasma membrane of susceptible genotypes and could directly affect the plasma membrane $H^+$-ATPase [39–42]. In the case of AK-toxins, the configuration at C-8 and C-9 was critical for phytotoxicity [43].

Recently, some genes were discovered to play important biological and pathological roles in the pathotype of *A. alternata*. Two NADPH (nicotinamide adenine dinucleotide phosphate) oxidase genes (NoxA and NoxB) were identified, and NoxB was found to be essential for the aggressiveness and basal pathogenicity of *A. alternata* [44]. The gene *PEX6*, encoding a protein required for the import of matrix proteins into peroxisomes, has been characterized in *A. alternata*. It plays a role in ROS (reactive oxygen species)-induced resistance and fungal pathogenicity in the mandarin pathotype of *A. alternata* [45].

AK-toxins, AF-toxins, and ACT-toxins have a common component, EDA, in their structures [10,11,35]. In a previous study, based on the [2-$^{13}$C]-sodium acetate feeding study of the Japanese pear pathotype of *A. alternata* and $^{13}$C NMR spectrum analysis, it was demonstrated that AK-toxins are biosynthesized from acetic acid via EDA [46]. In another study, $^{3}$H-labeled EDA was added to a growing liquid culture of the strain of the Japanese pear pathotype and was efficiently converted to AK-toxins. These results confirmed that EDA is an intermediate for toxin pathways [15].

The gene cluster involved in HST biosynthesis of *A. alternata* pathogens was first isolated from the Japanese pear pathotype, including *AKT1*, *AKT2*, *AKT3*, *AKT4*, *AKTR*, and *AKTS1* [47,48]. Recently, another gene, *AKT7*, encoding a cytochrome P450 monooxygenase was found to have the function of limiting the production of AK-toxin [7]. The biosynthetic genes of AF-toxins and ACT-toxins were identified by genomic cosmid libraries of the two pathotypes screened with the *AKT* gene probes [38]. For the biosynthetic genes of AF-toxins (*AFT*-genes), eleven *AFT*-genes and five transposon-like sequences (*TLS-S1* to *TLS-S5*) were isolated [49]. Among them, *AFT1*, *AFT3*, and *AFTR* show strong similarity to *AKT1*, *AKT3*, and *AKTR*, respectively [49]. The biosynthetic pathway of ACT-toxins was also found to be regulated by several genes, including *ACTT1*, *ACTT2*, *ACTT3*, *ACTT5*, *ACTT6*, *ACTTS2*, and *ACTTS3* [9,50–52], and *ACTT1* and *ACTT2* were considered to be the highly homologous genes of *AKT1* and *AKT2*, respectively, in the Japanese pear pathotype [52]. *AKT1*, *AKT2*, and *AKT3* were identified as involved in the biosynthesis of EDA, a common component of AK-, AF-, and ACT-toxins in the Japanese pear pathotype, as well as their orthologs in the strawberry and tangerine pathotypes [53]. Recently, a transcriptional regulator ACTR was identified to contribute to the biosynthesis of ACT-toxins via the mediator gene *ACTS4* in *A. alternata* [54]. These three genes were clustered on small chromosomes of less than 2.0 Mb in three pathotypic strains. They are not required for growth but confer an advantage in colonizing certain ecological niches [49,51,55–57].

### 2.2. AAL-Toxins

*Alternaria* stem canker disease is a serious disease of tomato (*Lycopersicon esculentum* Mill.). The disease was first described in 1975 [58]. It caused dark brown to black cankers on the stems of some tomato cultivars by a pathogenic strain, *A. alternata* f. sp. *Lycopersici* [16,17]. AAL-toxins were the main causative agent of the disease produced by the above pathogen. The first AAL-toxin was isolated in 1981 and its chemical structures, TA and TB, analogues of sphingosine and sphinganine, respectively, were determined [59–61]. To date, five types of AAL-toxin-related molecules, TA and TB, TC, TD, and TE, have been identified. Each of these fractions consisted of a mixture of two structural isomers (Figure 2). TA and TB showed toxicity to detached tomato leaves at 10 $ng \cdot mL^{-1}$. The toxicity of TD and TE is over 100 times lower than that of the form TA. The activity of TC was lower than that of TA, but higher than that of TD and TE [62]. Unlike other HSTs produced by *A. alternata*, the AAL-toxins can attack many other weeds, crops, and at least 25 species of solanaceous plants in addition to the susceptible tomato host [63,64]. On the other hand, some crops (e.g., maize, wheat, and resistant tomato varieties) are tolerant to AAL-toxins. Thus, AAL-toxins have been considered as very low-dose herbicides against a variety of broadleaf weeds such as datura, pricklesida, and black nightshade [63,65]. In addition, AAL-toxins are also toxic to cultured mammalian cells. The $IC_{50}$ value for the most sensitive hepatoma line, H4TG, was 10 μg·mL$^{-1}$ [66]. Such fact did limit the development of AAL-toxins as herbicides compared with some common herbicides, such as glyphosate, that are less toxic to mammals with the $LD_{50}$ ranging from 800 to >5000 mg·kg$^{-1}$ body weight for different animal species [67]. Recently, some AAL-toxin analogues were synthesized and one of them showed significant phytotoxicity and low mammalian toxicity, giving them potential for being developed as safe and effective natural herbicides [68,69].

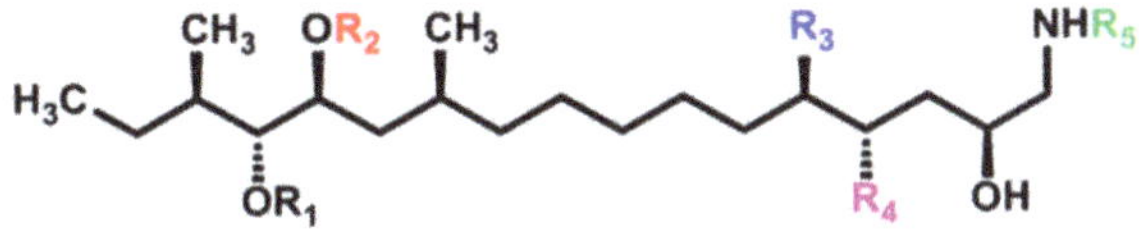


**AAL-toxins**

**TA$_1$: R$_1$=H, R$_2$= CO-CH$_2$-CH(COOH)-CH$_2$-COOH, R$_3$=OH, R$_4$ =OH, R$_5$=H**
**TA$_2$: R$_1$=CO-CH$_2$-CH(COOH)-CH$_2$ -COOH, R$_2$=H, R$_3$=OH, R$_4$=OH, R$_5$=H**
**TB$_1$: R$_1$=H, R$_2$= CO-CH$_2$CH(COOH)-CH $_2$-COOH, R$_3$=H, R$_4$= OH, R$_5$=H**
**TB$_2$: R$_1$=CO-CH$_2$-CH(COOH)-CH$_2$-COOH, R$_2$=H, R$_3$=H, R$_4$=OH, R$_5$=H**
**TC$_1$: R$_1$=H, R$_2$=CO-CH$_2$-CH(COOH)-CH$_2$-COOH, R$_3$=H,R$_4$=H,R$_5$=H**
**TC$_2$: R$_1$=CO-CH$_2$-CH(COOH)-CH$_2$-COOH, R$_2$=H, R$_3$=H, R$_4$=H, R$_5$=H**
**TD$_1$: R$_1$=H, R$_2$=CO-CH$_2$-CH(COOH)-CH$_2$ -COOH, R$_3$=H, R$_4$=OH, R$_5$=COCH$_3$**
**TD$_2$: R$_1$=CO-CH$_2$-CH(COOH)-CH $_2$ -COOH, R$_2$ = H, R$_3$ = H, R$_4$= OH, R$_5$=COCH$_3$**
**TE$_1$: R$_1$=H, R$_2$=CO-CH$_2$-CH(COOH)-CH$_2$-COOH, R$_3$=H, R$_4$=H, R$_5$=COCH$_3$**
**TE$_2$: R$_1$=CO-CH$_2$-CH(COOH)-CH$_2$-COOH, R$_2$=H, R$_3$=H, R$_4$=H, R$_5$=COCH$_3$**

**Figure 2.** Chemical structures of AAL-toxins.

When susceptible tomato leaves were treated with AAL-toxins, the accumulation of two amines, ethanolamine (EA) and phosphoethanolamine (PEA), occurred. This implies that AAL-toxins could interfere with amine metabolism [70]. When the $^{14}C$ label of EA was fed to susceptible leaf disks treated with AAL-toxins, there was a strong inhibition of the uptake of EA into phosphatidylethanolamine (PtdEA). This phenomenon suggests possible biochemical targets of AAL-toxins, which could be enzymes involved in the phospholipid pathway [71].

Based on their chemical structure, AAL-toxins are analogous sphinganine mycotoxins (SAMTs). The SAMTs cause competitive inhibition of ceramide synthase, suppressing the conversion of sphinganine, phytosphingosine, and other free sphingoid bases into complex ceramides. The resulting accumulation of free sphingoid bases acts as a second message that activates programmed cell death (PCD) transduction pathways [72,73]. When sensitive tomato tissues were treated with AAL-toxins, sphinganine and phytosphingosine accumulated in the tissue [74]. However, this phenomenon can be avoided by ceramide supplementation, suggesting that an imbalance of ceramide is critical for triggering cell death [61,75]. Further studies have shown that both jasmonic acid (JA) and ethylene can promote AAL-toxin-induced PCD in tomato leaves by interfering with sphingolipid metabolism [76]. AAL-toxin-induced PCD is associated with ceramide signaling and cell cycle disruption. The final physiological effects of AAL-toxins are the development of necrotic lesions on fruits and leaves, the inhibition of in vitro development of calli, pollen, roots, and shoots, and the reduction of protoplast and suspension cell viability [77].

Previous studies on feeding with labeled precursors showed that glycine and the methyl group of methionine were directly incorporated into AAL-toxins. The oxygen groups in the tricarboxylic acid moieties of AAL-toxins were derived from $H_2O$. The hydroxyl groups of the lipid backbone of the AAL-toxins were derived from molecular oxygen [78]. The AAL-toxin biosynthetic gene *ALT1* was identified, which encodes a type I PKS. ALT1 consists of seven domains that include α-ketoacyl synthase (KS), acyltransferase (AT), dehydratase (DH), methyl transferase (MT), β-ketoacyl reductase (KR), enoyl reductase (ER), and acyl carrier protein (ACP) [78]. Recently, a genomic BAC library of the tomato pathotype was screened using the *ALT1* probe. A 120-kb genomic region includes at least 13 genes involved in the biosynthesis of AAL-toxins. In addition to *ALT1*, the *ALT2*, *ALT3*, *ALT6*, and *ALT13* genes were also identified, encoding cytochrome P450 monooxygenase, aminotransferase, short-chain dehydrogenase/reductase, and Zn(II)2Cys6 transcription factor, respectively. *ALT* genes are located on a single small chromosome of about 1.0 Mb in the tomato pathotype strain [9].

### 2.3. ACR-Toxins

*Alternaria* brown spot disease of rough lemon was first discovered in South Africa [79]. The pathotype RLP (rough lemon) of *A. alternata* is the culprit. It can infect common citrus root species such as rough lemon (*Citrus jambhiri* Lush.) and rangpur line (*C. limonia* Osbeck) in some citrus growing areas [15,18,19]. The virulence of *A. alternata* RLP is due to the production of ACR-toxins, which may also be called ACRL-toxins [18,31,80,81]. ACR-toxins contain five compounds with different chain lengths, all of which have an $\alpha$-pyrone group (Figure 3). The main form of ACR-toxins (ACR-toxin I, MW = 496) consists of an $\alpha$-dihydropyrone ring in a polyalcohol with 19 carbon atoms [10,31]. ACR-toxins can cause brown necrosis on rough lemon leaves at 0.1 $\mu g \cdot mL^{-1}$, but did not affect mandarins and other non-hosts even at 1000 $\mu g \cdot mL^{-1}$ [18].

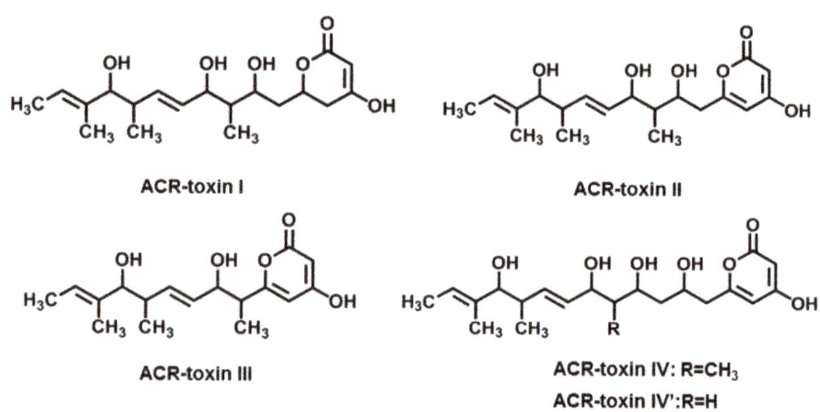

**Figure 3.** Chemical structures of ACR-toxins.

The target site of ACR-toxins is the mitochondrion, leading to mitochondrial dysfunction in rough lemon. ACR-toxins not only caused the uncoupling of mitochondrial oxidation phosphorylation, but also led to the exit of the cofactor $NAD^+$ from the TCA (tricarboxylic acid) cycle [82]. The ACRS (ACR-toxin sensitivity gene), which confers sensitivity to ACR-toxins in citrus species, was identified in the mitochondrial genome of rough lemon [83]. The sensitivity was controlled by the post-transcriptional modification of the ACRS transcript.

The rough lemon pathotype strain also carried a small chromosome of 1.2–1.5 Mb, and the presence of this chromosome was associated with ACR-toxin production and rough lemon pathogenicity [51]. Several ACRT genes responsible for the biosynthesis of ACR-toxins were identified by sequence analysis of the 1.5 Mb chromosome. *ACRTS1*, *ACRTS2*, and *ACRTS3* were characterized, encoding a putative hydroxylase, a putative reducing polyketide synthase type I (PKS), and a putative cyclase, respectively. All genes were closely related to ACR-toxin production and pathogenicity [84,85]. These genes are unique to the producers of ACR-toxins of the rough lemon pathotype [86].

### 2.4. AM-Toxins

Apple cultivars such as Indo and Delicious are highly susceptible to a pathogenic strain of *A. alternata* f. sp. *mali* that can cause severe economic losses, especially in Japanese orchards [20,21]. In 1974, AM-toxins were first isolated from *A. mali*, the apple pathotype of *A. alternata* that causes apple leaf spot disease, and structural studies were conducted. AM-toxins have three distinct types (I, II, and III. Figure 4) and are produced and released by both germinating conidia and cultured mycelia of the strain. Each toxin is a four-membered cyclic depsipeptide. AM-toxin I is the most abundant among AM-toxins, causing necrosis on leaves of highly susceptible apple cultivars at concentrations of $10^{-8}$ M [20,21,87].

The plasma membrane and chloroplasts are two targets of AM-toxins for susceptible apple cells [88]. Similar to AK-toxins, AM-toxins can also cause plasma membrane invagination and electrolyte loss. However, the effect of AM-toxins on Japanese pear was weaker than that of AK-toxins [88]. Membrane fragments and vesicles appeared in the chloroplasts, which had emerged from grana lamellae within 3 h after toxin treatment. Chloroplast disorganization was accompanied by a decrease in chlorophyll content and inhibition of photosynthetic $CO_2$ assimilation [89]. The photosynthetic activity of chloroplasts was inhibited. This phenomenon suggests that the chloroplast is a primary target of AM-toxins [38,41,90].

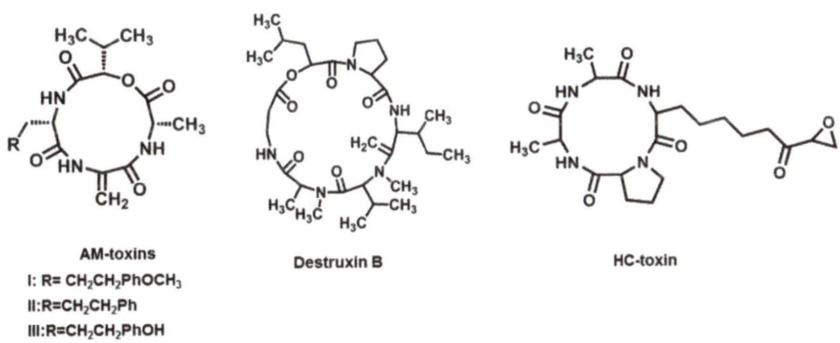


**Figure 4.** Chemical structures of AM-toxins, destruxin B, and HC-toxin.

AM-toxins belong to the cyclic peptides that are generally synthesized via non-ribosomal pathways by non-ribosomal peptide synthetases (NPRS) [91]. The AMT1, AMT2, AMT3, and AMT4 genes have been successfully isolated encoding proteins resembling enzymes involved in the secondary metabolism and modification of amino acids [92,93]. In 2007, a bacterial artificial chromosome (BAC) was isolated containing four AMT genes and other genes that are upregulated in AM-toxin-producing cultures, suggesting that genes for AM-toxin biosynthesis are clustered in the genome. It also revealed that the AMT genes are located on a conditionally dispensable (CD) chromosome of <with a size of 1.8 Mb in the strain [94].

### 2.5. Destruxin B

Black spot disease of *Brassica* spp. such as *B. campestris* and *B. napus* is caused by *A. brassicae* (Berk.) Sacc. The symptoms of the disease are lesions with grayish, brownish, or blackish centers and chlorotic margins on all above-ground parts of the plant, especially leaves, stems, and even siliques, resulting in huge economic losses in yields of about 40–60% [22,23]. The oil and protein content of the seeds is also significantly reduced, especially in *B. campestris*. Destruxin B, a cyclic peptide, is an HST that was first isolated from *A. brassicae* (Figure 4). Subsequent studies revealed that the sensitivity of *B. campestris* species to destruxin B was variable and that the order of sensitivity to destruxin B was similar to that of the pathogen. It did not cause symptoms in nine plant genera that are not hosts of *A. brassicae* [23,95]. Some researchers suggested that destruxin B may contribute to the aggressiveness of *A. brassicae* by conditioning host tissues and thereby determining host susceptibility [96].

However, in black-spot-disease resistant species (*Sinapis alba*), destruxin B could be converted into a less toxic product, hydroxydestruxin B. Essentially, hydroxydestruxin B was further biotransformed into the $\beta$-D-glucosyl derivative. Remarkably, it was observed that hydroxydestruxin B induced the biosynthesis of phytoalexins in black spot disease resistant species, but not in susceptible species [97].

In addition to phytotoxicity, destruxin B also exhibits a variety of biological activities. For example, significant cytotoxic effects were observed in L1210 leukemia cells and spleen lymphocytes treated with destruxin B [98]. It also showed suppressive effects on hepatitis B virus surface antigen and has been suggested as a potential candidate for the development of new anti-hepatitis agents [99,100]. Destruxin B was found to be a specific, dose-dependent, and reversible inhibitor of vacuolar ATPase, which maintains acidity in vacuolar organelles [101].

The biosynthetic pathway of destruxins, including destruxin B, was described in the fungus *Metarrhizium anisopliae*. Previously, destruxin B was thought to be biosynthesized from protodestruxin by N-methylation [102]. Feeding experiments with isotopically labeled precursors in *M. anisopliae* showed that methionine was involved in the incorporation of $^{13}C$ into the N-methyl group of MeVal and MeAla residues. Acetates were involved in the biosynthesis of the—CH (OH)-COOH fragment of the hydroxy acid moiety, proline, and isoleucine [103].

### 2.6. HC-Toxin

When northern corn leaf spot disease was first noted in the US in 1938, it was found that *Cochliobolus carbonum* was the key pathogenic strain and could produce HC-toxin [104]. In the 1970s, Pringle and co-workers purified and partially determined the structure of HC-toxin, indicating that it was a peptide containing Ala and Pro in the ratio of 2:1 [24]. Several years later, the complete structure was established, which was cyclo (D-Pro-L-Ala-D-Ala-L-Aeo) (Figure 4), with Aeo standing for 2-amino-9,10-epoxi-8-oxodecanoic acid [25]. In 2013, HC-toxin was also found in the culture filtrates of *A. jesenskae* that was isolated from seeds of *Fumana procumbens* [26]. HC-toxin could inhibit the root growth of susceptible maize (genotype *hm1/hm1*) at 0.5–2 $\mu g \cdot mL^{-1}$. The concentration needed to affect resistant maize (genotype *Hm1*/-) was 100-fold higher. The epoxide group of Aeo was critical for HC-toxin toxicity, and other amino acid residues also apparently played important roles in determining the bioactivity [105]. Besides phytotoxicity, HC-toxin also showed cytostatic activity against mammalian cells. The site of action of HC-toxin was histone deacetylase (HD), an enzyme that reversibly deacetylates the core histones (H3 and H4) [106].

HC-toxin production in *C. carbonum* was controlled by a complex locus, *TOX2*, that extended over 540 kb and contained several multicopy genes. The *TOX2* locus includes *HTS1, TOXA, TOXC, TOXD, TOXE, TOXF*, and *TOXG* genes, which encoded a nonribosomal peptide synthetase, a member of the major facilitator superfamily of transporters, a fatty acid synthase beta subunit, a predicted short-chain alcohol dehydrogenase, a pathway-specific transcription factor, a putative branched chain amino acid aminotransferase, and an alanine racemase, respectively [26,107]. *A. jesenskae* had high-scoring orthologs of all known genes involved in HC-toxin biosynthesis from *C. carbonum*. Based on genomic sequencing, *AjTOX2* was considered as a major gene involved in the biosynthesis of HC-toxin in *A. jesenskae*. The genes for HC-toxin biosynthesis were duplicated in these two fungi and the encoded orthologous proteins shared 75–85% amino acid identity [26].

### 2.7. Maculosin

Spotted knapweed (*Centaurea maculosa*) is a significant threat as a weed species in North America, particularly in the northwestern United States and southwestern Canada [27,28]. Its invasion of rangelands, roadsides, and pastures has resulted in a decline in forage production of about 70% and major losses in the millions of dollars. In 1984, an infected black-leaved orchid was found in Silver Bow County (Montana, USA), and *A. alternata* was identified as the causal agent. Although seven diketopiperazines were isolated and identified (Cyclo(-L-Pro-L-Tyr-), Cyclo(-L-Pro-L-Phe-), Cyclo(-L-Pro-D-Phe-), Cyclo—Pro-Hle-), Cyclo(-Pro-Val-), Cyclo(-Pro-Leu-), and Cyclo(-Pro-Ala-)) from the liquid culture of the orchid pathogenic strain of *A. alternata*, maculosin (Cyclo(-L-Pro-L-Tyr-), Figure 5) was established as a major HST of spotted knapweed because it exhibited high toxicity to spotted knapweed at 10 μM but no toxicity to other test plants even at 1 mM [108]. Thus, it has the potential to be developed as a safe and environmentally friendly bioherbicide against knapweed.

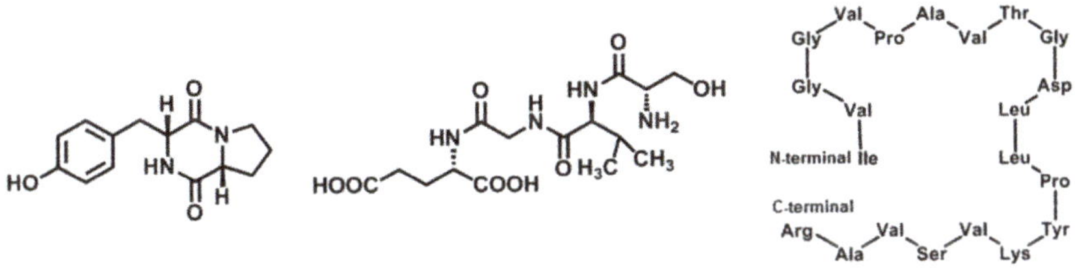


**Figure 5.** Chemical structures of maculosin, AS-I toxin, and ABR-toxin.

The target site of maculosin is the chloroplasts, since within 24 h of treatment with maculosin there is a progressive decay of the chloroplasts. The core component of maculosin activity is the diketopiperazine ring, which contains proline. Subsequently, the binding component of maculosin was identified as three large molecular weight proteins, one of which was thought to be ribulose-1,5-biphosphate carboxylase (RuBPcase) [109]. Maculosin is also a potent blocker of the delayed-rectifying potassium channel in guinea pig myocytes. It can increase alkaline phosphatase expression, induce differentiation, and exert antibacterial and antioxidant effects [110,111]. To date, there is no report on the biosynthetic pathway of maculosin. A systematic, in-depth study has yet to be conducted.

### 2.8. AS-I Toxin

In 1997, two phytotoxins were isolated from the culture filtrate of *A. alternata* that are pathogenic to sunflowers [29]. The chemical structure of one toxin was deduced using chemical and physicochemical methods as tetrapeptide Ser-Val-Gly-Glu and named as AS-I toxin (Figure 5). AS-I toxin can cause chlorosis or necrosis on leaves, inhibit seed germination of sunflowers, and lead to mild toxicity on tobacco and zucchini leaves, but has no toxic effect on other plants. These phenomena suggest that AS-I toxin is an HST [29,38]. The mode of action, target, and biosynthetic pathway for AS-I toxin are still not clear, so there is a wide research scope for this HST.

### 2.9. ABR-Toxin

Most HSTs are low-molecular-weight compounds and were discovered in liquid cultures. In 2008, some researchers indicated that the spore suspensions of *A. brassicae* can cause gray leaf spot disease on *Brassica* plants. After collecting spore germination fluid (SGF) on leaves, a fraction with a high molecular weight (above 10 kDa) and toxicity to host leaves was separated by ultrafiltration. Next, a new toxin was purified from that fraction by chromatography and named ABR-toxin. Further investigation showed that ABR-toxin was a protein toxin that loses its toxicity when treated at 60 °C or with proteinase K for 15 min. The isoelectric point of ABR-toxin was about 7.0 and the molecular weight was 27.5 kDa. It contains 21 amino acid residues (Ile-Val-Gly-Gly-Val-Pro-Ala-Val-Thr-Gly-Asp-Leu-Leu-Pro-Tyr-Lys-Val-Ser-Val-Ala-Arg) with an unblocked N-terminus (Figure 5). Biological activations showed that ABR-toxin at a concentration of 0.5–1 $\mu g{\cdot}mL^{-1}$ could induce symptoms on *Brassica* leaves, but a concentration greater than 50 $\mu g{\cdot}mL^{-1}$ had no effect on non-host leaves. ABR-toxin at a concentration of 0.5–1 μg·mL$^{-1}$ mixed with non-pathogenic spores of *A. alternata* could lead to symptoms similar to those caused by *A. brassicae* infection. The above results show that ABR-toxin not only triggered the initial colonization of host plants, but also showed a relationship with disease development that was different from that of destruxin B [23]. Currently, there are very few studies on ABR-toxin, so further detailed studies need to be conducted.

## 3. Non-Host-Selective Toxins

So far, less attention has been paid to NHSTs of *Alternaria* compared to HSTs. However, the role of NHSTs in virulence is more complex than that of HSTs. An in-depth exploration of NHSTs may reveal new and unexpected aspects for applications in many fields. Here, we detected 50 NHSTs from six families of *Alternaria*, including pyranones, quinones, tertramic acid, cyclic peptides, macrolides, and phenols (Table 2).

**Table 2.** NHSTs produced by *Alternaria* species.

| Family | Toxins | *Alternaria* Species | References |
|---|---|---|---|
| Pyranones | Radicinin | *A. radicina* | [112] |
| | Radicinol | *A. radicina, A. chrysanthemi* | [112,113] |
| | 3-epiradicinol | *A. chrysanthemi, A. longipipes* | [113,114] |
| | Deoxyradicinin | *A. helianthi* | [114] |
| | Pyrenocine A | *A. helianthi* | [115] |
| | Pyrenocine B | *A. helianthi* | [115] |
| | Solanapyrones A | *A. solani* | [116] |
| | Solanapyrones B | *A. solani* | [116] |
| | Solanapyrones C | *A. solani* | [116] |
| | Solanapyrones P | *A. tenuissima* | [117] |
| | Alternariol | *A. tenuis* | [118] |
| | Alternariol 9-methyl ether | *A. tenuis* | [118] |
| | Altenuene | *A. tenuis* | [118] |
| Quinones | Altertoxin I | *A. tenuis* | [119] |
| | Altertoxin II | *A. tenuis* | [119] |
| | Altertoxin III | *A. alternata* | [120] |
| | Altertoxin IV | *A. tenuissima* | [121] |
| | Altertoxin V | *A. tenuissima* | [122] |
| | Altertoxin VI | *A. tenuissima* | [122] |
| | Altertoxin VII | *Alternaria* sp. PfuH1 | [123] |
| | Alterlosins I | *A. alternata* | [124] |
| | Alterlosins II | *A. alternata* | [124] |
| | Alteichin | *A. eichorniae* | [120] |
| | Stemphyperylenol | *A. alternata* | [125] |
| | Stemphyltoxin III | *A. alternata* | [125] |
| | Altersolanol A | *A. solani* | [126] |
| | Altersolanol B | *A. solani* | [126] |
| | Altersolanol C | *A. solani* | [127] |
| | Altersolanol E | *A. solani* | [127] |
| | Altersolanol F | *A. solani* | [127] |
| | Macrosporin | *A. solani* | [126] |
| | Bostrycin | *A. eichhorniae* | [128] |
| | 4-Deoxybostrycin | *A. eichhorniae* | [128] |
| | Physcion | *A. porri* | [129] |
| | Erythroglaucin | *A. porri* | [129] |
| | Alterporriol B | *A. porri* | [130] |
| | Alterporriol K | *Alternaria* sp. ZJ9-6B | [130] |
| | Alterporriol L | *Alternaria* sp. ZJ9-6B | [130] |
| | Alterporriol M | *Alternaria* sp. ZJ9-6B | [130] |
| | Alterporriol T | *Alternaria* sp. XZSBG-1 | [131] |

**Table 2.** *Cont.*

| Family | Toxins | *Alternaria* Species | References |
|---|---|---|---|
| Tertramic acid | Tenuazonic acid | *A. alternata, A. longipes, A. tenuissima* | [132] |
| | 3-acetyl-5-isopropyltetramic acid | *A. tenuis* | [133] |
| | 3-acetyl-5-isobutyltetramic acid | *A. tenuis* | [133] |
| Cyclic peptides | Tentoxin | *A. alternata, A. citri, A. longipes, A. mali, A. porri, A. tenuis* | [134–139] |
| Macrolides | Brefeldin A | *A. carthami, A. zinnia* | [140,141] |
| | 7-Dehydrobrefeldin A | *A. carfhami* | [141] |
| | Aldaulactone | *A. dauci* | [142] |
| Phenolics | Zinniol | *A. zinnia, A. dauci, A. tagetica, A. solani, A. porri, A. carthami, A. macrospora, A. cichorii* | [143–145] |
| | α -Acetylorcinol | *A. tenuissima, A. brassicicola, A. dauci* | [146] |
| | *p*-Hydroxybenzoic acid | *A. tagetica, A. dauci* | [146,147] |

### 3.1. Pyranones

Pyranone is an important natural product that has attracted considerable attention due to its intriguing stereoisomeric structure and impressive bioactivity [148]. Simple pyranones and dibenzopyranones are the major groups of the pyranone family produced by *Alternaria* spp.

#### 3.1.1. Simple Pyranones

Pyranones without a benzene ring structure are defined as simple pyranones [6]. For the NHSTs of *Alternaria*, we have described here ten simple pyranones (Figure 6).

Radicinin was first found from *Stemphylium radicinum* [149] and then also isolated from *A. radicina*, including its analogue radicinol [112]. So far, many new simple pyranones NHSTs have been found in *Alternaria* spp. Radicinol and 3-epiradicinol have been isolated from other strains, such as *A. chrysanthemi*, which causes leaf spot disease in *Leucanthemum maximum* [113]. Further, 3-Epiradicinol is also found in *A. longipipes*. Deoxyradicinin was found in *A. helianthi*, an aggressive pathogen of sunflower (Figure 6a) [114].

Pyrenocine A and pyrenocine B were first described as products of the onion pink root fungus *Pyrenochaeta terrestris* [150]. They were then also found in the liquid medium of *A. helianthi* isolated from *Helianthus tuberosus* leaves with necrotic lesions. Pyrenocine A proved to be lethal to both isolated protoplasts and whole leaf tissue of *Helianthus* (Figure 6a) [115].

Solanapyrones A-C were isolated in 1983 from the phytopathogenic fungus *A. solani*, the causal agent of early potato blight [116]. Recently, solanapyrone P was discovered from *A. tenuissima*, an endophytic fungus in *Salvia przewalskii* (Figure 6a) [117].

Most compounds from this family showed lower phytotoxicity than radicinin [151]. Radicinin was found to be toxic to *Coix lachryma-christi* at 0.3 $\mu g \cdot leaf^{-1}$ [152]. It caused a 25% inhibition in root growth of carrot seedlings at a concentration of 10 $\mu g \cdot mL^{-1}$ [153]. In the structure of radicinin, the $\alpha$, $\beta$-unsaturated carbonyl group at C-4, a free secondary hy-droxyl group at C-3, and the stereochemistry of the same carbon and the unsaturation of the propenyl side chain play key roles to exhibit activity [151]. Due to its targeting activity against the host plant and the fact that it shows no toxicity to zebrafish embryos, radicinin has the potential to be developed as a natural bioherbicide [151]. As another bioactivity, radicinin exhibits antifungal, insecticidal, and antibiotic activity against Gram-positive bacteria, including *Staphylococcus aureus* and *Clostridium* sp. [152,154].

Radicinol showed anticancer activity in various cancer cells due to modulating both tumor suppressor protein (p53) and antiapoptic protein (BCL-2), which in turn increased the expression of caspase-3 [155].

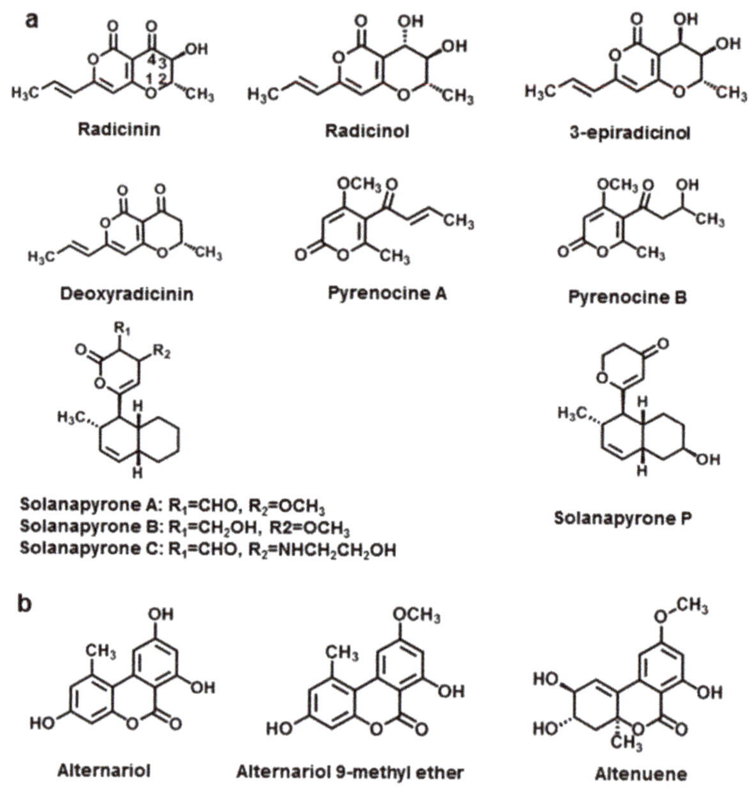


**Figure 6.** Chemical structures of *Alternaria* NHSTs belonging to simple pyranones (**a**) and dibenzopyranones (**b**) families.

Pyrenocine A caused leaf necrosis in the leaf injury bioassay and inhibited the growth of many plants, especially greater foxglove and autumn crocus. Pyrenocine A and B induced significant electrolyte loss in the leaf tissue of bermuda grass. However, pyrenocine B showed much weaker phytotoxic activity than pyrenocine A [156]. Moreover, pyrenocine A exhibited cytotoxicity against cancer cells with an $IC_{50}$ value of 2.6–12.9 μM [157]. Pyrenocine B inhibited the gene presentation of primary dendritic cells (DCs) in mice [158].

Solanapyrone A and B showed phytotoxicity in chickpea, resulting in stem death. Solanapyrone A was more toxic than solanapyrone B [159]. Solanapyrone A, C, and P showed antibacterial activities against various bacteria such as *Bacillus subtilis*, *B. megaterium*, *Clostridium perfringens*, *Micrococcus tetragenus*, and *Escherichia coli* with minimum inhibitory concentrations (MIC) ranging from 12.5 to 100 $\mu g \cdot mL^{-1}$ [117]. Solanapyrone A can also inhibit mammalian DNA polymerase $\beta$ and $\lambda$ activities in vitro, with $IC_{50}$ values of 30 μM and 37 μM, respectively [160].

Among these simple pyranones, the biosynthetic pathway of solanapyrone A was also discovered. Feeding experiments with [1-$^{13}$C], [1, 2-$^{13}$C] acetates, and [S$^{13}$CH$_3$] methionine showed that solanapyrones were biosynthesized from an octaketide consisting of one acetyl-CoA, seven malonyl-CoA, and two one-carbon units from methionine [161]. Recently, a gene cluster for solanapyrone biosynthesis containing six genes, *SOL1–SOL6*, was identified for the first time in *A. solani*, suggesting that solanapyrone biosynthesis requires eight acetates and one S-adenosylmethionine (SAM) as precursors [162]. Of these genes, *SOL1* encodes a polyketide synthase that initiates the solanapyrone biosynthetic pathway, and

*SOL5* encodes a Diels alderase that catalyzes both the oxidation and subsequent cyclization of the immediate precursor compound of solanapyrone A [163].

3.1.2. Dibenzopyranones

The dibenzopyranone skeleton is found in many natural products and biologically active molecules. Dibenzopyranone is of great importance as an intermediate for several interesting bioactive compounds [6,164]. In this section, the three major dibenzopyranones produced by *Alternaria* are presented, namely alternariol (AOH), alternariol-9-methyl ether (AME), and altenuol (ALT) (Figure 6b).

AOH, AME, and ALT are structurally related mycotoxins produced by different *Alternaria* strains, such as *A. tenuis* [118,132]. AOH and AME were first isolated and described in 1953 [165], while ALT was discovered in 1971 [118]. Subsequent studies showed that these three compounds are present in a wide range of vegetables, fruits, mushrooms, cereals, grapes, and feeds [132,166,167]. AOH and AME are considered to be the most important *Alternaria* toxins because they are produced in relatively large amounts by most species and account for up to 20% of crude extracts of *Alternaria* isolates, while ALT accounts for only 1–3% of extracts [119,168].

AOH and AME possess broad cytotoxicity, genotoxicity, and can induce oxidative stress [169–173]. In vitro, AOH and AME showed cytotoxicity to Henrietta Lacks's cervical cancer cell line HeLa cells [119]. Further studies revealed that AOH was cytotoxic to human colon carcinoma cell lines [174] and Caco-2 cells [172]. It effectively inhibited DNA relaxation and stimulated DNA cleavage activities of topoisomerase I, II$\alpha$, and II$\beta$ [170] and had mutagenic activity in mammalian cell lines [175]. AOH was also able to induce autophagy and senescence in murine macrophages and alter the morphology and cytokine secretion of murine and human macrophages [176,177]. In 1992, it was suggested that AOH and AME on cereals may be the most important factors for the increased incidence of human esophageal cancer in Linxian County, China [178]. Both AOH and AME appear to be highly mutagenic in the assay of *B. subtilis* and *E. coli* ND -160 [179]. Due to their widespread occurrence and high toxicity, the European Food Safety Authority (EFSA) has set the threshold of toxicological concern (TTC) for AOH and AME at 2.5 ng·kg$^{-1}$ body weight per day [180]. ALT is most acutely toxic in female mice with a $LD_{50}$ > 50 mg·kg$^{-1}$ body weight, compared to AOH and AME with a $LD_{50}$ > 400 mg·kg$^{-1}$ body weight [181]. Recently, ALT was reported to exhibit cytotoxic activity against HCT116 cell lines with an $IC_{50}$ value of 3.13 μM, and thus has the potential to be developed as a new antitumor drug candidate [182].

As for phytotoxic activity, AOH possessed a significant cytotoxic activity in soybean cells with an $EC_{50}$ value of 4.69 μM. It was suggested that the phenolic hydroxyl group played a key role in the toxicity to soybean cell culture [183]. AOH inhibited root growth of *Pennisetum alopecuroides*, *Medicago sativa*, and *Amaranthus retroflexus* at 1000 μg·mL$^{-1}$ [184]. AME inhibited the electron transport chain of spinach chloroplasts with an $IC_{50}$ value of 29.1 μM, and inhibited the growth of *Synechococcus* by directly interacting with one or more of the electron carriers involved in the electron transport chain [185]. Although there are reports of genotoxic, estrogenic, and mutagenic effects in laboratory animals, the toxicity of AOH and AME to humans and animals is low. Thus, these compounds represent a new lead structure and have the potential to be developed as new herbicides for weed control [185].

AOH, AME, and ALT are all polyketide-derived compounds. Due to their structural similarity, the biosynthetic pathway of these compounds should be of importance. The biosynthetic pathway of AOH was first studied in detail in 1961, which suggested that AOH could be synthesized by head–tail condensations of acetate units [186]. Further studies revealed that the formation of AOH occurs by the polycondensation of malonate, which is formed by the carboxylation of acetate [187]. Later, an enzyme, alternariol-O-methytransferase from *A. alternata*, was isolated that converts AOH to AME [188]. In 2019, the gene cluster for the biosynthesis of AOH and several derivatives of *A. alternata* was found. The gene cluster contains *PSKI, OMTI, MOXI, SDRI*, and *DOXI*, which encode O-methyltransferase, FAD-dependent monooxygenase, short-chain dehydrogenase, putative extradiol dioxygenase, and estradiol dioxygenase, respectively. Production begins with PKSI assembling an acetyl-CoA, together with six malonyl-CoA, to form the heptaketide AOH. AOH is further converted to AME by the methyltransferase OMTI. Next, 4-hydroxy-AME is catalyzed as an intermediate by the monooxygenase MOXI, followed by the opening of the lactone ring by SDRI to form altenusin. Finally, the formation of ALT from altenusin was catalyzed by DOXI for the rotation of the C-ring and lactonization [189].

### 3.2. *Quinones*

Quinones are an important species that interact with biological systems to promote many beneficial agents or even induce toxicities [190]. Among *Alternaria* toxins, there are three groups of quinones, including perylenequinones, anthraquinones, and bianthraquinone derivatives that have been isolated so far. In this section, twelve perylenequinones, ten anthraquinones, and five bianthraquinones, as well as their unique bioactivities, are presented.

#### 3.2.1. Perylenequinone Derivatives

Perylenequinones are a class of aromatic polyketides characterised by a highly conjugated pentacyclic core that gives them their potent bioactivity [191]. Here, twelve perylenequinones produced by *Alternaria* are presented, including altertoxin I–VII, alterlosin I and II, alteichin, stemphyperylenol, and stemphyltoxin III.

There are many types of altertoxins (ATXs) (Figure 7). We have described seven types of ATXs from *Alternaria* spp. ATX I and ATX II were first isolated from *A. tenuis* in 1973 and ATX III was isolated from *A. alternata* in 1983 [119,120]. The correct structure of ATX I was elucidated in 1983 [120]. ATX IV was isolated from the fermentation broth of an endophytic strain of *A. tenuissima* living in the stem of *Tribulus terrestris* [121]. ATXV and VI were isolated from the fermentation broth of *A. tenuissima* QUE1Se, which inhabits the stem tissue of *Quercus emoryi* [122]. Recently, ATX VII was isolated from the endophytic fungus *Alternaria* spp. PfuH1 of patchouli (*Pogostemon cablin*). Further studies showed that all of them are perylene derivatives, which can also be produced by other *Alternaria* spp. including *A. mali* and *A. eichorniae*. Although ATXs were produced in very low amounts by only a few species, they were important *Alternaria* toxins due to their high toxicity [168]. Among them, ATX II was the most potent [175,192].

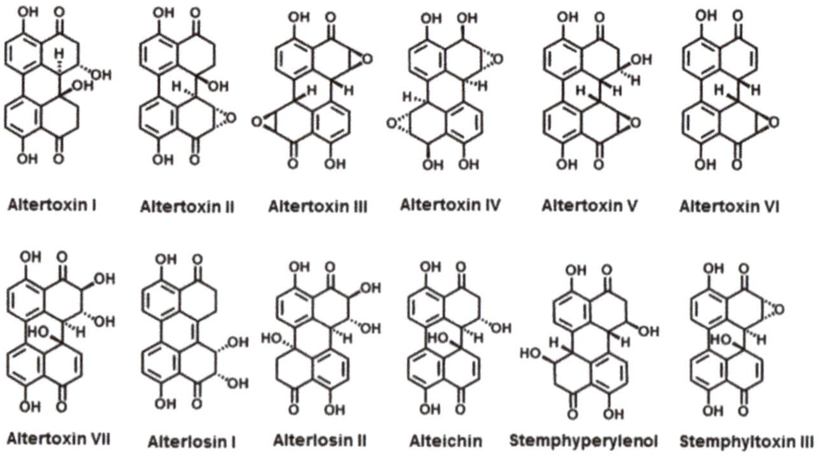


**Figure 7.** Chemical structures of *Alternaria* NHSTs belonging to perylenequinone family.

ATXs showed many activities; in particular, ATX I–III showed significant cytotoxicity, mutagenicity, and possibly carcinogenicity. In an Ames test, ATX I–III proved to be clearly mutagenic in TA98, TA100, and TA1537, with a ranking of ATX I < ATX II < ATX III [193]. ATX I and II were found to be highly toxic to the HeLa cells, with $IC_{50}$s of 20 and 0.5 μg·mL$^{-1}$, respectively [119]. ATX I–III were all cytotoxic to Chinese hamster V79 cells at concentrations greater than 5, 0.02, and 0.2 μg·mL$^{-1}$, respectively [175,194]. ATX IV showed cytotoxicity to human osteosarcoma cell lines (MG-63) and human hepatocellular carcinoma cell lines (SMMC-7721), with an $IC_{50}$ at 14.81 and 22.87 μg·mL$^{-1}$, respectively [195]. ATX V and VI showed the ability to inhibit HIV-1 viral replication in A3.01-infected cells. ATX V showed higher activity and could completely inhibit HIV-1 virus replication at concentrations of 0.5 μM. Thus, they have the potential to be developed as potent anti-HIV drugs [122]. ATX VII showed antibacterial activities against *S. agalactiae* with MIC values of 17.3 μg·mL$^{-1}$ [123].

As the major NHSTs of *Alternaria*, the biosynthetic pathway of ATXs was revealed. Based on the feeding experiment with $^{13}$C-labelled precursors, ATX I was used as an example of the biosynthetic pathway of ATXs. Five acetate molecules were found to be used for the synthesis of octalone analogues and tetralone analogues. ATXs were synthesised by the oxidative coupling of two molecules of tetralone analogues [120].

Alterlosins (ALS) include two compounds (Figure 7), ALS I and II. They were first isolated in 1989 from a host-selective strain of *A. alternata*, which is pathogenic on spotted knapweed. Both exhibited reasonable phytotoxicity, and ALS II was more potent than ALS I. ALS II was able to cause necrotic lesions on knapweed, lettuce, and Johnson grass at $10^{-4}$ M [124].

Alteichin (ALTCH) was isolated from *A. eichorniae* (Figure 7), a fungal pathogen of water hyacinth [120,196]. ALTCH was shown to have antifungal activity against *Valsa ceratosperma* and caused growth inhibition in lettuce seedlings [120]. Further studies revealed that ALTCH at a concentration of 0.1 mg·mL$^{-1}$ could induce necrotic spots on the leaves of water hyacinth, tomato thistle, wheat, sunflower, and barley within 12 h. The target of ALTCH can act directly on the plant cell and cause structural changes in plant membranes [196].

Stemphyperylenol and Stemphyltoxin III could be found in the culture of *Stemphylium botryosum* and *A. alternata* (Figure 7) [124,125]. Based on the bioactivity studies, stemphyperylenol is a toxin for finger millet [197]. Stemphyltoxin III showed an in vitro antibacterial activity against *B. subtifis*, *B. cereus*, and *E. coli*, as well as phytotoxic activity (Arnone et al., 1986). SOTTX-III was also mutagenic against Ames *S. typhimurium* TA98 and TA1537 [125,198].

#### 3.2.2. Anthraquinone Derivatives

Anthraquinones (9,10-dioxoanthracenes), with the rigid planar tricyclic aromatic system anthracene, form an important class of valuable natural products [199]. There are many *Alternaria* NHSTs belonging to this family (Figure 8a).

Altersolanol A-C, E, F, and macrosporine could be isolated from *A. solani*, a pathogen of solanaceous plants. Altersolanol A and B occurred only in the culture filtrate, while the others could be isolated from either the culture filtrate or mycelia [126,127,200]. Bostrycin and 4-deoxybostrycin were isolated from the culture filtrate of *A. eichhorniae* [128]. Physcion and erythroglaucin were isolated from *A. porri* [129].

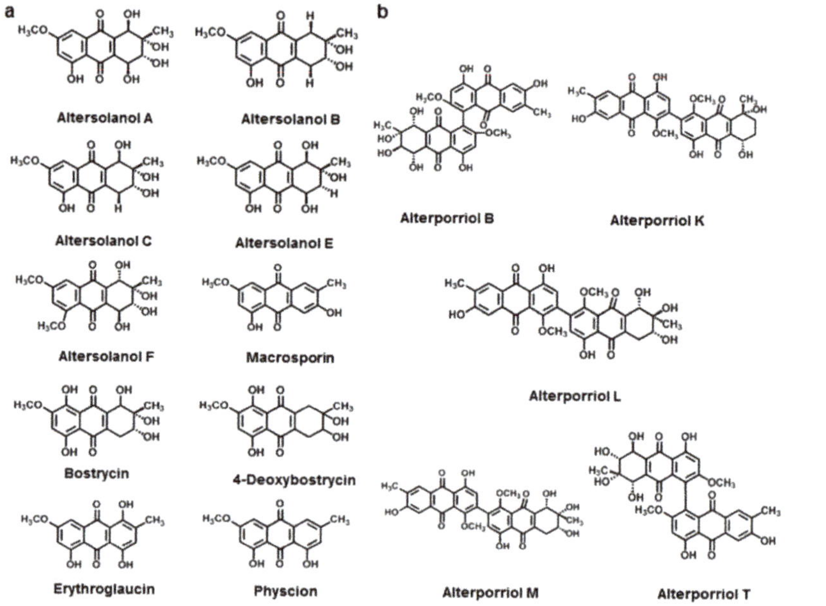


**Figure 8.** Chemical structures of *Alternaria* NHSTs belonging to anthraquinone (**a**) and bianthraquinone (**b**) families.

In bioactivity studies, altersolanols A and B showed an inhibitory effect on lettuce and stone-leek seedlings [201]. Altersolanol A-C, E, and F could act as electron transport inhibitors in the mitochondria of tobacco [202]. Altersolanol A could also cause necrosis and twisting on tomato leaves [203]. Besides phytotoxicity, altersolanols A–C and E showed antimicrobial activity against *S. aureus*, *B. subtilis*, *M. luteus*, and *Pseudomonas aeruginosa*. Altersolanols F showed obvious inhibitory activity against HCT-116 and HeLa cell lines with the $IC_{50}$ values of 3.026 and 8.094 μM, respectively [131]. Recent studies showed that altersolanol A exhibited cytotoxicity in vitro against 34 human cancer cell lines with an $IC_{50}$ ($IC_{70}$) value of 0.005 $\mu g \cdot mL^{-1}$ (0.024 $\mu g \cdot mL^{-1}$). Altersolanol A was a kinase inhibitor that induced cell death by apoptosis via the cleavage of caspase-3 and -9 and a decrease in the expression of anti-apoptotic proteins [195,204].

Macrosporine exhibited antibacterial and phytotoxic activity, inhibiting *Candida albicans*, *B. subtilis*, and *S. aureus* at a dose of 200 $\mu g \cdot disc^{-1}$, and induced significant necrosis by singlet oxygenation in plants [205].

Bostrycin and 4-deoxybostrycin showed antibacterial activity against *B. subtilis*. Bostrycin was also able to inhibit the growth of *Mycobacterium tuberculosis* in vitro and inhibit the activity of effector protein tyrosine phosphatase (MptpB) secreted by Mtb. In addition, bostrycin also acted as an antitumor agent against various cancer cell lines [206–209]. Both toxins showed a phytotoxic effect on water hyacinth at a concentration of 7 and 30 $\mu g \cdot mL^{-1}$, respectively [128].

Physcion had various pharmacological properties such as anti-inflammatory, antimicrobial, and antitumor effects, including cytotoxic activity in HeLa, A549, HL-60, and SW680 cells [210]. Physcion showed no mutagenicity in an Ames assay with TA100 and TA2638 [211]. The phytotoxic activity of physcion showed that it inhibited root and hypocotyl growth less at $7.0 \times 10^{-4}$ M in green amaranth and timothy [212].

Erythroglaucine showed a DPPH (1,1-diphenyl-2-picrylhydrazyl) radical scavenging property with an $IC_{50}$ value of 62 $\mu g \cdot mL^{-1}$ [213].

Based on the incorporation experiment of $^{13}C$-labelled sodium acetate and acetate, ageolanol A, macrosporin, and other similar pigments of *A. porri* were formed by eight acetates, which were condensed in a head-to-tail process to generate a linear octaketide. Subsequent cyclization and enolization, decarboxylation, and oxidation produced the final anthraquinone analogues [214,215].

#### 3.2.3. Bianthraquinone

Many *Alternara* spp. can produce alterporriol, a member of the bianthraquinone derivatives (Figure 8b). Alterporriol B was first described in *A. porri* in 1984. To date, many alterporriols have been discovered in *Alternaria*. Alterporriol K, L, and M were obtained from the extracts of *Alternaria* sp. ZJ9-6B and showed moderate cytotoxic activity against MDA-MB-435 and MCF-7 cells with $IC_{50}$ values ranging from 13.1 to 29.1 μM [130]. Alterporriol T was found in *Alternaria* sp. XZSBG-1 and showed an inhibition of α-glucosidase with an $IC_{50}$ value of 7.2 μM [131].

Some evidence suggests that preanthraquinones serve as precursors for a number of dimers. Alterporriols are homodimers composed of two alterolanols. Alterporriol A, for example, is formed by the oxidative coupling of a macrosporin and an alterolanol A. Other alterporriols can also be biosynthesized by the same pathways as alterporriol A [214].

### *3.3. Tertramic Acids*

Although they were isolated in the early 20th century, the various biological functions of tetramic acids (2,4-pyrrolidinediones) were not discovered until the 1960s [216]. Tenuazonic acid, 3-acetyl-5-isopropyltetramic acid, and 3-acetyl-5-isobutyltetramic acid are three classical analogues produced by *Alternara* [133].

Tenuazonic acid (TeA, (5S)-3-acetyl-5[(2S)-butan-2-yl]-4-hydroxy-1,5-dihydro-1H-pyrrol-2-one, Figure 9), an amide metabolite originally isolated from the culture filtrate of *A. tenuis*, is the simplest compound of the tetramic acids [217,218]. The structure and absolute configuration of TeA were elucidated after TeA was degraded by ozonolysis followed by acid hydrolysis [219]. Subsequently, TeA was also found in other species, such as *Phoma sorghina*, *Magnaporthe oryzae*, *Aspergillus* spp., and *Alternaria* spp., especially in *A. alternata*, *A. longipes*, and *A. tenuissima* [81,132,220–224]. Since its first isolation from cotton, TeA has been found in various vegetable, fruit, and crop plants contaminated with *Alternaria* [225–227].

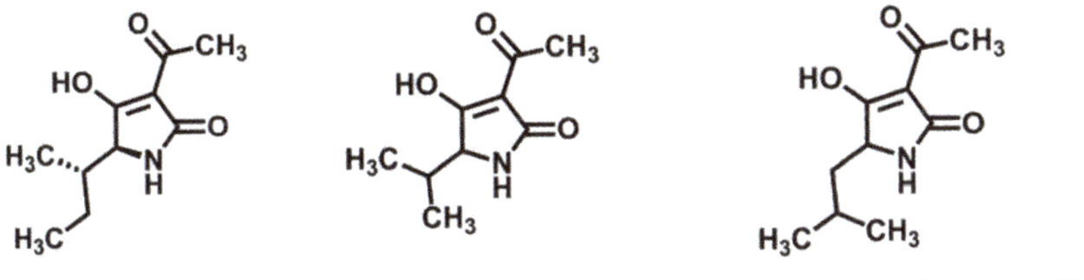


**Figure 9.** Chemical structures of *Alternaria* NHSTs belonging to the tertramic acids family.

TeA has long been reported to be toxic in animals, exhibiting antibacterial, antiviral, anticancer, and phytotoxicity effects [224,228–231]. The oral median lethal dose for male and female mice is 182 or 225 mg·kg$^{-1}$ and 81 mg·kg$^{-1}$ body weight, respectively [228,232]. It is also toxic to chicken embryos [233]. TeA inhibits protein biosynthesis by inhibiting the release of the polypeptide from the ribosome [234]. EFSA has evaluated the toxicity of TeA and set the threshold of toxicological concern at 1500 ng·kg$^{-1}$ body weight per day [235]. The first study on the effect of TeA on plant cells and seedlings was published in 1974. TeA could not only cause a necrotic spot on rice leaves, but also showed a striking stunting effect on the seedling growth of rice plants, mung beans, radishes, and turnips, as well as on the growth of cells of soybean and rice plants grown in suspension [236]. In the last two decades, an increasing number of articles have reported its phytotoxicity. TeA showed an inhibitory activity against 4-hydroxyphenylpyruvate dioxygenase (HPPD) with an $IC_{50}$

of 18 μM [237] and plant plasma membrane (PM) $H^+$-ATPase [238]. TeA was also able to inhibit the elongation of seedling roots and shoots [239–241], and resulted in a significant increase in multi-nucleolus of *Vicia faba* root tip cells at 400 $\mu g \cdot mL^{-1}$ [242]. Qiang et al. found a crude extract named AAC-toxin containing 5% TeA produced by *A. alternata*, the natural pathogen of *Ageratina adenophora*, a common noxious weed worldwide. Further purification of the AAC-toxin and subsequent bioassays showed that TeA was primarily responsible for herbicidal activity. It exhibited broad spectrum weed activity. Thus, TeA had the potential to be used as a bioherbicide in cotton fields [224,243–246].

Detailed studies on the main mechanism of action of TeA phytotoxicity revealed that TeA is a novel inhibitor of photosystem II (PSII), disrupting electron flow beyond the primary quinone acceptor, $Q_A$, by interacting with the D1 protein in the PSII reaction centers. The pyrrole ring, which contains an N-C=O group, is a core component of photosynthetic inhibitory activity [224,247,248]. TeA can induce a chloroplast-derived ROS burst that causes a range of irreversible cell damage, including chlorophyll degradation, lipid peroxidation, plasma membrane rupture, chromatin condensation, DNA cleavage, and organelle disinfection, eventually leading to rapid cell destruction and leaf necrosis in host plants [248]. TeA can also trigger the EXECUTER (EX) protein-dependent $^1O_2$ pathway leading to cell death in *Arabidopsis* seedlings [249]. A recent study suggests that cell death triggered by TeA is an essential requirement for the pathogen *A. alternata* to successfully infect host plants. Production of ROS was critical for pathogen invasion, proliferation, and disease symptom formation during infection. TeA significantly increased the ability of the pathogen to undergo invasive hyphal growth and spread [250].

Most tetramic acids are naturally derived from hybrid PKS and nonribosomal peptide synthetases (NRPS) that come from polyketides and α-amino acids [224,251,252]. Thus, TeA was also expected to be a product of a PKS–NRPS hybrid enzyme [224,253]. Previous experiments with radioactive precursors showed that *A. tenuis* first used L-isoleucine and two acetate molecules to synthesize N-acetoacetyl-L-isoleucine. Subsequently, TeA was formed by the cyclization of N-acetoacetyl-L-isoleucine [133,216,224,254]. Recently, the TeA biosynthetic gene *TAS1* was discovered from *M. oryzae*. *TAS1* encodes the TeA biosynthetic enzyme TAS1, which is a NPRS–PKS hybrid protein consisting of a C (condensation)-A (adenylation)-PCP (peptidyl carrier protein)-KS (ketosynthase) domain structure [252,255]. It was found that the C-A-PCP domain of TAS1 condenses L-isoleucine and acetoacetyl-CoA to yield N-acetoacetyl-L-isoleucine, while the KS domain recognizes the N-acetoacetyl-L-isoleucine hybrid to initiate the cyclization reaction to produce TeA [252,255]. In 2020, the mechanism of cyclization to form the tetramic acid ring was illustrated by the KS domain of TAS1 in the course of TeA biosynthesis. TAS1-KS contains a conserved catalytic triad Cys179-His322-Asn376. The substrate N-acetoacetyl-L-isoleucine was transferred from the PCP domain to Cys 179 via a thioester bond. The substrate was positioned by a hydrogen bond to Ser 324, and then the methylene proton was abstracted by His-322, which triggered a nucleophilic attack on the thioester carbonyl to give TeA. Asn376 could stabilize the conformation of cis-N-acetoacetyl-L-isoleusin for the nucleophilic attack to form TeA [252,256].

Much like the above biosynthetic pathway of TeA, the addition of L-isoleucine could stimulate the production of 3-acetyl-5-isopropyltetramic acid and 3-acetyl-5-isobutyltetramic acid. In this biosynthetic pattern, it is possible and useful to obtain tetramic acids with different side chains at the 5-position by growing the organism in media fed with different L-amino acids. Gatenbeck and co-workers added $^{14}C$-carboxyl-labeled L-valine or L-leucine to the culture media of *A. tenuis*. From the culture extracts, the 5-isopropyl and the 5-isobutyl derivatives of the tetramic acids were prepared and purified, i.e., 3-acetyl-5-isopropyltetramic acid (3-AIPTA) and 3-acetyl-5-isobutyltetramic acid (iso-TeA, Figure 9) [133].

Based on a bioassay, 3-AIPTA showed phytotoxicity to a wide range of plants. It inhibited the root and shoot length of seedlings and eventually killed seedlings of both monocotyledonous and dicotyledonous weeds. 3-AIPTA was able to inhibit PSII electron transport rates and the growth of algal cells [257]. Further studies indicated that 3-AIPTA had the same target and lethal mechanism as TeA on weeds, but the herbicidal effect was much weaker compared to TeA [258].

3-Acetyl-5-isobutyltetramic acid, also called iso-tenuazonic acid (iso-TeA), was an isomer of TeA. Because of its similar chemical structure to TeA (Figure 9), the two toxins were thought to have similar toxicological relevance. Iso-TeA showed remarkable toxic effects on *Artemia salina*, with a mortality rate of 68.9% compared to 73.6% for TeA [259]. Iso-TeA also showed antibacterial effect on *B. megaterium* [230]. It also showed significant phytotoxicity, such as the inhibition of rice root growth with an $ID_{50}$ (50% inhibitory dose) of 0.28 mM and marked browning of rice leaves at 10 mM [260].

### 3.4. Cyclic Peptides

Cyclic peptides exhibit remarkable biological activities due to their condensed structures [261]. In this section, we have introduced tentoxin and its competing derivatives, all of which belong to this family.

Tentoxin (TEN, Figure 10) is a secondary metabolite produced by several *Alternaria* species, including *A. alternata*, *A. citri*, *A. longipes*, *A. mali*, *A. porri*, and *A. tenuis* [134–139]. Based on the analysis of the acidic hydrolysis products and spectroscopic properties of the compound, it was found that tentoxin is a cyclic tetrapeptide containing glycine, L-leucine, N-methyl-L-alanine, and N-methy-L-dehydrphenylalany. The complete structure of tentoxin is cy-clo[*N*-methyl-L-alanyl-L-leucyl-(Z)-α,β-dehydro-*N*-methylphenylalanylglycyl] [262,263]. In addition to tentoxin, dihydrotentoxin (DHT) and isotentoxin (isoTEN) have also been isolated as metabolites from *Alternaria* species [253]. Tentoxin can be found in many products, including wheat, sorghum, fruit, and barley [264,265]. Therefore, the EFSA applied the toxicological threshold of concern (TTC) approach to TEN in its preliminary risk assessment, which was set at 1500 $ng \cdot kg^{-1}$ body weight per day [180,263].

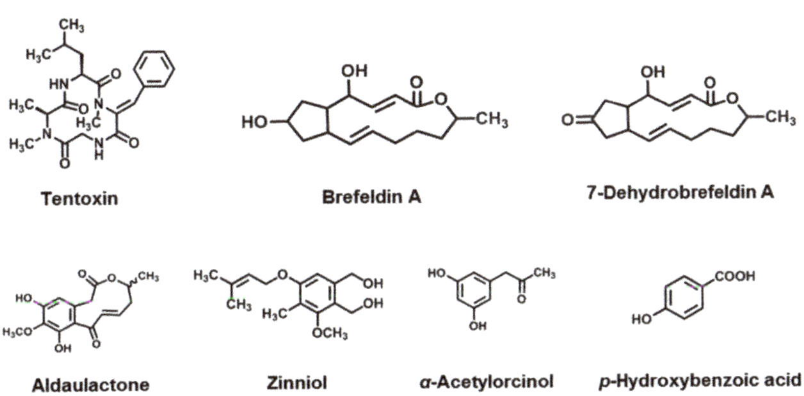


**Figure 10.** Chemical structures of *Alternaria* NHSTs belonging to the cyclic peptide (Tentoxin) macrolides (Brefeldin A, 7-dehydrobrefeldin A, and aldaulactone) and phenolics (Zinniol, α-acetylorcinol, and *p*-hydroxybenzoic acid) families.

As a phytotoxin, tentoxin was found to induce chlorosis in germinating seedlings of some dicotyledonous plants, but not in maize, tomatoes, and members of the Brassicaceae and Poaceae families. This was explained by the fact that sensitive species might possess a specific receptor site for tentoxin, resulting in the selective disruption of chloroplast function, reduction in the levels of chloroplast-specific lipids and proteins, and ultrastructural changes in chloroplasts [137,266,267]. Further studies indicated that tentoxin is a specific, non-competitive inhibitor of photophosphorylation and the site of action is associated with chloroplast F1-ATPase (CF1) [268,269]. Interestingly, ATP hydrolysis and synthesis were inhibited at a low dose of tentoxin, while ATPase activity was stimulated at high concentrations [270]. In 2002, the crystal structure of CF1 in complex with tentoxin showed that the binding site was located in a cleft at the $\alpha\beta$-subunit [269]. Recently, the study of the cryo-EM structure of CF1 in complex with tentoxin indicated that the cyclic ring of tentoxin with the charged or polar residues ($\beta$Asp83, $\beta$Thr82, $\alpha$Arg297, and $\alpha$Tyr271) and its isobutyl and phenyl moieties interact with the hydrophobic residues ($\alpha$Ile63, $\alpha$Leu65, and $\alpha$Val75) between the $\alpha$- and $\beta$-subunit, leading to a decrease in enzyme activity [271]. In addition, tentoxin also exhibits independent effects on plant metabolism, such as stomatal movements, ion uptake and translocation, and internal ion concentrations [272].

Previous studies indicated that tentoxin appeared on day 5 after inoculation with *A. alternata*, increased rapidly, and reached a maximum between days 9 and 12. After 14 days of inoculation, the synthesis decreased [273]. Methionine is the carbon donor in the biosynthesis of tentoxin and its precursor dihydrotentoxin [273,274]. In 1994, the tentoxin synthetase was isolated, which is a polyfunctional multienzyme with an integrated methyltransferase activity that contains active SH groups. The precursor amino acids were bound to the enzyme, then N-methylation and peptide extension occurred. Finally, dihydrotentoxin was formed by cyclization and then released to be converted into tentoxin [273]. Some researchers reported that the NRPS gene *CmNps3* was responsible for tentoxin biosynthesis in *C. miyabeanus*, and predicted that the gene *AaNps3* might be involved in tentoxin biosynthesis in *Alternaria* species [275]. Recently, two genes for tentoxin biosynthesis, a NPRS gene (*TES*) and a cytochrome P450 gene (*TES1*), were found in *A. alternata* [276]. *TES* encodes a protein of 5161 amino acids. *TES1* was closely associated with *TES* in a 5′-end-to-5′-end arrangement and was predicted to be involved in dehydrophenylalanine biosynthesis. Furthermore, a detailed analysis of TES revealed that it has a typical modular NRPS organization and consists of four modules with N-methyltransferase domains in both the second and fourth modules. The arrangement of the domains is A-T-C-A-M-T-C-A-T-C-A-M-T-C. TES assembles four precursor amino acids, Gly, Ala, Leu, and DPhe. The N-methylation of Ala and DPhe occurred in the N-methyltransferase domains, respectively. The condensation domain was located in the termination module of TES, which is responsible for the formation of intramolecular macrocyclization and final tentoxin release [276]. These findings are helpful for further studies on NRPS proteins in fungi and the mechanism of DPhe biosynthesis.

### 3.5. Macrolides

Brefeldin A (BFA) and its analogues 7-dehydrobrefeldin A (7-oxo-BFA) belong to the macrolide family, which possess antibiotic properties (Figure 10). Previously, BFA was isolated from *Penicillium* species and later found in *Alternaria* spp. such as *A. carthami* and *A. zinnia* [140,141]. 7-Oxo-BFA is another macrolide that is a potent phytotoxin of *A. carthami* [141]. BFA was particularly active. It could cause the rapid appearance of large necrotic patches and a 70% reduction in chlorophyll content when *Xanthium occidentale* leaves were treated with $10^{-4}$ M BFA [141]. One $\mu g \cdot mL^{-1}$ BFA was sufficient to inhibit both the germination and growth of tobacco pollen tubes and also cause the collapse of Golgi stacks [277]. Further studies showed that the Golgi stacks were the common target of BFA and 7-oxo-BFA. 7-Oxo-BFA was a more potent destroyer of the Golgi stacks than BFA [278]. As a macrolide, BFA also exhibited other important bioactivities, including antifungal, cytostatic, antimitotic, antiviral, and anticancer activities [279,280]. In most mammalian cells, 1–10 $\mu g \cdot mL^{-1}$ BFA not only inhibited secretion [281,282], but also caused profound morphological changes, including the decay of the Golgi apparatus and redistribution of Golgi enzymes into the endoplasmic reticulum [278]. BFA showed high cytotoxicity against HL-60, KB, Hela, MCF-7, and Spc-A-1 cell lines ($IC_{50}$ = 1.0–10.0 $ng \cdot mL^{-1}$) [280]. There is no report on the biosynthetic pathway of BFA and 7-oxo-BFA so far.

Aldaulactone was a 10-membered benzenediol lactone molecule and was firstly purified from *A. dauci*, which was toxic to a large range of dicotyledonous plants, especially carrot. It was also indicated that aldaulactone was involved in both fungal pathogenicity and plant resistance mechanisms. In phytotoxicity, aldaulactone was toxic to carrot cells, inducing a delay in embryonic development and a decrease in cell viability [142]. The biological function and biosynthetic pathway of aldaulactone has not yet been defined.

### 3.6. Phenols

Zinniol (Figure 10), a member of the phenol family, was first isolated from *A. zinnia* (Starratt, 1968) and later detected in culture filtrates of *A. dauci* [143], *A. tagetica* [144], *A. solani*, *A. porri*, *A. carthami*, *A. macrospora*, and *A. cichorii* [145]. Zinniol showed a broad phytotoxic spectrum that could cause necrotic leaf damage [144,283,284]. Zinniol could act specifically on a certain class of plant calcium channels, but its target is not comparable to calcium channel blockers [285]. The two hydroxymethyl groups of zinniol are essential for its phytotoxic activity [143]. However, a few studies suggested that zinniol is not markedly phytotoxic to embryogenic cellular cultures of *Daucus carota* [286] and the leaves of *Tagetes erecta* [287] at physiological concentrations. In addition, Zinniol also showed cytotoxic activity in rat embryonic fibroblasts with an $IC_{50}$ of 264 $\mu g \cdot mL^{-1}$ [284]. In general, there is still much room for research for this potential natural product.

α-Acetylorcinol is a resorcinol derivative that was first isolated from *C. lunata* in 1977 [288], and has also been reported as a secondary metabolite from various *Alternaria* spp., including *A. tenuissima*, *A. brassicicola*, and *A. dauci* [146]. It exhibited phytotoxic activity in many plants. It can induce necrosis to *Sida Spinosa*, *Chenopodium album*, *Ipomoea* sp., *Datura stramonium*, *Sorghum bicolor*, *S. halepense* [289], and *Nioctiana alata* [146]. α-Acetylorcinol also showed antifungal activity against *Trichophyton rubrum and A. fumigatus* [290].

There were many reports on the production of *p*-hydroxybenzoic acid by fungi, such as *A. tagetica* [147] *and A. dauci* [146]. It was also produced by *Epichloë bromicola* and *Diaporthe gulyae*, which were the phytopathogens of *Elymus tangutorum* and sunflower, respectively [291]. *p*-Hydroxybenzoic acid could inhibit the germination and root length of *Rumex crispus* [292]. Additionally, it also exhibited antibacterial [293], antioxidant [294], antifungal [295], antialgal [296], antimutagenic [297], and estrogenic activity [298].

## 4. Summary and Outlook

*Alternaria* is a ubiquitous genus in many ecosystems, consisting of saprophytic, pathogenic, and even endophytic species. Thus, they are a rich source of secondary metabolites. The production of various HSTs and NHSTs can be considered as a crucial reason for the survival of these fungi. In this review, we have listed only some parts of the toxins of the known *Alternaria* species. The great structural diversity, high potency, and exclusive mechanisms of action make these toxins extremely attractive for the discovery of their bioactivity. Many *Alternaria* toxins exhibit excellent herbicidal, antimicrobial, antitumor, and other bioactive properties. Some of them can be directly developed into drugs or pesticides, while others can serve as lead compounds for the discovery of new drugs or pesticides. However, several challenges must be overcome for their successful development as drug or pesticide candidates in the future. First, the biological activities and modes of action of most toxins are still unclear. Second, some active crude extracts need further purification to discover the exact active components. Third, the content of many toxins in *Alternaria* is low, so a deep exploration of their biosynthetic pathways is needed to increase the yield of the useful bioactive parts.

**Author Contributions:** Writing—original draft preparation, H.W., Y.G., Z.L., L.G., R.L. and Y.Z.; writing—review and editing, S.C., H.W., Y.G., H.M.K., Z.L. and S.Q. All authors have read and agreed to the published version of the manuscript.

**Funding:** This research was funded by Jiangsu Agriculture Science and Technology Innovation Fund (No. CX(21)3093) and the Postgraduate Research & Practice Innovation Program of Jiangsu Province (No. KYCX19_0617).

**Institutional Review Board Statement:** This article does not contain any studies with human or animal subjects.

**Informed Consent Statement:** Not applicable.

**Data Availability Statement:** All data presented in this study are contained in main text.

**Acknowledgments:** We thank Baojun Xu, who invited us to submit a paper to this special issue.

**Conflicts of Interest:** The authors have declared no conflict of interest.

## References

1. Thomma, B.P.H.J. *Alternaria* spp.: From general saprophyte to specific parasite. *Mol. Plant Pathol.* **2003**, *4*, 225–236. [CrossRef] [PubMed]
2. Lawrence, D.P.; Gannibal, P.B.; Peever, T.L.; Pryor, B.M. The sections of *Alternaria*: Formalizing species-group concepts. *Mycologia* **2013**, *105*, 530–546. [CrossRef] [PubMed]
3. Lee, H.B.; Patriarca, A.; Magan, N. *Alternaria* in food: Ecophysiology, mycotoxin production and toxicology. *Mycobiology* **2015**, *43*, 93–106. [CrossRef] [PubMed]
4. Woudenberg, J.H.C.; Seidl, M.F.; Groenewald, J.Z.; de Vries, M.; Stielow, J.B.; Thomma, B.P.H.J.; Crous, P.W. *Alternaria* section *Alternaria*: Species, formae speciales or pathotypes? *Stud. Mycol.* **2015**, *82*, 1–21. [CrossRef]
5. Zwickel, T.; Kahlm, S.; Rychlik, M.; Muller, M. Chemotaxonomy of mycotoxigenic small-spored *Alternaria* fungi-do multitoxin mixtures act as an indicator for species differentiation? *Front. Microbiol.* **2018**, *9*, 1368. [CrossRef] [PubMed]
6. Lou, J.; Fu, L.; Peng, Y.; Zhou, L. Metabolites from *Alternaria* fungi and their bioactivities. *Molecules* **2013**, *18*, 5891–5935. [CrossRef] [PubMed]
7. Takaoka, S.; Kurata, M.; Harimoto, Y.; Hatta, R.; Yamamoto, M.; Akimitsu, K.; Tsuge, T. Complex regulation of secondary metabolism controlling pathogenicity in the phytopathogenic fungus *Alternaria alternata*. *New Phytol.* **2014**, *202*, 1297–1309. [CrossRef]
8. Howlett, B.J. Secondary metabolite toxins and nutrition of plant pathogenic fungi. *Curr. Opin. Plant Biol.* **2006**, *9*, 371–375. [CrossRef]
9. Tsuge, T.; Harimoto, Y.; Akimitsu, K.; Kodama, M.; Akagi, Y.; Egusa, M.; Yamamoto, M.; Otani, H. Host-selective toxins produced by the plant pathogenic fungus *Alternaria alternata*. *FEMS Microbiol. Rev.* **2013**, *37*, 44–66. [CrossRef]
10. Nakashima, T.; Ueno, T.; Fukami, H.; Taga, T.; Masuda, H.; Osaki, K.; Otani, H.; Kohmoto, K.; Nishimura, S. Isolation and structure of AK-toxin I and II, host specific phytotoxic metabolites produced by *Alternaria alternata* Japanese pear pathotype. *Agric. Biol. Chem.* **1985**, *49*, 807–815. [CrossRef]
11. Nakatsuka, S.; Ueda, K.; Goto, T.; Yamamoto, M.; Nishimura, S.; Kohmoto, K. Structure of AF-toxin II, one of the host-specific toxins produced by *Alternaria alternata* strawberry pathotype. *Tetrahedron Lett.* **1986**, *27*, 2753–2756. [CrossRef]
12. Mikihiro, Y.; Fumio, N.; Syoyo, N.; Keisuke, K. Studies on host-specific AF-toxins produced by *Alternaria alternata* strawberry pathotype causing *Alternaria* black spot of strawberry (3) Use of toxin for determining inheritance of disease reaction in strawberry cultivar morioka-16. *Ann. Phytopath. Soc. Jpn.* **1985**, *51*, 530–535.
13. Pegg, K.G. Studies of strain of *Alternaria citri* pierce, the causal organism of brown spot of *Emperor mandarn*. *Queensl. J. Agric. Anim. Sci.* **1966**, *23*, 15–28.
14. Timmer, L.W.; Solel, Z.; Orozco-Santos, M. Alternaria brown spot of mandarins. In *Compendium of Citrus Diseases*, 2nd ed.; Timmer, L.W., Garnsey, S.M., Graham, J.H., Eds.; Amer Phytopathological Society: Saint Paul, MI, USA, 2000; pp. 19–21.
15. Akimitsu, K.; Peever, T.L.; Timmer, L.W. Molecular, ecological and evolutionary approaches to understanding *Alternaria* diseases of citrus. *Mol. Plant Pathol.* **2003**, *4*, 435–446. [CrossRef]
16. Grogan, R.; Kimble, K.; Misaghi, I. A stem canker disease of tomato caused by *Alternaria alternata* f. sp. *lycopersici*. *Phytopathology* **1975**, *65*, 880–886. [CrossRef]
17. Gilchrist, D.G. Production and nature of a host-specific toxin from *Alternaria alternata* f. sp. *lycopersici*. *Phytopathology* **1976**, *66*, 165–171. [CrossRef]
18. Kohmoto, K.; Scheffer, R.; Whiteside, J. Host-Selective Toxins from *Alternaria citri*. *Phytopathology* **1979**, *69*, 667–671. [CrossRef]
19. Kohmoto, K.; Akimitsu, K.; Otani, H. Correlation of resistance and susceptibility of citrus to *Alternaria alternata* with sensitivity to host-specific toxins. *Phytopathology* **1991**, *81*, 719–722. [CrossRef]
20. Okuno, T.; Ishita, Y.; Sawai, K.; Matsumoto, T. Characterization of alternariolide, a host-specific toxin produced by *Alternaria mali* Roberts. *Chem. Lett.* **1974**, *3*, 635–638. [CrossRef]
21. Hoffmeister, D.; Keller, N. Natural products of filamentous fungi: Enzymes, genes, and their regulation. *Nat. Prod. Rep.* **2007**, *24*, 393–416. [CrossRef]
22. Tewari, J.; Bains, P. Phytotoxins produced by *Alternaria brassicae* and bioassay of destruxin B. In *Toxins in Plant Disease Development and Evolving Biotechnology*; Upadhyay, R., Ed.; Science Publishers: Enfield, NH, USA, 1997; Volume 2, pp. 21–35.
23. Parada, R.Y.; Sakuno, E.; Mori, N.; Oka, K.; Egusa, M.; Kodama, M.; Otani, H. *Alternaria brassicae* produces a host-specific protein toxin from germinating spores on host leaves. *Biochem. Cell Biol.* **2008**, *98*, 458–463. [CrossRef] [PubMed]
24. Pringle, R.B. Amino acid composition of the host-specific toxin of *Helminthosporium carbonum*. *Plant Physiol.* **1971**, *48*, 756–759. [CrossRef]
25. Kawai, M.; Rich, D.H.; Walton, J.D. The structure and conformation of HC-toxin. *Biochem. Biophys. Res. Comm.* **1983**, *111*, 398–403. [CrossRef]

26. Wight, W.D.; Labuda, R.; Walton, J.D. Conservation of the genes for HC-toxin biosynthesis in *Alternaria jesenskae*. *BMC Microbiol.* **2013**, *13*, 165. [CrossRef] [PubMed]
27. Harris, P.; Cranston, R. An economic evaluation of control methods for diffuse and spotted knapweed in Western Canada. *Can. J. Plant Sci.* **1979**, *59*, 375–382. [CrossRef]
28. Bobylev, M.M.; Bobyleva, L.I.; Strobel, G.A. Synthesis and bioactivity of analogs of maculosin, a host-specific phytotoxin produced by *Alternaria alternata* on spotted knapweed (*Centaurea maculosa*). *J. Agric. Food Chem.* **1996**, *44*, 3960–3964. [CrossRef]
29. Liakopoulou-Kyriakides, M.; Lagopodi, A.L.; Thanassoulopoulos, C.C.; Stavropoulos, G.S.; Magafa, V. Isolation and synthesis of a host-selective toxin produced by *Alternaria alternate*. *Phytochemistry* **1997**, *45*, 37–40. [CrossRef]
30. Feng, B.; Nakastuka, S.; Goto, T.; Tsuge, T.; Nishimura, S. Biosyntheses of host-selective toxins produced by *Alternaria alternata* pathogens, I: (8*R*, 9*S*)-9,10-epoxy-8-hydroxy-9-methyl-deca-(2*E*, 4*Z*, 6*E*)-trienoic acid as a biological precursor of AK-toxins. *Agric. Biol. Chem.* **1990**, *54*, 845–848. [CrossRef]
31. Gardner, J.M.; Kono, Y.; Tatum, J.H.; Suzuki, Y.; Takeuchi, S. Structure of the major component of ACRL-toxins, host-specific pathotoxic compound produced by *Alternaria citri*. *Agric. Biol. Chem.* **1985**, *49*, 1235–1238. [CrossRef]
32. Foyer, C.H.; Noctor, G. Redox sensing and signalling associated with reactive oxygen in chloroplasts, peroxisomes and mitochondria. *Physiol. Plant.* **2003**, *119*, 355–364. [CrossRef]
33. Namiki, F.; Yamamoto, M.; Nishimura, S.; Nakatsuka, S.; Goto, T.; Kohmot, K.; Otani, H. Studies on host-specific AF-toxins produced by *Alternaria alternata* strawberry pathotype causing *Alternaria* black spot of strawberry. 4. Protective effect of AF-toxin II on AF-toxin I-induced toxic action and fungal infection. *Ann. Phytopath. Soc. Jpn.* **1986**, *52*, 428–436. [CrossRef]
34. Maekawa, N.; Yamamoto, M.; Nishimura, S.; Kohmoto, K.; Kuwada, K.; Watanabe, Y. Studies on host-specific AF-toxins produced by *Alternaria alternata* strawberry pathotype causing *Alternaria* black spot of strawberry. (1) Production of host-specific toxins and their biological activities. *Ann. Phytopathol Soc. Jpn.* **1984**, *50*, 600–609. [CrossRef]
35. Kohmoto, K.; Itoh, Y.; Shimomura, N.; Kondoh, Y.; Otani, H.; Kodama, M.; Nishimura, S.; Nakatsuka, S. Isolation and biological activities of two host-specific toxins from the tangerine pathotype of *Alternaria alternata*. *Phytopathology* **1993**, *83*, 495–502. [CrossRef]
36. Otani, H.; Kohmoto, K.; Nishimura, S.; Nakashima, T.; Ueno, T.; Fukami, H. Biological activities of AK-toxins I and II, host-specific toxins from *Alternaria alternata* Japanese pear pathotype. *Ann. Phytopathol Soc. Jpn.* **1985**, *51*, 285–293. [CrossRef]
37. Park, P.; Ikeda, K. Ultrastructural analysis of responses of host and fungal cells during plant infection. *J. Gen. Plant Pathol.* **2008**, *74*, 2–14. [CrossRef]
38. Meena, M.; Samal, S. *Alternaria* host-specific (HSTs) toxins: An overview of chemical characterization, target sites, regulation and their toxic effects. *Toxicol. Rep.* **2019**, *6*, 745–758. [CrossRef] [PubMed]
39. Namiki, F.; Okamoto, H.; Katou, K.; Yamamoto, M.; Nishimura, S.; Nakatsuka, S.; Goto, T.; Kohmoto, K.; Otani, H.; Novacky, A. Studies on host-specific toxins produced by *Alternaria alternata* strawberry pathotype causing *Alternaria* black spot of strawberry (5) Effect of toxins on membrane potential of susceptible plants by means of electrophysiological analysis. *Ann. Phytopathol. Soc. Jpn.* **1986**, *52*, 610–619. [CrossRef]
40. Otani, H.; Tomiyama, K.; Okamoto, H.; Nishimura, S.; Kohmoto, K. Effect of AK-toxin produced by *Alternaria alternata* Japanese pear pathotype on membrane potential of pear cells. *Ann. Phytopathol. Soc. Jpn.* **1989**, *55*, 466–468. [CrossRef]
41. Otani, H.; Kohmoto, K.; Kodama, M.; Nishimura, S. Role of host-specific toxins in the pathogenesis of *Alternaria alternata*. In *Molecular Strategies of Pathogens and Host Plants*; Patil, S.S., Ouchi, S., Mills, D., Vance, C., Eds.; Springer: Berlin, Germany, 1991; Volume 3, pp. 139–149.
42. Otani, H.; Kohmoto, K.; Kodama, M. *Alternaria* toxins and their effects on host plants. *Can. J. Bot.* **1995**, *73*, 453–458. [CrossRef]
43. Uemura, I.; Miyagawa, H.; Ueno, T. Asymmetric total synthesis of AK-toxins. *Terahedron* **2002**, *58*, 2351–2358. [CrossRef]
44. Morita, Y.; Hyon, G.; Hosogi, N.; Miyata, N.; Nakayashiki, H.; Muranaka, Y.; Inada, N.; Park, P.; Ikeda, K. Appressorium-localized NADPH oxidase B is essential for aggressiveness and pathogenicity in the host-specific, toxin-producing fungus Japanese *Alternaria alternata* pear pathotype. *Mol. Plant Path.* **2013**, *4*, 365–378. [CrossRef] [PubMed]
45. Wu, P.; Chen, C.; Choo, C.; Chen, Y.; Yago, J.; Chung, K. Proper functions of peroxisomes are vital for pathogenesis of *Citrus* brown spot disease caused by *Alternaria alternata*. *J. Fungi* **2020**, *6*, 248. [CrossRef] [PubMed]
46. Nakatsuka, S.; Feng, B.; Goto, T.; Tsuge, T.; Nishiumra, S. Biosynthesis of host-selective toxins produced by *Alternaria alternata* pathogens. 2. Biosynthetic origin of (8*R*,9*S*)-9,10-epoxy-8-hydroxy-9-methyl-deca-(2*E*,4*Z*,6*E*)-trienoic acid, a precursor of AK-toxins produced by *Alternaria alternata*. *Phytochemistry* **1990**, *29*, 1529–1531. [CrossRef]
47. Tanaka, A.; Shiotani, H.; Yamamoto, M.; Tsuge, T. Insertional mutagenesis and cloning of the genes required for biosynthesis of the hostspecific AK-toxin in the Japanese pear pathotype of *Alternaria alternata*. *Mol. Plant-Microbe Interact.* **1999**, *12*, 691–702. [CrossRef]
48. Tanaka, A.; Tsuge, T. Structural and functional complexity of the genomic region controlling AK-toxin biosynthesis and pathogenicity in the Japanese pear pathotype of *Alternaria alternata*. *Mol. Plant-Microbe Interact.* **2000**, *13*, 975–986. [CrossRef]
49. Hatta, R.; Ito, K.; Hosaki, Y.; Tanaka, T.; Tanaka, A.; Yamamoto, M.; Akimitsu, K.; Tsuge, T. A conditionally dispensable chromosome controls host-specific pathogenicity in the fungal plant pathogen *Alternaria alternata*. *Genetics* **2002**, *161*, 59–70. [CrossRef] [PubMed]
50. Masunaka, A.; Tanaka, A.; Tsuge, T.; Peever, T.L.; Timmer, L.W.; Yamamoto, M.; Yamamoto, H.; Akimitsu, K. Distribution and characterization of AKT homologs in the tangerine. *Phytopathology* **2000**, *90*, 762–768. [CrossRef] [PubMed]

51. Masunaka, A.; Ohtani, K.; Peever, T.; Timmer, L.; Tsuge, T.; Yamamoto, M.; Yamamoto, H.; Akimitsu, K. An isolate that is pathogenic to both tangerines and rough lemon and produces two host-selective toxins, ACT- and ACR-toxins. *Phytopathology* **2005**, *95*, 241–247. [CrossRef] [PubMed]
52. Miyamoto, Y.; Masunaka, A.; Tsuge, T.; Yamamoto, M.; Ohtani, K.; Fukumoto, T.; Gomi, K.; Peever, T.; Akimitsu, K. Functional analysis of a multicopy host-selective ACT-toxin biosynthesis gene in the tangerine pathotype of *Alternaria alternata* using RNA silencing. *Mol. Plant-Microbe Interact.* **2008**, *21*, 1591–1599. [CrossRef] [PubMed]
53. Imazaki, A.; Tanaka, A.; Harimoto, Y.; Yamamoto, M.; Akimitsu, K.; Park, P.; Tsuge, T. Contribution of peroxisomes to secondary metabolism and pathogenicity in the fungal plant pathogen *Alternaria alternata*. *Eukaryot. Cell* **2010**, *9*, 682–694. [CrossRef]
54. Li, L.; Ma, H.; Zheng, F.; Chen, Y.; Wang, M.; Jiao, C.; Li, H.; Gai, Y. The transcription regulator ACTR controls ACT-toxin biosynthesis and pathogenicity in the tangerine pathotype of *Alternaria alternata*. *Microbiol. Res.* **2021**, *248*, 126747. [CrossRef] [PubMed]
55. Covert, S.F. Supernumerary chromosomes in filamentous fungi. *Curr. Genet.* **1998**, *33*, 311–319. [CrossRef] [PubMed]
56. Ito, K.; Tanaka, T.; Hatta, R.; Yamamoto, M.; Akimitsu, K.; Tsuge, T. Dissection of the host range of the fungal plant pathogen *Alternaria alternata* by modification of secondary metabolism. *Mol. Microbiol.* **2004**, *52*, 399–411. [CrossRef] [PubMed]
57. Miyamoto, Y.; Isshiki, Y.; Honda, A.; Masunaka, A.; Tsuge, T.; Yamamoto, M.; Ohtani, K.; Fukumoto, T.; Gomi, K.; Peever, T.; et al. Function of genes encoding acyl-CoA synthetase and enoyl-CoA hydratase for host-selective ACT-toxin biosynthesis in the tangerine pathotype of *Alternaria alternata*. *Phytopathology* **2009**, *99*, 369–377. [CrossRef] [PubMed]
58. Sugahara, S.; Ito, Y.; Sakurai, Y.; Narikawa, T.; Sakata, Y. Varietal difference of the resistance to stem canker caused by *Alternaria alternata* in tomato. *Res. Bull. Aichi Agric. Res. Cent.* **1989**, *21*, 170–175.
59. Bottini, A.; Gilchrist, D.; Phytotoxins, I. A 1-aminodimethylheptadecapentol from *Alternaria alternata* f. sp. *lycopersici*. *Tetrahedron Lett.* **1981**, *22*, 2719–2722. [CrossRef]
60. Bottini, A.T.; Bowen, J.R.; Gilchrist, D.G. Phytotoxins. II. Characterization of a phytotoxic fraction from *Alternaria alternata* f. sp. *lycopersici*. *Tetrahedron Lett.* **1981**, *22*, 2723–2726. [CrossRef]
61. Brandwagt, B.F.; Mesbah, L.A.; Takken, F.L.W.; Laurent, P.L.; Kneppers, T.J.A.; Hille, J.; Nijkamp, H.J.J. A longevity assurance gene homolog of tomato mediates resistance to *Alternaria alternata* f. sp. *lycopersici* toxins and fumonisin B1. *Proc. Natl. Acad. Sci. USA* **2000**, *97*, 4961–4966. [CrossRef]
62. Caldas, E.D.; Jones, A.D.; Ward, B.; Winter, C.K.; Gilchrist, D.G. Structural characterization of three new AAL-toxins produced by *Alternaria alternata* f. sp. *lycopersici*. *J. Agric. Food Chem.* **1994**, *42*, 327–333. [CrossRef]
63. Abbas, H.K.; Duke, S.O.; Paul, R.N.; Riley, R.T.; Tanaka, T. AAL-toxin, a potent natural herbicide which disrupts sphingolipid metabolism of plants. *Pestic. Sci.* **1995**, *43*, 181–187. [CrossRef]
64. Mesbah, L.A.; Van Der Weerden, G.M.; Nijkamp, H.J.J.; Hille, J. Sensitivity among species of *Solanaceae* to AAL toxins produced by *Alternaria alternata* f.sp. *lycopersici*. *Plant Pathol.* **2000**, *49*, 734–741. [CrossRef]
65. Duke, S.O.; Dayan, F.E. Modes of action of microbially-produced phytotoxins. *Toxins* **2011**, *3*, 1038–1064. [CrossRef] [PubMed]
66. Shier, W.T.; Abbas, H.K.; Mirocha, C.J. Toxicity of the mycotoxins fumonisins B1 and B2 and *Alternaria alternata* f. sp. *lycopersici* toxin (AAL) in cultured mammalian cells. *Mycopathologia* **1991**, *116*, 97–104. [CrossRef] [PubMed]
67. Van Bruggen, A.H.C.; He, M.M.; Shin, K.; Mai, V.; Jeong, K.C.; Finckh, M.R.; Morris, J.G. Environmental and health effects of the herbicide glyphosate. *Sci. Total Environ.* **2018**, *616*, 255–268. [CrossRef]
68. Abbas, H.K.; Tanaka, T.; Shier, W.T. Biological activities of synthetic analogues of *Alternaria alternata* toxin (AAL-toxin) and fumonisin in plant and mammalian cell cultures. *Phytochemistry* **1995**, *40*, 1681–1689. [CrossRef]
69. Duke, S.O.; Dayan, F.E. Clues to new herbicide mechanisms of action from natural sources. *ACS Sym. Ser.* **2013**, *1141*, 203–215.
70. Gilchrist, D.G. Mycotoxins reveal connections between plants and animals in apoptosis and ceramide signaling. *Cell Death Differ.* **1997**, *4*, 1312–1317. [CrossRef] [PubMed]
71. Orolaza, N.; Kawakita, K.; Doke, N. Inhibitory effect of AL-toxin produced by *Alternaria alternata* tomato pathotype on the biosynthesis of phosphatidylethanolamine in tomato leaves susceptible to the fungus. *Ann. Phytopath. Soc. Jpn.* **1992**, *58*, 719–725. [CrossRef]
72. Shi, L.; Bielawski, J.; Mu, J.; Dong, H.; Teng, C.; Zhang, J.; Yang, X.; Tomishige, N.; Hanada, K.; Hannun, Y.A.; et al. Involvement of sphingoid bases in mediating reactive oxygen intermediate production and programmed cell death in *Arabidopsis*. *Cell Res.* **2007**, *17*, 1030–1040. [CrossRef]
73. Shao, Z.; Zhao, Y.; Liu, L.; Chen, S.; Li, C.; Meng, F.; Liu, H.; Hu, S.; Wang, J.; Wang, Q. Overexpression of FBR41 enhances resistance to sphinganine analog mycotoxin-induced cell death and *Alternaria* stem canker in tomato. *Plant Biotechnol. J.* **2020**, *18*, 141–154. [CrossRef]
74. Abbas, H.K.; Tanaka, T.; Duke, S.O.; Porter, J.K.; Wray, E.M.; Hodges, L.; Sessions, A.E.; Wang, E.; Merrill, A.H.; Riley, A.R.T. Fumonisin-induced and AAL-toxin-induced disruption of sphingolipid metabolism with accumulation of free sphingoid gases. *Plant Physiol.* **1994**, *106*, 1085–1093. [CrossRef]
75. Markham, J.E.; Hille, J. Host-selective toxins as agents of cell death in plant-fungus interactions. *Mol. Plant Pathol.* **2001**, *1*, 229–239. [CrossRef]
76. Zhang, L.; Jia, C.; Liu, L.; Li, C.; Wang, Q. Involvement of jasmonates and ethylene in *Alternaria alternata* f. sp. *lycopersici* toxin-induced tomato cell death. *J. Exp. Bot.* **2011**, *15*, 5405–5418. [CrossRef]
77. Ismaiel, A.; Papenbrock, J. Mycotoxins: Producing fungi and mechanisms of phytotoxicity. *Agriculture* **2015**, *5*, 492–537. [CrossRef]

78. Caldas, E.D.; Sadilkova, K.; Ward, B.L.; Jones, A.D.; Winter, C.K.; Gilchrist, D.G. Biosynthetic studies of fumonisin B-1 and AAL toxins. *J. Agric. Food Chem.* **1998**, *46*, 4734–4743. [CrossRef]
79. Doidge, E. A study of some *Alternarias* infecting citrus in South Africa. *S. Afr. Dept Agric. Sci. Bull.* **1929**, *69*, 99–112.
80. Nishimura, S.; Tatano, S.; Miyamoto, Y.; Ohtani, K.; Fukumoto, T.; Gomi, K.; Tada, Y.; Ichimura, K.; Akimitsu, K. A zinc-binding citrus protein metallothionein can act as a plant defense factor by controlling host-selective ACR-toxin production. *Plant Mol. Biol.* **2013**, *81*, 1–11. [CrossRef]
81. Nishimura, S.; Kohmoto, K. Host-specific toxins and chemical structures from *Alternaria* species. *Annu. Rev. Phytopathol.* **1983**, *21*, 87–116. [CrossRef]
82. Akimitsu, K.; Kohmoto, K.; Otani, H.; Nishimura, S. Host-specific effect of toxin from the rough lemon pathotype of *Alternaria alternata* on mitochondria. *Plant Physiol.* **1989**, *89*, 925–931. [CrossRef]
83. Ohtani, K.; Yamamoto, H.; Akimitsu, K. Sensitivity to *Alternaria alternata* toxin in citrus because of altered mitochondrial RNA processing. *Proc. Natl. Acad. Sci. USA* **2002**, *99*, 2439–2444. [CrossRef] [PubMed]
84. Izumi, Y.; Kamei, E.; Miyamoto, Y.; Ohtani, K.; Masunaka, A.; Fukumoto, T.; Gomi, K.; Tada, Y.; Ichimura, K.; Peever, T.; et al. Role of the pathotype-specific *ACRTS1* gene encoding a hydroxylase involved in the biosynthesis of host-selective ACR-toxin in the rough lemon pathotype of *Alternaria alternata*. *Phytopathology* **2012**, *102*, 741–748. [CrossRef]
85. Izumi, Y.; Ohtani, K.; Miyamoto, Y.; Masunaka, A.; Fukumoto, T.; Gomi, K.; Tada, Y.; Ichimura, K.; Peever, T.; Akimitsu, K. A polyketide synthase gene, *ACRTS2*, is responsible for biosynthesis of host-selective ACR-toxin in the rough lemon pathotype of *Alternaria alternata*. *Mol. Plant-Microbe Interact.* **2012**, *25*, 1419–1429. [CrossRef] [PubMed]
86. Akimitsu, K.; Ohtani, K.; Shimagami, T.; Katsumoto, M.; Igarashi, C.; Tanaka, S.; Matsuoka, S.; Mochizuki, S.; Tsuge, T.; Yamamoto, M.; et al. Citrus as a molecular contact point for co-evolution of *Alternaria* pathogens. *Physiol. Mol. Plant Pathol.* **2016**, *95*, 93–96. [CrossRef]
87. Ueno, T.; Nakashima, T.; Uemota, M.; Fukami, H.; Lee, S.; Izumiya, N. Mass spectrometry of *Alternaria mali* toxins and related cyclodepsipeptides. *Biol. Mass Spectrom.* **1977**, *4*, 134–142. [CrossRef] [PubMed]
88. Park, P.; Nishimura, S.; Kohmoto, K.; Otani, H.; Tsujimoto, K. Two action sites of AM-toxin I produced by apple pathotype of *Alternaria alternate* in host cell: An ultrastructural study. *Can. J. Bot.* **1981**, *59*, 301–310. [CrossRef]
89. Kohmoto, K.; Otani, H. Host recognition by toxigenic plant-pathogens. *Experientia* **1991**, *47*, 755–764. [CrossRef]
90. Li, Y.; Aldwinckle, H.; Sutton, T.; Tsuge, T.; Kang, G.; Cong, P.; Cheng, G. Interactions of apple and the *Alternaria alternata* apple pathotype. *Crit. Rev. Plant Sci.* **2013**, *32*, 141–150. [CrossRef]
91. Keller, N.; Turner, G.; Bennett, J. Fungal secondary metabolism-biochemistry to genomics. *Nat. Rev. Microbiol.* **2005**, *3*, 937–947. [CrossRef]
92. Johnson, R.D.; Johnson, L.; Itoh, Y.; Kodama, M.; Otani, H.; Kohmoto, K. Cloning and characterization of a cyclic peptide synthetase gene from *Alternaria alternata* apple pathotype whose product is involved in AM-toxin synthesis and pathogenicity. *Mol. Plant-Microbe Interact.* **2000**, *13*, 742–753. [CrossRef]
93. Harimoto, Y.; Tanaka, T.; Kodama, M.; Yamamoto, M.; Otani, H.; Tsuge, T. Multiple copies of AMT2 are prerequisite for the apple pathotype of *Alternaria alternata* to produce enough AM-toxin for expressing pathogenicity. *J. Gen. Plant Pathol.* **2008**, *74*, 222–229. [CrossRef]
94. Harimoto, Y.; Hatta, R.; Kodama, M.; Yamamoto, M.; Otani, H.; Tsuge, T. Expression profiles of genes encoded by the supernumerary chromosome controlling AM-toxin biosynthesis and pathogenicity in the apple pathotype of *Alternaria alternata*. *Mol. Plant-Microbe Interact.* **2007**, *20*, 1463–1476. [CrossRef] [PubMed]
95. Bains, P.S.; Tewari, J.P. Purification, chemical characterization and host-specificity of the toxin produced by *Alternaria-Brassicae*. *Physiol. Mol. Plant Pathol.* **1987**, *30*, 259–271. [CrossRef]
96. Buchwaldt, L.; Green, H. Phytotoxicity of destruxin B and its possible role in the pathogenesis of *Alternaria brassicae*. *Plant Pathol.* **1992**, *41*, 55–63. [CrossRef]
97. Pedras, M.S.; Zaharia, I.; Gai, Y.; Zhou, Y.; Ward, D. In planta sequential hydroxylation and glycosylation of a fungal phytotoxin: Avoiding cell death and overcoming the fungal invader. *Proc. Natl. Acad. Sci. USA* **2001**, *98*, 747–752. [CrossRef] [PubMed]
98. Morel, E.; Païs, M.; Turpin, M.; Guyot, M. Cytotoxicity of cyclodepsipeptides on murine lymphocytes and on L 1210 leukemia cells. *Biomed. Pharmacother.* **1983**, *37*, 184–185. [PubMed]
99. Sun, C.; Chen, H.; Yeh, S. Suppressive effects of metabolites from *Alternaria brassicae* on the hepatitis B surface antigen. *Planta Med.* **1994**, *60*, 87–88. [CrossRef] [PubMed]
100. Chen, H.; Chou, C.; Sun, C.; Yeh, S. Suppressive effects of destruxin B on hepatitis B virus surface antigen gene expression in human hepatoma cells. *Antivir. Res.* **1997**, *34*, 137–144. [CrossRef]
101. Bandani, A.R.; Amiri, B.; Butt, T.M.; Gordon-Weeks, R. Effects of efrapeptin and destruxin, metabolites of entomogenous fungi, on the hydrolytic activity of a vacuolar type ATPase identified on the brush border membrane vesicles of *Galleria mellonella* midgut and on plant membrane bound hydrolytic enzymes. *Biochim. Biophys. Acta* **2001**, *1510*, 367–377. [CrossRef]
102. Suzuki, A.; Tamura, S. Isolation and structure of protodestruxin from *Metarrhizium anisopliae*. *Agric. Biol. Chem.* **1972**, *36*, 896–898. [CrossRef]
103. Jegorov, A.; Sedmera, P.; Matha, V. Biosynthesis of destruxins. *Phytochemistry* **1993**, *33*, 1403–1405. [CrossRef]
104. Ullstrup, A.J. Inheritance of susceptibility to infection by *Helminthosporium maydis* race 1 in maize. *J. Agric. Res.* **1941**, *63*, 331–334.

105. Kim, S.D.; Knoche, H.W.; Dunkle, L.D.; Mccrery, D.A.; Tomer, K.B. Structure of an amino acid analogue of the host-specific toxin from *Helminthosporium carbonum*. *Tetrahedron Lett.* **1985**, *26*, 969–972. [CrossRef]
106. Walton, J.D. Host-selective toxins: Agents of compatibility. *Plant Cell* **1996**, *8*, 1723–1733.
107. Pitkin, J.W.; Nikolskaya, A.; Ahn, J.H.; Walton, J.D. Reduced virulence caused by meiotic instability of the TOX2 chromosome of the maize pathogen *Cochliobolus carbonum*. *Mol. Plant-Microbe Interact.* **2000**, *13*, 80–87. [CrossRef]
108. Stierle, A.C.; Cardellina, J.H.; Strobel, G.A. Maculosin, a host-specific phytotoxin for spotted knapweed from *Alternaria alternata*. *Proc. Natl. Acad. Sci. USA* **1988**, *85*, 8008–8013. [CrossRef]
109. Park, S.H.; Strobel, G.A. Cellular protein receptors of maculosin, a host-specific phytotoxin of spotted knapweed (*Centaurea maculosa* L.). *Biochim. Biophys. Acta-Gen. Subj.* **1994**, *1199*, 11–19. [CrossRef]
110. Lopes, S.C.D.N.; Fedorova, A.; Castanho, M.A.R.B. Cholesterol modulates maculosin's orientation in model systems of biological membranes relevance towards putative molecular recognition. *Steroids* **2004**, *69*, 825–830. [CrossRef]
111. Paudel, B.; Maharjan, R.; Rajbhandarib, P.; Aryalc, N.; Azizc, S.; Bhattaraic, K.; Barald, B.; Malla, R.; Bhattarai, H. Maculosin, a non-toxic antioxidant compound isolated from *Streptomyces* sp. KTM18. *Pharm. Biol.* **2021**, *59*, 933–936. [CrossRef]
112. Robeson, D.J.; Gray, G.R.; Strobel, G.A. Production of the phytotoxins radicinin and radicinol by *Alternaria chrysanthemi*. *Phytochemistry* **1982**, *21*, 2359–2362. [CrossRef]
113. Sheridan, H.; Canning, A.M. Novel radicinol derivatives from long-term cultures of *Alternaria chrysanthemi*. *J. Nat. Prod.* **1999**, *62*, 1568–1569. [CrossRef]
114. Tal, B.; Robeson, D.J.; Burke, B.A.; Aasen, A.J. Phytotoxins from *Alternara helianthi* and the structures of deoxyradicinol and radianthin. *Phytochemistry* **1985**, *24*, 729–731. [CrossRef]
115. Tal, B.; Robeson, D.J. The production of pyrenocines A and B by a novel *Alternaria* species. *Z. Naturforsch. C* **1986**, *41*, 11–12. [CrossRef]
116. Ichihara, A.; Tazaki, H.; Sakamura, S. Solanapyrones A, B and C, phytotoxic metabolites from the fungus *Alternaria solani*. *Tetrahedron Lett.* **1983**, *24*, 5373–5376. [CrossRef]
117. Wang, X.; Luo, X.; Xiao, J.; Zhai, M.; Yuan, Y.; Zhu, Y.; Crews, P.; Yuan, C.; Wu, Q. Pyrone derivatives from the endophytic fungus *Alternaria tenuissima* SP-07 of Chinese herbal medicine *Salvia przewalskii*. *Fitoterapia* **2014**, *99*, 184–190. [CrossRef] [PubMed]
118. Pero, R.W.; Owens, R.G.; Dale, S.W.; Harvan, D. Isolation and identification of a new toxin, altenuene, from the fungus *Alternaria tenuis*. *Biochim. Biophys. Acta* **1971**, *230*, 170–179. [CrossRef]
119. Pero, R.W.; Posner, H.; Blois, M.; Harvan, D.; Spalding, J.W. Toxicity of metabolites produced by the "*Alternaria*". *Environ. Health Perspect.* **1973**, *4*, 87–94. [CrossRef]
120. Okuno, T.; Natsume, I.; Sawai, K.; Sawamura, K.; Furusaki, A.; Matsumoto, T. Structure of antifungal and phytotoxic pigments produced by *Alternaria* sps. *Tetrahedron Lett.* **1983**, *24*, 5653–5656. [CrossRef]
121. Wu, W.; Yue, G.; Huang, Q.; Sun, L.; Zhang, W. A new compound from an endophytic fungus *Alternaria tenuissima*. *J. Asian Nat. Prod. Res.* **2014**, *16*, 777–782. [CrossRef] [PubMed]
122. Bashyal, B.P.; Wellensiek, B.P.; Ramakrishnan, R.; Faeth, S.H.; Ahmad, N.; Gunatilaka, A.A.L. Altertoxins with potent anti-HIV activity from *Alternaria tenuissima* QUE1Se, a fungal endophyte of *Quercus emoryi*. *Bioorg. Med. Chem.* **2014**, *22*, 6112–6116. [CrossRef]
123. Kong, F.; Yi, T.; Ma, Q.; Xie, Q.; Zhou, L.; Chen, J.; Dai, H.; Wu, Y.; Zhao, Y. Biphenyl metabolites from the patchouli endophytic fungus *Alternaria* sp. PfuH1. *Fitoterapia* **2020**, *146*, 104708. [CrossRef]
124. Stierle, A.C.; Cardellina, J.H.; Strobel, G.A. Phytotoxins from *Alternara alternata*, a pathogen of spotted knapweed. *J. Nat. Prod.* **1989**, *52*, 42–47. [CrossRef]
125. Stack, M.E.; Mazzola, E.P. Stemphyltoxin III from *Alternaria alternata*. *J. Nat. Prod.* **1989**, *52*, 426–427. [CrossRef] [PubMed]
126. Stoessl, A. Some metabolites of *Alternaria solani*. *Can. J. Chem.* **1969**, *47*, 767–776. [CrossRef]
127. Yagi, A.; Okamura, N.; Haraguchi, H.; Abo, T.; Hashimoto, K. Antimicrobial tetrahydroanthraquinones from a strain of *Alternaria solani*. *Phytochemistry* **1993**, *33*, 87–91. [CrossRef]
128. Charudattan, R.; Rao, K.V. Bostrycin and 4-deoxybostrycin: Two nonspecific phytotoxins produced by *Alternaria eichhorniae*. *Appl. Environ. Microbiol.* **1982**, *43*, 846–849. [CrossRef]
129. Suemitsu, R.; Iwai, J.; Kawaguchi, K.; Haitani, N.; Kitagawa, N. Isolation and identification of erythroglaucin (1, 4, 5-trihydroxy-7-methoxy-2-methylanthraquinone) from the mycelium of *Alternaria porri* (Ellis) Ciferri. *Agric. Biol. Chem.* **1977**, *41*, 2289–2290.
130. Huang, C.; Pan, J.; Chen, B.; Yu, M.; Huang, H.; Zhu, X.; Lu, Y.; She, Z.; Lin, Y. Three bianthraquinone derivatives from the mangrove endophytic fungus *Alternaria* sp ZJ9-6B from the South China Sea. *Mar. Drugs* **2011**, *9*, 832–843. [CrossRef]
131. Chen, B.; Shen, Q.; Xun Zhu, X.; Lin, Y. The Anthraquinone derivatives from the fungus *Alternaria* sp. XZSBG-1 from the Saline Lake in Bange, Tibet, China. *Molecules* **2014**, *19*, 16529–16542. [CrossRef]
132. Ostry, V. *Alternaria* mycotoxins: An overview of chemical characterization, producers, toxicity, analysis and occurrence in foodstuffs. *World Mycotoxin J.* **2008**, *1*, 175–188. [CrossRef]
133. Gatenbeck, S.; Sierankiewicz, J. Microbial production of tenuazonic acid analogues. *Antimicrob. Agents Chemother.* **1973**, *3*, 308–309. [CrossRef] [PubMed]
134. Saad, S.M.; Halloin, J.M.; Hagedorn, D.J. Production, purification, and bioassay of tentoxin. *Phytopathology* **1970**, *60*, 415–418. [CrossRef] [PubMed]

135. Liebermann, B.; Oertel, B. Bildung und isolierung des phytotoxins tentoxin aus *Alternaria alternata*. *J. Basic Microbiol.* **1983**, *23*, 503–511. [CrossRef]
136. Kono, Y.; Gardner, J.M.; Takeuchi, S. Nonselective phytotoxins simultaneously produced with hostselective ACTG-toxins by a pathotype of *Alternaria citri* causing brown spot. *Agric. Biol. Chem.* **1986**, *50*, 2401–2403.
137. Edwards, J.V.; Lax, A.R.; Lillehoj, E.B.; Boudreaux, G.J. Structure-activity relationships of cyclic and acyclic analogues of the phytotoxic peptide tentoxin. *J. Agric. Food Chem.* **1987**, *35*, 451–456. [CrossRef]
138. Suemitsu, R.; Sano, T.; Yamamoto, M.; Arimoto, Y.; Morimatsu, F.; Nabeshima, T. Structural elucidation of alterporriol B, a novel metabolic pigment produced by *Alternaria porri* (Ellis) *ciferri*. *Agric. Biol. Chem.* **1984**, *48*, 2611–2613. [CrossRef]
139. Duke, S.O. Tentoxin effects on variable fluorescence and P515 electrochromic absorbance changes in tentoxin-sensitive and -resistant plant. *Plant Sci.* **1993**, *90*, 119–126. [CrossRef]
140. Tietjen, K.G.; Matern, U. Induction and suppression of phytoalexin biosynthesis in cultured cells of safflower, *Carthamus tinctorius* L.; by metabolites of *Alternaria carthami* Chowdhury. *Arch. Bioch. Biophy.* **1984**, *229*, 136–144. [CrossRef]
141. Vurro, M.; Evidente, A.; Andolfi, A.; Zonno, M.C.; Giordano, F.; Motta, A. Brefeldin A and $\alpha$, $\beta$-dehydrocurvularin, two phytotoxins from *Alternaria zinniae*, a biocontrol agent of *Xanthium occidentale*. *Plant Sci.* **1998**, *138*, 67–79. [CrossRef]
142. Courtial, J.; Hamama, L.; Helesbeux, J.J.; Lecomte, M.; Renaux, Y.; Guichard, E.; Voisine, L.; Yovanopoulos, C.; Hamon, B.; Oge, L.; et al. Aldaulactone-an original phytotoxic secondary metabolite involved in the aggressiveness of *Alternaria dauci* on carrot. *Front. Plant Sci.* **2018**, *9*, 502. [CrossRef]
143. Barasch, I.; Mor, H.; Netzer, D.; Kashman, Y. Production of zinniol by *Alternaria dauci* and its phytotoxic effect on carrot. *Physiol. Plant Pathol.* **1981**, *19*, 7–16. [CrossRef]
144. Cotty, P.; Mishagi, I.; Hine, R. Production of zinniol by *Alternaria tagetica* and its phytotoxic effect on *Tagetes erecta*. *Phytopathology* **1983**, *73*, 1326–1328. [CrossRef]
145. Cotty, P.; Mishagi, I. Zinniol production by *Alternaria* species. *Phytopathology* **1984**, *74*, 785–788. [CrossRef]
146. Leyte-Lugo, M.; Richomme, P.; Poupard, P.; Pena-Rodriguez, L.M. Identification and quantification of a phytotoxic metabolite from *Alternaria dauci*. *Molecules* **2020**, *25*, 4003. [CrossRef]
147. Gamboa-Angulo, M.M.; García-Sosa, K.; Alejos-González, F.; Escalante-Erosa, F.; Delgado-Lamas, G.; Peña-Rodríguez, L.M. Tagetolone and tagetenolone: Two phytotoxic polyketides from *Alternaria tagetica*. *J. Agric. Food Chem.* **2001**, *49*, 1228–1232. [CrossRef] [PubMed]
148. Avula, S.K.; Das, B.; Csuk, R.; Al-Rawahi, A.; Al-Harrasi, A. Recent advances in the stereoselective total synthesis of natural pyranones having long side chains. *Molecules* **2020**, *25*, 1905. [CrossRef] [PubMed]
149. Clarke, D.D.; Nord, F.F. Radicinin: A new pigment from *Stemphylium radicinum*. *Arch. Biochem. Biophys.* **1953**, *45*, 469–470. [CrossRef]
150. Sato, H.; Konoma, K.; Sakamura, S. Phytotoxins produced by onion pink root jungus, *Pyrenochaeta terrestris*. *Agric. Biol. Chem.* **1979**, *43*, 2409–7506.
151. Masi, M.; Freda, F.; Clement, S.; Cimmino, A.; Cristofaro, M.; Meyer, S.; Evidente, A. Phytotoxic activity and structure-activity relationships of radicinin derivatives against the invasive weed buffelgrass (*Cenchrus ciliaris*). *Molecules* **2019**, *24*, 2793. [CrossRef] [PubMed]
152. Nakajima, H.; Ishida, T.; Otsuka, Y.; Hamasaki, T.; Ichinoe, M. Phytotoxins and related metabolites produced by *Bipolaris coicis*, the pathogen of Job's tears. *Phytochemistry* **1997**, *45*, 41–45. [CrossRef]
153. Solfrizzo, M.; Vitti, C.; Girolamo, A.D.; Visconti, A.; Logrieco, A.; Fanizzi, F.P. Radicinols and radicinin phytotoxins produced by *Alternaria radicina* on carrots. *J. Agric. Food Chem.* **2004**, *52*, 3655–3660. [CrossRef]
154. Santoro, E.; Mazzeo, G.; Marsico, G.; Masi, M.; Longhi, G.; Superchi, S.; Evidente, A.; Abbate, S. Assignment through chiroptical methods of the absolute configuration of fungal dihydropyranpyran-4-5-diones phytotoxins, potential herbicides for buffelgrass (*Cenchrus ciliaris*) biocontrol. *Molecules* **2019**, *24*, 3022. [CrossRef] [PubMed]
155. Giridharan, P.; Verekar, S.A.; Gohil, A.R.; Mishra, P.D.; Khanna, A.; Deshmukh, S.K. Antiproliferative activity of hamigerone and radicinol isolated from *Bipolaris papendorfii*. *Biomed Res. Int.* **2014**, *2014*, 890904. [CrossRef] [PubMed]
156. Kim, J.C.; Choi, G.J.; Kim, H.T.; Kim, H.J.; Cho, K.Y. Pathogenicity and pyrenocine production of *Curvularia inaequalis* isolated from zoysia grass. *Plant Dis.* **2000**, *84*, 684–688. [CrossRef] [PubMed]
157. Myobatake, Y.; Kamisuki, S.; Tsukuda, S.; Higashi, T.; Chinen, T.; Takemoto, K.; Hachisuka, M.; Suzuki, Y.; Takei, M.; Tsurukawa, Y.; et al. Pyrenocine A induces monopolar spindle formation and suppresses proliferation of cancer cells. *Bioorg. Med. Chem.* **2019**, *27*, 115–149. [CrossRef]
158. Shishido, T.; Hachisuka, M.; Ryuzaki, K.; Miura, Y.; Tanabe, A.; Tamura, Y.; Kusayanagi, T.; Takeuchi, T.; Kamisuki, S.; Sugawara, F.; et al. EpsinR, a target for pyrenocine B, role in endogenous MHC-II-restricted antigen presentation. *Eur. J. Immunol.* **2014**, *44*, 3220–3231. [CrossRef]
159. Hamid, K.; Strange, R.N. Phytotoxicity of solanapyrones A and B produced by the chickpea pathogen *Ascochyta rabiei* (Pass.) Labr. and the apparent metabolism of solanapyrone A by chickpea tissues. *Physiol. Mol. Plant Pathol.* **2000**, *56*, 235–244. [CrossRef]
160. Mizushina, Y.; Kamisuki, S.; Kasai, N.; Shimazaki, N.; Takemura, M.; Asahara, H.; Linn, S.; Yoshida, S.; Matsukage, A.; Koiwai, O.; et al. A plant phytotoxin, solanapyrone A, is an inhibitor of DNA polymerase $\beta$ and $\lambda$. *J. Biol. Chem.* **2002**, *277*, 630–638. [CrossRef] [PubMed]

161. Oikawa, H.; Yokota, T.; Sakano, C.; Suzuki, Y.; NaYa, A.; Ichihara, A. Solanapyrones. Phytotoxins produced by *Alternaria solani*: Biosynthesis and isolation of minor components. *Biosci. Biotenchnol. Biochem.* **1998**, *62*, 2016–2022. [CrossRef]
162. Kasahara, K.; Miyamoto, T.; Fujimoto, T.; Oguri, H.; Tokiwano, T.; Oikawa, H.; Tokiwano, T.; Oikawa, H.; Ebizuka, Y.; Fujii, I. Solanapyrone synthase, a possible Diels-Alderase and iterative type I polyketide synthase encoded in a biosynthetic gene cluster from *Alternaria solani*. *ChemBioChem* **2010**, *11*, 1245–1252. [CrossRef] [PubMed]
163. Kim, W.; Park, C.M.; Park, J.J.; Akamatsu, H.O.; Peever, T.L.; Xian, M.; Gang, D.; Vandemark, G.; Chen, W. Functional analyses of the Diels-Alderase gene *sol5* of *Ascochyta rabiei* and *Alternaria solani* indicate that the solanapyrone phytotoxins are not required for pathogenicity. *Mol. Plant-Microbe Interact.* **2015**, *28*, 482–496. [CrossRef]
164. Pan, L.; Dong, J.; Xie, D.; Li, Y.F.; Liu, Q. Synthesis of 2-(Trifluoromethyl)-dibenzopyranones with Rhodium(III)-catalyzed Formal anti-Michael Addition as Key Step. *Adv. Synth. Catal.* **2018**, *360*, 958–964. [CrossRef]
165. Raistrick, H.; Stickings, C.E.; Thomas, R. Studies in the biochemistry of microorganisms. 90. Alternariol and alternariol monomethyl ether, metabolic products of *Alternaria tenuis*. *Biochem. J.* **1953**, *55*, 421. [CrossRef] [PubMed]
166. Scott, P.M. Analysis of agricultural commodities and foods for *Alternaria* mycotoxins. *J. Aoac. Int.* **2001**, *84*, 1809–1817. [CrossRef] [PubMed]
167. Gruber-Dorninger, C.; Novak, B.; Nagl, V.; Berthiller, F. Emerging mycotoxins: Beyond traditionally determined food contaminants. *J. Agric. Food Chem.* **2017**, *65*, 7052–7070. [CrossRef] [PubMed]
168. Schade, J.; King, A. Analysis of the major *Alternaria* toxins. *J. Food Product.* **1984**, *47*, 978–995. [CrossRef]
169. Wollenhaupt, K.; Schneider, F.; Tiemann, U. Influence of alternariol (AOH) on regulator proteins of cap-dependent translation in porcine endometrial cells. *Toxicol. Lett.* **2008**, *182*, 57–62. [CrossRef] [PubMed]
170. Fehr, M.; Pahlke, G.; Fritz, J.; Christensen, M.O.; Boege, F.; Altemoller, M.; Podlech, J.; Marko, D. Alternariol acts as a topoisomerase poison, preferentially affecting the II$\alpha$ isoform. *Mol. Nutr. Food Res.* **2009**, *53*, 441–451. [CrossRef]
171. Solhaug, A.; Vine, L.L.; Ivanova, L.; Spilsberg, B.; Holme, J.A.; Pestka, J.; Collins, A.; Eriksen, G.S. Mechanisms involved in alternariol-induced cell cycle arrest. *Mutat. Res.-Fund. Mol. Mutagen.* **2012**, *738*, 1–11. [CrossRef] [PubMed]
172. Fernandez-Blanco, C.; Juan-Garcia, A.; Juan, C.; Font, G.; Ruiz, M.J. Alternariol induce toxicity via cell death and mitochondrial damage on Caco-2 cells. *Food Chem. Toxicol.* **2016**, *88*, 32–39. [CrossRef]
173. Vila-Donat, P.; Fernández-Blanco, C.; Sagratini, G.; Font, G.; Ruiz, M.J. Effects of soyasaponin I and soyasaponins-rich extract on the alternariol-induced cytotoxicity on Caco-2 cells. *Food Chem. Toxicol.* **2015**, *77*, 44–49. [CrossRef] [PubMed]
174. Bensassi, F.; Gallerne, C.; el Dein, O.S.; Hajlaoui, M.R.; Bacha, H.; Lemaire, C. Cell death induced by the *Alternaria* mycotoxin alternariol. *Toxicol. In Vitro* **2012**, *26*, 915–923. [CrossRef]
175. Fleck, S.; Burkhardt, B.; Pfeiffer, E.; Metzler, M. *Alternaria* toxins: Altertoxin II is a much stronger mutagen and DNA strand breaking mycotoxin than alternariol and its methyl ether in cultured mammalian cells. *Toxicol. Lett.* **2012**, *214*, 27–32. [CrossRef] [PubMed]
176. Solhaug, A.; Torgersen, M.; Holme, J.; Lagadic-Gossmann, D.; Eriksen, G. Autophagy and senescence, stress responses induced by the DNA-damaging mycotoxin alternariol. *Toxicology* **2014**, *326*, 119–129. [CrossRef]
177. Solhaug, A.; Wisbech, C.; Christoffersen, T.E.; Hult, L.O.; Lea, T.; Eriksen, G.S.; Holme, J.A. The mycotoxin alternariol induces DNA damage and modify macrophage phenotype and inflammatory responses. *Toxicol. Lett.* **2015**, *239*, 9–21. [CrossRef] [PubMed]
178. Liu, G.; Qian, Y.; Zhang, P.; Dong, W.; Qi, Y.; Guo, H. Etiological role of *Alternaria alternata* in human esophageal cancer. *Chin Med. J.* **1992**, *105*, 394–400.
179. An, Y.; Zhao, T.; Miao, J.; Liu, G.; Zheng, Y.; Xu, Y.; Van Etten, R.L. Isolation, identification, and mutagenicity of alternariol monomethyl ether. *J. Agric. Food Chem.* **1989**, *37*, 1341–1343. [CrossRef]
180. European Food Safety Authority. Panel on Contaminants in the Food Chain. Scientific opinion on the risks for animal and public health related to the presence of *Alternaria* toxins in feed and food. *EFSA J.* **2011**, *9*, 2407–2504. [CrossRef]
181. Siegel, D.; Feist, M.; Proske, M.; Koch, M.; Nehls, I. Degradation of the *Alternaria* mycotoxins alternariol, alternariol monomethyl ether, and altenuene upon bread baking. *J. Agric. Food Chem.* **2010**, *58*, 9622–9630. [CrossRef]
182. Xiao, J.; Zhang, Q.; Gao, Y.; Tang, J.; Zhang, A.; Gao, J. Secondary metabolites from the endophytic botryosphaeria dothiadea of melia azedarach and their antifungal, antibacterial, antioxidant, and cytotoxic activities. *J. Agric. Food Chem.* **2014**, *62*, 3584–3590. [CrossRef]
183. De Souza, G.D.; Mithofer, A.; Daolio, C.; Schneider, B.; Rodrigues-Filho, E. Identification of *Alternaria alternata* mycotoxins by LC-SPE-NMR and their cytotoxic effects to soybean (*Glycine max*) cell suspension culture. *Molecules* **2013**, *18*, 2528–2538. [CrossRef]
184. Tang, J.; Huang, L.; Liu, Y.; Toshmatov, Z.; Zhang, C.; Shao, H. Two phytotoxins isolated from the pathogenic fungus of the invasive weed *Xanthium italicum*. *Chem. Biodivers.* **2020**, *7*, e2000043. [CrossRef] [PubMed]
185. Demuner, A.J.; Barbosa, L.C.; Miranda, A.C.M.; Geraldo, G.C.; da Silva, C.M.; Giberti, S.; Bertazzini, M.; Forlani, G. The fungal phytotoxin alternariol 9-methyl ether and some of its synthetic analogues inhibit the photosynthetic electron transport chain. *J. Nat. Prod.* **2013**, *76*, 2234–2245. [CrossRef] [PubMed]
186. Thomas, R. Studies in the biosynthesis of fungal metabolites. *Biochem. J.* **1961**, *80*, 234–240. [CrossRef]
187. Gatenbeck, S.; Hermodsson, S. Enzymic synthesis of the aromatic product alternariol. *Acta Chem. Scand.* **1965**, *19*, 65–71. [CrossRef]

188. Hiltunen, M.; Söderhäll, K. Alternariol-O-methyltransferase from *Alternaria alternata*: Partial purification and relation to polyketide synthesis. *Exp. Mycol.* **1992**, *16*, 44–51. [CrossRef]
189. Wenderoth, M.; Garganese, F.; Schmidt-Heydt, M.; Soukup, S.T.; Ippolito, A.; Sanzani, S.M.; Fischer, R. Alternariol as virulence and colonization factor of *Alternaria alternata* during plant infection. *Mol. Microbiol.* **2019**, *1*, 131–146. [CrossRef]
190. Kumagai, Y.; Shinkai, Y.; Miura, T.; Cho, A.K. The chemical biology of naphthoquinones and its environmental implications. *Annu. Rev. Pharmacol. Toxicol.* **2012**, *52*, 221–247. [CrossRef]
191. Hu, J.; Sarrami, F.; Li, H.; Zhang, G.; Stubbs, K.; Lacey, E.; Stewart, S.; Karton, A.; Piggott, A.; Chooi, Y. Heterologous biosynthesis of elsinochrome A sheds light on the formation of the photosensitive perylenequinone system. *Chem. Sci.* **2019**, *10*, 1457–1465. [CrossRef]
192. Stack, M.E.; Mazzola, E.P.; Page, S.W.; Pohland, A.E.; Highet, R.J.; Tempesta, M.S.; Corley, D.G. Mutagenic perylenequinone metabolites of *Alternaria alternata*: Altertoxins I, II, and III. *J. Nat. Prod.* **1986**, *49*, 866–871. [CrossRef]
193. Stack, M.E.; Prival, M.J. Mutagenicity of the *Alternaria* metabolites altertoxins-I, altertoxins-II, and altertoxins-III. *Appl. Environ. Microbiol.* **1986**, *52*, 718–722. [CrossRef]
194. Fleck, S.C.; Sauter, F.; Pfeiffer, E.; Metzler, M.; Hartwig, A.; Köberle, B. DNA damage and repair kinetics of the *Alternaria* mycotoxins alternariol, altertoxin II and stemphyltoxin III in cultured cells. *Mutat. Res. Genet. Toxicol. Environ. Mutagen.* **2016**, *798–799*, 27–34. [CrossRef]
195. Zhang, N.; Zhang, C.; Xiao, X.; Zhang, Q.; Huang, B. New cytotoxic compounds of endophytic fungus *Alternaria* sp. isolated from *Broussonetia papyrifera* (L.) Vent. *Fitoterapia* **2016**, *110*, 173–180. [CrossRef] [PubMed]
196. Robeson, D.; Strobel, G.; Matusumoto, G.K.; Fisher, E.L.; Chen, M.; Clardy, J. Alteichin: An unusual phytotoxin from *Alternaria eichorniae*, a fungal pathogen of water hyacinth. *Experientia* **1984**, *40*, 1248–1250. [CrossRef] [PubMed]
197. Hradil, C.M.; Hallock, Y.F.; Clardy, J.; Kenfield, D.S.; Strobel, G. Phytotoxins from *Alternaria cassia*. *Phytochemistry* **1989**, *28*, 73–75. [CrossRef]
198. Davis, V.M.; Stack, M.E. Mutagenicity of stemphylotoxin-III, A metabolite of *Alternaria alternata*. *Appl. Environ. Microbiol.* **1991**, *57*, 180–182. [CrossRef] [PubMed]
199. Malik, E.M.; Muller, C.E. Anthraquinones as pharmacological tools and drugs. *Med. Res. Rev.* **2016**, *36*, 705–748. [CrossRef]
200. Okamura, N.; Mimura, K.; Yagi, A. Altersolanol-related compounds from the culture liquid of *Alternaria solani*. *Phytochemistry* **1996**, *42*, 77–80. [CrossRef]
201. Suemitsu, R.; Yamadaa, Y.; Sanoa, T.; Yamashitaa, K. Phytotoxic ativities of Altersolanol A, B and dactylariol, and activities of altersolanol A against Some microorganisms. *Agric. Biol. Chem.* **1984**, *48*, 2383–2384.
202. Haraguchi, H.; Abo, T.; Fukuda, A.; Okamura, N.; Yagi, A. Mode of phytotoxic action of altersolanols. *Phytochemistry* **1996**, *43*, 989–992. [CrossRef]
203. Evidente, A.; Rodeva, R.; Andolfi, A.; Stoyanova, Z.; Perrone, C.; Motta, A. Phytotoxic polyketides produced by *Phomopsis foeniculi*, a strain isolated from diseased *Bulgarian fennel*. *Eur. J. Plant Pathol.* **2011**, *130*, 173–182. [CrossRef]
204. Mishra, P.D.; Verekar, S.A.; Deshmukh, S.K.; Joshi, K.S.; Fiebig, H.H.; Kelter, G. Altersolanol A: A selective cytotoxic anthraquinone from a *Phomopsis* sp. *Lett. Appl. Microbiol.* **2015**, *60*, 387–391. [CrossRef] [PubMed]
205. Trigos, A.; Mendoza, G.; Espinoza, C.; Salinas, A.; Fernandez, J.J.; Norte, M. The role of macrosporin in necrotic spots. *Phytochem. Lett.* **2011**, *4*, 122–125. [CrossRef]
206. Yuan, P.; He, L.; Chen, D.; Sun, Y.; Ge, Z.; Shen, D.; Lu, Y. Proteomic characterization of *Mycobacterium tuberculosis* reveals potential targets of bostrycin. *J. Proteom.* **2020**, *212*, 103576. [CrossRef] [PubMed]
207. Lin, W.; Fang, L.; Liu, J.; Cheng, W.; Yun, M.; Yang, H. Inhibitory effects of marine fungal metabolites from the South China Sea on prostate cancer cell line DU-145. *Int. J. Intern. Med.* **2008**, *35*, 562–564.
208. Chen, W.; Hou, J.; Guo, Y.; Yang, H.; Xie, C.; Lin, Y.; She, Z. Bostrycin inhibits proliferation of human lung carcinoma A549 cells via downregulation of the PI3K/Akt pathway. *J. Exp. Clin. Cancer Res.* **2011**, *30*, 17. [CrossRef]
209. Jie, J.; Shi, L.; Yue, S.; Wang, M.; Zhang, J. Bostrycin inhibits growth of tongue squamous cell carcinoma cells by inducing mitochondrial apoptosis. *Transl. Cancer Res.* **2020**, *9*, 3926–3936. [CrossRef]
210. Qin, X.; Peng, Y.; Zheng, J. In vitro and in vivo studies of the electrophilicity of physcion and its oxidative metabolites. *Chem. Res. Toxicol.* **2018**, *31*, 340–349. [CrossRef]
211. Mueller, S.O.; Schmitt, M.; Dekant, W.; Stopper, H.; Schlatter, J.; Lutz, W.K. Occurrence of emodin, chrysophanol and physcion in vegetables, herbs and liquors. Genotoxicity and anti-genotoxicity of the anthraquinones and of the whole plants. *Food Chem. Toxicol.* **1999**, *37*, 481–491. [CrossRef]
212. Anke, H.; Kolthoum, I.; Laatsch, H. Metabolic products of microorganisms. 192. The anthraquinones of the *Aspergillus glaucus* group. II. Biological activity. *Arch. Microbiol.* **1980**, *126*, 231–236. [CrossRef] [PubMed]
213. Wang, S.; Li, X.; Teuscher, F.; Li, D.; Diesel, A.; Ebel, R.; Proksch, P.; Wang, B. Chaetopyranin, a benzaldehyde derivative, and other related metabolites from *Chaetomium globosum*, an endophytic fungus derived from the marine red alga *Polysiphonia urceolata*. *J. Nat. Prod.* **2006**, *69*, 1622–1625. [CrossRef]
214. Ohnishi, K.; Tanabe, H.; Hayashi, S.; Suemitsu, R. Biosynthesis of Alterporriol-A by *Alternara porri*. *Biosci. Biotechnol. Biochem.* **1992**, *56*, 42–43. [CrossRef]
215. Kang, S.; Pandey, R.; Lee, C.; Sim, J.; Jeong, J.; Choi, B.; Jung, M.; Ginzburg, D.; Zhao, K.; Won, S.; et al. Genome-enabled discovery of anthraquinone biosynthesis in *Senna tora*. *Nat. Commun.* **2020**, *11*, 5875. [CrossRef] [PubMed]

216. Royles, B.J.L. Naturally occurring tetramic acids: Structure, isolation, and synthesis. *Chem. Rev.* **1995**, *95*, 1981–2001. [CrossRef]
217. Rosett, T.; Sankhala, R.H.; Stickings, C.E.; Taylor, M.E.U.; Thomas, R. Studies in the biochemistry of microorganisms. Metabolites of *Alternaria tenuis* auct: Culture filtrate products. *Biochem. J.* **1957**, *67*, 390–400. [CrossRef]
218. Kumari, A.; Singh, K. Evaluation of prophylactic efficacy of cinnamaldehyde in murine model against *Paradendryphiella arenariae* mycotoxin tenuazonic acid-induced oxidative stress and organ toxicity. *Sci. Rep.* **2021**, *11*, 19420. [CrossRef]
219. Stickings, C.E. Studies in the biochemistry of micro-organisms. 106. Metabolites of *Alternaria tenuis* auct.: The structure of tenuazonic acid. *Biochem. J.* **1959**, *72*, 332–340. [CrossRef]
220. Iwasaki, S.; Muro, H.; Nozoe, S.; Okuda, S.; Sato, Z. Isolation of 3,4-dihydro-3,4,8-trihydroxy-2(2H)-naphthalenone and tenuazonic acid from *Pyricularia oryzae* Cavara. *Tetrahedron Lett.* **1972**, *1*, 13–16. [CrossRef]
221. Steyn, P.S.; Rabie, C.J. Characterization of magnesium and calcium tenuazonate from *Phoma sorghina*. *Phytochemistry* **1976**, *15*, 1977–1979. [CrossRef]
222. Montemurro, N.; Visconti, A. Alternaria metabolites-chemical and biological data. In *Alternaria Biology, Plant Diseases and Metabolites*; Chelkowski, J., Visconti, A., Eds.; Elsevier Science: Amsterdam, The Netherlands; London, UK; New York, NY, USA; Tokyo, Japan, 1992; pp. 449–541.
223. Ebbole, D. Magnaporthe as a model for understanding host-pathogen interactions. *Annu. Rev. Phytopathol.* **2007**, *45*, 437–456. [CrossRef]
224. Chen, S.; Qiang, S. Recent advances in tenuazonic acid as a potential herbicide. *Pestic. Biochem. Phys.* **2017**, *143*, 252–257. [CrossRef]
225. Siegel, D.; Rasenko, T.; Koch, M.; Nehls, I. Determination of the *Alternaria* mycotoxin tenuazonic acid in cereals by high-performance liquid chromatography-electrospray ionization ion-trap multistage mass spectrometry after derivatization with 2, 4-dinitrophenylhydrazine. *J. Chromatogr. A* **2009**, *1216*, 4582–4588. [CrossRef] [PubMed]
226. Gross, M.; Curtui, V.; Ackermann, Y.; Latif, H.; Usleber, E. Enzyme immunoassay for tenuazonic acid in apple and tomato products. *J. Agric. Food Chem.* **2011**, *59*, 12317–12322. [CrossRef]
227. Motoyama, T. Secondary metabolites of the rice blast fungus *Pyricularia oryzae*: Biosynthesis and biological function. *Int. J. Mol. Sci.* **2020**, *21*, 8698. [CrossRef]
228. Miller, F.; Rightel, W.; Sloan, B.; Ehrlich, J.; French, J.; Bartz, Q.; Dixon, G. Antivial activity of tenuazonic acid. *Nature* **1963**, *2000*, 1338–1339. [CrossRef] [PubMed]
229. Kaczka, E.A.; Gitterman, C.O.; Dulaney, E.L.; Smith, M.C.; Hendlin, D.; Woodruff, H.; Folkers, K. Discovery of inhibitory activity of tenuazonic acid for growth of human adenocarcinoma-1. *Biochem. Biophys. Res. Commun.* **1964**, *14*, 54–57. [CrossRef]
230. Gitterman, C.O. Antitumor, cytotoxic, and antibacterial activities of tenuazonic acid and congeneric tetramic acids. *J. Med. Chem.* **1965**, *8*, 483–486. [CrossRef] [PubMed]
231. Zhao, H.; Cui, Z.; Gu, Y.; Liu, Y.; Wang, Q. The phytotoxicity of natural tetramic acid derivatives. *Pest Manag. Sci.* **2011**, *67*, 1059–1061. [CrossRef]
232. Smith, E.R.; Fredrickson, T.N.; Hadidian, Z. Toxic effects of the sodium and the *N*, *N*′-dibenzylethylenediamine salts of tenuazonic acid. *Cancer Chemother. Rep.* **1968**, *52*, 579–585.
233. Davies, N.D.; Diner, U.L.; Morgan-Jones, G. Tenuazonic acid production by *Alternaria alternata* and *Alternaria tenuissima* isolated from cotton. *Appl. Environ. Microbiol.* **1977**, *34*, 155–157. [CrossRef] [PubMed]
234. Shigeura, H.T.; Gordon, C.N. The biological activity of tenuazonic acid. *Biochemistry* **1963**, *2*, 1132–1137. [CrossRef]
235. Asam, S.; Rychlik, M. Potential health hazards due to the occurrence of the mycotoxin tenuazonic acid in infant food. *Eur. Food Res. Technol.* **2013**, *236*, 491–497. [CrossRef]
236. Umetsu, M.; Chiba, S.; Ogawa, S.; Nakao, T. Immune responses in mycoplasma pneumoniae infections. *Uirusu* **1974**, *24*, 157–163. [CrossRef] [PubMed]
237. Meazza, G.; Scheffler, B.E.; Tellez, M.R.; Rimando, A.M.; Romagni, J.G.; Duke, S.O.; Nanayakkara, D.; Khan, I.A.; Abourashed, E.A.; Dayan, F.E. The inhibitory activity of natural products on plant *p*-hydroxyphenylpyruvate dioxygenase. *Phytochemistry* **2002**, *59*, 281–288. [CrossRef]
238. Bjork, P.K.; Rasmussen, S.A.; Gjetting, S.K.; Havshoi, N.W.; Petersen, T.I.; Ipsen, J.O.; Larsen, T.O.; Fuglsang, A.T. Tenuazonic acid from Stemphylium loti inhibits the plant plasma membrane $H^+$-ATPase by a mechanism involving the C-terminal regulatory domain. *New Phytol.* **2020**, *226*, 770–784. [CrossRef] [PubMed]
239. Zonno, M.; Vurro, M. Effect of fungal toxins on germination of *Striga hermonthica* seeds. *Weed Res.* **1999**, *39*, 15–20. [CrossRef]
240. Marfori, E.C.; Kajiyama, S.I.; Fukusaki, E.I.; Kobayashi, A. Phytotoxicity of the tetramic acid metabolite trichosetin. *Photochemistry* **2003**, *62*, 715–721. [CrossRef]
241. Tylkowska, K.; Grabarkiewicz-Szczesna, J.; Iwanowska, H. Production of toxins by *Alternaria alternata* and *A. radicina* and their effects on germination of carrot seeds. *Seed Sci. Technol.* **2003**, *31*, 309–316. [CrossRef]
242. Zhou, B.; Qiang, S. Effect of tenuazonic acid produced by *Alternaria alternata* on mironucleus and karyokinesis of *Vicia faba* root tip cells. *Chin. J. App. Environ. Biol.* **2007**, *13*, 803–806.
243. Qiang, S.; Summerell, B.A.; Li, Y. Pathogenicity of *Alternaria alternata* on Crofton weed (*Eupatorium adenophorum*). In Proceedings of the 17th Asian-Pacific Weed Science Society Conference, Bangkok, Thailand, 22–27 November 1999; pp. 556–561.

244. Qiang, S.; Wan, Z.; Dong, Y.; Li, Y. Phytotoxicity of crude metabolites produced by *Alternaria alternata* to Crofton weed. In *Sustainable Management towards the 21 Century in China, Proceedings of the 6th Weed Science Conference of China, Nanning, China, 1 March 1999*; Guangxi Nationality Press Nanning: Nanning, China, 1999.
245. Chen, S.; Dai, X.; Qiang, S.; Tang, Y. Effect of a nonhost-selective toxin from *Alternaria alternata* on chloroplast-electron transfer activity in *Eupatorium adenophorum*. *Plant Pathol.* **2005**, *54*, 671–677. [CrossRef]
246. Zhou, B.; Wang, H.; Meng, B.; Wei, R.; Wang, L.; An, C.; Chen, S.; Yang, C.; Qiang, S. An evaluation of tenuazonic acid, a potential biobased herbicide in cotton. *Pest Manag. Sci.* **2019**, *75*, 2482–2489. [CrossRef]
247. Chen, S.; Xu, X.; Dai, X.; Yang, C.; Qiang, S. Identification of tenuazonic acid as a novel type of natural photosystem II inhibitor binding in $Q_B$-site of *Chlamydomonas reinhardtii*. *Biochim. Biophys. Acta* **2007**, *1767*, 306–318. [CrossRef] [PubMed]
248. Chen, S.; Yin, C.; Qiang, S.; Zhou, F.; Dai, X. Chloroplastic oxidative burst induced by tenuazonic acid, a natural photosynthesis inhibitor, triggers cell necrosis in *Eupatorium adenophorum* Spreng. *Biochim. Biophys. Acta* **2010**, *1797*, 391–405. [CrossRef] [PubMed]
249. Chen, S.; Strasser, R.J.; Qiang, S. In vivo assessment of effect of phytotoxin tenuazonic acid on PSII reaction centers. *Plant Physiol. Biochem.* **2014**, *84*, 10–21. [CrossRef]
250. Shi, J.; Zhang, M.; Gao, L.; Yang, Q.; Kalaji, H.; Qiang, S.; Strasser, R.; Chen, S. Tenuazonic acid-triggered cell death is the essential prerequisite for *Alternaria alternata* (Fr.) Keissler to infect successfully host *Ageratina adenophora*. *Cells* **2021**, *10*, 1010. [CrossRef]
251. Schobert, R.; Schlenk, A. Tetramic and tetronic acids: An update on new derivatives and biological aspects. *Bioorg. Med. Chem.* **2008**, *16*, 4203–4221. [CrossRef]
252. Mo, X.; Gulder, T.A.M. Biosynthetic strategies for tetramic acid formation. *Nat. Prod. Rep.* **2020**, *38*, 1555–1566. [CrossRef]
253. Collemare, J.; Billard, A.; Böhnert, H.; Lebrun, M.H. Biosynthesis of secondary metabolites in the rice blast fungus *Magnaporthe grisea*: The role of hybrid PKS-NRPS in pathogenicity. *Mycol. Res.* **2008**, *112*, 207–2015. [CrossRef]
254. Stickings, C.; Townsend, R. Studies in the biochemistry of micro-organisms. 108. Metabolites of *Alternaria tenuis* Auct.: The biosynthesis of tenuazonic acid. *Biochem. J.* **1961**, *78*, 412–418. [CrossRef] [PubMed]
255. Yun, C.; Motoyama, T.; Osada, H. Biosynthesis of the mycotoxin tenuazonic acid by a fungal NRPS–PKS hybrid enzyme. *Nat. Commun.* **2015**, *6*, 8758. [CrossRef]
256. Yun, C.; Nishimoto, K.; Motoyama, T.; Shimizu, T.; Hino, T.; Dohmae, N.; Nagano, S.; Osada, H. Unique features of the ketosynthase domain in a nonribosomal peptide synthetase–polyketide synthase hybrid enzyme, tenuazonic acid synthetase 1. *J. Biol. Chem.* **2020**, *295*, 11602–11612. [CrossRef] [PubMed]
257. Chen, S.; Zhou, F.; Yin, C.; Strasser, R.; Yang, C.; Qiang, S. Application of fast chlorophyll a fluorescence kinetics to probe action target of 3-acetyl-5-isopropyltetramic acid. *Environ. Exp. Bot.* **2011**, *71*, 269–279. [CrossRef]
258. Chen, S.; Yin, C.; Strasser, R.; Govindjee; Yang, C.; Qiang, S. Reactive oxygen species from chloroplasts contribute to 3-acetyl-5-isopropyltetramic acid-induced leaf necrosis of *Arabidopsis thaliana*. *Plant Physiol. Biochem.* **2012**, *52*, 38–51. [CrossRef] [PubMed]
259. Qin, J.; Zhang, Y.; Hu, L.; Ma, Y.; Gao, J. Cytotoxic metabolites produced by *Alternaria* No.28, an endophytic fungus isolated from Ginkgo biloba. *Nat. Prod. Commun.* **2009**, *4*, 1473–1476. [CrossRef]
260. Lebrun, M.H.; Nicolas, L.; Boutar, M.; Gaudemer, F.; Ranomenjanahary, S.; Gaudemer, A. Relationship between the structure and the phytotoxicity of the fungal toxin tenuazonic acid. *Phytochemistry* **1988**, *27*, 77–84. [CrossRef]
261. Wong, C.; Lam, H.; Song, T.; Chen, G.; Li, X. Synthesis of constrained head-to-tail cyclic tetrapeptides by an imine-induced ring-closing/contraction strategy. *Angew. Chem. Int. Ed.* **2013**, *52*, 10212–10215. [CrossRef] [PubMed]
262. Meyer, W.L.; Kuyper, L.F.; Phelps, D.W.; Cordes, A.W. Structure of the cyclic tetrapeptide tentoxin. Crystal and molecular structure of the dihydro derivative. *J. Chem. Soc. Chem. Commun.* **1974**, *184*, 339–340. [CrossRef]
263. Liu, Y.; Rychlik, M. Development of a stable isotope dilution LC−MS/MS method for the *Alternaria* toxins tentoxin, dihydrotentoxin, and isotentoxin. *J. Agric. Food Chem.* **2013**, *61*, 2970–2978. [CrossRef]
264. Rodríguez-Carrasco, Y.; Mañes, J.; Berrada, H.; Juan, C. Development and validation of a LC-ESI-MS/MS method for the determination of *Alternaria* toxins alternariol, alternariol methyl-ether and tentoxin in tomato and tomato-based products. *Toxins* **2016**, *8*, 328. [CrossRef]
265. De Sa, S.V.M.; Monteiro, C.; Fernandes, J.O.; Pinto, E.; Faria, M.A.; Cunha, S.C. Emerging mycotoxins in infant and children foods: A review. *Crit. Rev. Food Sci. Nutr.* **2021**, 1–15. [CrossRef]
266. Halloin, J.M.; Hagedorn, D.J. Effects of tentoxin on enzymic activities in cucumber and cabbage cotyledons. *Mycopathologia* **1975**, *55*, 159–162. [CrossRef]
267. Schadler, D.L.; Steele, J.A.; Durbin, R.D. Some effects of tentoxin on mature and developing chloroplasts. *Mycopathologia* **1976**, *58*, 101–105. [CrossRef] [PubMed]
268. Steele, J.A.; Uchytil, T.F.; Durbin, R.D.; Bhatnagar, P.; Rich, D.H. Chloroplast coupling factor 1: A species-specific receptor for tentoxin. *Proc. Natl. Acad. Sci. USA* **1976**, *73*, 2245–2248. [CrossRef] [PubMed]
269. Groth, G. Structure of spinach chloroplast F-1-ATPase complexed with the phytopathogenic inhibitor tentoxin. *Proc. Natl. Acad. Sci. USA* **2002**, *99*, 3464–3468. [CrossRef] [PubMed]
270. Santolini, J.; Haraux, F.; Sigalat, C.; Moal, G.; Andre, F. Kinetic analysis of tentoxin binding to chloroplast F-1-ATPase-A model for the overactivation process. *J. Biol. Chem.* **1999**, *274*, 849–858. [CrossRef] [PubMed]
271. Yang, J.; Williams, D.; Kandiah, E.; Fromme, P.; Chiu, P.L. Structural basis of redox modulation on chloroplast ATP synthase. *Commun. Biol.* **2020**, *3*, 842. [CrossRef] [PubMed]

272. Klotz, M.G. The action of tentoxin on membrane processes in plants. *Physiol. Plantarum* **1988**, *74*, 575–582. [CrossRef]
273. Ramm, K.; Ramm, M.; Liebermann, B.; Reuter, G. Studies of the biosynthesis of tentoxin by *Alternaria alternata*. *Microbiology* **1994**, *140*, 3257–3266. [CrossRef]
274. Liebermann, B.; Ramm, K. N-methylation in the biosynthesis of the phytotoxin tentoxin. *Phytochemistry* **1991**, *30*, 1815–1817. [CrossRef]
275. De Bruyne, L.; Van Poucke, C.; Di Mavungu, D.J.; Zainudin, N.A.I.M.; Vanhaecke, L.; De Vleesschauwer, D.; Turgeon, B.G.; De Saeger, S.; Hofte, M. Comparative chemical screening and genetic analysis reveal tentoxin as a new virulence factor in *Cochliobolus miyabeanus*, the causal agent of brown spot disease on rice. *Mol. Plant Pathol.* **2016**, *17*, 805–817. [CrossRef]
276. Li, Y.; Han, W.; Gui, X.; Wei, T.; Tang, S.; Jin, J. Putative nonribosomal peptide synthetase and cytochrome P450 genes responsible for tentoxin biosynthesis in *Alternaria alternata* ZJ33. *Toxins* **2016**, *8*, 234. [CrossRef] [PubMed]
277. Rutten, T.L.M.; Knuiman, B. Brefeldin A effects on tobacco pollen tubes. *Eur. J. Cell Biol.* **1993**, *61*, 247–255.
278. Driouich, A.; Jauneau, A.; Staehelin, L.A. 7-Dehydrobrefeldin A, a naturally occurring brefeldin A derivative, inhibits secretion and causes a cis-to-trans breakdown of Golgi stacks in plant cells. *Plant Physiol.* **1997**, *113*, 487–492. [CrossRef] [PubMed]
279. Harri, E.; LoeMer, W.; Singh, H.; Stahlin, H.; Tamm, C. Die constitution von brefeldin A. *Helv. Chem. Acta* **1963**, *46*, 1235–1243.
280. Wang, J.; Huang, Y.; Fang, M.; Zhang, Y.; Zheng, Z.; Zhao, Y.; Su, W. Brefeldin A, a cytotoxin produced by *Paecilomyces* sp. and *Aspergillus clavatus* isolated from *Taxus mairei* and *Torreya grandis*. *FEMS Immunol. Med. Microbiol.* **2002**, *34*, 51–57. [CrossRef] [PubMed]
281. Misumi, Y.; Miki, K.; Takatsuki, A.; Tamura, G.; Ikehara, Y. Novel blockade by brefeldin A of intracellular transport of secretory proteins in cultured rat hepatocytes. *J. Biol. Chem.* **1986**, *261*, 1398–1403. [CrossRef]
282. Fujiwara, T.; Oda, K.; Yokota, S.; Takatsuki, A.; Ikehara, Y. Brefeldin A causes disassembly of the golgi-complex and accumulation of secretory proteins in the *endoplasmic reticulum*. *J. Biol. Chem.* **1986**, *263*, 18545–18552. [CrossRef]
283. Gamboa-Angulo, M.M.; Escalante-Erosa, F.; García-Sosa, K.; Alejos-González, F.; Delgado-Lamas, G.; Peña-Podríguez, L.M. Natural zinniol derivatives from *Alternaria tagetica*. Isolation, synthesis, and structure-activity correlation. *J. Agric. Food Chem.* **2002**, *50*, 1053–1058. [CrossRef] [PubMed]
284. Berestetskii, A.O.; Yuzikhin, O.S.; Katkova, A.S.; Dobrodumov, A.V.; Sivogrivov, D.E.; Kolombet, L.V. Isolation, identification, and characteristics of the phytotoxin produced by the fungus *Alternaria cirsinoxia*. *Appl. Biochem. Microbiol.* **2010**, *46*, 75–79. [CrossRef]
285. Thuleau, P.; Graziana, A.; Rossignol, M.; Kauss, H.; Auriol, P.; Ranjeva, R. Binding of the phytotoxin zinniol stimulates the entry of calcium into plant protoplasts. *Proc. Natl. Acad. Sci. USA* **1988**, *85*, 5932–5935. [CrossRef]
286. Lecomte, M.; Hamama, L.; Voisine, L.; Gatto, J.; He´lesbeux, J.-J.; Séraphin, D.; Peña-Rodriguez, L.M.; Richomme, P.; Boedo, C.; Yovanopoulos, C.; et al. Partial resistance of carrot to *Alternaria dauci* correlates with in vitro cultured carrot cell resistance to fungal exudates. *PLoS ONE* **2014**, *9*, e101008. [CrossRef] [PubMed]
287. Qui, J.A.; Castro-Concha, L.A.; García-Sosa, K.; Miranda-Ham, M.L.; Peña-Rodríguez, L.M. Is zinniol a true phytotoxin? Evaluation of its activity at the cellular level against *Tagetes erecta*. *J. Gen. Plant Pathol.* **2010**, *76*, 94–101. [CrossRef]
288. Nukina, M.; Marumo, S. α-Acetylorcinol from *Cochliobolus lunata*. *Agric. Biol. Chem.* **1977**, *41*, 717. [CrossRef]
289. Venkatasubbaiah, P.; Baudoin, A.B.A.M.; Chilton, W.S. Leaf spot of hemp dogbane caused by *Stagonospora apocyni*, and its phytotoxins. *J. Phytopathol.* **1992**, *135*, 309–316. [CrossRef]
290. Peng, W.; You, F.; Li, X.L.; Jia, M.; Zheng, C.; Han, T.; Qin, L.P. A new diphenyl ether from the endophytic fungus *Verticillium* sp. isolated from *Rehmannia glutinosa*. *Chin. J. Nat. Med.* **2013**, *11*, 673–675. [CrossRef] [PubMed]
291. Andolfi, A.; Boari, A.; Evidente, M.; Cimmino, A.; Vurro, M.; Ash, G.; Evidente, A. Gulypyrones A and B and phomentrioloxins B and C produced by diaporthe gulyae, a potential mycoherbicide for Saffron Thistle (*Carthamus lanatus*). *J. Nat. Prod.* **2015**, *78*, 623–629. [CrossRef] [PubMed]
292. Chaves, N.; Sosa, T.; Al, J.C.; Escudero, J.C. Identification and effects of interaction phytotoxic compounds from exudade of *Cistus ladanifer* leaves. *J. Chem. Ecol.* **2001**, *27*, 611–621. [CrossRef] [PubMed]
293. Cho, J.Y.; Moon, J.H.; Seong, K.Y.; Park, K.H. Antimicrobial activity of 4-hydroxybenzoic acid and trans 4-hydroxycinnamic acid isolated and identified from rice hull. *Biosci. Biotechnol. Biochem.* **1998**, *62*, 2273–2276. [CrossRef]
294. Merkl, R.; Hrádková, I.; Filip, V.; Šmidrkal, J. Antimicrobial and antioxidant properties of phenolic acids alkyl esters. *Czech J. Food Sci.* **2010**, *28*, 275–279. [CrossRef]
295. Deba, F.; Xuan, T.D.; Yasuda, M.; Tawata, S. Herbicidal and fungicidal activities and identification of potential phytotoxins from *Bidens pilosa* L. var. *radiata* Scherff. *Weed Biol. Manag.* **2007**, *7*, 77–83. [CrossRef]
296. Zuo, S.; Zhou, S.; Ye, L.; Ma, S. Synergistic and antagonistic interactions among five allelochemicals with antialgal effects on bloom-forming *Microcystis aeruginosa*. *Ecol. Eng.* **2016**, *97*, 486–492. [CrossRef]
297. Ceylan, O.; Alic, H. Antibiofilm, antioxidant, antimutagenic activities and phenolic compounds of *Allium orientale* BOISS. *Brazilian Arch. Biol. Technol.* **2015**, *58*, 935–943. [CrossRef]
298. Pugazhendhi, D.; Pope, G.S.; Darbre, P.D. Oestrogenic activity of *p*-hydroxybenzoic acid (common metabolite of paraben esters) and methylparaben in human breast cancer cell lines. *J. Appl. Toxicol.* **2005**, *25*, 301–309. [CrossRef] [PubMed]

MDPI
St. Alban-Anlage 66
4052 Basel
Switzerland
www.mdpi.com

*Journal of Fungi* Editorial Office
E-mail: jof@mdpi.com
www.mdpi.com/journal/jof

www.ingramcontent.com/pod-product-compliance
Lightning Source LLC
Chambersburg PA
CBHW061827210926
51850CB00006B/829
*9783036588100*